Darwin's Reach

Darwin's Reach
21st Century Applications
of Evolutionary Biology

Norman Johnson

CRC Press
Taylor & Francis Group
Boca Raton London

CRC Press is an imprint of the
Taylor & Francis Group, an **informa** business

First edition published 2022
by CRC Press
6000 Broken Sound Parkway NW, Suite 300, Boca Raton, FL 33487–2742

and by CRC Press
2 Park Square, Milton Park, Abingdon, Oxon, OX14 4RN

© 2022 Taylor & Francis Group, LLC

CRC Press is an imprint of Taylor & Francis Group, LLC

ISBN: 978-1-138-58742-7 (hbk)
ISBN: 978-1-138-58739-7 (pbk)
ISBN: 978-0-429-50396-2 (ebk)

DOI: 10.1201/9780429503962

Typeset in Times
by Apex CoVantage, LLC

Contents

SECTION II Food

SECTION III Environment

SECTION IV Society

Foreword

We live in a world riddled with contradictions around science and its applications. On the positive side, as we begin to emerge from the worst global pandemic in over a century, surely one must be in awe of scientists having developed, rigorously safety-tested, and mass-produced vaccines to a new disease in barely a year's time. Many of these vaccines were not simple "tweaks" on existing ones but instead used an entirely new approach. Science for the win! On the negative side, misinformation about science has proliferated at such a rate that half or more of people in some communities in the United States are simply declining to take this vaccine. They would literally rather face a disease shown to be potentially deadly than trust the science and scientists that accomplished this feat.

Evolutionary biology in particular has been a topic that has divided and angered nonscientists. One of the most magnificent and influential observations in all of biology is that all forms of life on Earth—bacteria, plants, animals, fungi—not only share various characteristics but are actually directly related. Yes, if we could draw a long enough family tree, the reader would be on the same tree with seagulls, sunflowers, and their own gut bacteria. This conclusion is "science fact" with as much evidence as any science fact—no observation argues against it while tons argue for it—yet, again, half or more of people in some communities in the United States have simply decided to take this fact as untrue, or at best, unproven. Worse yet, this science fact is taken by many people as an affront—an insult to humanity. I am often asked if anything more can be done to convince people to accept this science. We evolutionary biologists face the challenges of not only wanting people to appreciate the importance of our subject but even to accept its truth.

To this end, I am thrilled that Dr. Norman Johnson has stepped in, yet again, to lay out why and how evolutionary biology concepts are being used to help humanity and our planet. Some readers may be familiar with Dr. Johnson from his previous excellent book, *Darwinian Detectives*. I have known Dr. Johnson since 1992, but our shared history goes back further. We were both undergraduates at the College of William and Mary in Virginia, and we both studied with Professor Bruce Grant there (discussed by Dr. Johnson in Chapter 13). Professor Grant inspired us both to devote our lives to studying evolutionary genetics through a course bearing that title. However, the inspirational part of this course was not simply the interesting content but also the manner in which Professor Grant described the backstories of how the research results came about—evolutionary genetics, and indeed all science, is a human endeavor. Researchers were inspired by personal motivations and histories, by their own creative ideas, and through persevering through false starts and wrong tentative conclusions.

Since I have known Norman, he has always thought of science as such a human endeavor, and he has always loved to tell "the stories of the people behind the discoveries" rather than simply laying out the conclusions. I recall many stimulating hallway discussions with Norman when we were graduate students about science and its players, including even some of the specifics mentioned in this book (e.g., Lewontin and Hartl on the use of DNA fingerprinting in law enforcement). I am thrilled, and entirely unsurprised, that he takes exactly this "story of the people behind the science" approach in this delightful new book.

A Native American proverb purportedly says, "Tell me the facts and I will learn. Tell me the truth and I will believe. But tell me a story and it will live in my heart forever." Consistent with his in-person conversations and with this time-tested pedagogical approach, Dr. Johnson provides history and context to each discovery, often including direct personal interviews with the players, along with explaining the scientific principles that apply. He even references the science or news media coverage of the time when appropriate. Citations from classic works like Emily Dickinson's poems, to classic science fiction like *Star Trek*, and to recent movies like *The Martian* work their way into the text to further engage and connect with the reader. In short, he portrays all science as the human endeavor that it is and writes to the reader as a person.

Equally importantly, Dr. Johnson's book appropriately emphasizes throughout the importance of the scientific results described—knowing these answers matters not just to the scientists conducting the work but also to humanity and the proper care of our planet. Chapter 3 explains how research on an insect bacterium can help curb the spread of Zika and dengue, Chapter 5 shows a great example of how genetic mapping studies of Crohn's disease has revealed possible treatments for it, and Chapter 12 shows the dramatic economic consequences of ocean acidification. Other insights emerge throughout, such as that the dogma for why human babies are so helpless still often taught in schools is probably wrong (Chapter 7). And there are some quick fun observations, like that the caffeine we love functions in plant leaves as an insecticide. Between the engaging, storied approach and the incredible number of applications discussed of evolutionary research affecting humanity, this book will surely spark thousands of dinner-time and cocktail party "Did you know . . .?" conversations. It has already done so in my household.

In the Epilogue, Dr. Johnson bemoans that many stories did not get sufficient attention or were not even presented. Alas, this fact will always apply to science because it constantly builds upon itself. While being rigorous, this book is also amazingly current, with numerous references to ongoing studies and to results that came out this year. I have no doubt that there will be room for a companion book or update within just a few years to capture some of these continually emerging stories.

Back to the original question—how do we convince people to accept science? The answer lies within this book: scientific truths, personal stories, engaging prose, and clear applications to let the reader know why they should care. Dr. Johnson has done both the readers and the scientific community a huge service with this book, and it will serve as an outstanding reference to both professionals and the public. Educators often quote Dobzhansky's famous "Nothing in biology makes sense except in the light of evolution" line, but this book lays out why that is true in a compelling manner for everyone. Those people reading this book are in for a treat!

Mohamed Noor
Dean of Natural Sciences and Professor of
Biology at Duke University

Preface

This book is about applied evolution – the application of the principles of and information about evolutionary biology to diverse practical matters.

To some extent, applied evolution has existed for an exceptionally long time. The reciprocal exchange between evolution and the applied science of breeding animals and plants literally started with Charles Darwin, the pigeon breeder and author of *Variation Under Domestication* as well as *The Origin of Species*. This exchange continues to present day; for instance, evolutionary geneticists are developing statistical models that take advantage of the tremendous power of complete genome sequences to determine how breeding designs would improve livestock. From time to time, earlier generations of biologists applied evolution to some aspect of human health. The notable mid-20th-century contributions Peter Medawar and George Williams made toward an evolutionary theory of aging quickly come to mind. Yet these early efforts have been piecemeal; there was little coherence across these applications until fairly recently.

Applied evolution of today differs from that of the nascent attempts of yesteryear, both in being more coherent and in having a much wider scope. In the 1990s, George Williams and Randy Neese, followed by others, developed Darwinian (evolutionary) medicine, applying the principles of evolutionary biology to the broader question "Why do we get sick?" In the time since, evolutionary medicine is becoming a separate discipline, with its own various institutes, centers, and journals. Today's applied evolution is more than just evolutionary medicine and agriculture. Evolution also is being applied to environmental concerns: Which of these different populations of a beetle likely to go extinct, and how can we best conserve them? Is this species of sea urchins likely to be able to adapt to the changing environments arising from global warming? Evolution's reach extends still further. It now also includes forensic biology and the law, as its principles are being used in criminal and civil court proceedings. Ideas from evolutionary biology can be used to inform policy regarding foreign affairs and national security. For instance, the evolution of large horns in some beetle species shares an underlying framework with the arms races between nations that have led to dangerous escalations of weapon stockpiles.

This last decade has given researchers two game-changing tools and a major shift in thinking that pervade much of biology. First, high-throughput sequencing has drastically increased the speed at which DNA sequences can be generated and opened up research to any organism on the planet. Coupled with advances in computation, high-throughput sequencing is changing how many evolutionary biologists conduct their research programs. They can now address topics that were only contemplated a decade ago. For instance, consider the Neanderthal genome studies. Biologists not only can conclusively demonstrate that we interbred with Neanderthals, but they can estimate how much Neanderthal DNA individuals have. They can even infer the action of selection at particular parts of the Neanderthal genome that came into our own. These sequencing advances are important not just to biologists; in addition, they have also reduced the price tag to the extent that getting one's genome sequenced is within the budgets of middle-income people. Already, millions of people have had parts of their DNA assessed through consumer genomics kits to infer their ancestry and/or to assess their risk of obtaining a disease. Knowledge of evolutionary biology informs both ancestry genetic tests and disease risk. The principles of evolutionary biology also underlie the methodologies used to map disease and other traits to genetic variants.

Perhaps even more importantly, the editing tool CRISPR/Cas9 allows researchers to alter gene sequences specifically and with relative ease. CRISPR editing already is empowering researchers who are interested in the gene function for just about any organism. These scientists can address questions such as which sequence changes will cause their organism of choice to have greater tolerance to drought. This technology is also being developed to control pest species. Evolutionary principles are important in ensuring that such measures are safe and effective.

As high-throughput sequencing and CRISPR genome editing are the tools of 2010s biology, the concept of the decade is the microbiota. This is the collection of microbes – bacteria, archaea, fungi, and viruses – that live on and within the bodies of larger organisms, including humans. Recent research shows that the microbiota is dramatically more important to our health and even our behavior than what was thought even a decade ago. We really are ecosystems, and as such, ecosystem ecology and evolution are of essence to the study of our health. Recent research is also showing that the microbiotas of other organisms are also important to their ability to respond to challenges.

In addition to these new technologies and discoveries, we face challenges new and old. Global climate change has intensified over the last decade. The principles of evolution are central to addressing whether organisms can continue to adapt to rising temperatures and increasingly acidic oceans. Other human activity has placed a large number of species in peril. While the looming extinction crisis is grave, it is not hopeless. Evolutionary biology—combined with ecology and genetics—is vital to these essential conservation efforts. The COVID-19 pandemic has reminded us of the power of infectious disease to cause misery and death. Here, we will see how evolutionary principles are used to track the spread of viruses and develop vaccines. Evolutionary biology has also been vital to the development of drug regimens to treat HIV-AIDS.

The application of evolutionary biology also entails possible misapplications wherein science is perverted to justify dubious or even evil ends. The most extreme historical examples include Lysenko's perversion of genetics for ideological goals in the Soviet Union and the use of evolution to justify genocide, brutality, and "racial" superiority in Nazi Germany. But there are less extreme versions of such misapplications, some of which persist today. For instance, some authors have misused information about genetic variation among human populations to make unsubstantiated, racist claims. These misapplications do not need to arise from nefarious motivations. The much-hyped paleo diet, regardless of its health benefits, is not well supported by the evidence about what our ancestors ate and the speed to which we can adapt to dietary changes.

In addition to informing various audiences about the importance of these evolutionary applications, I hope this book encourages dialogue among different researchers studying different evolutionary applications, with the aim of a better understanding of the nascent science of evolutionary applications. Perhaps, in the not-too-distant future, such dialogue may lead to mutual illumination across the disparate branches of this young field. Perhaps lessons learned in a research program using evolutionary approaches to better treat cancer could then be transferred to research in attempts to halt the spread of an invasive species, such as the cane toad.

In thinking about evolutionary applications as an emerging field, I see at least three main types of evolutionary applications. First, investigating the evolution of an organism is useful because that organism's biology matters to the practical problem. For example, determining whether a doctor accused of infecting his partner with HIV samples from a patient depends on learning about the evolutionary relationships of the partner's viral load among other viral samples. Another type of application involves searching for parallels between an evolved system and an aspect of human culture. An example is similarities between arms races in conflicts between human societies and arms races that occur within and between nonhuman animals. The third type of application, which is central to evolutionary medicine, is viewing human health from the perspective of human beings as evolved organisms that live within an ecosystem.

Applied evolution is not only interdisciplinary but also multidisciplinary. Consequently, no one can be an expert on all of the fields related to applied evolution. This book is for experts in one field that pertains to applied evolution who are interested in other aspects. It is also for students at the undergraduate and graduate levels. It is they who will most likely forge and deepen the new connections. Finally, the book is for the nonexperts, the general readers, who are interested in following how evolution affects their lives. One of the public relations challenges that evolutionary biology faces is that most people do not see it being all that relevant to their daily lives. Even many who accept evolution do not grasp how far Darwin's reach goes. I hope to help change that perception with this book.

This book is divided into four sections (in order): Health, Food, Environment, and Society. The arrangement is somewhat arbitrary; each section can be read independently of the others, though chapters within sections will build on one another. Due to space limitations, not every subject affected by evolutionary applications can be covered. For instance, I will not cover much about the exciting work in uses of evolutionary principles in computation. Although several chapters will discuss the microbiota, no one chapter will be devoted to it.

This book cannot touch on all topics related to evolutionary applications. Moreover, science is a continuing process. Accordingly, I will provide updates and other information in a blog found at: https://wordpress.com/page/darwinsreach.wordpress.com/home

Acknowledgments

This book started with discussions I had with my editor Chuck Crumly at the Society for the Study of Evolution meetings in Portland, Oregon, in 2017. I thank Chuck for his patience, encouragement, and guidance over the years. I also thank Ana Lucia Eberhart and Sherry Thomas at Taylor and Francis for their excellent work in getting the book ready for production. I thank Aruna Rajendran and her team for their outstanding work in copyediting and producing the book. Michael DeGregori illustrated several excellent figures for the book. I also thank Mohamed Noor for graciously writing a thoughtful foreword to the book.

Three experiences stand out in the development of this book. First, starting more than a decade ago, I co-organized and then co-ran a working group at the National Evolutionary Synthesis Center (NESCent) on Communicating Human Evolution. Many ideas about the book came from interactions with various members of the group. I thank the members of the group, most notably my co-organizers Weezie Mead and Jim Smith.

Rich Kliman, the editor of *The Encyclopedia of Evolutionary Biology*, invited me to be a section editor on applied evolution. Soliciting and reading entries in applied evolution has helped me gain more expertise in a broad range of topics in applied evolution, a skill that has been extremely useful in writing this book. I thank Rich and the contributors to the section.

More recently, I participated in a small meeting/workshop on cancer and evolution organized by Jeff Townsend and Jason Somarelli. From this experience, I learned much more about cancer and the interaction of cancer and evolutionary biology. I also have enjoyed the pleasure of working with extraordinary researchers working at this intersection. I thank Jason and Jeff and the participants.

Numerous people have helped me during the course of the book. The following have helped through correspondence and/or interviews: Lynn Adler, Cathie Aime, Craig Albertson, Tim Anderson, Jake Barnett, Eve Beaury, Regina Baucom, Dan Blumstein, Dan Bock, Amy Boddy, Ana Caicedo, Vincent Cannataro, Ben Chan, Andy Conith, Graham Coop, James Crall, Holly Dunsworth, Doc Edge, David Enard, Bruce Grant, Dan Hartl, David Hillis, Vaishali Katju, Morgan Kelly, Sophia Kimmg, Rob Kulathinal, Joe Lachance, Carol Lee, Cinnamon Mittan, Randy Neese, Ben Normark, Evan Palmer-Young, Bret Payseur, Melissa Pespini, Loren Rieseberg, Lynnette Sievert, Jason Somarelli, Paul Turner, Mike Wade, Jack Werren, Andrew Whitehead, Melissa Wilson, and Kelly Zamudio.

In addition, many have given advice on drafts of chapters and/or the prospectus. These include those in the preceding paragraph as well as Courtney Babbitt, Wendy Curtis, Hannah Broadley, Julie Froehlig, Lian Gao, Rodger Gwiazdowski, Diane Kelly, Rich Kilman, Barbara King, Duncan Irschick, Emily Monosson, Josh Moyer, Dina Navron, Jeff Podos, Sarah Purvis, Suzanne Veyrat, Abby Vander Linden, Marlene Zuk, and the UMass OEB BAM(P) reading group. I am also grateful for the support from family and friends. Thank you!

Introduction: What Is Evolution and Why Is It Important?

SUMMARY

Within a population, individuals differ. These differences affect their ability to survive and reproduce. Offspring resemble their parents. Given these conditions, biologists expect the composition of the population to change after repeated rounds of reproduction and screening (selection). This is the essence of evolution by natural selection, a process by which organisms adapt to their environment. Chemists have used an analogous process (directed evolution) to create new enzymes and other molecules: herein, they go through cycles of new variants and screening those variants for desired characteristics. This is but one application of the principles of evolution, the overarching theme of the book. This chapter, which serves as an introduction to evolution, addresses several topics that include the difference between acclimation and adaptation and the importance of the environment (including the vast microbial world that organisms interact with) in discussions of evolution. Evolution is not just a process; it also has a historical aspect, which can be important in applications of evolution. All organisms are related, but some are more closely related than others. This nested hierarchy of relatedness and how it can be applied in medicine and other realms is discussed.

"Biology has this one process that is responsible for all this glorious complexity we see in nature"—Frances Arnold.[1]

On October 3, 2018, Frances Arnold was awarded the Nobel prize for chemistry for research in what is called directed evolution.[2] Arnold, the Linus Pauling Professor of Chemical Engineering at Cal Tech, won the prize for developing and using methods that apply the principles of evolution to create new molecules for use in various capacities.

Arnold's method has two essential components: inducing changes in the genes that coded for enzymes in bacteria and then screening the bacteria, picking those that performed the task the best to start the next round. The induced changes are mutations. Like mutations in the natural world, they are the source for variation. A screening process goes on in nature, too. It's called natural selection. Repeated cycles of directed evolution and natural selection each can lead to profound changes over time.

One application of directed evolution is in the manufacture of sitagliptin, a drug used to manage type II diabetes by smoothing out the peaks and valleys of blood glucose levels. A key step in the standard production of this drug currently uses a rhodium-based catalyst. This is nasty stuff; its use causes the release of substantial toxic waste. Researchers at Codexis, a protein engineering company, used directed evolution to generate an enzyme that could replace this rhodium-based catalyst and can work under the harsh conditions involved in drug production[3]. With the evolved

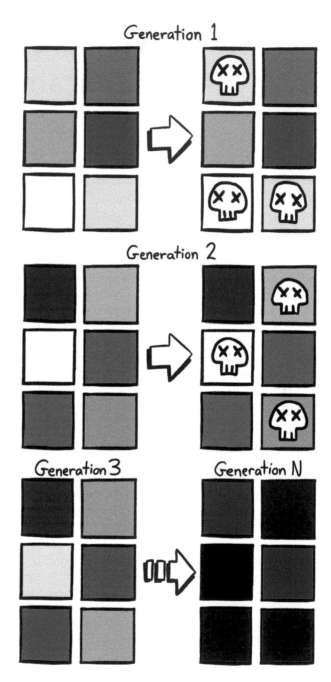

Evolution by natural selection arises as a repeated cycle of reproduction and differential survival of genotypes. We assume that the colors of parents and their offspring are correlated, though not necessarily perfectly. In this case, lighter squares are more likely to survive than darker ones. Eventually, the population becomes uniformly dark. Note that in a different environment, darker squares could be favored. (Note that evolution by natural selection can also act via differential reproduction.)

Source: Illustrated by Michael DeGregori.

enzyme, sitagliptin production is not only more efficient but also more environmentally friendly because it generates less solvent waste.

Directed evolution often takes advantage of a feature of biochemistry: enzymes typically catalyze more than one reaction. A given enzyme is often most efficient at catalyzing a given reaction but will also catalyze other reactions less well. These "promiscuous" reactions are important in several respects. First, they are excellent starting material for direct evolution of novel functions. As directed evolution proceeds, weak and inefficient reactions can become stronger and more efficient. Alternatively, drugs that are enzyme based could have side effects due to these other promiscuous reactions. Directed evolution could reduce the extent to which the enzyme catalyzes these reactions; by making the enzymes more selective, directed evolution can minimize side effects of drug[4]. The same principle occurs in the natural world; because enzymes and other molecules have multiple functions, changes that occur via evolution by natural selection can have side effects, and these side effects can propel further evolution.

While there are profound similarities shared by the screening that occurs in nature and the screening that Arnold and other biologists use in directed evolution, the two types of screening do differ in one key respect. In their experiments, Arnold and her colleagues set a specific task for the bacteria to perform. In nature, the task is not determined by anyone; it is simply the existential task of survival and reproduction. Individuals in natural populations vary in numerous ways, some of which are obvious while others are subtler. If the differences help particular individuals to survive the multitude of slings and arrows they encounter, such individuals are more likely to transmit genes to the next generation. Their offspring, accordingly, should also survive and reproduce better in those environments.

Natural selection usually leads to organisms having a better fit with their environment.[5] Biologists call the features that this process generates adaptations. The resistance to drought of a cactus in a desert is an adaptation. The rapid flight of hummingbirds and their ability to hover are adaptations. The fleeing behaviors of a mouse trying to evade an avian predator are adaptations, as are the keen vision and rapid acceleration of the hawk trying to catch the mouse. Bacteria that live in hot springs have enzymes that can withstand very high temperatures, in contrast to the less heat-tolerant enzymes of bacteria that live in milder environments. Those are adaptations too. All are generated by mutation and natural selection.

Although natural selection and evolution are related processes, they are not the same. Evolution by natural selection requires that the traits be passed on from parent to offspring. This does not mean that the offspring need to exactly resemble their parents. They just have to be more like their parents than other members of the population. The resemblance of parents and offspring and the mechanisms behind this resemblance are called heredity.[6] Note that without heredity, selection can still take place. It just would not result in evolutionary change.

Without selection, heritable differences still evolve over the course of the generations, just from random changes. Biologists call this random process genetic drift. These changes that accumulate over time due to this random genetic drift would not affect how well individuals survive and reproduce, but they still are important in other realms. For instance, most of the differences that biologists use to track the evolutionary history of organisms arise from changes generated by random genetic drift.

Both in nature and in the experimental setting of directed evolution, selection can be complex. The success of a particular genetic variant is context dependent. Certain variants perform well in a given environment but not well in others. For example, consider a small mammal that faces visual predators. One would expect individuals whose coloration more closely matches the background to be more protected from predators and thus have a higher survival rate than those whose coloration does not blend as readily. Light coloration should be selected for if the background is light; in contrast, dark coloration should be selected for if the background is dark.

Recent work by Hopi Hoekstra, an evolutionary biologist at Harvard, demonstrates this example of the context dependence of evolution.[7] Hoekstra has spent the last 20 years investigating the genetic basis of coloration in deer mice of the genus *Peromyscus* and found few key pigmentation genes affect the coloration patterns. One of these is the signaling gene *Agouti*.

Led by her postdoc, Rowan Barrett, Hoekstra's team added wild mice to field enclosures in the Sand Hills of Nebraska. Some of these enclosures were light in color and some neighboring enclosures were dark. Consistent with their expectations, Barrett and Hoekstra found the dark mice survived better in dark enclosures and the light mice survived better in light enclosures. This led to changes in the overall distribution of color in the dark and light enclosures, in opposite directions. But the changes seen were not just in the color phenotypes but also in the underlining genetic makeup of the mice populations. Specifically, there were changes in the frequencies of specific nucleotides at *Agouti*, and these changes were in opposing directions in the light and dark enclosures.

Context is not limited to the external environment: the effect of a genetic change can also depend on the particular genetic backgrounds. A genetic change in one genetic environment might be advantageous, but that same change might be deleterious in a different genetic environment. Writing with Philip Romero, Francis Arnold discussed the complexities in both directed evolution and evolution in the natural world.[8] One such complexity they discuss involves protein stability. Enzymes are proteins; as such, they require a certain amount of stability to function. Changes that improve the efficiency of the reaction the enzyme performs often decrease the stability of the enzyme. If the initial enzyme is more stable, changes that increase efficiency at the expense of stability can occur. But if the initial enzyme were already near the edge of stability, such changes would be much less likely to evolve.

Directed evolution is useful in determining the relationships between changes in genotypes and changes in outward traits (phenotypes) of interest; as such, this applied science can benefit basic science in genetics and evolutionary biology. A similar virtuous cycle has long existed between evolutionary biologists and breeders of animals and plants.[9] Charles Darwin, himself a pigeon breeder, used information from breeders in the development of and arguments for his evolutionary theories during the middle of the 19th century. Our understanding of inbreeding and its evolutionary consequences came from Sewall Wright, a pioneer in evolutionary biology who started his academic career during the 1920s in the employ of the US Department of Agriculture. Breeders and evolutionary genetics are still working together to develop new methodologies for crop and livestock improvement in this age of genome sequences.

While individuals are products of evolutions, the process of evolution does not happen to an individual; it is populations, not individuals, that evolve. If you spent several weeks in the Rocky Mountains, or some other high elevation area, you would likely acquire physiological changes, such as an increase in red blood cell number. This is acclimation, not evolution. The capacity to acclimate, though, is an adaptation. Individuals are able to temporarily change their physiologies in response to changes in their environment because natural selection over numerous generations favored such flexibility; individuals that had flexible responses to the changes inherent with increased altitude were better able to survive and reproduce. Human populations that have lived at high elevation for several generations, however, have evolved adaptations for the reduced oxygen concentrations they experience.[10] Note that the adaptations of humans who lived at high elevation are different from the acclimation that individuals who live at low elevations experience when they go uphill. We'll dive deeper into these adaptations to high altitude in Chapter 4.

Acclimation is not just important for human medicine; how organisms respond via acclimation will be relevant to how they respond to climate change and other environmental challenges. It is also relevant to other practical challenges; for instance, an understanding of acclimation responses could be useful in determining how well pesticides will work to protect crops from insect pests.

Although it is populations that evolve, evolution can occur within individuals in the cases where populations exist within the individual. Cancer is an evolutionary process where populations of cells evolve within the individual animal. Similarly, bacterial cells within the microbiota of a larger organism can evolve. In this sense, a human body serves as an ecosystem where microbial populations can enter, evolve, and exit.

Evolution is not just selection and the other processes that take place in populations. It also has a historical aspect. Evolution is about dinosaurs and the transitions whales made as they went from being land mammals to a return to the seas. It is about which particular beetle species is most closely related to another beetle. Evolution goes on continuously and can be observed in nature and in the lab, but it is also steeped in history that goes back to the start of life, about 4 billion years ago.

Let's jump back to London in 1837. Here, Charles Darwin, age 28, was writing the travelogue about his experiences as a naturalist and scholar on the expedition of the HMS Beagle that had ended the previous year. The notes that Darwin had already published about that five-year voyage would soon make him a local celebrity. But Darwin at that time was also in flux; the experiences he saw traveling on the Beagle and in excursions in South America were changing his worldview about the origins of new life forms. He was ready to make the radical leap toward what we now call evolution. Writing in his notebook, Darwin drew a sketch of a tree-like figure along with the words: "I think".[11] Darwin's tree diagram was a representation of the connectedness of (presumably all known) life forms. A couple decades later, Darwin published his *Origin of Species*. Here, he had one figure. It was also a tree, another representation of the connectedness of life. The most revolutionary aspect of this diagram was not so much the argument that organisms could change via natural populations, but that the life that we know—bumblebees, turnips, yeasts, oak trees, mice, octopuses, and most certainly, humans—had a shared, common ancestor and had evolved differences from that common ancestor.

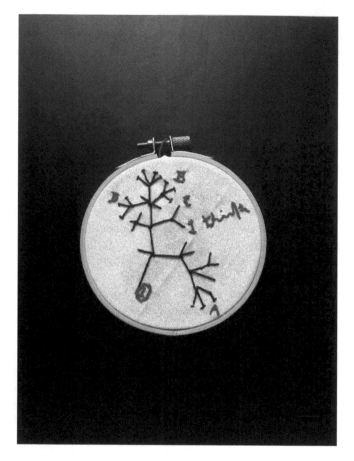

Cross-stitch of Darwin's "I think" drawing.

Source: Courtesy of Sarah Emel.

Following Darwin, evolutionary biologists have used trees to represent evolutionary relationships. These trees highlight a crucial principle about evolution: all organisms are related, but some are more related than others. Family relationships serve as a useful analogy here. Let's say that Ann and Barbara are sisters and that they have a first cousin, Charles. Ann and Barbara are more related to each other than they are to Charles because they share common parents with each other but only share common grandparents with Charles. More distantly related cousins share still more distant ancestors. The same works with species. Species that are more closely related share more recent common ancestors than do more distantly related species. A key point is that relationships in evolutionary biology are nested. Humans, bonobos. and chimpanzees form a group (what biologists call a clade) that is closely related to other apes. Apes, another clade, are nested within the clade of primates. Primates are nested within the clade of mammals. Mammals are nested within vertebrates. Vertebrates are nested within the Metazoa (animals). We can continue drawing wider boxes to include all known life.

In this way, evolutionary biology is much like geology and astronomy. These fields "deal not only with general processes and mechanisms, but also unique historical particulars".[12] These historical particulars—what the Canadian evolutionary biologist Ryan Gregory calls "evolution's path"—is also important with respect to applications of evolution.

The shared evolutionary history lies at the base of animal models used in medicine and in the testing of nonhuman animals for products to be used in humans.[13] Animals that share a closer common ancestor with us are more likely to share common physiology, though there are exceptions. The nested hierarchy of relatedness can be applied here; often, researchers begin with trials of medicines using rodents before moving to more closely related animals. Medical researchers need to keep in mind that the animals they use can differ in some fundamental respects from humans. For instance, mice and humans differ in the structure and function of their telomeres (ends of chromosomes). These differences limit the mouse model in some aspects of the study of cell senescence and cancer.[14]

Upon extreme and acute stress, the hearts of some people will misfire. This is why heart attacks often increase during severe earthquakes. Other mammals show a similar phenomenon—capture myopathy—when subject to the extreme stress of being handled. A collaboration between the UCLA researchers Dan Blumstein, a behavioral ecologist, and Barbara Natterson-Horowitz, a cardiologist,[15] has used data about capture myopathy and the patterns of evolutionary relationships in a wide array of mammals to assess the factors that make certain animals more susceptible to capture myopathy. Preliminary analysis suggests that risk factors include large size, big brains, and long life spans. Such analysis can lead to a better understanding of capture myopathy, which could be useful in mitigating its existence in wildlife management and in agriculture. There could also be clinical benefits for humans with cardiac stresses.

Collaborating with the science journalist Katryn Bowers, Natterson-Horowitz had documented the large number of cases where humans and other mammals shared similar diseases and other ailments. In their book, *Zoobiquity*, they note:

> Western lowland gorillas die from a terrifying condition in which the body's biggest and most critical artery, the aorta, ruptures. Torn aortas also killed Lucille Ball, Albert Einstein, and the actor John Ritter, and strike thousands of less famous human beings every year.[16]

The similarity does not end at cardiovascular ailments. Natterson-Horowitz and Bowers also recounted how cancers in humans and dogs behave in similar ways; for instance, osteosarcoma, a form of bone cancer, often strikes teenage humans as they undergo a growth spurts and large dogs at a comparable state of development. Variants at the gene *BRCA1* predispose humans and English springer spaniels to increased risk of breast cancer. Instances of self-harm, self-medication, and other attributes that we have traditionally thought of being unique to humans are also found in other animals. We are not nearly as special as we would like to think we are.

That all life shares common ancestry means that there is no such thing as a more evolved or a less evolved organism. We think of bacteria as being simple, but they have been evolving on the planet for almost four billion years. In this time, they evolved complex biochemical machinery. While there are trends toward gains of new traits in certain lineages during evolution, the reduction and even loss of traits is pervasive through evolution.[17] For instance, as the ancestors of whales returned to a marine lifestyle, they lost their limbs. Presumably, such traits hindered survival and reproduction in the oceans. What happens on a given path of evolution is dependent on the environments that the organisms face.

The environments that organisms face along their evolutionary path include other organisms. Pathogens and parasites have always been part of our evolutionary path. These pathogens and parasites have evolved in response to us, while we have evolved in response to them. This mutual evolutionary circle, called coevolution, can be very potent because it is self-reinforcing. Evolutionary responses to physical features of the environment should taper off if the environments stop changing. But with coevolution, the circle never stops. As humans adapt to the pathogens and parasites, they adapt to us.

Coevolution is also important with respect to the microbiota. The human body contains about as many microbial cells as it does human cells. As the British science writer Ed Yong demonstrates,[18] we indeed contain multitudes—these microbial cells are in profound ways parts of us, and not just connected to us. Our human cells and our microbial cells have coevolved in ways that we are just beginning to figure out.

Although each branch of the tree of life has its own particular historical path, regular patterns do emerge. To some extent evolution is predictable: certain outcomes are more likely than others under particular sets of conditions. The paleontologist and evolution advocate par excellence Stephen Jay Gould famously posed the question: if we were to replay the tape of life, would we get the same result? Three decades ago in his classic book *Wonderful Life*, Gould[19] stressed the quirkiness, the path-dependency, the contingencies of evolution; he envisioned each play of the tape leading toward rather, and perhaps even strikingly, different outcomes.

More recently, the pendulum has swung—mind you, not completely, but noticeably—toward the predictability of evolution. In his recent book *Improbable Destinies*, the Washington University evolutionary biologist Jonathan Losos[20] marshals the evidence demonstrating that evolutionary outcomes can often be predictable.

Guppies from Trinidad illustrate just how predictable evolution can be. Here, guppies that live in areas where predators are common differ in notable respects from guppies of the same species that do not regularly encounter predators. Where predators are absent, male guppies are vibrantly colorful in contrast to drab females. Where predators are common, the males are as drab as the females. Guppies who face predators are smaller; they live fast and die young, compared with those that do not face this threat. The guppies even behave differently: those from predator-rich environments are skittish, while those living where predators are absent boldly go where their counterparts fear to tread. But wait! These changes can be readily reproduced experimentally. Expose a population of guppies that evolved in a predator-free environment to predators and wait a few generations; it does not take that long. They will become smaller. The males will lose their vibrant colors. They will get more skittish. They evolve faster lifestyles, maturing earlier and dying younger (even if they are not eaten). These guppies will evolve to be like those who lived with predators. This can be done in reverse too. Take guppies that faced predators and place them in an environment without predators. Again, wait a few generations. The males will evolve vibrant colors. They will get larger. They will become bolder. This is repeatable evolution in action!

Repeated outcomes of evolutionary paths are a good indicator of the action of natural selection. In the pages that follow, we will see examples of such repeated outcomes that have likely been driven by natural selection. Different populations of fish exposed to toxic petroleum-based compounds independently evolved responses in the same gene, though not at precisely the same location of the gene. The ability to continue to be able to digest lactose (the sugar in milk) into adulthood has evolved in multiple human populations from changes at the same gene but at different locations of the gene.

Nearly half a century ago, the Russian-born evolutionary biologist Theodosius Dobzhansky argued "nothing in biology makes sense except in the light of evolution".[21] Because organisms evolve, one cannot fully understand the patterns and processes of biology without referring to evolution. In the pages that follow, I show that evolutionary processes and paths matter much to our everyday world, in areas that range from medicine to food to the environment to our society.

NOTES

1. Arnold is quoted in Gibney et al. (2018, p. 176)
2. Directed evolution is also known as "test tube" and *in vitro* evolution. Frances Arnold shared the prize with George Smith at the University of Missouri–Columbia and Sir Gregory Winter, a Fellow of Trinity College of Cambridge University. Their work was in developing phage display, a technique that uses evolutionary principles to produce antibodies.
3. See Savile et al. (2010) for details of the use of directed evolution to produce sitagliptin.
4. Hammer et al. (2017).
5. There are exceptions; evolution by natural selection is not guaranteed to increase the fit of the organisms to their environments, but these exceptions are beyond the scope of this introduction.
6. See Zimmer (2018) for a detailed treatment of the history of the study of heredity as well as for the current state of our knowledge about human heredity. See also Johnson (2018).
7. Barrett et al. (2019b).
8. See Romero and Arnold (2009).
9. See Johnson (2016) for more information about the contributions evolutionary biologists have made to breeding programs. We will return to this subject in Chapter 8.
10. Lachance and Tishkoff (2013) review cases of adaptations in human populations, including those involving high elevations.
11. See Quammen (2018) for discussion about Darwin's representation of the history of life as trees. Quammen's book also discusses complications to this tree metaphor. Some organisms, especially microbes, can sometimes exchange genes with rather distant relatives. Thus, branches can come together. This complication, while extremely interesting, does not alter the fact that all known life is related.
12. Gregory (2008, p. 50).
13. Kelly (2016).
14. Hanahan and Weinberg (2011).

15. Blumstein et al. (2015).
16. For more details, see Natterson-Horowitz and Bowers (2012). The quote is on p. 7.
17. Johnson et al. (2012) provide a summary of case examples of trait loss and decay. They also further discuss the broader implications of pervasive trait loss throughout evolution.
18. Yong (2016) is arguably the best recent book on the microbiome and its role in our evolution as well as the evolution of other animals.
19. Gould (1989).
20. Losos (2017).
21. Dobzhansky (1973).

Section I

Health

1 Paging Dr. Darwin
Evolutionary Medicine in the 21st Century

SUMMARY

A major problem in medicine is antibiotic resistance: as bacteria evolve resistance to antibiotics, the effectiveness of this essential treatment soon declines. This chapter begins with a new treatment designed to inhibit the evolution of resistance with (bacterio)phages, viruses that attack bacteria. In adapting to a specific type of phage, the bacteria are forced to evolve less resistance to antibiotics. Bacteria that cannot be resistant to both the phage and antibiotics are an example of a tradeoff, a fundamental concept in evolution. The chapter then discusses other ways of using evolutionary principles to limit the evolution of antibiotic resistance. The applications of evolution to antibiotic resistance are a subset of the larger field of evolutionary medicine. A central question within evolutionary medicine is why we get sick and remain vulnerable to disease. Some explanations include the more rapid evolution of pathogens and parasites compared to our own, that some symptoms of illness (such as fever) are actually defenses against infectious disease, mismatches between the environments we live in and those our ancestors evolved in, and that natural selection favors traits that enhance reproduction and not necessarily our health or our happiness.

"Evolutionary biology offers a framework for organising the diverse facts in medicine, and a way to understand why the body is vulnerable to disease."—Randolph Nesse.[1]

FIGHTING RESISTANCE IS NOT FUTILE

In 2012, Dr. Ali Khodadoust, then a 76-year-old ophthalmologist from New Haven, Connecticut, had coronary bypass surgery. Soon after surgery, complications ensued: Khodadoust developed a bacterial infection coming from the graft the surgeons had used on his aorta. The infectious agent was the bacterium *Pseudomonas aeruginosa*, one commonly found in hospitals. *Pseudomonas* usually is not harmful, but it can be for people who have deficient or compromised immune systems. Due to his open-heart surgery, Khodadoust's immune system was likely impaired.[2]

Despite multiple treatments with antibiotics, Khodadoust's infections kept reoccurring.[3] The *Pseudomonas* quickly had evolved resistance to the antibiotic ciprofloxacin. Khodadoust became septic. His doctors determined that the bacterium was not completely resistant to a different antibiotic (ceftazidim), which they gave him intravenously. After several weeks, the infection was sufficiently reduced. But soon after, it came back again. By 2015, Khodadoust seemed to have exhausted his options.

DOI: 10.1201/9780429503962-2

3

Khodadoust's story is just one example of the growing threat of antibiotic-resistant bacteria. Almost as soon as one antibiotic becomes commonly used, bacteria evolve resistance to it. Bacteria replicate quickly and generate large numbers, attributes that are conducive for very rapid evolution. Moreover, bacteria have the capacity to transfer genes across to even relatively distant relatives through a variety of mechanisms. This process—lateral gene transfer—also contributes to the ability of bacteria to evolve antibiotic resistance. In particular, *P. aeuruginosa* is adept at evolving resistance, having many genetic mechanisms of resistance available to it.[4]

How can we mitigate this resistance threat? One idea arises from the natural history of bacteria. Like just about every other life form, bacteria have natural enemies. Usually, the biggest threats to bacteria are bacteriophages—"phages" for short—viruses that specifically target bacteria. The word "phage" comes from the Greek "phagen", in English, "to devour". And phages are proficient at devouring their bacterial hosts! Many phages look like space capsules. They dock at a receptor site of the bacterial cell and insert their DNA. This phage DNA then is replicated many times. Here the bacterium does the heavy lifting: the phage DNA is replicated by the machinery the bacterium uses to replicate its own DNA. How devious! These phage DNA segments are packaged into capsules, which then bust through the cell, leaving devastation. The cycle is then ready to repeat, only with many phage capsules instead of just one. With just a few such cycles, bacterial infections can be wiped out.

What if one of these phages could be used to thwart the antibiotic-resistant bacteria infecting Khodadoust and others like him? This idea—phage therapy—has a long and murky history. Phages were used to treat bacterial infections from almost as soon as they were discovered in the early part of the twentieth century. Outside of the countries dominated by the former Soviet Union, phage therapy was largely abandoned for many decades. As we will see, this treatment, with some new twists, has been making a comeback in recent years.

The French Canadian virologist Felix d'Herelle dominates the early history of phages and phage therapy.[5] Well-connected, peripatetic, and largely self-taught, d'Herelle had an interest in the transmissible particles that could eat through bacteria, causing what he called "taches vierges" or "clear plaques" on bacterial lawns. In the summer of 1915, an outbreak of severe hemorrhagic dysentery swept through French troops stationed outside of Paris. While stationed at the nearby Pasteur Institute, d'Herelle investigated the cause of the dysentery. He noticed that these transmissible particles would also devour the *Shigella* bacteria that had caused the soldiers' dysentery. Perhaps, he thought, these particles—which we now know as phages—could be used to treat infections.

Also working at the Pasteur Institute was a virologist from the then Soviet Republic of Georgia, George Eliava. The younger virologist caught d'Herelle's enthusiasm for phages and phage therapy, leading to him to establish the Bacteriophage Institute in the Georgian city of Tbilisi. Phage therapy studies flourished at Tbilisi. Unfortunately, Eliava did not. Because he had angered Stalin's friends, Eliava was named an "enemy of the people". He was executed in 1937.

Research on and applications of phage therapy continued during the second half of the twentieth century in the Soviet Union and the associated Eastern Bloc. Phages were even used as a preventative measure in schools, military bases, and other areas where large numbers of people were in close association due to the enhanced risk of rapid outbreaks in such localities.

With the advent of the effective antibiotics, the interest in phage therapy waned outside of the Soviet Union. Still, as an ever-growing list of antibiotics lost their effectiveness due to bacteria evolving resistance, phage therapy is staging a revival. One of the limitations of phage therapy is that bacteria can also evolve resistance to the phage. This limitation is mitigated, though not completely, by the prospect of phages evolving to counteract the bacterial resistance. Such coevolutionary arms races are frequent in nature.

Ben Chan and Paul Turner, who are evolutionary biologists at Yale University, have developed a new approach to using phage therapy that takes explicit advantage of bacteria evolving in response to the phage. They identified a particular phage that could make the bacteria less resistant to antibiotics as the bacteria evolve responses to it.[6] This phage, known as OMKO1, was found in water samples from Dodge Pond in East Lyme, Connecticut—not far from Yale's New Haven campus.

Here's how it works. The OMKO1 phage devours the bacteria. Variation exists within the bacteria in how vulnerable they are to the phage. Over time, the bacteria that are best able to withstand the phage persist and divide, leaving a population of bacteria that are more resistant to the phage. Nothing surprising is happening—this is simply the evolution of resistance to the phage. But there is a twist here. The bacteria that are resistant to the phage have different efflux pumps than those that are sensitive. The efflux pumps are protein-based structures located on the bacterium's cytoplasmic membrane that transport chemicals. While making the bacteria more resistant to the phage, this change in the efflux pumps also makes the bacterial cells less resistant to several mainstream antibiotics. The bacteria face a tradeoff that traps them in a box: they can be resistant to the phage or they can be resistant to the antibiotic, but they cannot be resistant to both. This double whammy of the phage and the antibiotics ought to keep the *Pseudomonas* infection in check.

Chan and Turner and their team then used the phage to treat ophthalmologist Ali Khodadoust. They applied a cocktail of the phage and the antibiotic ceftazidime at the site of the infection.[7] Soon after being given the phage cocktail, the bacteria infecting him became less resistant to several antibiotics. With reduced resistance, the bacteria could be brought under control by antibiotic treatment. Khodadoust's infection was cleared, likely allowing him a couple extra years of higher quality life and the ability to resume his ophthalmology work. He died in March 2018 at the age of 82.[8]

Surgery patients are not the only ones who are susceptible to invasion by *Pseudomonas* and other opportunistic bacteria. So are people who have cystic fibrosis, a genetic disease that causes a buildup of mucus in the lungs.

Cystic fibrosis is due to mutations in a particular gene known as the cystic fibrosis transmembrane conductance regulator. Yes, that's a mouthful. Hence, the gene often goes by the moniker CFTR. But its full name provides clues as to what the gene does. The word "transmembrane" means "across a membrane". Our cells are replete with membranes. The membrane here is the outer one of the cells that provide a protective lining of our lungs. Conductance sounds like something involved with electricity. In fact, it is the movement of ions, charged atoms. Here, the ions are chloride ions, one

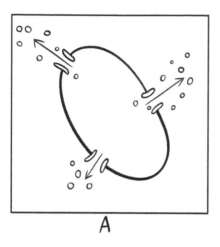
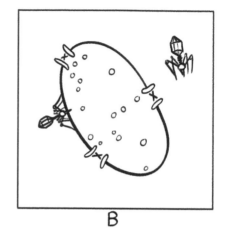

A   B

In the absence of the phage, the bacteria are resistant to the antibiotic in part because their open efflux pumps allow for easy removal of the antibiotic. In the process of adapting to the phage, the bacteria evolve efflux pumps that are more closed. This change makes the bacteria less resistant to the antibiotic.

Source: Illustrated by Michael DeGregori.

of the components that make up ordinary table salt (sodium chloride). So this gene provides information to produce a protein—an ion channel protein—that rests inside the membrane of cells lining the lungs. This ion channel protein regulates the balance of ions by pumping chloride ions out. A person who has two mutated versions of this gene will have cells that are less adept at transporting those chloride ions out of the lungs. As a consequence, their lungs will become salty, leading to a buildup of mucus in the lungs.[9]

One cystic fibrosis patient, a 22-year-old woman, came to Chan's team in October 2017 to see if the phage would work for her.[10] Over the next two months, the team examined whether the OMKO1 phage could work on her infection. Spoiler alert: it could! After they received approval from the FDA and other regulatory bodies, the team treated her, giving her the phage cocktail that she could self-administer with a nebulizer. Even after just two days, her infection also quickly responded, and she had more energy. Although the bacteria had not completely cleared after a ten-day course of the phage cocktail, it did become sensitive to an array of antibiotics. With this regained antibiotic sensitivity, the infection could be tamed.

During December 2018, almost exactly a year after the woman with cystic fibrosis had received the phage cocktail, I met with Paul Turner and Ben Chan in Turner's office at Yale.[11] It was a chilly, overcast day in New Haven, with snow threatening to fall, but not very successfully. Over a lunch of burritos, we chatted.

Prior to his studies of phage therapy, Turner had more than two decades of experience examining basic questions of evolutionary biology using viruses. Some of the questions he had examined involved game theory showing that viruses (which completely lack anything like what we would call consciousness) still can engage in strategies with other viruses. Other projects examined how microbes evolve to specialize

Ben Chan searching for phage.
Source: Courtesy of Paul Turner.

Paul Turner in his lab.
Source: Courtesy of Paul Turner.

or generalize in resource use and how new species of microbes can evolve. Turner told me that while he had focused on basic science for most of his career, he also has had a long-standing interest in medical applications of this work. This interest in part grew during the 1990s as he watched the ravages of the AIDS epidemic when he was a graduate student working in southern California.

Chan, too, started in basic research. His doctoral research had been on poison dart frogs. He had been enraptured by the microbial world after a project on the skin microbiome of frogs and its relevance to resistance to the chytrid fungus that has threatened many frog species. After finishing his PhD, Chan worked at a private company generating phage products for agricultural uses. Turner had recruited Chan to join his team initially for a different project. Chan soon convinced Turner that evolutionary principles, as we have already seen, could be applied to phage therapy in clinical settings.

We also discussed why phage therapy had been abandoned in the West. Turner and Chan gave several possible reasons. It was partly because most of the early work had been done in the Soviet Union, partly because much of the early work had been overhyped, partly because it is easier to make money by producing antibiotics. It's very difficult to make money off phages. Another factor was just inertia. Regardless of why phage therapy had been ignored throughout much of the twentieth century, it seems a promising tool to be used now, especially when bacteria evolve resistance to traditional antibiotics.

Another promising strategy for thwarting antibiotic resistance in disease-causing bacteria is to stop or limit their ability to cause disease, rather than killing them. Selection requires changing the survival or the reproductive output of certain individuals. Hence, drugs that do not harm the bacteria should not cause selection for resistance. In contrast, it's the ability to cause disease—what is often called "virulence"—and not the presence of the bacteria per se that is of interest with regard to human health. If a drug could reduce or eliminate the virulence of bacteria, it would be useful in managing the disease. Moreover, the drug should remain useful for a long time because evolution of resistance to it should be limited.

The initial phase of infection often involves bacteria adhering to the host cells.[12] For example, some strains of E. coli that cause urinary tract infections generate pili, appendages that attach to epithelial cells of the urogenital tract. Without these pili, the bacteria could easily be washed away by the flow of urine through the tract. To produce these pili, the bacteria ramp up expression of a group of genes, the pap pilus operon. Some of the most virulent bacteria have multiple copies of this operon, enabling rapid formation of pili upon contact with host cells. One way to limit adhesion of these bacteria is to use structures—pilicides—that mimic subunits of the pili. As these pilicides are added to the developing pilus, they prevent it from further elongation and render it nonfunctional. Researchers have achieved success with this technique in mice; by reducing the bacteria's ability to adhere to cells, the pilicides have greatly reduced bladder infections.

Virulence factors are not limited to the factors that produce pili. Some bacteria, like Bacillus anthracis (the one that causes anthrax) produce toxins that hinder the cellular functions of their hosts. Still others, such as the Pseudomonas aeruginosa we met at the start of this chapter, produce biofilms, wherein colonies of bacteria shield themselves by secreting chemicals that produce an outer protective coating. Just as with the pilicides, researchers are increasingly able to target these other virulence factors, leading to making them less adept at causing disease.

At first glance, one would expect the loss of virulence not to result in selection for resistance of the drug. After all, the drug does not kill the bacteria; instead, it only alters its state. But wait a minute. What if the drugs that reduce virulence alter the survival and/or reproductive output of the bacteria?[13] If that were the case, then being resistant to the drug may confer a selective advantage, and resistance would be expected to evolve. Whether this happens depends on how virulence works, specifically in the relationship between the virulence factor and the ability of bacteria to reproduce (its fitness). Some of these are beneficial to the bacteria by increasing survival or reproduction; others are non-beneficial and either decrease an aspect of bacterial fitness or have no effect. If the virulence factors are non-beneficial, it should be easier to prevent evolution toward virulence. Further research examining the fitness benefits and costs to different virulence factors in a variety of environments is warranted.

A third way biologists have applied evolutionary principles to the problem of antibiotic resistance involves examining how the resistance evolves. Mutation, the generation of new variants, is an essential part of the evolutionary process. Various chemicals, such as ethylmethylsulfonate, and X-rays make DNA replication more error prone. For nearly a century, geneticists have used these agents to produce mutants and to study the mutation process. More recently, geneticists have also used transposable elements, which are segments of DNA that can move from one location to another within the genome, for a similar purpose. When a transposable element lands within particular regions of a gene, it can disrupt the gene's function. Mutagenesis with transposable elements not only avoids the use of harmful chemicals or radiation but also allows the researchers greater precision in their targeting and easier identification of the affected site.

Such a mutagenesis screen using transposable elements was used to detect mutations that either increased or decreased susceptibility to ciprofloxacin within *Pseudomonas aeruginosa*. Remarkably, more than one hundred different genes affect susceptibility! The large number of genes itself points to why antibiotic resistance is such a vexing problem—there are so many ways in which it can occur.[14] The identities of the mutations are also important in that they provide more information about the biology of resistance, which could eventually be used in clinical settings. Many of the mutations that increased susceptibility to the antibiotic play roles in DNA replication and repair. One of the mutations that decreased susceptibility (increased resistance) was in an efflux regulator, the same type of gene involved in the phage study by Turner and Chan. Others were in genes that transport iron ions. Iron transport and iron metabolism appear increasingly important in antibiotic resistance.

Researchers can also track the evolution of resistance to antibiotics in an experimental setting. Alex Wong and his colleagues at Carleton College in Ottawa, Canada, did this by placing *P. aeruginosa* in culture media that replicated the lungs of people with cystic fibrosis.[15] To mimic the sticky conditions of a cystic fibrosis lung, they added mucin, a gel-like glycoprotein substance. Thus, the bacteria were facing challenges similar to those that they would face adapting to a lung in someone with cystic fibrosis. In some of these cultures, the researchers also added ciprofloxacin.

The genetic changes in those cultures with the antibiotic in this study were almost completely different from those observed in the study that generated mutations that caused resistance to the antibiotic. Unlike the mutagenesis experiment, the experimental evolution approach adds selection. Here, as Wong and colleagues note, "selection acts as an extra sieve that will weed out slow-growing mutants that, while they confer resistance, are out-competed on the way to fixation by other mutations conferring higher fitness".[16] Consistent with selection operating in the treatments with antibiotics, these treatments had genotypes with more mutations than those from treatments without the antibiotic.[17]

Parallel evolution—the same changes happening multiple times—was observed during the adaptation to the antibiotic. This parallel evolution provides information about the key traits that selection operated on during the adaptation to the antibiotic, which then could be used in drug development and other applications. Of the 24 genotypes from populations evolved with the antibiotic, 20 had mutations in *nfxB*, a gene that regulates efflux pumps. Nine genotypes had mutations in *gyrB* and four genotypes had *gyrA* mutations; these two genes encode subunits of DNA gyrase, the enzyme that helps to coil and uncoil DNA. In addition, six of the genotypes had changes at the gene *orfN*, which plays a role in the production of flagellum, hair-like appendages that bacteria use to move. So, efflux pump regulation, DNA gyrase activity, and flagellum production appear important in the resistance to the antibiotic under these conditions.

That these mutations in these genes appeared so many times also suggests that these genes substantially increase resistance to the antibiotic, which further experiments confirm. These mutations also have little cost in the absence of the antibiotic. As such, they are unlikely to be outcompeted by bacteria sensitive to the antibiotic when the antibiotic is removed. The authors suggest that the lack of costs likely comes about from additional "second-site" mutations that compensate for any costs associated with the changes that lead to resistance. In addition to the parallel evolution occurring at the "usual suspect" genes, there were also several changes that only occurred in a single replicate. The authors speculate that some of those odd-ball changes may be second site changes compensating for the costs incurred by the changes that yielded the resistance. Hence, these also should be investigated further.

Wong and colleagues raise another clinical implication of these results: if cost-free resistance is commonplace, then we should not expect sensitivity to return after the antibiotic is no longer used because the resistant bacteria carry no burden.

Costs to antibiotic resistance do appear common: a survey of studies found that most mutations resistant to an antibiotic had reduced competitive ability compared to sensitive bacteria in the absence of the antibiotic.[18] But commonplace is not universal: resistance mutations vary considerably in the costs, with some having no detectable costs. One interesting finding is that those mutations that provided the largest resistance effect tended to have large costs and those with modest resistance effect were more likely to be apparently cost free.

The finding that resistance mutations of modest effect have little or no costs has a possibly profound clinical application. Selection is likely to drive such mutations to high frequency when the concentrations of the antibiotic are high enough to retard growth of sensitive bacteria but not so high as to kill them. Such sublethal concentrations of antibiotics are often found in compartments of the respiratory systems of people who have been on long-term antibiotic regimes, such as those with cystic fibrosis. An experimental study in laboratory culture confirms this intuition: resistant genotypes evolved in the presence of sublethal concentrations of ciprofloxacin and appeared to have little or no cost.[19] The resistance appears to be due to mutations in the two genes that encode the subunits of DNA gyrase. This study also found the troubling result of cross-resistance: these mutations also gave the bacteria some resistance to other antibiotics.

The propensity of bacteria to quickly evolve antibiotic resistance continues to vex the medical community. The approaches presented here provide promise in the struggle to maintain the effectiveness of antibiotics. They rely on a deep understanding of evolutionary principles along with the acquisition of relevant empirical evidence. They are but a few of the applications emerging from evolutionary medicine.

WHY WE GET SICK

Evolutionary medicine began in earnest during the early 1990s by the interaction of a medical doctor (Randy Nesse) and an evolutionary biologist (George Williams). Then a practicing psychiatrist, Nesse had a desire to understand the evolutionary origins of disease and had Williams on his radar as potential collaborator due to their shared interest in the evolution of aging.[20] As an undergraduate at the University of Michigan, Nesse had written a term paper arguing that aging—the increase in the rate of mortality with age—evolved because it directly favored groups of individuals. That is, groups of individuals who had higher mortality when old would have a selective advantage over groups where individuals did not age. This notion is almost certainly incorrect, as selection on individuals staying alive for as long as possible would outweigh any group benefit. Evolutionary biologists at Michigan had directed Nesse to the seminal paper that George Williams wrote in 1957 on the subject. Here, Williams had proposed that some genetic variants could be beneficial for a person while young, but come at the cost of being harmful when that person became older.[21] Because younger people are more likely to have offspring in the future than are their elders, natural selection is more potent on variants that affect the survival of younger people. Accordingly, natural selection should favor these variants that improve survival when young even though they hurt survival in old age. Although much debate still exists regarding the exact mechanisms of aging, the central ideas proposed by Williams and extended by others are still held in high regard. One demographer commenting on the core principle noted, "evolution coddles you when you are young and forsakes you when you are old".[22]

After meeting him in 1985, Nesse found a kindred spirit in Williams. Though they were coming from different backgrounds, both men were searching for the evolutionary basis of disease. As Nesse would tell me: "I was looking for an evolutionary biologist who wanted to find the reasons for disease, he was looking for a doctor to do the same. It was a wonderful collaboration."[23]

After writing a review article for the *Quarterly Review of Biology*[24], Nesse and Williams' wrote the 1994 book, *Why We Get Sick*. Here, they encapsulated the central organizing question of evolutionary medicine—"why, in a body of such exquisite design, are there a thousand flaws and frailties that make us more vulnerable to disease?"[25] Shouldn't evolution by natural selection reduce these vulnerabilities as much as possible? The answer is that a variety of aspects of evolution unwittingly conspire against us. It's not a unitary factor; as a result, we should not expect single answers to most common ailments. Still, knowledge about these conspirators can lead to a better understanding of disease; and as we saw with the phage therapy studies, this understanding can potentially lead to clinical benefits.[26]

One of these conspirators is infection. From viruses to bacteria to fungi and other microscopic eukaryotes to nematodes, we have a large and diverse array of pathogens and parasites that make a living feeding off of us. In doing so, they often cause pestilence or at least make us miserable. Yes, we have evolved measures against these pathogens. But as we saw with antibiotic-resistant bacteria, these pathogens usually can evolve countermeasures faster than we evolve defenses because of their short generation time and immense potential reproductive output. They can run the evolutionary race faster than we can! In this book, we will focus on evolutionary applications concerning viral disease (Chapter 2) and vector-borne diseases (Chapter 3)

It's not just the infectious agents that cause disease. In an ironic twist, the very genetic defenses we have evolved against the disease agents can in themselves compromise our health. In Africa and in other areas where malarial infections are common, humans have evolved genetic changes in their hemoglobin, the molecule that transports oxygen in the blood. One of these is a change in the sixth position in the beta chain of hemoglobin; instead of the amino acid glutamic acid, these individuals have the amino acid valine. This amino acid change leads to slight shift in the electric charge of the molecule. Individuals with one copy of the genetic variant (or allele) have red blood cells that have a subtle difference in shape. This change is enough to give these heterozygotes (people with two different alleles) an increased resistance to malaria with little cost. Unfortunately, individuals with two copies of the allele (homozygotes) have sickle cell anemia. Other changes to hemoglobin in response to malarial defense also have deleterious effects.[27] We will discuss this further in Chapter 3.

Recall that people with cystic fibrosis have two altered versions of the *CFTR* gene. What about people with just one copy? Do these heterozygotes have an evolutionary advantage? There has been much speculation and a few studies investigating this possibility, including the idea that having one copy of the altered *CFTR* gene may be protective against cholera, though nothing definitive has been found.[28]

In addition to genetic responses that evolved to prevent infection, we have also evolved responses that kick in upon infection in order to fight off the pathogen. These so-called plastic responses can make us feel miserable but may be useful. Nesse and Williams suggested fever as an example of such a response, as elevated temperatures may help fight infection.[29] They noted that various animals, from cold-blooded lizards to juvenile rabbits (which are unable to generate fevers) will seek out warmer temperatures when infected. Moreover, while adult rabbits can increase their body temperature in response, they are less likely to survive infection if given drugs that block their ability to generate fevers. Clinical investigation of this hypothesis has been slow in the making, but one recent study found that fever could improve survival of patients with severe sepsis; those with elevated temperatures had greater survival than those with temperatures closer to normal.[30] As Nesse and Williams noted, fever comes with its own costs. In addition to the subjective malaise, even moderate fever increases metabolic rates and is associated with temporary male sterility. As the severity of the fever increases, so does the costs, which can include neurological damage. Deciding whether to treat fever, thus, requires full consideration of both the benefits and costs of fever. With more research on both the benefits and costs, however, medical practitioners may someday be in a better position to make such informed decisions.[31]

Sometimes, these plastic responses to infection can be detrimental. Dozens of millions died across the world from the 1918–19 influenza pandemic. This immense death toll may be partly due to overreactions of immune systems of those who died. Pathologists examining bodies of the flu's victims found evidence of extensive inflammation, "red, engorged lungs that were hard to the touch and seeped a watery, bloody fluid", as described by the British science journalist Laura Spinney in her recent book *Pale Rider*.[32] This extensive inflammation is consistent with a "cytokine storm", a fanatical overreaction by the triggered immune system that caused far more damage than the virus itself.[33] As we will see in Chapter 2, overreactions by the immune system also appear to be involved with more severe complications seen with infection with the new coronavirus, SARS-COV-2.

Another reason why we get sick is that our bodies evolved under different environments than the ones we exist in today. *There is now a mismatch between our bodies and our environments.* For most of our evolutionary history, we did not have ready access to foods rich in calories and sugar. We also did not live in large cities until the last few thousand years. We have different social networks than we did in the past. These social changes have brought new diseases and new stressors. Related to the concept of mismatch is the notion of historical legacies. The traits that evolved in response to new environments did not do so in vacuum. Instead, they evolved from traits that existed before. In Chapter 4, we will further discuss the effects of mismatch and the limitations of ascribing our ailments to mismatch.

A quarter century since the groundbreaking work of Nesse and Williams, evolutionary medicine has grown into a field of its own, with all of the trappings of academia—textbooks[34], college and even high school courses, and professional societies and their corresponding journals. I asked Nesse what he thought was the biggest advance in evolutionary medicine. He replied:

The biggest advance is not a specific conclusion, it is a change in perspective from that of a mechanic to that of an engineer. Almost all medical research has been about how the body works, what is broken, and how to fix it. Aspects of bodies that leave it vulnerable to problems were casually attributed to mutations or limitations of natural selection. George and I encouraged people to ask a new question. Instead of asking why one person gets sick, we encouraged asking why natural selection has left all members of a species vulnerable to a disease. We encouraged seeking answers to that question not just in mutation, migration, and drift, but also in the exigencies of coevolution, mismatch, tradeoffs, [and] reproduction at the expense of health, and defenses.[35]

Reflecting on the successes and lack thereof over the last quarter century of evolutionary medicine, Nesse notes that the breakthroughs have largely come from work on "coevolution, antibiotic resistance, and somatic evolution in tumors". Research that is based on the phylogenetic patterns of similarity and differences of humans and nonhuman animals has, thus far, not been as successful. Nesse generalizes, "research based on cells evolving fast turns out to be especially tractable and practical, compared with work on larger questions about why bodily traits are the way they are".

Nesse is careful to note there are historical precedents to these questions. Evolutionary medicine is built upon an edifice of study by evolutionary biologists, medical doctors, functional biologists, and natural theologians that stretches back centuries. He continues, "What is new is encouraging people to take the question seriously, and to consider evidence for and against multiple alternative explanations."

Evolutionary theory thus provides a perspective for generating hypotheses about our vulnerabilities. A quarter century before his involvement in evolutionary medicine, George Williams had cautioned evolutionary biologists from being too quick to accept hypotheses that a trait or a phenomenon because it was favored by natural selection. He noted "adaptation is a special and onerous concept that should be used only when it is really necessary".[36] Leaping to conclusions about the adaptive functions of traits is easy, and being too eager to accept adaptive explanations for traits is a seductive trap. Not all of evolution is adaptive; indeed, much of it is maladaptive.[37] As part of the maturation of evolutionary medicine, increasingly more care has gone into rigorous formulation and testing of hypotheses.

Medical doctors have much to do in terms of learning about evolution and its relationship to disease, but the burden of learning is not just on them; there is much that evolutionary biologists will need to do as well. Randy Nesse notes

> There is a tendency in many evolutionary circles to study disease in the abstract, without recognizing the potential major practical implications. There is also a tendency to lack respect for physicians, often considering them to be simple mechanics who don't know much science. That fails to recognize how deeply many doctors are interested in the origins of disease. Along these lines, it's hard for outsiders to appreciate the pressures on doctors to make decisions on the spot using whatever knowledge they have in the best interests of their patients.

I wonder what Darwin would think about the applications of evolutionary biology to medicine. When he was barely in his teens, young Charles Darwin had been in a program of medicine at Edinburgh University. Charles had no stomach for blood and no interest in surgery, which led him to dropping out. This decision did not sit well with Darwin's father, Robert, a well-respected country doctor. Robert fretted about his son, at one point telling him that "you are good for nothing but shooting, dogs, and rat catching, and you will be a disgrace to yourself and all your family". As a young man, Robert also had been pushed into going into medicine despite a similar aversion to surgery but had gotten over it.[38] Robert's father, grandfather to Charles, Erasmus Darwin also was a medical doctor and polymath of considerable note. His studies of microorganisms and speculations about the origins of life helped inspire Mary Shelley to write *Frankenstein*. Erasmus was also a confidant of the early geologist James Hutton, whose ideas about the Earth being old and being subject to slow gradual forces would later influence the younger Darwin's ideas about evolution. In addition, Erasmus had speculated on what would be known as evolution, notably that mammals may have had a single common origin.[39] I would like to think that Erasmus, Robert, and Charles each would be pleased about these applications of evolution to medicine.

NOTES

1. Nesse (2008).
2. Zimmer (2016).
3. Sano et al. (2002).
4. Chan et al. (2018).
5. Moradall et al. (2017).
6. The history of phage therapy and d'Herelle's role in it comes from Chanishvili (2012).
7. Chan et al. (2016).
8. Monosson (2018).
9. We will return to cystic fibrosis in more detail in Chapter 5. For details on the disease, see Trivedi (2020).
10. For details on the use of phage therapy on the young woman with cystic fibrosis, see Monosson (2018).
11. The meeting took place on December 13, 2018. All quotations from Turner or Chan are from this meeting.
12. Rasko and Sperandio (2010).
13. Allen et al. (2014).
14. Breidenstein et al. 2008.
15. Wong et al. (2012).
16. Wong et al. (2012, p. 6)
17. The antibiotic treatment also showed another telltale sign of selection that we will discuss in subsequent chapters. The sequence of nucleotides in DNA provides information for the sequence of amino acids in the protein encoded by that DNA. Due to the nature of the genetic code, some mutations in DNA change the amino acid code, while others do not. Those that change the amino acid are more likely to be affected by natural selection than those that do not. Accordingly, if there are lots more mutations that change the amino acid than those that do not change the amino acid, selection for change is very likely going on. In the replicates with the antibiotic present, the mutations that changed the amino acid were much more common than those that did not.
18. Melnyk et al. (2014).
19. Jorgensen et al. (2013).
20. Background taken from e-mail correspondence with Randy Nesse, February 2019.
21. Williams (1957).
22. Wachter (1997, p. 3).
23. From e-mail correspondence with Randy Nesse, February 2019.
24. Williams and Nesse (1991).
25. Nesse and Williams (1994). The quote is on p. 3.
26. See Nesse (2011) for an updated list of the various factors that cause vulnerability to disease in humans.
27. Sabeti (2008).
28. Beckett (1997).

29. Nesse and Williams (1994).
30. Kushimoto et al. (2013).
31. In reviewing a draft of this chapter in October 2019, Randy Nesse informed me that a large study of patients in the ICU found neither benefits nor costs to lowering fever. He adds that these ICU patients had access to the best medical interventions, so this study may not be a good test of the utility of fever in the natural environment. Fever remains an under-studied topic. During production of the book, a thoughtful review of the costs and benefits of fever in the context of the pandemic caused by SARS-CoV-2 appeared (Wrotek et al. 2021). Here, the authors suggested that in many cases, the benefits of fever (including the public health aspects) outweigh the risks. Stay tuned!
32. Spinney (2017). The quote is from p. 192.
33. Morens and Fauci (2007).
34. Gluckman et al. (2016).
35. This quote and subsequent quotes from Nesse come from written responses to questions I asked him in March 2018.
36. Williams (1966).
37. See Brady et al. (2019) and Masel and Promislow (2016).
38. Browne (1995).
39. Stott (2013).

2 Going Viral
Evolution and Viral Diseases

SUMMARY

As we have seen with the pandemic COVID-19 (caused by the virus SARS-CoV-2), viruses continue to affect human health. The chapter begins with a comparison of the COVID-19 pandemic with the influenza pandemic a century before. After a discussion of the nature of coronaviruses in general and SARS-CoV-2 in particular, we discuss the likely ways this virus spilled over into humans. We then explore how evolution has been used to track SARS-CoV-2. Next, we turn to the evolution of the properties of SARS-CoV-2, including the new, more infectious Delta variant. We also investigate the genetic factors that can affect susceptibility to COVID, including the intriguing prospect that we obtained both predisposing and protective variants from Neanderthals. Next, we consider the history of vaccines starting with smallpox and ending with COVID-19. The principles of evolution can be used to address the best strategies for vaccination schedules given limited resources. Finally, we turn to the AIDS pandemic with discussions about how evolution has been used to trace the evolution of HIV, the virus that causes AIDS, and how the evolutionary principles inform treatment.

"Ecology exposes us to new epidemics, but evolution is even worse: the diseases we already know change even as we attempt to come to grips with them."—Marta Wayne and Ben Bolker.[1]

"The fact is that we are really no better prepared for a bad outbreak today than we were when Spanish flu killed tens of millions of people a hundred years ago. The reason that we haven't had another experience like that isn't because we have been especially vigilant. It's because we have been lucky."—Michael Kinch.[2]

A TALE OF TWO PANDEMICS

Though often called the Spanish flu, the influenza pandemic that killed tens of millions across the world and about two-thirds of a million in the United States during 1918 and 1919 did not begin in Spain. It did not hit Spain harder than other countries. Instead, that pandemic is thus named because the best information about the spread of that pandemic came from Spain. Because Spain did not participate in the war waging over much of Europe, the Spanish media did not face the same wartime censorship that most other European countries and the United States imposed.[3]

Somewhere between 1 and 2 percent of those infected with the 1918 influenza virus died, but death was not the only toll from this pandemic. Many of the infected had long-lasting, even permanent, complications, such as the loss of smell and taste. Some, like the American author and journalist Katherine Ann Porter, had impaired

DOI: 10.1201/9780429503962-3

color vision, likely because the infection had spread to the optic nerve. In her short story "Pale Horse, Pale Horse", Porter reflected on the washed-out colors of her world postinfection: "Sitting in a long chair; near a window, it was in itself a melancholy wonder to see the colourless sunlight slanted on the snow, under a sky drained of its blue."[4] Detailed record keeping during the buildup to the US involvement in World War II would reveal subtle, lifelong effects on fetuses developing while their mothers were infected. Recruits born in 1919 whose mothers were infected were ever so shorter—1.3 millimeters or 1/20 of an inch—than their counterparts born to uninfected mothers.[5]

World leaders were not immune to this pandemic. US President Woodrow Wilson and several of the other heads of state conducting peace negotiations in the aftermath of what would later be known as World War I would catch it in the spring of 1919. Wilson subsequently had a massive stroke in October 1919, leaving him largely bedridden and unable to govern. Medical experts agree that Wilson's infection with influenza earlier that year likely contributed to the stroke. Wilson's stroke not only affected domestic governance, but also international relations. Because the infirm Wilson was unable to marshal support for the United States to sign the Treaty of Versailles, Germany was subject to much harsher terms. In addition, a healthy Wilson may have been able to garner support for US entry into the League of Nations. Imagine the counterfactual histories that could be written!

Subsequent influenza epidemics have appeared since 1918, but none took anywhere as high a toll. The pandemics of 1957 and 1968 each only killed about a 100,000 Americans—and at a time when the United States had a much larger population than in 1918.[6] In 2009, a strain of flu that was of the same class—H1N1—as the 1918 flu emerged. Sometimes called the swine flu because its components came together in pigs before it jumped into humans, this virus alarmed medical experts due to its similarity to the deadly 1918 flu. Fortunately, the 2009 flu was not as deadly as feared; although it officially was called a pandemic because it caused numerous cases across the world and had not been discovered before, it was not more lethal than a typical seasonal flu.[7]

Flus give medical experts many reasons to worry about their potential for becoming deadly pandemics. Michael Kinch, who we saw in one of the epigraphs, notes that flu can evolve rapidly.[8] Although we have vaccines for flus, this evolution can render vaccines useless. This evolution is why a previous year's flu shot may not be effective on this year's flu and why vaccine developers devote considerable time to attempting to predict the evolutionary tracks viruses will take. It is also why people can get the flu multiple times: their immune systems may not recognize the new strain. Kinch notes that some flus, like the H5N1 avian flu, are very deadly—much more so than even the 1918 one.

But as we have learned, influenza viruses are not the only ones that have pandemic potential. SARS (severe acute respiratory syndrome) and MERS (Middle Eastern respiratory syndrome) are coronaviruses that caused outbreaks in the each of the first two decades of this century: SARS in East Asia and MERS in the Arabian Peninsula. The viruses that caused these diseases were even more lethal than the 1918 flu, but fortunately the outbreaks were contained before becoming pandemics.

The scares from SARS and MERS kept coronaviruses on the radar screens of medical and public health experts. A 2010 review paper on coronaviruses concluded: "Based on available evidence, it seems that the question of emergence of another pathogenic human coronavirus from bat reservoirs might be more appropriately expressed as 'when' than as 'if.'"[9] In fact, in a January 2019 review paper, Nathan Grubaugh of Yale University and his colleagues imagined a new coronavirus causing an outbreak of illness.[10] In their hypothetical scenario, the first case was a 22-year-old man in Miami, Florida, who had been exposed to numerous birds and other wildlife while golfing at nearby resorts. Among other things, this review highlighted some of the methods genomics, epidemiology, and evolutionary biology could use to track the disease and halt its spread. In their hypothetical scenario, the outbreak happily was contained. Unfortunately, the efforts to contain a novel coronavirus in the real world in 2020 failed.

This novel coronavirus started spreading rapidly in southern China in the waning weeks of 2019. It caused a respiratory illness, mild in most people but severe in others, that would soon be called COVID-19. In the weeks that followed, the virus continued marching through China. Moreover, as its human hosts traveled outside China, the virus took footholds in country after country. On March 11, 2020, the World Health Organization (WHO) declared COVID-19 a pandemic.[11] By this time, the virus had reached 114 countries, with 118,000 known cases and 4.300 deaths worldwide. That was only the beginning.

In the subsequent days and weeks, businesses and schools in the United States closed as the number of cases of the virus skyrocketed. State and local governments across the United States imposed a patchwork of increasingly stringent measures— eventually including stay-at-home orders—to try to contain the spread of the virus. Then came the rush of deaths. We lost John Prine, the folk-country singer. We lost Ellis Marsalis, the jazz bandleader. We lost mothers, daughters, fathers, sons, friends, and colleagues. In this rush of deaths, we lost people we knew and people we did not know. After a summer lull and relaxation of restrictions, viral spread surged again in both Europe and the United States during the fall of 2020. By mid-November, the US death toll had reached a grim milestone of a quarter million dead, with thousands being added to the list each day.[12]

Like the pandemic of a century ago, the mortality rate of those infected with this coronavirus is about 1 or 2 percent on average.[13] In the COVID-19 pandemic mortality increases sharply with age, a pattern distinct from that of the 1918 influenza virus, where young adults and those in early middle age were hit hardest.[14] Like the 1918 flu, death was not the only adverse outcome: many of those affected have survived the virus but with long-lasting cardiovascular, respiratory, and neurological complications. Describing these so-called long-haulers, Jennifer Possick, the medical director of the Winchester Chest Clinic at Yale New Haven hospital, noted "the symptoms that they have span every organ system".[15] Again like the influenza pandemic of a century ago, Covid wears many faces.

Unlike in 1918—decades before we knew what genes were, much less how to decipher them—we have the capacity to rapidly sequence DNA. We knew what the virus was within days of the initial reports coming out of Wuhan. Knowing the identity of the virus and having its sequence allows for diagnostic testing. As we will explore later, knowing the genetics and biology of the virus also is a tremendous asset in the development of vaccines.

We should have been better prepared. The failure has been most acute in the United States. "The story of COVID-19 in the US is one of the strangest paradoxes of the whole pandemic",[16] wrote Richard Horton, editor-in-chief of *The Lancet* and author of *The COVID-19 Catastrophe*. He continued: "No other country in the world has the concentration of scientific skill, technical and productive capacity possessed by the US. It is the world's scientific superpower bar none. And yet this colossus of science utterly failed to bring to bring its expertise successfully to bear on the policy and politics of the nation's response." While critical of the responses of China, the United Kingdom, and other countries, Horton rightfully but sadly arraigned the responses of the US government in the early stages of the pandemic. A full accounting of the failings of the Trump administration and some state and local governments is beyond the scope of this book, but much of the failure can be attributed to a lack of preparedness and a deep state of denial.[17]

Still, there are tools that have been used and can be used to combat this 21st-century pandemic that were not available to our predecessors living in 1918. These tools also allow us to prevent and/or contain future outbreaks, preventing them from becoming pandemic. Many of these involve evolution and the allied fields of genetics and ecology. Given COVID-19's tremendous toll,[18] limiting the likelihood and/or scope of future pandemics is greatly desired.

A CLOSER LOOK AT SARS-COV-2 AND OTHER CORONAVIRUSES

Coronaviruses are one of several kinds of RNA viruses. As such, they only use RNA and do not use DNA as their genetic material. They naturally infect both humans and nonhuman animals, including livestock.[19] Four genera of coronaviruses exist: *Alphacoronavirus*, *Betacoronavirus*, *Gammacoronavirus*, and *Deltacoronavirus*. The first two genera, which are the ones that can infect humans, only infect mammals. Prior to the early 2000s, coronaviruses were not generally seen as a serious threat to human health. They are responsible for about a quarter of the cases of what is typically called the common cold. In addition, opportunistic infections of coronaviruses can be dangerous to people with compromised immune systems, while harmless to those with intact immune systems.

The SARS outbreak in 2002–3 alerted the world to the dangers of coronaviruses. The mortality rate from SARS was about 10 percent, though it varied by country and demographic factors. To put this in context, the mortality rate for most seasonal flus is around one-tenth of 1 percent, a hundred-fold lower.[20] The mortality rate of COVID-19 is intermediate between SARS and seasonal influenza. A decade later, a second lethal coronavirus-associated disease appeared in Saudi Arabia: MERS, which had a similar mortality rate to SARS. Both SARS and MERS were contained, though there have been periodic outbreaks of MERS.

The coronavirus that causes the disease COVID-19 was officially named SARS-CoV-2, reflecting its close evolutionary relationship with the virus that caused the original SARS back in 2002–4. In fact, SARS-CoV and SARS-CoV-2 are the two human coronaviruses that are each other's closest relative found in humans. The naming and classification of viruses today, like the modern practice of taxonomy of the macroscopic organisms, is based on evolutionary practices.[21]

People infected with SARS-CoV-2—even those who are not showing symptoms—shed respiratory droplets as they breathe, talk, sing, and engage in other activities. These respiratory droplets containing the virus can then infect others. The epidemiology of SARS-CoV-2 shows that most transmission is within close contact, which guided the recommendation of maintaining at least six feet of separation from others. Increasing evidence supports a limited role of longer-range transmission in indoor areas with poor ventilation where the droplets can linger in the air. Nonetheless, SARS-CoV-2 does not spread as easily through the air as some other viruses, like the one that causes measles.[22] The ability to spread through air is higher in the new variants of the virus as compared with the one that first spread across the world.

How infectious is SARS-CoV-2? Epidemiologists use a term—the basic reproductive number or R—to express the infectiousness of a virus. R is the number of individuals that the average infected person will infect. A reproductive number of two means that each infected person will infect two other people. Suppose a person (call her Anne) transmits a virus to two people, Bob and Catherine, and they each spread it to two friends, and so on. Unchecked, the virus should spread quickly, approximately doubling with every transmission cycle.[23] The larger R, the more likely and the faster the virus will spread. Viruses with R values above 1.0 will spread, while those with R values less than 1.0 will die out.[24]

R values are not constant and will vary, depending on the virus and the population. A virus that readily spreads through the air will likely infect more people per cycle than one requiring contact with bodily fluids for transmission. R could be reduced through hygienic measures (such as washing hands), wearing face coverings, and through social distancing: these measures reduce R by making transmission more difficult. Moreover, as a novel virus sweeps through a population, individuals that are infected usually become immune. This immunity does not just protect the immune individuals but also is beneficial for the population. With fewer susceptible individuals, the virus has more difficulty spreading: consequently, its R declines with the susceptible pool of hosts.[25] Vaccinating people with an effective vaccine also reduces R by reducing the pool of the susceptible.

Epidemiological studies put out in the early days of the COVID-19 spread in Wuhan estimated that the R prior to intervention was about 2.2. This estimate came with much uncertainty, with confidence intervals that ranged from 1.4 to 3.9.[26] Preliminary estimates from King County, Washington state, estimate R was closer to 3 during late February 2020, prior to social distancing and other measures to contain the spread. With the containment measures, R fell to approximately 1 by mid-March.[27] Various websites have tracked R in localities as the viral spread has waxed and waned.[28]

SARS-CoV-2 is more infectious than flu viruses; the R values for the 1918 and other pandemic influenzas were only about 1.8. Seasonal flus, with R values around 1.25, are less infectious still.[29] In contrast, the version of the SARS-CoV-2 virus that swept across the world in 2020 is less infectious than smallpox or measles, which have reproductive numbers of about 6 and above 10, respectively.[30] Viruses like smallpox or measles are difficult to control even with vaccines, unless the vaccines are highly

A

B

The reproductive rate (R) is a metric of how quickly the virus is spreading. When R is 2 (as in top panel A), each infected person infects an average of two people. When R is 3 (as in bottom panel B), each infected person infects an average of three people. If R is greater than 1.0, the virus should be expected to increase exponentially. The larger R, the faster the expected rate of the increase in viral counts. If R is less than 1.0, the virus should be contained.

Source: Illustrated by Michael DeGregori.

effective and most people get vaccinated. The COVID-19 pandemic should be easier to contain with vaccines.

So how does a virus like SARS-CoV-2 infect people? Consider what happens when we inhale. After leaving the trachea, sometimes known informally as the wind-pipe, air moves the bronchial tubes of the lungs. These bronchial tubes are lined with cells containing cilia, fine hair-like structures that push away debris as the inhaled air passes through the tubes. Many viruses, including influenzas and both SARS-CoV and SARS-CoV-2, attack these cells. Here, the viruses are provided a safe haven from where they can be fruitful and multiply. But in order to thrive within the cells, the virus must be able to enter the cells. How?

A typical route for viral entry is receptors that cells have on their surfaces. Whether and how well a virus can get into cells depends on the interaction between the receptors that cells have and the structure of the virus. Viruses differ in which receptors they use. Both the original SARS virus and the SARS-CoV-2 get into cells via angiotensin-converting enzyme-2 (ACE2) receptors in the cell membrane.[31] In contrast, influenza viruses exploit glycoproteins that contain sialic acids, which are various acidic sugars whose structure contains a ring with nine carbons.

ACE2 receptors vary in their structure across mammals: this variation can be predictive of determining which animals are likely to transmit SARS-CoV-2. For example, cats and ferrets, which contract this virus easily, have ACE2 receptors that differ by only two amino acids from the human one.[32] Ferrets, thus, should make excellent model systems for research on not just SARS-CoV-2 but also other coro-naviruses that could be threats to human health. Likewise, different animals vary in the nature of the sialic acids on the glycoproteins, and this variation affects how well different influenza viruses can get into cells of different animals.

These receptors did not evolve for viruses to use them. Instead, they evolved because they are important for numerous different functions of our bodies. The viruses just exploit these receptors for their own use. Among other functions, ACE2 modulates blood pressure: this protein is not just a receptor but also an enzyme that converts angiotensin II (which contracts blood vessels) to angiotensin (which dilates blood vessels).[33] Not surprisingly, ACE2 is highly expressed in tissues that are rich in capillaries. The prevalence of ACE2 in these tissues likely explains why COVID-19 can have such diverse effects and can target so many organ systems. That the virus subverts an enzyme whose activity tends to reduce blood pressure could explain why hypertension is a risk factor for severe COVID-19.

A critical feature of whether a specific SARS-CoV-2 particle can infect a specific host cell is the spike protein. It is this protein that gives viral particles of coronavi-ruses the characteristic crown-like shape that led to the name of these viruses. As the virus tries to enter the cell, the spike protein is cut into two subunits: S1 and S2. A region of S1 that is called the "receptor binding domain" binds to the peptide domain of the ACE2. Successful entry requires both cleavage of spike into S1 and S2 as well as proper matching between S1's receptor binding domain and the host cell's receptor binding domain.

Why COVID-19 varies so much in severity is still not yet fully known, but a few pieces of the puzzle are coming into view. First, the virus is extremely good at usurp-ing its host cells: the cells infected with SARS-CoV-2 can have an extraordinary amount of viral RNA, meaning that the cells have become factories for the virus.[34] Second, the infection can in some cases throw the immune system out of balance. Our immune system does not rely on a single type of response; instead, it has several that ordinarily work together. One major response is the release of interferons, various signaling proteins that direct other nearby cells to respond to a threat. They activate natural killer cells and other immune cells to attack the virus. Another major response is the release of chemokines that go into the circulatory system for a longer-range response. SARS-CoV-2 infection sometimes shuts down the short-range interferon response while leaving the long-range chemokine response intact. Some virus experts, such as Benjamin tenOever at Mount Sinai Medical School, have hypothesized that severe COVID-19 may be due to this imbalance of immune responses.[35] As immune systems age, they are likely to be more sensitive to perturbations, which could explain why severe COVID-19 is much more common in older people.

EVOLUTION AND CORONAVIRUS SPILLOVER

Evolutionary principles and studies can be useful in tracking the source of viruses. Most viruses that get into humans originated in other animals. Contact between humans and these animals allows for the possibility of spillover[36] of the virus. What animal was the original host of the virus? When and where did it leap into humans? Can we infer anything from sequence information about which viruses are most likely to be deadly were they to jump into our species? Information addressing these questions can inform surveillance practices.

SARS emerged in late 2002 from the Guangdong Province in southeast China, bordering the Pacific and not far from Hong Kong. It is about 1,000 kilometers south of Wuhan, the original epicenter of SARS-CoV-2. In his journalistic account of the SARS epidemic, Karl Taro Greenfield, then editor of *Time Asia*, paints a picture of Guangdong in the first years of the 21st century. There and then, in that booming "Era of Wild Flavor", the newly affluent residents and tourists could go to markets where they could get an extraordinary variety of exotic animals to eat. Describing the market, Greenfield wrote:

> A startling musky smell overwhelmed me as I walked between the stalls. I realized it was the combination of the feces of a thousand different animals mingled with their panicked breaths. The range and scope of wildlife on display was a zoological chart brought to life. There were at least a dozen types of dogs, from Saint Bernard to Labrador. There had to be at least as many different breeds of house cat. There were raccoon dogs, badgers, civets, squirrels, deer, boars, rats, guinea pigs, pangolins, muskrats, ferrets, wild sheep, mountain goats, bobcats, mountain lions, three types of monkeys, horses, ponies, bats, and one camel out in the parking lot. And these were just the mammals.[37]

The combination of such diverse animals together was a recipe for spillover disaster. Indeed, the first people to get SARS had had exposure to these exotic animals. Phylogenetic analysis of viruses collected from animals in the area pointed to transmission between civets, a small and elongated cat, and humans. Civets in the area even showed antibodies to the virus.[38]

But is this the whole story? Virus transmission can sometimes be complicated, wherein virus circulates freely in one species causing little or no disease.
Epidemiologists call this the reservoir host. The reservoir host allows for a continuous supply of virus that can spread to humans or to other animals. Then there are
species that can get the virus from a reservoir host and, in turn, transmit the virus to
humans. This is the intermediate (or amplifying) host. An important consideration
is that viruses can evolve within the intermediate host; sometimes this evolution can
make the virus more amenable for living in humans.

Are civets the reservoir host or the intermediate host of the SARS virus? When
injected with the virus, civets get fevers and become lethargic. Moreover, wild civets
seldom display antibodies to the SARS virus. These observations suggest that civets
are not the reservoir host—the ultimate source—of the virus. They also led biologists to search for the reservoir host.[39]

Substantial evidence, including phylogenetic data, supports bats—specifically,
horseshoe bats of the genus *Rhinolophus*—as the reservoir host of the SARS virus.[40]
These small bats—all weigh less than an ounce (28 grams)—that have horseshoe-
shaped noses are hunted for food in southeast Asia and other places across the world.
Many of these bats sampled in Guangdong, China, where SARS likely emerged,
display antibodies to the virus, and some have active virus. Not only are the viruses
taken from these bats genetically diverse, but their diversity encompasses the variation seen in viruses taken from humans and civets. Viruses from humans and civets
are but one branch within the diversity of viruses taken from bats. Hence, the direction of transmission of the SARS virus was Horseshoe bats -> civets -> humans.
Given that civets were commonly found at Guangdong markets and that the early
SARS cases were animal handlers, the markets are a likely source for the spillover.

A similar picture is emerging about the origins of SARS-CoV-2, but the story is
less clear. Given the origin of the original SARS virus, early attention turned again
to meat markets. Suspicion grew upon the news that virus samples taken from the
surfaces of a seafood and wildlife market in Wuhan City were found to have strikingly similar sequences to virus RNA taken from the first patients in Wuhan. One
might be tempted to conclude that the market was the direct source of the virus,
but caution is warranted. Writing in an April 2020 perspective for *Cell*, Yong-Zhen
Zhang and Edward Holmes noted: "Unfortunately, the apparent lack of direct animal
sampling in the market may mean that it will be difficult, perhaps even impossible,
to accurately identify any animal reservoir at this location."[41]

Early studies also found that horseshoe bats in China had viruses that have similar
sequences to those of SARS-CoV-2. The similarity was sufficiently strong to suggest
that these bats were part of the chain of transmission of the virus. Research by Alice
Latinne at EcoHealth Alliance and her colleagues strengthened the connection to
bats.[42] As in the case with the original SARS virus, SARS-CoV-2 also forms a cluster
that is within the diversity of viruses of horseshoe bats in the genus *Rhinolophus*.
These researchers also showed that bats in this genus are prone to exchanging viruses
with other species.

As of spring 2021, there is still much uncertainty about the identity of the intermediate host of SARS-CoV-2. A leading candidate is the pangolin because these small
mammals have coronaviruses that cluster with SARS-CoV-2.[43] Sometimes called

spiny anteaters, pangolins are indeed covered with scales, but they are not anteaters or closely related to anteaters. In fact, they are not closely related to anything but themselves. They belong to the mammalian order Pholidota, one that contains only the eight species of pangolins. They are only called spiny anteaters because they eat ants (and termites). Prized for their scales and skin as well as for their meat, pangolins are in high demand. One estimate puts the number of pangolins that have been smuggled since the start of this century at nearly a million. Moreover, they are easily captured; when threatened, pangolins will roll into a ball. Further sampling, especially near Wuhan, will be needed to fully address the role of pangolins in the spillover that led to SARS-CoV-2. It is clear, however, that they have coronaviruses that are related to SARS-CoV-2, which could spill over into humans. They ought to be a priority for surveillance regardless of whether they were the intermediate host of SARS-CoV-2.

At least two of the three coronaviruses that cause serious disease in humans have bats—indeed, a single genus of horseshoe bats—as their reservoir host. MERS, the third pathogenic coronavirus, appears to have jumped from camels to humans directly, though bats cannot be ruled out as a reservoir host.[44] Bats have also been reservoir hosts for numerous other deadly viruses that have jumped into humans, including Ebola and Nipah.[45] Bats are likely a major source of spillover viruses because they contain so many viruses: compared with other mammals, bats have more virus. One possible reason for the high viral load of bats is that there are so many species of bats; because species usually exchange viruses with related species, groups of mammals that are more species-rich should maintain more viruses. Another plausible explanation has to do with unique features of bat's biology due to adaptations related to flight.[46] The extra metabolic demands that come with flying have altered the mitochondria of bats as well as the nuclear genes that interact with mitochondrial genes. These changes have altered the immune systems of bats, allowing them to be more tolerant of intercellular invaders such as most viruses. These adaptations, however, appear to have come with the tradeoff of leaving bats more vulnerable to extracellular infections, such as the one that causes white-nose syndrome.

Regardless of why bats harbor more viruses, it is clear that they do. They also frequently exchange viruses between different species, giving viruses opportunities to evolve in new directions, which can lead to spillover to humans. Coronaviruses are also particularly prone to switch hosts. These facts should be considered in virus surveillance programs. Interesting, although the three pathogenic coronaviruses are all betacoronaviruses, alphacoronaviruses are even more prone to host switching than betacoronaviruses.[47] This finding suggests that we should also prioritize alphacoronaviruses in surveillance programs.

Are there features of a coronavirus that predict whether it is likely to cause serious disease if it enters into humans? Analysis of the differences between the pathogenic and the nonpathogenic coronaviruses in humans reveals a few signature traits.[48] One is an insertion of four amino acids in the spike protein that is found in the pathogenic but not in the non-pathogenic coronaviruses. This change evolved independently in both SARS viruses and in the MERS virus. Another signature trait is in the nuclear localization signal domain of the N protein. This part of the protein directs it to the nucleus of host cells. The pathogenic viruses have nuclear localization signal domains that are more positively charged than the nonpathogenic viruses. Information coming from these phylogenetic analyses thus allows biologists to predict pathogenicity potential, which will be useful in surveillance programs. Viruses that are more likely to become pathogenic should receive higher priority.

USING EVOLUTION TO TRACK VIRUSES

Once a pathogenic virus gets into humans, evolutionary principles and data can be useful in tracking how it spreads in human populations. Because the coronavirus evolves over time, phylogenetics methods can be used to track its transmission patterns. Such information can provide insight on how the virus is spread and could conceivably inform containment protocols. Evolutionary studies can also be used to determine whether and how the virus is evolving in human populations. Let's look at these approaches with respect to coronaviruses and with SARS-Cov-2 in particular.

One early epicenter of the virus was in the New York City area. On March 3, 2020, the first case of community-acquired infection was identified in New Rochelle, a commuter suburb of New York City located about 25 miles (40 kilometers) north of Manhattan. In an early step to control the virus, Governor Andrew Cuomo set up a containment area around this small city of 77,000 starting on March 12.[49] Unfortunately, by this time, there was too much community spread outside of the New Rochelle containment area to stop the virus.

Matthew Maurano at NYU's Institute for System Genetics and colleagues analyzed the sequence of viruses from people in New York City and the surrounding area early in the outbreak there.[50] They identified more than 100 introductions of the virus to the NYC area, most of which occurred during late February. Most of these introductions came from people who had traveled from Europe, not Asia. Evidently, the virus had been festering in Europe for a considerable time. Such a mode of transmission demonstrates that the Trump administration's partial travel ban from China was likely ineffective at containing the virus, as it had many possible paths to take.

Another study led by Trevor Bedford at the Fred Hutchinson Cancer Research Center in Seattle, Washington, looked at patterns of community spread in Washington state, another early epicenter.[51] Most of these infections were from a single introduction, most likely during the last week of January or the first week of February. The virus was already circulating well before the alarm came and restrictions were put in place.

The first confirmed case in the United States was a man (WA1) who came from Wuhan to Washington on January 15 and had tested positive on January 19. Was the community spread detected by Bedford's study linked to this individual? Possibly. The data are consistent with this scenario, as the viruses from the Washington state group do appear to be direct descendants of WA1. But because the rate of evolution of the virus is much slower than the rate at which it is transmitted, it is also possible that the cluster in Washington state was descended from a related virus. WA1 could still be a side branch. Samples from British Columbia interdigitate with those from Washington state, suggesting that the cluster could have come from Canada. A different group of researchers also studying the outbreak reached similar conclusions.[52]

There has been speculation that the virus was spreading in Washington state and other parts of the country before January 2020.[53] Results from Bedford's study show that this is exceedingly unlikely. In addition to the phylogenetic analysis, they also examined collections of samples from the Seattle flu study. Of 5,270 samples taken from January 1 to February 20, all were negative for SARS-CoV-2. The first positive was on February 21. Assuming that there would be a delay of a couple of weeks between the time that community spread was seeded and when people with flu symptoms came into clinics, this result is consistent with the timing, based on the phylogenetic analysis.

With the sequencing of viral genomes now being reasonably cheap and quick, outbreaks of SARS-CoV-2 and other viruses can be traced almost in real time. The combination of sequencing and phylogenetics will augment viral surveillance going forward.

EVOLVING VIRUSES

As Wayne and Bolker noted in the epigraph, the evolution of the diseases we know can complicate our efforts to control them. Influenza viruses evolve rapidly. With this rapid evolution, influenza viruses can evade the immunity we obtained from prior exposure. Sometimes, as with the virus that brought on the 1918 pandemic, a new pathogenic virus can emerge. What about the evolution of SARS-CoV-2? Should we be worried about this disease becoming worse?

Being Darwinian creatures with short generation times and high mutation rates, viruses will evolve. As we saw earlier, the accumulation of mutations is useful to us tracking the virus. The question is whether the virus will evolve in such a way that it changes its characteristics, such as increasing its infectivity or in alterations to the severity of the symptoms it causes.

The speed at which a virus would be expected to evolve is dependent on several factors, including the mutation rate. All other things being equal, an increased mutation rate would permit faster evolution. Compared with influenza, coronaviruses evolve more slowly because they have a lower mutation rate. Unlike influenza viruses, coronaviruses have a gene that encodes a protein (nsp 14) that proofreads the replication of the virus RNA. Like having someone else read over what you wrote should reduce typographical errors, the addition of nsp 14 reduces the number of mutations that coronaviruses make each time they replicate.[54] The proofreading function is far from perfect, but it is sufficient to reduce the mutation rate several fold.

The viral population size also affects evolutionary rates. The larger the population size, the more mutations that are produced and the more efficient selection is. With large populations, mutations that are advantageous for the host are more likely to increase in frequency and become fixed. Large population sizes also make selection more efficient at weeding out mutations that are deleterious to the virus. Thus, the large number of active cases of the virus during much of calendar year 2020 gave the virus more opportunity to evolve by natural selection than if the virus had been checked at low numbers.

The largely unknown variable is the type of selection pressure that is placed on the virus. We should expect to see selection favoring the virus becoming better able at transmission. But predicting how the virulence of the virus will evolve is more difficult. Variants that make SARS-CoV-2 more virulent leading to more cases of severe COVID-19 could be favored, but so could variants that make it less virulent.

The question of what is most likely depends on whether changes in virulence make the virus better able at infection. Moreover, changes in virulence could arise through neutral evolution, wherein a variant that alters virulence—in either direction—rises in frequency simply due to random genetic drift.

Through the early fall of 2020, SARS-CoV-2 had little genetic variation.[55] Molecular evolution analysis pointed to strong purifying selection weeding out mutations that are deleterious to the virus during this stage of the pandemic. Movement of the virus to new localities and containment measures also subjected the virus to bottlenecks. Additional analysis also found only scant evidence for selection making the virus more infectious.[56]

From the initial report of the virus till the early fall of 2020, there was only a single change in the virus for which we had evidence that its rise was driven by selection. Early in the pandemic, nearly all variants of the virus had spike proteins with the amino acid aspartate (D) in the 614th position.[57] There was also a rare variant with the amino acid glycine (G) in that position. As the pandemic took hold during the late winter and early spring of 2020, the variant with glycine rose in frequency and soon displaced the variant with aspartate as the most common variant. By that summer, the glycine variant predominated across the world. This change is denoted as D614G, in accordance with the standard practice of having the abbreviation of the original amino acid, followed by the position, followed by the abbreviation of the new amino acid.

Bette Korber, a researcher at the Los Alamos National Laboratory who had worked on HIV evolution before working on SARS-CoV-2, found statistical evidence that the increase of the G variant is repeated and that it is unlikely this is due to chance.[58] These changes happening repeatedly in different places suggests that selection, and not just genetic drift, is responsible for the changes. In addition to the statistical evidence, Korber and her colleagues found functional differences: the new G variant is associated with a higher viral load in patients and is more easily detected in tests. It is not, however, associated with how severe the disease is. It is also not more resistant to antibodies. In fact, it may increase susceptibility to antibodies.

Although D614G appears to be associated with increased infectivity, some researchers are expressing caution. In a commentary in the same issue of *Cell* where the Korber article was published, Nathan Grubaugh noted that the "higher detection of SARS-CoV-2 RNA in oral and nasal swabs might not be a direct reflection of transmission detection. In addition, much transmission likely happens in the presymptomatic stage, and we don't know how these differences compare".[59] Furthermore, Lucy Van Dorp and colleagues found no association of D614G with increasing infection, though stressed that different methods could lead to different results.[60]

Although the D614G change likely increased the infectivity of the virus, it also may have made the virus more susceptible to vaccines. This change in the spike protein alters its three-dimensional shape (conformation), making it easier to attach to the ACE2 receptors. With better attachment of spike to the receptors, the transmissibility of the virus increases. But this increased transmissibility comes at a price. The altered conformation leaves the receptor binding domain exposed, which makes the virus more vulnerable to attacks from neutralizing antibodies generated by prior exposure or through vaccines.[61] Sometimes, tradeoffs work in our favor.

Since the fall of 2020, things have changed: several variants of the virus have risen in frequency, and some of these appear to have been driven by selection. The concern is that these variants are either more infectious, more deadly, and/or more resistant to antibodies generated by natural infection or vaccines. One notable variant known as B.1.1.7 (also known as Alpha) first emerged in England in September 2020 and became the dominant lineage there over the course of that fall.[62] This variant, which contains eight mutations in the spike protein and nine elsewhere, is considerably more infectious than its predecessor. One likely candidate mutation to explain the increased infectiousness of Alpha is a mutation in the spike's receptor binding domain (N501Y), which increases the ability of the virus to bind to the ACE2 receptor. The link of infectiousness to that mutation is not yet certain; there may be others. Regardless of the specific causal mutation or mutations, the greater ability of Alpha to spread increases the threshold of immunity needed (by prior infection or vaccines) to halt its increase. By the spring of 2021, Alpha became the predominant variant in the United States.[63] Fortunately, the aggressive vaccinations in the United States (as discussed later) during that spring forestalled a major surge of the virus.

How did Alpha and these other more infectious or more deadly variants emerge? A likely hypothesis is that they arose via evolution within a few immunocompromised people who had a chronic SARS-CoV-2 infection. A study that traced the evolution of the virus in an immunocompromised individual provided support for the hypothesis.[64] This patient was treated with the plasma from people who had recovered from the infection (convalescent plasma therapy). In response to the convalescent plasma treatment, the virus in this patient evolved a deletion that apparently helped it escape the antibodies in the plasma. This same deletion is found in Alpha and other variants.

There is also concern about cross-transmission between humans and other animals because it may lead to mutations—some of which may lead to changes in infectivity, changes in virulence, or changes that allow the virus to escape immunity (from prior infections or vaccination). In Denmark and in other locations, humans and minks have been transmitting SARS-CoV-2 to each other.[65] Associated with this cross-transmission are changes in the sequence of the spike protein, though whether or how these changes will affect infectivity, virulence, or immunity is unknown. As a precaution, several million mink have been culled to prevent further cross-infection.

Although it is wise to be concerned about the evolution of viruses after cross-infection, these genetic changes need not necessarily lead to a worse disease. A year after the original SARS outbreak, a few people who had interacted with civets were infected with SARS-CoV that had been adapting to civets. Unlike most cases of the original SARS, where symptoms are usually severe and often fatal, these people had only mild symptoms.[66]

WHAT GENETIC VARIANTS AFFECT COVID SUSCEPTIBILITY AND WHY?

People are not equal in their susceptibility to infections whether it be viral or bacterial. Some of that variation is linked to environment and behavior and some may be luck, but abundant data points to the existence of considerable genetic variation for susceptibility to infectious disease in general. What about COVID-19?

Blood type influences the propensity for getting severe COVID: several studies have shown that having type A blood increases one's risk.[67] In contrast, type O blood is protective. It's not too surprising that blood type would affect COVID risk, as it has affected the propensity for getting other infectious diseases. For instance, blood group affects susceptibility to hepatitis B, a virus that causes nearly a million deaths each year and chronically infects a quarter of a billion. Especially in areas where hepatitis B is highly prevalent, type O blood carries a higher risk of hepatitis B and type B is protective.[68]

Associations between blood type and infectious disease are to be expected because blood type is linked to the immune system. Blood type is based on sugar-like molecules that are attached to the membranes of red blood cells: type A blood has the A sugar, while type B blood has the B sugar. Rounding out the types, type AB has both sugars and type O has neither. These sugars on the red blood cells act as antigens, molecules that are recognized by antibodies. People produce antibodies in their blood serum corresponding to antigens *that they lack*. (Type A blood and anti-A antibodies would lead to adverse consequences.) Thus, people who have type O blood generate both anti-A and anti-B antibodies; people with Type B blood generate just anti-A antibodies; people with Type A blood generate just anti-B antibodies; and people with Type AB generate neither antibody. As an aside: people with type O blood are sometimes called universal donors; because their blood lacks both the A and B antigens, small amounts of their blood as with an emergency blood transfusion are less likely to adverse consequences than small amounts of other blood. Still, the best course in non-emergency cases is for people to receive blood of their own blood type. Getting back to the main point: people with different antibodies circulating in their blood are prone to vary somewhat in the propensity to acquire certain infections.

The situation is more complicated than that. Most people also have the ABO antigens in their bodily fluids, but some do not. Those that do—secretors—have at least one functional copy of the gene *FUT2*, which encodes an enzyme called alpha (1,2) fuscosyltransferase. Those that lack functional copies are called nonsecretors. In most populations, about one-fifth are nonsecretors. Molecular evolution studies show that *FUT2* has the signature of balancing selection and that different point mutations are responsible for the phenotypes in different populations.[69] Along with these patterns of sequence evolution is evidence that secretors and nonsectors vary in disease susceptibility. For instance, nonsecretors are less prone to certain noroviruses that often cause diarrhea and other intestinal disorders. Interestingly, secretors and non-secretors vary in the composition of their microbiota.[70] Nonsecretors also appear to have some protection for COVID-19, though the mechanism by which this protection works is still unknown.[71]

Another set of genes has also been implicated in COVID-19 susceptibility. A study that looked at the association between markers and disease susceptibility (genome-wide association or GWAS) found a 50,000 base pair region on chromosome 3 that affects one's likelihood of getting severe COVID.[72] Having one variant of this region approximately doubled the risk. The genes in this region have historically been inherited as a unit, as there has been insufficient recombination over the course of history to break down the association. We cannot pinpoint the casual factor: any one genetic variant (or even multiple variants acting in concert) could be the culprit. One possible gene, SLC6A20, is of interest because its product (the sodium-imino acid transporter 1) interacts with ACE2, the key receptor that SARS-CoV-2 exploits. But there are also other candidates that look good. These include the C-X-C motif chemokine receptor 6—which regulates the cellular immune responses to influenza and other viruses.

What is the evolutionary origin of this region? Svante Paabo at the Max Planck Institute for Evolutionary Anthropology has proposed an intriguing explanation: it is from humans interbreeding with Neanderthals.[73] Paabo is one of the leading researchers in the genomics of Neanderthals and other archaic humans.[74] His lab was the first to obtain a complete genome sequence from Neanderthals—given the degradation of DNA from samples that are tens of thousands of years old, that is no easy feat. One of the striking findings that has come out of the 2010s was incontrovertible evidence of Neanderthals mating with humans, leading to most (but not all) contemporary human populations having a couple of percent of Neanderthal DNA in their genome. We will discuss the interbreeding of humans and Neanderthals in much more detail in Chapter 17.

Working with Hugo Zeberg, Paabo found that the risk factor for chromosome 3 was prevalent in Neanderthals; one Neanderthal from Croatia is homozygous for the risk factor! They calculated that the probability that the Neanderthal haplotype was inherited from the common ancestor—instead of through introgression—to be less than one in a thousand. Thus, this is strong evidence that the risk factor came from introgression from Neanderthals. Though rare in East Asia, the Neanderthal haplotype is most common in South Asia. It is especially common in Bangladesh; almost two-thirds of the population there carry at least one copy. Intriguingly, people of Bangladeshi ancestry living in the United Kingdom are at about twice the risk of dying from COVID-19 compared with the population average.

The finding that Neanderthal ancestry could affect susceptibility to a new virus had been anticipated. Two years before the appearance of COVID-19, the evolutionary biologists David Enard and Dmitri Petrov had written "it would be interesting to study whether the presence of Neanderthal at VIPs [proteins that interact with viruses] still leads to variable susceptibility to modern viruses in modern humans".[75] Their remarks came in a paper they had written showing that introgressions of Neanderthal DNA that showed signs of being under positive selection were frequently associated with proteins that interacted with viruses (VIPs).[76] This association is particularly prevalent for proteins that interact with RNA, rather than DNA, viruses. Interestingly, compared with DNA viruses, RNA viruses are more likely to jump from one species to another.

I asked David Enard what he thought about whether Neanderthal ancestry at that region on chromosome 3 modulates COVID-19 severity. He said that there was a very real chance that it did.[77] He continued:

> The result fits well with previous evidence that RNA viruses . . . have driven abundant
> adaptive introgression from Neanderthals to Eurasian modern humans, and evidence

that several bat species are reservoirs of potentially pathogenic coronaviruses in the geographic locations where Zeberg and Paabo found the Neanderthal haplotype at higher frequencies. This suggests a hypothesis that the frequency of Neanderthal ancestry increased in these locations due to an ancient coronavirus, or at least another virus associated with similar host genetic susceptibility.

Enard did caution that the result could come from population stratification wherein the association between the disease severity and the genetic marker could be confounded by ancestry and suggested that further controls were needed.

Enard has also found that the proteins that interact with viruses show two seemingly contradictory patterns.[78] They generally evolve more slowly, which is evidence of selection weeding out deleterious changes that are indicative of functional constraint. But they also show frequent signatures of adaptive evolution. Nearly a third of the amino acid changes that are adaptive are in these proteins. These results suggest that viruses and other pathogens are important drivers of evolution, especially at the level of the proteins we have. Related to coronaviruses specifically, one protein called aminopeptidase N (ANPEP) shows extensive adaptive evolution across mammals. This protein is used by human coronavirus 229E, an alphacoronavirus that usually causes mild colds but can cause more serious disease in people with compromised immune systems.

VACCINES

In late fall 2020, glimmers of hope appeared with news of effective vaccines against SARS-CoV-2. Both Pfizer and Moderna released vaccines deemed safe and extremely effective in clinical trials.[79] Soon after, limited supplies of the vaccines were deployed to high priority individuals, such as health care personnel and individuals older than 65, in both the United Kingdom and the United States. Amusingly, one of the early UK recipients was an 81-year-old man named William Shakespeare; reports of his vaccination set off a flurry of cheery puns as the world took in the encouraging news.[80] Developed and tested in less than a year, these vaccines are an extraordinary and welcomed feat!

Vaccination remains a highly effective means of protection from infectious pathogens, including viruses. It serves a dual purpose. First, vaccinations reduce the risk and/or severity of sickness in the individuals who are vaccinated. They also can reduce the spread of the contagion in populations. As more people are vaccinated, fewer individuals in the population are susceptible to the virus. Hence, vaccinations should reduce the R. Because the spread of the virus is checked through the reduction of R, people who are not immunized benefit. This feature is sometimes called herd immunity.[81]

The idea behind vaccinations is to stimulate the immune system to generate a response against the infectious agent. This is achieved by presenting individuals with a harmless or less harmful stimulus that mimics the harmful agent. This agent provokes not just an immediate immune response but also induces the immune system to store in molecular memory information about the agent that would result in a quicker and more robust response upon subsequent exposure. The central premise of vaccines is that the immune system would not just respond to the original agent (the vaccine) but would also response to the related, disease-causing virus.

"Smallpox was an infectious disease caused by either of two virus variants, *Variola major* and *Variola minor*." Stephen Pinker, the cognitive psychologist and linguist, calls this his favorite sentence.[82] Pinker found this sentence on Wikipedia and uses it in his recent book *Enlightenment Now* to extol the progress humanity has made thanks to science and other enlightenment values. What Pinker likes about the sentence is the "was" part: smallpox was an infectious disease. It is no longer so, but before it was eradicated, smallpox likely killed half a billion people during the last century (1880–1979) it plagued us.[83] But no longer; we eliminated it thanks to vaccines.

Smallpox has also played a major part in our history. It likely helped lead to the decline of the Roman Empire during the second century. Prior to the smallpox outbreak, Rome was a vibrant city of perhaps as many as a million people. At that time, Roman citizens could travel from present-day England to present-day Egypt on Roman roads.[84] Incidentally, the high population densities in Rome and other Roman cities combined with the ease of travel that Roman roads provided were ingredients in allowing smallpox and other infectious diseases to maintain themselves and spread. The Roman Empire would not fall for a couple of centuries yet, so it is difficult to say that it caused the fall. But the toll smallpox took certainly sapped the strength of Rome and set the conditions for its fall.

Columbus and then the Spanish conquistadors who followed brought smallpox to the Americas. Lacking immunity, the Indigenous Americans were ravaged by the disease, first in the Caribbean and then in the North America mainland. Some estimates point to a decline of more than 90 percent of the Indigenous American population.[85]

In some cases, European settlers actively engaged in biological warfare against the Indigenous Americans. One such example hits close to home. I live in Amherst, Massachusetts, a town named after the man who would become Lord Jeffrey Amherst. During the 1760s, the British under Amherst's command fought the French in a campaign to acquire what is now Canada. Several Indigenous American nations had joined in with the French. After an attack on Fort Pitt (now Pittsburgh), officers under Amherst's command suggested sending blankets infected with smallpox to the Indigenous Americans.[86] Amherst, who had advocated total war against the French and especially the Indigenous Americans, agreed. Exactly what was done and how well it worked is unclear. Historian Elizabeth Fenn noted that such biological warfare using smallpox was not uncommon during the 18th century in the North American colonies

Smallpox is a former disease, and *Variola* is extinct in the wild.[87] It is a former disease thanks to the deployment of vaccines, coordinated by Donald Ainslie Henderson, who led the World Health Organization Smallpox Eradication Unit from 1967 to 1977. Henderson, known as D.A. even to his wife, accomplished this feat in large part due to his tireless dedication and his extraordinary gift of organization.[88] That *Variola* lacks an animal reservoir and that most infected people display symptoms also helps the effort.[89] But such a program of containment and then eradication would not be possible without vaccines.

Smallpox was the first disease for which a vaccine was developed. In the 1790s, the English physician Edward Jenner developed a smallpox vaccine and systematically demonstrated its effectiveness. Although Jenner predates Darwin's theory of evolution, his vaccine is based on a principle of evolutionary relatedness. Jenner's vaccine was actually cowpox, a virus found in cows that is related to smallpox. Cowpox exposure primes the human immune response to smallpox but has little or no ill effects. In fact, the word vaccine comes from the Latin word for cow. Legend is that Jenner observed that the skin of the young women who milked cows were generally smooth and not marked by the pitted scars that were the remnants of smallpox.[90] Exposure to cows seemed to provide these young women immunity to smallpox.

Controversy about vaccinations is also very old; it even predates the practice. Before Jenner developed a vaccine for smallpox, people had used a similar but cruder technique called inoculation.[91] Here, the pustules of actual smallpox were rubbed into an incision made in the skin of the person being inoculated. If inoculation went according to plan, those inoculated would experience a substantially milder version of the disease.

One of the earliest successful uses of inoculation was also highly contentious. During a smallpox outbreak in 1721, a young physician named Zabdiel Boylston performed a series of inoculations that sparked heated debate in the city.[92] Interestingly, the primary proponent of inoculation and Boylston's patron was an older and established theologian: Cotton Mather. A generation earlier, Mather had been among the judges of the Salem Witch Trials. By the start of the 1720s, public opinion had soured on the religious frenzy of the past and Mather had wanted to restore his reputation as a man of learning. Although Mather thought inoculation as a measure that would improve the health of Bostonians, he likely was acting out of mixed motives as he probably was also looking at it as a means of redemption. In contrast, one of the most ardent opponents of inoculation was a brash, young newspaper publisher named James Franklin. Barely out of his teens, Franklin was primarily interested in making money and notoriety by whipping up controversy. And yes, this Franklin was the older brother of the Benjamin Franklin, who in later life would become a champion of inoculation.

As stated earlier, the reason influenza vaccines from the year before do not work as well is evolution. Influenza viruses evolve quickly; in doing so, they can evolve to escape from the vaccine. Moreover, developing flu vaccines requires predicting how the virus will evolve because vaccines are developed and produced several months before they are used. Similar processes are likely to go on with subsequent development of coronavirus vaccines. Fortunately, coronaviruses do not evolve as rapidly as influenza viruses.

Prior to COVID-19, the 2017/8 influenza season was one of the deadliest in the United States in recent years, rivaling and perhaps even exceeding the 2009 "swine flu" pandemic. Nearly 50 million people (about one in seven) in the US caught the flu. Nearly a million were hospitalized. About 80,000 died. Part of the reason that the flu was so bad was that the vaccine that season was not very effective: only 38 percent.[93]

Why was the 2017/8 season so bad? Why was the vaccine much less effective than usual? The likely answer is in how flu vaccines are typically mass-produced. The vaccine is a virus that has been killed. But before it is killed, it needs to be replicated in mass quantities. This production involves multiple generations of the virus being grown in chicken eggs. As the virus completes multiple generations in eggs, it can adapt to the conditions in the eggs. This egg adaptation can change the way the immune system views the virus—what is known as its antigenic properties—and thus its effectiveness as a vaccine against the circulating virus. When the virus is of a particular type—H3N2, commonly found in chickens—the egg adaptation appears to be stronger, and hence there is more of a mismatch between the vaccine developed and the viruses that are circulating.[94]

The vaccines developed by Moderna and Pfizer against SARS-CoV-2 are unusual. Instead of being a live but harmless virus (like the smallpox virus) or a killed virus (like most flu viruses), these coronavirus vaccines are just naked RNA. There are no viral particles. The design of these vaccines, which target the spike protein, was developed extraordinarily quickly, even before there were cases in the United States.[95] Part of the reason why these vaccines could be developed at these extraordinary speeds is that researchers had knowledge gained from 15 years of investigations of the original SARS virus, giving researchers a foothold to tackle the problem. Once the sequence of SARS-CoV-2 was in hand, they could use the principles of comparative biology to make educated guesses about what part of the virus to tackle.

The rapid vaccination of Americans during the spring of 2021 likely prevented a resurgence of the virus even as Alpha and other variants rose in frequency. Vaccinations took away the fuel of the virus. Assuming sufficient people get vaccinated, Americans can soon resume more normal life without fear of the pandemic. Vaccination campaigns in the United Kingdom, Israel, and other countries have also blunted the pandemic. Elsewhere in the world, notably in India, the pandemic rages on. Worldwide intensive vaccination will be required to curb the pandemic and reduce the likelihood of new troubling variants.

The Moderna and Pfizer vaccines were designed to be given in a two-dose regimen, with doses spaced weeks apart. Early in the vaccination effort, when supplies were limited, Canada and the United Kingdom delayed the second dose to some in order to get more people with first doses, as the first dose provided substantial immunity on its own. With more people immune in this so-called dose-sparing strategy, the spread of the virus would be muted. In addition, fewer people would be hospitalized. Some have expressed concern that the dose-sparing strategy would leave a considerable fraction of people with partial immunity. Would a large proportion of partially immune people lead to selection for variants that escape the immune response? Sarah Cobey and her colleagues presented the argument that given limited vaccines, the dose-sparing strategy would actually impede the evolution of resistance to the vaccine.[96] The partial immunity of those with a single dose limits the spread of the virus; this in turn helps slow down the rate of variant generation. As of May 2021, vaccine supply outstrips demand in the United States. Other rich countries also have or will soon have an abundance of vaccine. But shortages are likely to persist in other places in the world. Spreading out doses seems to be an effective strategy there.

HIV/AIDS: THE OTHER PANDEMIC

In the fall of 1981, doctors in the United States were alerted to clusters of young gay men with unusual symptoms. These previously healthy men were coming into clinics with *Pneumocystis carinii* pneumonia and other opportunistic infections, ones that are generally only seen in patients with compromised immune systems. This syndrome, which would soon be named AIDS (acquired immunodeficiency syndrome), attracted the attention of an immunologist named Anthony Fauci.[97] Now Director of the National Institute of Allergies and Infectious Diseases and the public face of the public health effort at combatting the COVID-19, Fauci first made his mark managing the public health battle against the other pandemic of our times.

Caused by HIV (human immunodeficiency virus), AIDS has killed at least 37 million worldwide.[98] Despite this horrific toll, there are reasons to be optimistic about AIDS. Infections in the United States are now less than a third of what they were at the peak. Once an almost certain death sentence, various treatments have made the disease manageable, particularly if it is treated early. Research in evolutionary biology has shed light on the origins of HIV and its movement across the world. It also can provide insight into treatments.

In 1992, *Rolling Stone* magazine published a provocative hypothesis for HIV's origin. The claim was that the virus originated in monkeys and that it had entered into humans from contamination during the development of polio vaccines in the Congo during the late 1950s.[99] Although most biomedical researchers were highly skeptical of the hypothesis, some were more receptive to its plausibility. Among those more open to the idea was Frank Lily, a geneticist who was on the HIV/AIDS commission.[100]

This hypothesis did attract the attention of William Hamilton, who was at the time among the world's most notable evolutionary biologists.[101] As a graduate student in the early 1960s, Hamilton had worked out the foundations of the theory of kin selection, the notion that natural selection could favor variants of genes that promote individuals to take actions to help a relative's reproductive success even at a cost to itself.[102] Not long afterward, Hamilton developed a mathematical theory for the antagonistic pleiotropy model for the evolution of aging, extending the work of George Williams. In the 1980s, Hamilton had promoted the importance of host-parasite coevolution to explain both the existence of sex and for female preferences of particular traits in their male counterparts.

By the middle of the 1990s, Hamilton was now approaching 60 and looking for new challenges. After reading accounts of the development of the polio vaccine, Hamilton embarked on an expedition to investigate this hypothesis. Writing to his friends Yura and Blanka Ulehla on Christmas Eve 1999, Hamilton described the rationale for the project:

> The idea is that, if it is indeed true that Hilary Koprowski's chimpanzee-bonobo testing camp for his live oral polio vaccines at Kisangani (then Stanleyville) in the (then) Belgian area of Africa was the source of an accidental contamination of the vaccines which were fed to a million Africans in the late 50s, it should be possible to find an appropriate wild SIV (simian immunodeficiency virus) in chimps of the area where Koprowski's assistants collected the animals for his testing camp. . . . It is to these precise areas that I now intend to go together with two young and athletic Canadians and two litres of special RNA buffer solution.[103]

Feverish from malaria, Hamilton returned to London at the end of January. He was admitted to University College Hospital London, where he was given treatment for malaria. On March 7, 2000, Hamilton died. To what extent the malaria infection factored into Hamilton's death is still unclear.

The hypothesis that HIV emerged from development or testing of polio vaccines during the 1950s in Central America has been shown to be false based on several lines of evidence.[104] One of these is from dating of the origin of HIV-1. Molecular evolution studies estimate that HIV-1 entered human populations around the start of the 20th century and rule out an origin after the 1930s. Moreover, samples of HIV-1 taken from 1959 and 1960 from the Democratic Republic of Congo show substantial genetic diversity, which is inconsistent with origin in the 1950s.[105]

Where did chimpanzees acquire the virus? Examination of the sequences of SIV (simian immunodeficiency virus, the nonhuman primate equivalent of HIV) and other viruses from different primates point to it arising from recombination between a preexisting chimpanzee virus and one from sooty mangabeys (*Cercocebus atys*), monkeys found in the forests of West Africa.[106] Somehow, the combination of parts of the chimpanzee virus and parts of the sooty mangabey virus produced a deadly virus. Chimpanzees infected with SIV get sick and have reduced mortality. Moreover, SIV infection in chimpanzees decimates the same T cells as HIV does in humans. AIDS was around as a chimpanzee disease before HIV.

HIV jumped into a human early in the 20th century somewhere in Africa. The most likely reason for this spillover is someone hunting chimpanzees, but we do not know much more than that.[107] For about half a century, HIV was present in low numbers in southern Africa. Around 1960, the viral spread took off. One plausible cause for the outbreak was the availability of inexpensive needles used in medical facilities.[108]

In *And the Band Played On*, an account of the early days of the AIDS crisis in America, journalist Randy Shilts suggested that HIV arrived in the United States with the various parades centered around the 1976 Bicentennial.[109] This scenario is unlikely: molecular evolution studies estimate that HIV came to the United States considerably earlier (most likely 1966–1972),[110] though it is quite possible that the bicentennial parades accelerated the spread of the virus in the United States. Most likely, the virus had come to the United States through Haiti. A likely reason why the virus had spread from the Congo to Haiti is that because both are French-speaking countries; many Haitians were working in the Congo during the 1960s.[111]

The first successful drug used against AIDS, AZT (also known as zidovudine) was developed in 1987.[112] This drug—like most antivirals—targets an aspect of the biology of the virus. HIV is a retrovirus, and as such it replicates like the RNA viruses (influenza and the coronaviruses), but retroviruses have an added twist. They can convert their RNA back into DNA and insert that DNA into the genomes of their hosts. The enzyme they use to transcribe RNA information into DNA information is called reverse transcriptase. AZT inhibits reverse transcriptase by mimicking the nucleotides used in nucleic acids.

Though this treatment was a major advance at the time, success was limited because HIV evolved resistance to the drug, typically in a matter of months to a couple of years. Combination therapy uses multiple drugs that block reverse transcriptase at different steps as well as drugs that inhibit the enzymes that HIV uses to break down proteins. With the combination treatments, the virus has more difficulty in evolving resistance, which has led to much improved health and increased survival of those infected with HIV.

A topic of debate about the genetic basis of evolutionary adaptation appears to have ramifications for HIV treatment. In the classic view of adaptation, a mutation that is favored by natural selection—let's say that it tweaks an enzyme in a way that allows the organism to survive better—and that mutation increases in frequency, sweeping through the population. In recent years, evolutionary biologists have had an increased appreciation for adaptation that arises from standing genetic variation (variation present in the population). As before, the variants that are selectively favored increase in frequency, sweeping through the population.

Both views of adaptation involve sweeps, but the sweeps are different. In the classic view of the sweep starting from a single mutation—what are called hard sweeps—variation is quickly lost in the region of the chromosome near the favored mutation. Unless there is recombination, the favored mutation will carry variants linked to it. In the sweeps from standing genetic variation, which are called soft sweeps, there are multiple copies of favored variants; each of these sweeps a different set of linked variants with it, allowing for much more variation existing after the sweep.[113] The depletion of variation after a hard sweep makes future adaptation less likely; in contrast, future adaption is not as constrained after a soft sweep.

Pleuni Pennings, now at San Francisco State University, examined the sequences of HIV in patients collected over time. Here, Pennings and her colleagues saw both soft sweeps and hard sweeps occurring in HIV patients who had evolved drug resistance.[114] Finding both types of sweeps in comparable frequencies suggest that the adaptative mutations were occurring about once a viral generation. If they were much more common, then soft sweeps should predominate. If they were much less common, then variation-reducing hard sweeps should predominate.

Following up on this study, Pennings' lab looked at patients who had undergone different treatment regimens to examine the relationship between how often hard and soft sweeps occurred and the general success of the treatment.[115] They found that hard sweeps and a loss of genetic diversity predominated in the viruses of patients who had a better type of treatment (such as in many of the newer combination therapies). In contrast, soft sweeps and a maintenance of genetic diversity were much more common in treatment regimens that had poor outcomes (such as AZT alone).

One possible application coming out of this research is the development of a biomarker to see how well soft sweeps are predictive of poor outcomes. Development of a biomarker could be useful because sequencing is much easier and quicker than waiting months to see changes in clinical symptoms.

CODA

With spillover coming from deforestation, deforestation creates more forest edges, and these edges lead to more contact between humans and wildlife. This is especially pronounced if the forest area is fragmented. Bats, host to many spillover viruses, including both SARS-CoV-1 and SARS-CoV-2, also are more likely to feed near humans when their habitat is disturbed. Wildlife trade entails people going into forests to collect the wildlife, enhancing the likelihood of spillover. Remember the epigraph by Wayne and Bolker: ecology sets up the opportunity for new spillovers of potentially pathogenic viruses.

Reducing deforestation, enhanced money for monitoring wildlife trade, increasing early detection and control—programs that combined cost tens of billions a year—are all ways to minimize the prospect of another bad pandemic.[116] Although these programs have some costs, these costs are dwarfed by the trillions lost and the death toll that has come from COVID-19. Prudent action now can minimize the risk of subsequent deadly and costly pandemics.

During the production of this book, a few things have changed regarding the COVID-19 pandemic. During the summer of 2021, the Delta variant (also known as B.1.617.2) of the SARS-CoV-2 virus supplanted previous variants in the United States. This variant, which is a cousin but not direct descendant of the Alpha variant, is about 50 percent more transmissible than Alpha and about twice as transmissible as the SARS-CoV-2 that sparked the first global wave of the pandemic. Combined with a modest ability to evade immunity of previously infected and vaccinated individuals, this superior transmissibility led to a subsequent wave of infections across much of the globe. Delta infections may also be somewhat more potent than those of prior variants: some studies have found hospitalization rates are higher for those infected with Delta than those infected with Alpha. For now, vaccines are still highly protective against all major variants, including Delta. As of September 2021, no new variants that are able to outcompete Delta and are better at evading vaccine-induced immunity are at appreciable frequencies. Evolutionary genetic tools and principles will remain vital not only in tracking the evolution of new variants of this vaccine but also in devising policy to thwart the evolution of such variants.[117]

The origin of the virus remains unknown. During the summer 2021, simmering claims about the prospect of a lab origin for SARS-CoV-2 rose to the surface again. Although escape from a lab cannot be completely dismissed as an origin of the virus, the bulk of evidence argues against it. Published in September 2021 a *Cell* review authored by many notable evolutionary biologists who specialize in viruses concludes "the most parsimonious explanation for the origin of SARS-CoV-2 is a zoonotic event."[118] They note similarity between the early epidemiology of this virus and other viruses, such as the first SARS-CoV, that emerged from animal markets. They also discuss how the genetic and evolutionary patterns associated with this virus argue for a natural zoonotic origin.

Also in September 2021, researchers at the Pasteur Institute and the University of Laos posted a reprint documenting the isolating viruses from bats in Laos that are the closest to SARS-CoV-2. Notably, these viruses have receptor binding domains that, like those of SARS-CoV-2, can enter human cells through the ACE2 receptor. These viruses, however, lack the furin cleavage site present in SARS-CoV-2 which assists in entry through the receptor. Thus, these viruses probably are less efficient at entering human cells than SARS-CoV-2 is. The findings do not establish that SARS-CoV-2 evolved from bat viruses in Laos (or nearby in Southeast Asia), but they do present that as a plausible hypothesis.[119]

NOTES

1. Wayne and Bolker (2015, p. 7).
2. Michael Kinch is quoted in Bryson (2019, p. 334).

3. See Spinney (2017) for information on the 1918–19 pandemic.

4. Spinney (2017), p. 48.

5. *Ibid.*, p. 217.

6. Biggerstaff et al. (2014).

7. See Wayne and Bolker (2015).

8. See Bryson (2019).

9. Graham and Barle (2010). The quote is on p. 3142.

10. Grubaugh et al. (2019).

11. Branswell and Joseph (2020).

12. Wamsley (2020). During the long, brutal Covid winter predicted in the fall of 2020, hundreds of thousands more Americans died.

13. Yang et al. (2021) estimate that 1.39 percent of those infected die, with confidence limits of 1.04 and 1.77 percent. Estimating the true mortality rate is challenging given the difficulties of estimating the actual number of people who have been infected, which is much larger than those recorded as being infected. In addition, the number of deaths attributed to COVID-19 is an underestimate of the actual number of deaths caused by the virus. The underestimate of COVID-19 deaths is much smaller than the underestimate of the number of COVID-19 infections. Moreover, this mortality rate likely varied at different times during the pandemic.

14 Spinney (2017) noted that most older people exposed to the 1918 pandemic flu likely had partial or complete immunity from prior exposure to flu viruses from the past. We should not expect such immunity to be present for the coronavirus SARS-CoV-2, given that it came into an immunologically naïve population.

15. Bernstein (2020).

16. Horton (2020, p. 17).

17. Lipton et al. (2020).

18. In addition to the mortality and morbidity from the disease, the pandemic also has had a great economic burden. Dobson et al. (2020) estimate the global losses from the pandemic to be around 10 trillion dollars.

19. Cui et al. (2019).

20. The CDC reported an estimated 35.5 million people in the United States got sick with influenza, and there were an estimated 34,200 deaths during the 2018–19 flu season. See https:cdc.gov/flu/about/burden/2018-2019.html. Accessed December 11, 2020.

21. The Coronavirdae Study Group (2020, p. 536) noted the need to consider the genetic variability within viruses when classifying them. They also highlighted the importance of

> the question of how much difference to an existing group is large enough to recognize the candidate virus to previously identified viruses infecting the same host or established monophyletic groups of viruses, often known as genotypes or clades, which may or may not include viruses of different hosts.

22. CDC. Scientific brief: SARS-CoV-2 and potential airborne transmission. Updated October 5, 2020. http:www.cdc.gov/coronavirus/2019-ncov/scientific-brief-sars=cov-2. html. Accessed December 11, 2020.

23. An analogous version of R can be applied to cultural transmission. A famous series of television ads from the early 1980s illustrated the power of such transmission: here a young woman is talking about being so impressed with a shampoo that she told two friends. As she says the words, the screen splits in two. Then, the friends then told two friends and the screen splits again. And so on, as the screen keeps splitting. With just a few iterations of doubling (R equals 2), the number of people involved rapidly increases.

24. Wayne and Bolker (2015).

25. Because R will decline as more and more people are infected and then gain immunity, sometimes the value of R_0 (R-naught) is used to describe the value of R in an unexposed population. As more people are infected and then recover, R declines at roughly $R = R_0$ $(1 - I)$ where I is the proportion of immune individuals. When R goes below 1, the infection should be checked. This should happen when $I > 1 - (1/R_0)$. This assumes that infected people are equally likely to infect others. Substantial variation in spreading ability of different classes of people, such as super-spreaders spreading most cases of the virus, can complicate the analysis.
26. Li et al. (2020b).
27. See Thakkar et al. (2020) for a study showing how early containment efforts reduced the reproductive number of the virus in Washington State. More recently, Brauner et al. (2021) examined data from several countries to ascertain the effectiveness of different containment measures. Strict limits on gatherings and closing high-risk businesses (e.g., restaurants and bars) had large effects on reducing R. Stay-at-home orders provided only modest further reductions.
28. For instance, see covidactnow.org. As of December 11, 2020, R is 1.13 in Rhode Island, the state with the highest number of new cases per 100 K people. This means that the average infected person is infecting an average of 1.13 new people. So cases should continue to increase in that state. In contrast, R is 0.81 in Iowa. Accordingly, cases should decrease there.
29. Biggerstaff et al. (2014).
30. Wayne and Bolker (2015). Also note that while the original version of SARS-CoV-2 is less infectious than smallpox, at least one of the subsequent variants likely is at least as infectious as smallpox, if not more so. Unchecked and in an immune-naïve population, the R of the Delta variant is probably around 5 to 7.
31. Morens and Fauci (2020).
32. Shi et al. (2020).
33. Samavati and Uhal (2020).
34. Somers (2020).
35. Blanco-Melo et al. (2020).
36. Biologists also call such animal–human transmission of disease zoonosis. David Quammen (2012) has written an excellent book overviewing the various ways emerging viruses have spilled over into humans.
37. Greenfield (2009, p. 11).
38. Cui et al. (2019).
39. Wang and Eaton (2007).
40. Li et al. (2005).
41. Zhang and Holmes (2020).
42. Latinne et al. (2020).
43. See Quammen (2020) for an overview of pangolins and the possibility that they are the intermediate host. See Li et al. (2020a) for skepticism that pangolins are the intermediate host.
44. Cui et al. (2019).
45. Letko et al. (2020).
46. Brook and Dobson (2014).
47. Latinne et al. (2020).
48. Gussow et al. (2020).
49. Chappell (2020).
50. Maurano et al. (2020).

51. Bedford et al. (2020).
52. Worobey et al. (2020).
53. There is evidence that some individuals in the West Coast of the United States had antibodies that reacted to the virus (Basavaraju et al. 2020), but that is not direct evidence that they had SARS-CoV-2.
54. Abdelrahman et al. (2020).
55. Dearlove et al. (2020).
56. Van Dorn et al. (2020).
57. See Callaway (2020) for an overview. Biochemists use the abbreviation (D) for aspartate because (A) is used for alanine. The site is on the surface of the S protein and not at the receptor binding domain.
58. Korber et al. (2020).
59. Grubaugh et al. (2020). The quote is on p. 794.
60. Van Dorn et al. (2020)
61. Mansbach et al. 2021.
62. Davies et al. (2021).
63. Parker-Pope (2021).
64. Kemp et al. (2021).
65. Fischer (2020).
66. Song et al. (2005).
67. See, for instance, Fan et al. (2020) and Valenti et al. (2020).
68. Jing et al. (2020).
69. Ferrer-Admetlla et al. 2009.
70. Gampa et al. (2017).
71. Valenti et al. 2020.
72. The Severe Covid-19 GWAS Group (2020).
73. See Zeberg and Paabo (2020) for the study and Luo (2020) for commentary.
74. Paabo (2014).
75. Enard and Petrov (2018). The quote is p. 369.
76. The signature of positive selection here is the presence of a long stretch of DNA derived from Neanderthals found at appreciable frequency in a modern human population. Their rationale is that if sometime soon after Neanderthals mated with modern humans there was positive selection for the Neanderthal introgression in the human genome, then that haplotype would rapidly increase in frequency. If the rise in frequency is not rapid, then the haplotype is more likely to have been broken up by genetic recombination.
77. E-mail conversation with David Enard. December 2020.
78. Enard (2016).
79. See Hensely (2020) and Welland and Zimmer (2020). Both the Pfizer and the Moderna vaccines were more than 90 percent effective in clinical trials and in the early months of their use in the general population. Although their effectiveness has declined somewhat due the Delta variant and waning immunity, they remain highly effective.
80. Hassan (2020).
81. Wayne and Bolker (2015). Herd immunity can be reached through vaccinations, prior infection, or some combination of the above. Because vaccines in general, and in the Covid vaccines in particular, are far safer than the virus they are designed to immunized against, the desired goal is to reach herd immunity via vaccinations as much as possible.
82. Pinker (2018). The quote is on p. 64.
83. Foreword by Douglas Preston of Henderson (2009).
84. Kinch (2018).

85. Mann (2005).
86. Penn (2000).
87. As of 2020, the smallpox virus exists in only two locations: one laboratory in each of the United States and Russia.
88. Preston (2009).
89. Henderson (2009).
90. There is abundant controversy about how Jenner came to the discovery about exposure to cowpox as well as his exact role in developing the vaccine. Whether Jenner was first to use cowpox as a vaccine, he certainly was the first to systematically demonstrate that it worked and that it performed better than other methods of the time. See Boylston (2012), Kinch (2018), and Wolfe (2011) for more detail.
91. See Boylston (2012).
92. Coss (2016).
93. Even though the vaccine was only 38 percent effective, it still prevented about 7 million cases of the disease, about a 100,000 hospitalizations, and 8,000 deaths in the United States. The model predicts that the positive effects of the vaccine were most felt in young children (under the age of 5), reducing hospitalizations by two-fifths. Because this model only examined the direct benefits of immunization and did not consider secondary benefits through herd immunity, the figures may be underestimates of the true benefits. See Rolfes et al. (2019) for details.
94. Barr et al. (2018).
95. Wallace-Wells (2020).
96. Cobey et al. (2021)
97. Abbasi (2018).
98. *Ibid.*
99. Kolata (1992).
100. Dunlap (1995).
101. Biographic information about Hamilton can be found in Segerstrale (2013).
102. One example of a potential behavior that could arise from kin selection would be a squirrel gave an alarm call to help warn relatives of a predator's approach even though it put itself at risk. Another would be a bird that helped raise her sister's brood while forsaking or at least delaying her own chance for reproduction.
103. Segerstrale (2013). In the same letter, Hamilton noted:

> This town is the very place about which Joseph Conrad wrote "The Heart of Darkness" and even to some extent it's back to even more primitive than then when there were at least the big Belgian-run steam boats.

104. See Wolfe (2011) for further information.
105. Worobey et al. (2008).
106. Sharp and Hahn (2010).
107. Quammen (2012) has written a plausible narrative story of how this spillover could have happened.
108. Wolfe (2011).
109. Shiits (1987).
110. Gilbert et al. (2007).
111. Pepin (2011).
112. Tseng et al. (2014).
113. Hermisson and Pennings (2005).
114. Pennings et al. (2014).

115. Feder et al. (2016).
116. Dobson et al. (2020).
117. See Katella (2021) and Otto et al. (2021).
118. Holmes et al. (2021). The quote is on p. 4852.
119. Gale (2021) and Temmam et al. (2021).

FURTHER READING

Wayne, M. L. and B. Bolker. 2015. *Infectious Disease: A Very Short Introduction.* Oxford University Press.

Wolfe, N. 2011. *The Viral Storm: The Dawn of a New Pandemic Age.* Time Books.

3 Vectors of Disease

SUMMARY

Mosquitoes are the animal responsible for the most human deaths because of the diseases they carry, notably malaria. We first consider how malaria has shaped both human history and our genomes, including sickle cell anemia. Only a few mosquito species are disease vectors. Why? In addressing this question, we explore why some mosquitoes have specialized on humans. Here, we examine the neurobiology of smell and taste in mosquitoes and how these senses have evolved. We then turn to how malarial parasites have been controlled over the years. Here, we find that resistance to these agents often evolves. We discuss reasons for why some regions of the world have seen more resistance to evolution than others. We then consider ways mosquito populations have been controlled over the years. As with the antimalarials, evolution of resistance was an important reason why DDT lost favor. We examine sterile male technique, the use of the bacterium *Wolbachia* to generate incompatibility, and CRISPR-generated gene drives.

"Malaria is not a disease of the environment in the way that, say, asthma is, or certain kinds of cancer are. And yet its transmission depends upon an exacting set of environmental conditions. The protozoan parasite, despite all its sophisticated wiles and cunning, is more like a seed than a self-sufficient predator. Like a pip on the wind, it must alight in a fertile bed, be enveloped in the proper amount of moisture, and be bathed in the correct level of sunlight."—Sonia Shah (2010)[1]

"The mosquito has killed more people than any other cause of death in human history. Statistical extrapolation situates mosquito-inflicted deaths approaching half of all humans that have ever lived. In plain numbers, the mosquito has dispatched an estimated 52 billion people from a total of 108 billion throughout our relatively brief 200,000-year existence."—Timothy Winegard (2019)[2]

THE LETHAL MOSQUITO

Mosquitoes still kill. During the first two decades of the 21st century, the annual death toll attributable to mosquitoes and the diseases they carry has been about 2 million. Humans are extraordinarily good at killing other humans, but we are no match for the mosquito. Of animal taxa, mosquitoes are by far the number one killer of humans. Our self-inflicted death rate over that time period is *only* about half a million a year. Still, that is good enough for second place.[3]

Of course, the mosquito itself does not do the actual killing. The mosquito is the delivery system; it is the vector. The killing agents are biological packages inside the mosquitoes. Some are viruses like West Nile, yellow fever, Dengue, and Zika. Others are eukaryotic microbes like *Plasmodium*, the agent that causes malaria. These agents, however, would be unable to persist as disease-causers were it not for their mosquito vectors.

DOI: 10.1201/9780429503962-4

Today we think of malaria, influenza, and other infectious disease as being caused by an agent. But this "germ theory" view of disease is rather recent; it was not generally accepted until well into the 19th century. Accounts from the 1918–9 flu pandemic show that even the early 20th century was a period of transition where acceptance of the germ theory had not fully overcome prior beliefs.[4] The dominant belief prior to germ theory for malaria and many other diseases was that something was bad in the air. The word malaria literally comes "bad air": the disease was more common in Italy than in Northern Europe during the 18th and 19th centuries, and Italians would speak of their fevers as "mal'aria" ("bad air"). Horace Walpole, the English writer and art historian, is credited for bringing the word into the English language. Writing in 1740, he stated: "There is a horrid thing called the malaria, that comes to Rome every summer, and kills one."[5]

Conclusive demonstration that mosquitoes were vectors of disease came in the waning days of the 19th century. A trio of discoveries in 1897 were instrumental. First, the Indian-born British medical doctor Robert Ross showed that avian malaria required transmission by *Anopheles* mosquitoes. He speculated that these same mosquitoes were vectors of human malaria, a discovery confirmed the same year by the Italian biologist Giovanni Grassi. In that same year Robert Koch, famous for his postulates about demonstrating causes of disease, conclusively showed that human blood could be cleared of the malarial parasite with quinine. A few years later, the American doctor Walter Reed, following up on work by a Cuban doctor named Carlos Finlay, conclusively showed that *Aedes* mosquitoes were vectors of yellow fever. These discoveries slammed the door shut on the miasmatic theory.[6]

The discovery that *Aedes* mosquitoes were responsible for the transmission of yellow fever was a pivotal historical moment. Fresh off its victory in the Spanish American War and the acquisition of the Philippines and other territories in Asia, the United States was ever more eager to establish a more direct route to the Pacific. A canal through the Isthmus of Panama would save valuable time and expense on trips from the Eastern United States to Asia. The problem was that mountains were not the only barrier to constriction of the canal; so were diseases like yellow fever. The famed French engineer Ferdinand de Lesseps, who led the building of the Suez Canal, had failed miserably trying to recreate his Suez success in Panama. Yellow fever and other diseases killed a considerable fraction of his team and left the survivors too weak to build.[7]

The knowledge that the *Aedes* mosquito was the disease vector helped the American efforts to contain yellow fever. As Havana's chief medical sanitary officer, William Gorgas had exterminated that city's mosquitoes and consequently had eradicated yellow city from the city by 1902. When the United States had secured domain over a sliver of land through the newly independent country of Panama the following year,[8] Gorgas repeated the mosquito eradication strategies used in Cuba. By 1906, yellow fever was eliminated! In addition, the incidence of malaria had been cut to a mere tenth of what it had been. The removal of these disease threats allowed for successful completion of the Panama Canal by 1914. In turn, the canal's completion helped make the United States a world power.

This is but one example of the central argument that the historian Timothy Winegard in makes his recent and provocative book *Mosquito*. Not only have the

mosquitoes that transmit disease caused billions of deaths and untold sorrow throughout history, but they have also played pivotal roles in our history. As Winegard puts it:

> The mosquito sponsored both the rise and fall of ancient empires, she gave birth to
> independent nations while callously subduing and subjugating others. She has crippled
> and even destroyed economies. She has prowled the most momentous and pivotal bat-
> tles, menaced and slaughtered the greatest armies of her generations, and outmaneu-
> vered the most celebrated generals and military minds ever mustered to arms, slaying
> many of these men in the course of her carnage.[9]

According to Winegard, Alexander the Great's conquests in the 4th century before
the common era were halted as bouts of malaria sapped his strength. Later, malaria
likely killed Alexander at the age of 33. Three centuries later, Caesar and Pompey,
once friends and co-leaders of the Roman Republic, battled for control over it. One
fateful day, Caesar took his army across the Rubicon river, an action that would spark
a civil war and would eventually lead to Caesar's dictatorship and the emergence of the
Roman Empire. Unknown to Pompey, Caesar and his army were battling malaria during their cross of the Rubicon. Had he known, Pompey could have taken advantage and
prevented Caesar's rule. Moving forward a few centuries more, the fall of the Roman
Empire was hastened by malaria, as well as the smallpox outbreaks discussed in the
previous chapter.

Spanish and then other colonists brought *Anopheles* mosquitoes and their
Plasmodium to the Americas, where these parasites would continue their influence on historical events. In fact, malaria-carrying mosquitoes were instrumental in
bringing the Revolutionary War to an end.

The fall I took US history in high school coincided with the bicentennial of
the battle of Yorktown. There, in October 1781, the last major battle of the
Revolutionary War was fought; with help from the French, the American forces
under George Washington defeated the army commanded by Charles Cornwallis.
Being only a couple hours' drive from Yorktown, Virginia, we took a class trip to
the ceremony.[10] There, we saw US President Ronald Reagan and French President
Francois Mitterrand speak. What I did not know then was that the French were
not the only forces helping the Americans. Washington's forces had help from the
Anopheles mosquito and its malarial parasite. Malaria was endemic through much
of what would become the United States, particularly the South, during the mid-18th
century and most inhabitants—including Washington—had had several bouts with
it. These prior exposures seasoned the Americans; they were either less sick from
or nearly immune to subsequent infection. In contrast, the British had much less
prior exposure and consequently faced the full strength of malaria. As Cornwallis
reported to his superior,

> I have the mortification to inform your Excellency that I have been forced to give up
> the post. . . . The troops being much weakened by sickness. . . . Our numbers had been
> diminished by the Enemy's fire, but particularly by Sickness. . . . Our force diminished
> daily by sickness . . . to little more than 3,200 rank & file fit for duty.[11]

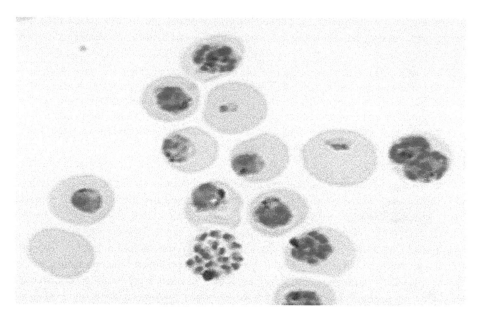

Cultured *Plasmodium falciparum* malarial parasites.

Source: Courtesy of John Tan and Tim Anderson.

The British were no match for the stinging guns of the mosquito loaded with *Plasmodium* bullets.

As we will see, mosquitoes have shaped our genomes as well as our history.

MALARIA'S GENETIC LEGACY

Humans, a few species of *Anopheles* mosquitoes, and the various *Plasmodium* parasites that malaria have engaged in a coevolution dance over the millennia. This coevolutionary dance has left several marks on the human genome as natural selection for resistance to malaria has altered the frequencies of genetic variants, though not uniformly in different populations. The best-known genetic adaptation to malaria is the sickle cell anemia allele. Although people with two copies of this allele have sickle cell anemia and almost always died young before modern medicine, those with one copy (those having the sickle cell trait) are protected from malaria. Although the exact mechanism for this protection is not fully known, the protection is strong. In fact, a recent study looking at genetic associations for malaria risk in malaria-endemic areas found that those with the trait were seven times less likely to get severe malaria.[12] This is a classic example of heterozygote advantage, a form of balancing selection wherein both alleles are maintained by selection because the heterozygote has higher reproductive success than either homozygote.

As we saw in Chapter 1, sickle cell anemia is due to a single genetic change: the alleles differ by just one nucleotide of DNA change that alters just one amino acid in the beta chain of hemoglobin.[13] Incidentally, there are mutations in hemoglobin that provide some protection from malaria without causing devasting illness in homozygotes. We would have been better off with that mutation, but unfortunately, the sickle cell anemia allele won the race. That quirk of evolutionary history has led to much death and suffering.[14]

The genetic characterization of sickle cell anemia was done in the mid-1950s; as such, it was one of the first diseases to be characterized. It is also an old, though not necessarily the oldest, genetic adaptation to malaria. As we will soon see, the sickle cell trait is far from the only adaptation to malaria.

Recent studies by Charles Rotimi and his group provide insight as to when and where the sickle cell allele originated. Now Director of the Trans-NIH Center for Research on Genomics and Global Health, Rotimi has long been interested in both the genetic and sociological dimensions of disease.[15] Born and raised in Benin City, Nigeria, Rotimi attended the University of Alabama for his doctoral studies. Here, he assessed whether foundry workers in Ohio particularly exhibited rates of stomach and lung cancer. They did not. Then, starting in the 1990s, Rotimi turned his attention toward investigating patterns of disease in African and African American families. He found that hypertension, diabetes, and coronary heart disease tended to run in some African American families but did not in others. This led to further investigations into the environmental as well as genetic factors of these diseases in people of West African descent. With the rise of genomics, Rotimi has forcefully argued for greater study of genomes from people of African descent. As he puts it in a recent interview, "African genomes, because they are the oldest, have among the highest degree of variation, and that you cannot fully understand population variation without studying and understanding these most ancestral human genomes." This is a point we will return to in Chapters 5 and 17.

Working with Daniel Shriner, Rotimi found strong evidence of a single origin of sickle anemia.[16] This discovery contradicted earlier speculation that sickle cell anemia may have evolved multiple times. The ancestral allele most likely evolved 7,300 years ago. But there is considerable uncertainty about that age: the confidence intervals stretch from 3,400 to 11,100 years ago. DNA analysis of 5,200-year-old Egyptian mummies has found the genetic signature of sickle cell anemia. Hence, this date sets a lower bound of its age.

Remember that the sickle cell allele is only advantageous in heterozygote form: the carriers are more fit than the noncarriers, but those with two alleles are very unfit because they have anemia. In such a situation of heterozygote advantage, we should expect the frequency of the mutation to rapidly increase when it is rare but then slow down its increase and eventually stop increasing once it has reached the equilibrium frequency. Shriner and Rotimi's simulations suggest that the time between the origin of the heterozygous-favored mutation and its approach to equilibrium should take about two millennia. This estimate assumes that the homozygous individuals who get sickle cell anemia do not survive long enough to have and rear children. Accordingly, the heterozygotes have about a 16 percent fitness advantage over noncarriers. If homozygous individuals can live long enough to bear an appreciable number of children, then this estimate may be altered.

Shriner and Rotimi's study also addressed whether the origin of the sickle cell allele coincided with Bantu expansion. About 5,000 years ago, the Bantu people started a dramatic expansion across Africa. Starting from equatorial West Africa (approximately the location of present-day Gabon) over the course of a few millennia, the Bantu expanded south and east into most of Africa. It is possible that the Bantu expansion and the sickle mutation started at the same time. But Shriner and Rotimi argue that the more likely scenario based on their data is that the allele arose long before the expansion started. In fact, they think that the allele already was near its equilibrium frequency when the expansion started.

Where did the mutation arise? There are several possibilities. It could have arisen in west-central Africa in rain forest near where the Bantu lived. But Shriner and Rotimi point out an even more intriguing possibility. If the mutation arose during the most likely time of 7,000 or so years ago, then perhaps it arose in what is now the Saharan desert. "The Sahara Desert?" I hear you ask incredulously. Well, during this time, the Sahara was not a desert. In this "Green Sahara" period, it was rather wet and could support a mosquito-laden forest. In any case, the Bantu people likely had a high frequency of the carriers as they started their invasion. As they moved south and east, the genetic immunity of the carriers likely gave them an advantage over other populations, thus fostering the continuation of their expansion.

But even before the sickle cell allele arose, genetic adaptations to malaria had evolved. Before we go into detail about these, we need to discuss the different types of *Plasmodium* and the malarias they cause.[17] An early mild human malaria was *Plasmodium malariae*, which was able to live in humans for long periods of time—even up to several decades—in a state akin to suspended animation. Being able to stay in humans indefinitely patiently waiting to be bitten by another mosquito was an advantage for the parasite, but that came with a tradeoff: *Plasmodium malariae* reproduced slowly in both humans and mosquitoes. Consequently, it eked out a meager existence of quiet desperation. It was replaced by another *Plasmodium*, *P. vivax*, which was able to better attach itself to human red blood cells. This parasite evaded the defenses of human immune systems better than its predecessor. Thus, it could replicate faster within humans, an evolutionary advantage. The malaria from *P. vixax* is of intermediate severity: definitively stronger than *malariae* but milder than some of the malarias that would come later.

Although *P. vivax* evolved in Africa, it is largely absent from the continent today. The reason is a genetic adaptation known as Duffy negativity. Duffy is the name of a specific cell surface receptor found on red blood cells. *P. vivax* had evolved mechanisms to exploit the Duffy receptor; thus, it can get into red blood cells of those who have the receptor. Individuals who do not express the Duffy receptor on their red blood cells—those that are Duffy negative—are immune to *vivax* malaria. Many thousand years ago, a mutation that altered the expression of this Duffy receptor evolved in Africa. People with this mutation expressed the receptor on most of their cells except for their red blood cells. This mutation rose in frequency to near fixation; today, a whopping 97 percent of people from West and Central Africa are Duffy negative and thus are basically unaffected by *vivax* malaria. This genetic adaptation most likely led to the downfall of *vivax* malaria in Africa, though it would subsequently reappear in Europe. While the Duffy mutation could stop *Plasmodium vivax*, it is unable to thwart the newer and more deadly *Plasmodium falciparum*, which is responsible for the most pathogenic form of malaria.

Duffy negativity does not come without cost; there are, as there usually are, tradeoffs. People who are Duffy negative are more likely to get asthma.[18] This genetic link may partially explain why people of African descent are more likely to acquire asthma, even after taking into socioeconomic and other factors. People who are Duffy negative also face a substantially greater risk of being infected by HIV. One study of African Americans found that those who were homozygous Duffy negative had a 40 percent increased risk of HIV infection.[19] Based on this result, the authors estimate that the Duffy mutation could explain about a tenth of the HIV infection load in Africa. Given that about half of those living with HIV (or about 20 million people, as of 2019) are in East and Southern Africa,[20] this entails that the Duffy mutation may have led to a couple million of additional people who are living with HIV. Interestingly, after infection, the Duffy mutation has a slight protective effect: AIDS progression is somewhat slower in people with the Duffy mutation.

HIV itself and malaria act synergistically.[21] Being infected with HIV increases one's risk for getting infected with malaria, likely because of the suppressed immune system associated with HIV infection. In turn, malaria infection increases the replication of HIV. Modeling studies show that this synergetic effect poses a threat for areas where one infection is at a highly prevalent and the other infection is either at a low rate or an unstable rate. Without the synergism, the rarer infection is likely to die out; with the synergism, this rarer infection is more likely to be maintained in the population.

Yet another genetic adaptation to malaria is G6PDD, a genetic condition that comes from having insufficient activity of glucose-6-phosphate dehydrogenase. While G6PDD provides some protection from some malarias, some environmental triggers can cause the premature breakdown of red blood cells, which can lead to anemia and feeling run down. Ironically, several antimalarial drugs can trigger the breakdown of red blood cells. An episode of the television series M*A*S*H has a subplot about G6PPD and its complications.[22] Because of a delivery snafu, the unit only has primaquine on hand to fight malaria. Doctors in the 1950s, when M*A*S*H is set, were aware that many people of African ancestry would not tolerate primaquine to what we now know is G6PPD. But what they did not know at the time was that many people whose ancestors came from southern Europe and the Middle East also had G6PPD. Corporal Klinger, by then the company clerk, is given primaquine and becomes ill. At first, the crew thinks he is malingering. But it turns out that he had developed a serious illness. Just before the credits roll, a medical note is posted noting that in the years since the Korean War biologists have discovered that G6PPD is common in people of Mediterranean and Middle Eastern ancestry, as Klinger was Lebanese American.

As we saw with COVID-19 and other viral diseases, blood groups affect susceptibility to disease. With malaria, having type B or especially type A blood increases the risk of severe disease, while type O is protective.[23] The mechanism for this differential susceptibility has to do with rosettes, which are clumps that occur when infected red blood cells surround themselves with uninfected ones. The antigen associated with type A blood is more prone to bind with PfEMP1 (*P. falciparum* erythrocyte membrane protein 1), a protein made by the parasite that leads to the adhesion of the cells. In standard Mendelian genetics, the parent from which one gets each genetic variant does not matter. But that is not the case here. The effect of blood type on malaria susceptibility is stronger if you inherited the non-O from your mother than if you get it from your father.

WHY ARE SOME MOSQUITOES VECTORS?

Mosquitoes are an ancient and diverse group of insects. When the first dinosaurs and first true mammals appeared in the Triassic, mosquitoes were already around. In fact, a molecular evolution study estimates that the split between the two major groups of mosquitoes—the Culicinae (which includes both *Culex* and *Aedes* mosquitoes) and the Anophelinae subfamilies (containing *Anopheles* mosquitoes and their relatives)—most likely occurred about 225 million years ago or so, right in the middle of the Triassic.[24] Mosquitoes vary considerably in size and in habitat, but most have specialized, elongated mouthparts that reflect the blood-feeding way of life of most female mosquitoes. It is this blood-feeding habit that predisposes mosquitoes to be vectors of disease.

Although there are 3,500 species of mosquitoes, only a small fraction of these feed on humans. Mosquito disease vectors almost always emerge from these human-feeding mosquitoes. For the mosquito to be a disease vector, it must be a suitable place for the pathogen or parasite disease-causing agent. There is also a Goldilocks dilemma. If the mosquito's immune system is too effective at squashing down replication of the pathogen or parasite, disease is unlikely. If the mosquito's immune system insufficiently checks the replication of the parasite, the mosquito may not live long enough to be an effective transmitter of disease.

If you took a phylogenetic tree of mosquitoes and shaded in those twigs representing disease-vector species, you would see a few shaded lines separated by vast spaces of the absence of shading. The closest relatives of vector mosquitoes usually do not carry disease. These related non-disease-bearing mosquitoes often look a whole lot like the vector species. They are hard to distinguish based just on morphological characteristics, complicating efforts to contain the mosquitoes that cause disease.

Evolutionary biologists refer to closely related and morphologically similar species as sibling species. Sibling species have figured in on debates about how we should think about species and how we should distinguish them. Starting in the 1930s and 1940s, several prominent evolutionary biologists promoted the idea that species should be based on whether populations could interbreed and produce fertile offspring. If when brought together, two populations could produce fertile offspring, they would be considered the same species. If they could not, they would be considered two separate species.[25] The German-born, American-immigrant ornithologist and evolutionary biologist Ernst Mayr was among the most forceful advocates of defining species based on their reproductive compatibility or lack thereof. Mayr also was keenly aware that sibling species were crucial to the argument that defining species based on morphology alone could be deceptive. He pointed to sibling species of *Anopheles* mosquitoes as an exemplar:

> Perhaps the most celebrated case of sibling species is that of the malaria-mosquito complex in Europe. According to the older literature, malaria in Europe is caused by

the malaria mosquito, *Anopheles maculilipennis*. A study of the distribution and ecology of this mosquito revealed all sorts of puzzling irregularities. *Anopheles* mosquitoes were found to be quite common in certain parts of Europe where malaria was absent.[26]

After further discussion of those irregularities, Mayr reveals that the *Anopheles* in Europe are six species that look extraordinarily similar, but only two (*A. labranchiae* and *A. bacharowi*) carry malaria. In addition to whether they transmit malaria, these species differ in egg size and shape as well as in habitat. When Mayr wrote these words, DNA markers were decades away from being available. But even then, evolutionary geneticists could reliably distinguish mosquito species based on morphological characteristics of their chromosomes that were visible under the microscope. Now, these mosquitoes can quickly be distinguished by examining their DNA. Quick determination of whether the mosquitoes observed in an area are species that are disease vectors makes surveillance much easier.

What attracts a few mosquito species, but not others, to humans? To examine the differences between those that specialize on humans and those that do not, we need to first look at the sensory systems of mosquitoes. Differences in the mammals they bite depend on what the mosquitoes smell and what they taste.

Mosquitoes smell with their antennae, the tip of their proboscis, and their maxillary palps.[27] These structures are coated with sensilla, sensory hairs that contain olfactory sensory neurons. Within each hair, there are proteins that bind to the trace chemicals in the aroma. These odorant-binding proteins will transport the odorants through a watery substance up to the membrane of the neuron. This neuronal membrane contains various receptors that recognize specific odorants. Recognition causes the neuron to fire. Changes in the genes that encode receptors or changes in genes that control the expression of these genes can result in changes in which chemicals—and thus, which potential hosts—the mosquito will respond to.

The odor profiles of humans and other animals are also affected by what types of bacteria on their skin.[28] Changes in the composition of the skin bacteria can affect how mosquitoes will respond. Interestingly, the extent of differences between skin profiles of different primates is largely disconnected to their evolutionary relationships. Although our closest living relatives are chimpanzees and bonobos, our skin odor profile is closer to those of lemurs and cows than they are to those of chimpanzees and bonobos. One reason that human sweat is distinct from our chimpanzee and bonobo relatives is that human sweat is especially rich in *Staphylococcus* bacteria. These bacteria prefer the excretion from eccrine glands, sweat glands that are common in humans but rare in other primates. These glands, which play a vital role in thermoregulation, are particularly common on our hands and feet.[29]

Differences in preference can evolve quickly. *Aedes aegypti*, now a vector for yellow fever and more recently Zika, originally lived in forested areas in sub-Sahara, where it bit nonhuman animals. The forest subspecies *A. a. formosus*, which is black in color, still resides in the African forest and shows no preference for human blood. The brown-colored domestic subspecies *A. e. aegypti* is found across the world and has a distinct preference for human blood. Although the two subspecies have remained separate in the wild, they will hybridize in the lab and produce fertile offspring.

An urban Senegal population is much more closely related to urban populations in Mexico than it is to a forest population nearby. The divergence between the urban Senegal and Mexican populations is so small that it suggests that the two populations split only a few hundred years ago.[30]

Carolyn McBride, then at The Rockefeller University, and her colleagues took advantage of the interfertility of these subspecies to get at the genetic basis of their difference in preference. Hybrid mosquitoes that preferred human odors expressed the odor receptor gene *Or4* much more than hybrid mosquitoes that did not have a preference. Moreover, this receptor gene was highly expressed in the antennae of females from the domestic subspecies. This gene encodes a receptor that recognizes sulcatone, a ketone that is much more common in human sweat than in the sweat in nonhuman animals. This chemical, which smells like citrus fruit, is commonly used to attract mosquitoes.[31]

The difference in *Or4* is almost certainly not the only difference between the forest and the domestic subspecies, nor is sulcatone the only chemical that distinguishes human and nonhuman smell to the mosquitoes. For instance, lactic acid, a chemical that our muscles build up during anaerobic exercise, also plays a role. Human skin emits much more lactic acid than the skins of other animals. The human-feeding domestic subspecies is especially attracted to lactic acid. Adding lactic acid to a previously unappealing animal odor will attract this mosquito, whereas removing lactic acid from an appealing human mélange will turn this mosquito off.[32] Still, identification of the change in *Or4* is a first step toward characterizing the genetic basis of the traits that make mosquitoes bite humans. The hope is that such characterization should provide insight as to why some mosquitoes evolved specialization to living on humans, setting them up to be likely candidates for disease transmission.

Mapping out how different species of mosquitoes respond to different odors can also be useful in developing attractants (to be used in traps, for instance) and repellants.[33] A recent study used genetically engineered *Anopheles coluzzii*[34] to examine the physiology behind how repellants affect this mosquito. While natural repellants activated specific receptor neurons in the antenna, DEET (N, N-diethyl-meta-toluamide) and other synthetic repellants did not.[35] Instead, DEET appears to act by masking human odors; with DEET, fewer human odorants reach the antenna, and thus humans are rendered "invisible" to the mosquito. Interestingly, the mode of DEET repulsion is different for *Culex* mosquitoes, wherein DEET activates a specific odorant receptor.

In addition to their olfactory sense, mosquitoes also use taste to make decisions about where and whether to feed and to lay eggs. Although biologists know much more about the physiology and the neurobiology of mosquito smell, they have recently made substantial progress unraveling how taste works in mosquitoes. Mosquitoes taste potential food or blood several times, with different organs each time.[36] Upon encountering a potential source of blood or food, a mosquito first tastes with its tarsi, the parts of its legs that are furthest away from the body. If it decides to continue feeding, the mosquito will taste again with the stylet, a needle-shaped mouthpart which the mosquito uses to bite and to ingest material. After ingestion, the mosquito will taste again with its cibarium, an internal organ where the food or blood is chewed. These taste organs all have sensilla. Like the sensilla in the olfactory system, these sensory hairs contain receptor neurons, here called gustatory receptor neurons. At each step, these receptor neurons will guide signals to the brain of the mosquito. Inputs from these receptor neurons will determine the mosquito's response.

While only female mosquitoes ingest blood, both males and females ingest nectar through their stylets. When feeding on blood, female mosquitoes take much larger meals than when they feed on nectar. The stylets of males and females also differ in morphology and in the receptors they contain. A recent study on *Aedes aegypti* finds that these mosquitoes have a different sense of taste for blood than they do for nectar.[37] This different sense of taste emerges because female stylets also respond to blood differently than they do nectar. Both blood and nectar contain sugar, but blood also contains sodium chloride and sodium hydrogen carbonate (the active ingredient in baking soda). The combination of sugar, salt, and baking soda—no one single compound— causes the responses in the receptor neurons in the stylet that signify blood.

Why did the human-feeding subspecies of *Aedes aegypti* evolve a taste for human blood? One explanation is that changing patterns of precipitation have caused the mosquito to spend more time in areas with a high density of humans because that is where they can easily find durable sources of water. A recent study led by Noah Rose in the McBride lab supports this hypothesis.[38] They looked at the biting preferences of mosquitoes collected from 27 sites across Africa and found that rainfall seasonality along with human population density predicted the preference. Not surprisingly, mosquitoes collected from areas with high human density had a taste for human blood. But in addition, mosquitoes from locations where precipitation was clustered in a season—where there were wet and dry seasons—also preferred human blood. These mosquitoes typically lay eggs in rock pools, tree holes, and other wet surfaces slightly above the water line. In dry areas, such sources of egg-laying sites are likely to have been constructed by humans. Thus, in these areas, selection would likely favor mosquitoes that are around humans. Such selection may have led mosquitoes to favor human blood. The study also found evidence of selection on a region of the first chromosome that contained *Or4*, the odorant receptor associated with a preference for human blood.

Although climate change is likely to affect Africa in the coming decades, Rose and colleagues note that the change is unlikely to dramatically affect seasonality of rainfall. Hence, the climatic trends are unlikely to greatly affect preference. In contrast, demographic projections portend that more and more people in Africa will live in cities. This increased urbanization portends that mosquitoes will evolve a preference for human blood across even more of Africa. And with this shift in preference, the risk for yellow fever and other diseases increases.

ANTIMALARIALS AND EVOLUTION OF RESISTANCE

Evolution thwarts our ability to combat malaria in two fundamental ways. First, the *Plasmodium* parasite has evolved resistance to the drugs used to control it. Second, mosquitoes have evolved resistance to the chemical agents used to kill them. There are some similarities in these two aspects. In both, there is a long history of overreliance on a single control measure and there are the corresponding setbacks as the organisms evolve resistance. We'll start with the efforts to control *Plasmodium*.

When I took chemistry as an undergraduate, Dr. Armstrong taught many of the chemistry labs and occasionally guest lectured. Well into his 70s at the time, Dr. Armstrong was a stickler for detail, but also an engaging and witty storyteller. If some of us were running behind in the completion of the lab exercise for the day, Armstrong would urge us stragglers not to dawdle as he wanted to go home, have a gin and tonic, and watch the evening news. This was in the era where the 6:30 network news evening show was a bigger deal than it is now. Armstrong's drink is related to malaria, as tonic water contains the first antimalarial: quinine. The mixed drink originated with British soldiers in India during the 19th century who were given quinine as an antimalarial. Quinine has an unpleasant taste so, the soldiers started putting it in a solution of water, sugar, and lime juice—tonic water. They then used this tonic as a mix for gin. Thus was born the gin and tonic.

Quinine comes from the bark of several species of *Cinchona* trees.[39] Getting the bark from these trees was difficult; hence, quinine was often in short supply. In fact, one of the reasons why the Union eventually prevailed in the US Civil War was that it had bigger and more reliable sources of quinine than the logistically challenged Confederacy. Successful chemical synthesis of quinine did not arrive until World War II; even today, it is most often obtained from bark. Moreover, in the doses typically given, quinine was not all that effective. The quinine in today's gin and tonic would provide little protection from malaria; even the quinine concentrations from older tonic water was barely sufficient. Higher doses were effective but costly. In addition, doses large enough to be therapeutic came often led to pounding in the ears, sweating, and difficulty seeing, among other side effects.

The expense and the side effects prohibited the widespread use of quinine. Consequently, malarial parasites faced only sporadic selective pressure to become resistant to quinine and did not evolve substantial resistance to it. The same is not true of the subsequent, more effective antimalarials.

Since World War II, progress in the control of malaria has taken a sawtooth path. Advances in treatments lead to substantial declines in malaria; unfortunately, these advances are rolled back as malarial parasites evolve resistance to the drugs. Chloroquine, which has gotten renewed attention recently thanks to speculation that it could be used as a treatment for COVID-19, had been effective at controlling malaria during the 1940s and 1950s. The evolution of resistance to chloroquine led to a resurgence in malaria, particularly in southeast Asia, as the Vietnam War intensified. In fact, malaria was among the most pressuring concerns of the US armed forces, as it infected about half a million of the US soldiers stationed in southeast Asia (roughly half the total force). The Walter Reed Army Institute undertook a massive search for new antimalarials but was unsuccessful in developing any during the Vietnam War era. Also interested in developing new drugs to fight malaria, the Chinese government launched a secret mission in 1967, then called the 523 project because it started on May 23, that was ultimately successful in generating perhaps the most useful antimalarial of the late 20th century: artemisinin.[40]

The 523 team was led by Youyou Tu, a chemist who had experience with synthesizing pharmaceutical chemicals from herbs used in traditional Chinese medicine. Her team was tasked with searching through the voluminous literature on herbs used in traditional Chinese medicine to find suitable candidates to then synthesize chemicals that could be used as antimalarials. After testing several chemicals that did not pan out, Tu and her team turned their attention to *Artemisia annua*, known as sweet wormwood. This sweet-smelling plant has a long history in traditional Chinese medicine, where it has gone by the name qinghao. In the Eastern Jin Dynasty of the 4th century CE, Ge Hong wrote about an extract from this plant that could be used to remedy the symptoms of malaria. From this plant, Tu's team synthesized and tested the active ingredient: artemisinin.[41] For this breakthrough, Tu eventually would share in the 2015 Nobel Prize in Physiology or Medicine.

Artemisinin is commonly used a short-term-acting drug against malaria. It quickly kills malarial parasites, but it has a short half-life and is quickly degraded. Exactly how artemisinin works is still being worked out, but it appears to act as a prodrug. That is, artemisinin is relatively inert under most circumstances, but will activate when triggered.[42] One trigger is the presence of a great deal of free heme, an iron-containing component of hemoglobin. When activated, artemisinin generates reactive oxygen compounds that cause damage to cells—including the parasite—in the environment. Artemisinin is thus a targeted drug; it only wreaks damage in certain parts of the body.[43]

Today, artemisinin usually is used in combination with other antimalarials, such as piperaquine, that act more slowly.[44] By combining drugs that act via different mechanisms, this artemisinin combination therapy takes an approach similar to that taken in treating HIV infection. The idea is resistance to multiple drugs should be more difficult for the parasite to evolve, especially when the drugs act on different aspects of the parasite's life cycle. Unfortunately, these defenses are breaking down: resistance to artemisinin started appearing in Cambodia and elsewhere in the early days of the 21st century. Evolution of resistance again poses a threat to controlling malaria.

Although they knew that resistance was increasing in Cambodia, the researchers did not have a good way of tracking that resistance. Assays are difficult and time consuming. So researchers searched for a biomarker—a genetic assay—for resistance. By 2014, they had developed such a biomarker: allelic variation at the gene *kelch13* was predictive in determining whether a parasite was resistant or not.[45] Having a biomarker makes surveillance much more efficient. These days, performing DNA assays is now drastically quicker and cheaper than tracking the symptoms and parasite loads of individuals over the course of several days.

Evolution was central to the generation of the biomarker. Figuring out which gene or genes caused resistance is not easy due to the type of population structure *Plasmodium* has. Related individuals are often clustered geographically together. Moreover, allelic variation at one locus can strongly correlate with allelic variation at another locus. The clustering of relatives and this correlation across loci—known as linkage disequilibrium—can lead to false positives and false negatives when linking natural genetic variation to traits. So the researchers first used directed evolution (see Introduction) to generate a highly resistant strain of the parasite in the lab. They then sequenced the resistant strain and found that it differed by seven changes from the sensitive progenitor. To determine which change or changes were responsible for resistance in natural settings, the researchers looked for associations of these changes with resistance in Cambodia. Six of the seven changes showed no association, but the genotype at *kelch13* was strongly associated with resistance. Artemisinin was much less effective at clearing the parasite from patients if their parasites had specific alleles of *kelch13*.

One particular allele (known as C580Y) of *kelch13* stands out because it is much more frequent than the other alleles. The name of the allele comes from it being a change in the 580th amino acid from a cysteine (C) to a tyrosine (Y). By the late 2010s, this variant made up about 85 percent of the resistant alleles in Southeast Asia and was still outcompeting the other resistant alleles and is increasing in frequency. Why is this allele spreading? Is it because this allele is much better at conveying resistance? Is it because the allele has low fitness costs?

Data support neither possibility. Hospital records show that C580Y is intermediate in resistance strength: it is better than some variants at slowing down clearance of the parasites but is significantly worse than other alleles. What about fitness costs? As we saw in the discussion of antibiotic resistance in Chapter 1, fitness costs are important in predicting the fates of resistance alleles. An allele that was intermediate in resistance but had low fitness costs in the absence of artemisinin could outperform other resistant alleles, even those that conveyed greater resistance. Tim Anderson's lab at the Texas Biomedical Research Institute in San Antonio used CRISPR editing to generate *Plasmodium* that differ only at known sites at *kelch13*. Anderson's group found that C580Y actually has relatively high fitness costs compared with other resistance alleles.[46] So what's going on?

The most likely explanation is that one of the parasites carrying the C580Y resistance allele had a mutation that ameliorated the fitness costs associated with C580Y. With this compensatory mutation, C580Y was able to outcompete the other resistance alleles and rapidly increase in frequency. Note that the C580Y allele that Anderson's group created by CRISPR editing would not have this compensatory mutation. The compensatory evolution here mirrors what we saw in Chapter 1 with some cases of antibiotic resistance.

I spoke with Tim Anderson about this. I have known Anderson since 1988 when we were graduate students at the University of Rochester. In all that time, I have never known anyone as knowledgeable and passionate about parasites. As a student, Anderson's dissertation project involved collecting and dissecting human roundworms in Guatemala.

Anderson told me that his lab plans to use long-term artificial selection experiments to test the feasibility of the compensatory evolution explanation for why C580Y has been increasing in frequency despite being only intermediate in resistance and being saddled with high fitness costs.[47] If compensatory evolution is common under the lab conditions, then this would support the scenario. Replicating compensatory evolution also allows its mechanisms and genetic basis to be studied more closely. For instance, is the compensatory evolution arising from one or a couple genetic changes of large effect or from many changes, each with modest effects?

While we were talking, Anderson brought up an interesting observation about malaria's resistance to drug: Although malaria is much more common in Africa than in southeast Asia, resistance to antimalarial drugs evolves much more often in southeast Asia than in Africa. Why? Anderson presented several hypothetical explanations, all of which may be operating. First, people in different areas respond differentially to malaria. In southeast Asia, malaria is rare enough that most people get it only one or a few times. There, it is an acute illness. And like most with most acute illnesses, people treat it by seeking medical help. This means using antimalarial drugs and the resulting selective pressure for resistance to those drugs. In Africa, malaria is a chronic illness. It is much more common; people get infected numerous times and are seasoned. There, people live with malaria and are less likely to get treatment. Consequently, selective pressure is much reduced.

Another plausible explanation for why malarial resistance is more common in southeast Asia is the difference in how intensely different malarial parasites are competing within the same host. In southeast Asia, people seldom are infected by more than one parasite at a time; and when they are, they almost always have just two different parasites. In Africa, however, being infected by more than one parasite is common; sometimes, people are infected by more than even two parasites. Recall that fitness costs are common: in an untreated person, resistant parasites should have lower fitness than sensitive parasites. In a host infected by just one parasite, this fitness cost is usually inconsequential. But if a host is infected by both a resistant and a sensitive parasite, the sensitive parasite is likely to outcompete the resistant one. Accordingly, the increased intra-host competition in Africa could be checking the evolution of resistance there.

A third possibility is differences in the population structure in southeast Asia and in Africa and how that affects compensatory evolution. In southeast Asia, the parasite populations are largely clonal. Variants at one position are in high linkage disequilibrium with those of others. In Africa, linkage disequilibrium is quickly lost. Compensatory evolution is more amenable in southeast Asia because it occurs more often when there is strong linkage disequilibrium.

Anderson's group also looked at the molecular evolution of samples of *Plasmodium* taken at different times as artemisinin resistance evolved in southeast Asia.[48] They found that even though the C580Y variant had increased to high frequency by the mid-2010s, many selectively favored variants had appeared earlier. Genetic diversity has decreased as this one variant has taken over. If we were to continue sampling the genotypes of resistant parasites, we would likely find that one variant would continue to fixation with a corresponding loss of genetic diversity at other sites along the genome. Recall the distinction between hard and soft sweeps we discussed in relation to HIV evolution. The pattern that we would expect in the future would look like a hard sweep—selection sweeping along a single variant—even though we know from having samples collected in the past that the sweep was soft (selection initially favored multiple variants).

This finding casts doubt about inferences made about the nature of the evolution of resistance to earlier antimalarials. The pattern of evolution of resistance to both chloroquine and to pyrimethamine[49] gives the appearance of being due to a single hard sweep for each drug. But given the time-course data on the evolution of resistance to artemisinin, can we say with confidence that the evolution of the resistance to the earlier antimalarials were hard sweeps? Could they be soft sweeps that hardened as the most successful variant outcompeted the others?

Recall that in the HIV evolution studies that those drug combinations to which resistance evolution involved hard sweeps had better outcomes for patients. The same is likely to be true with combinations of antimalarials. That a soft sweep occurred used in the evolution of resistance to artemisinin and likely occurred with the evolution of chloroquine and to pyrimethamine does not bode well for use of these drugs in combination therapy.

That the sweep in the evolution of resistance to artemisinin was soft, involving selection favoring multiple variants, not only indicates that artemisinin resistance evolves readily but also implies that the product of the effective size of the population of parasites infecting a body and the mutation rate toward mutations to artemisinin resistance is large. Artemisinin resistance likely has a large mutation target size—there are many ways to tweak *kelch13* such that resistance appears—which could make resistance to it relatively easy to evolve. The development of new antimalarials should take mutation target size into account: drugs with smaller likely mutation target sizes ought to be more resilient to the evolution of resistance.[50]

Pinpointing artemisinin resistance to variants of *kelch13* is not only useful for surveillance and population genetic applications but also has led to a better understanding of the mechanistic biology behind the resistance, which in turn could lead to therapeutic advances. Just after 2019 rolled over to 2020, a report broke showing the mechanism of artemisinin resistance.[51] While they reside in red blood cells, *Plasmodium* parasites are busily chewing up hemoglobin. As they degrade the hemoglobin, heme is released, which activates artemisinin, which kills the parasites. The parasites with the *kelch13* mutation do not break up hemoglobin as readily when they are in the young ring stage. As a result, less heme is released into the environment and less artemisinin is activated. With less of the toxin activated, the parasites are more likely to survive. The downside is that these parasites grow more slowly. They would likely lose if they were in competition with genotypes with intact *kelch13*. Pinpointing the mechanism by which artemisinin kills *Plasmodium* also has implications for the use of partner drugs to use with it. The other drug with artemisinin in the combination therapies often also targets the parasite's digestion of hemoglobin. Subsequent combinations wherein the other drug targets a different biochemical pathway may be less susceptible to the parasite evolving resistance.[52]

MOSQUITO CONTROL

In addition to efforts to control malaria by killing the parasite, people have also targeted the mosquito vector. For instance, Gorgas and his crew eradicated yellow fever from Havana, Cuba, by 1902 with aggressive measures against the *Aedes* mosquitoes.[53] They drained swamps and ponds and other containers of water to stop the mosquitoes from breeding. They erected nets. They used various fumigating agents, such as sulfur and pyrethrum. This was difficult work. But it did the job. Similar mosquito-control practices eradicated or contained mosquitoes and their diseases in Panama and elsewhere.

In *The Fever*, Sonia Shah describes the science of mosquito control during the first four decades of the 20th century as a "nuanced multidisciplinary field, enlisting the insights of engineers, entomologists, ecologists, clinicians, and anthropologists".[54] Then came a gigantic shift that would reduce these nuanced field with its complexities to what Shah called "a single brute question: how to coat interior walls with two grams of chemical per square foot".[55] This shift was the realization that dicholoro-diphenyltrichloroethane (commonly known as DDT) was a potent and apparently harmless killer of mosquitoes.[56]

DDT looked like a magic bullet. It was easy to apply. It apparently was toxic to insects and similar small, cold-blooded animals. It was durable; a single application could last for months. To postwar Americans fresh off their victories in Europe and Asia, it looked like just the weapon to use against mosquitoes. And like other magic bullets, it was prone to over- and misuse. Farmers and gardeners used it against insects broadly. Where it was used, insect populations collapsed.

This overuse led to Darwinian trouble. On Halloween 1948—only five years after the initial spray—the *New York Times* reported house flies in Orlando, Florida being resistant to DDT.[57] The Times report did not explicitly use the word "evolution", though it did mention the rapid turnover of generations of the fly as an explanation for the DDT resistance appearing in these flies and not (yet) in other animals.

Moreover, further research showed that DDT was not as harmless as had been believed or advertised. DDT is absorbed and stored in fat tissues, where it can accumulate. Moreover, with each step of the food chain, the DDT is further concentrated so, as the mouse that ate the grasshoppers gets the DDT that grasshopper absorbed, the bird that eats mice gets the DDT eaten by the mice as they ate grasshoppers and other insects. For many birds, the DDT they accumulate causes them to produce eggs with thinner shells; the thinner shells imperil the young developing in the egg. This was among the evidence that Rachel Carson[58] and others used to call for ending the use of DDT. A few years later, Canadian folk singer Joni Mitchell urged farmers to stop using the chemical so that we could enjoy the birds and bees. By the end of 1972, the EPA had banned DDT for most use in the United States.

In recent years, some have criticized Mitchell's and Carson's admonitions about the dangers of DDT given the resurgence of malaria in many countries, particularly in Africa. But the problem is that DDT had lost its effectiveness as a mosquito killer due to the evolution of resistance brought on by its overuse. As Winegard put it:

> With the benefit of looking back at the bygone DDT clouds from both sides now, Mitchell was right to reprimand farmers for paving paradise with the insecticide. It was the widespread, carpeting agricultural application of DDT that created environmental degradation and mosquito defiance, not its relatively limited and surgical use solely as a mosquito killer.[59]

With DDT no longer an option, other measures of mosquito control would—some old, some new—would come into use.

One mosquito control measure came out of work by Edward Knipling at the USDA on screwworm flies (*Cochliomyia hominivorax*). The larvae of these screwworm flies infest the flesh of cattle, sheep, and other livestock. They are particularly prone to attack wounds, where they eat both healthy and damaged tissue. In addition to causing distress, the screwworms can also carry disease. Knipling had the idea of releasing sterile males to control the screwworm population. If female screwworms mate with sterile males and are thus prevented from having offspring with fertile males, the reproductive potential of the population could be reduced. During the

1950s, Knipling was inspired by work of the evolutionary geneticist Hermann Joseph Muller, who had won the Nobel prize for showing that X-rays could induce mutations in *Drosophila* flies and that some of these mutations could cause sterility.[60]

Knipling first showed that this sterile male technique could be used to control screwworm populations in the lab. Then he used the technique to eradicate screwworms from Curaçao, a small island nation north of the coast of Venezuela.[61] Combined with other eradication methods, the sterile male technique had removed screwworms from the United States by 1966.[62]

If the sterile male technique can eradicate the screwworm from the United States, could it be used to control mosquitoes? In principle, it should. But in practice there are complications that need to be considered. As Knipling himself noted, there are attributes of populations that make them better or worse candidates for use of sterile male technique.[63] For instance, sterile male technique works best when the size of the target population is not too large because the technique works best when there are more released sterile males than wild males. Mosquito populations are typically large, thus necessitating the production of numerous sterile males to control them. Moreover, attributes of the sterile males need to be considered. Ideally, the radiation (or other sterilizing treatment) should only affect the fertility of the males. If the treatment also substantially reduces the ability of the males to mate with females, then the effectiveness of the treatment will suffer. Evolutionary genetic theory here can be applied to estimate how much the effectiveness will decline with a given loss of competitiveness. The reduction in competitive ability in irradiated sterile male mosquitoes has limited the usefulness of the sterile male technique in controlling their populations.

The exasperation with irradiation as a sterilizing tool combined with the pressing need to control certain mosquito populations has led biologists to consider alternatives. One such alternative involves a fascinating bacterium that lives inside the cells of numerous insects and other arthropods. This bacterium, *Wolbachia*, which was first discovered in *Culex* mosquitoes, is extraordinarily abundant. An estimated two-thirds of insect species—which translates into considerably more than a million species—have it. *Wolbachia* also has attracted attention due to the numerous effects these bacteria have on their hosts. They can kill males. They can convert males into females. They can cause females to become parthenogenic, enabling unmated females to lay eggs that hatch into females. The contagion in *The Walking Dead* has nothing on these microscopic critters that live in the cells of insects. They are truly, as Jack Werren at the University of Rochester and his colleagues put it, "master manipulators of invertebrates".[64]

One of the most fascinating things that *Wolbachia* can do is to make different strains and populations incompatible with one another. In nature, this incompatibility can be the impetus to the evolution of new species. But we can also exploit the incompatibility to produce the functional equivalent of sterile males. In *Wolbachia*-induced incompatibility, if the sperm lacks the type of *Wolbachia* present in the egg, the resulting embryo dies early in development.[65] Offspring dying early in development is at least as good as the males being sterile with respect to population. Moreover, unlike the sterile males created by irradiation, the males that lack *Wolbachia* or who have a different *Wolbachia* type to the females are generally quite viable and competitive in mating.

Wolbachia-induced incompatibility looks to be a promising new wrinkle on controlling mosquito populations through sterile male technique.[66] In this incompatibility approach, care must be taken to ensure that females and males of the same *Wolbachia*-type are not released together, as the matings between such would be compatible and would generate lots of offspring. The irradiation and incompatibility approaches can be combined. In a proof-of-principle study, researchers recently have used such a combined approach to control *Aedes albopictus* (an important vector of Zika and dengue viruses) in islands near Guangzhou, a city in southern China that has had substantial dengue transmission.[67] The success of this combined approach in Guangzhou should lead to further trials.

Another wrinkle on the sterile male technique is to use CRISPR gene editing. I recently chatted with Mike Wade at the University of Indiana, who is attempting to use males that have CRISPR constructs that target female fecundity genes. These experiments are now being done in flour beetles raised in the lab. But the potential is out there to use these for controlling other pests, such as mosquitoes.[68] Stay tuned!

GENE DRIVES

CRISPR gene editing is also being examined in yet another way to control pest populations—gene drives. This unusual idea is based on outlaw genes found in nature that violate Mendel's laws and spread in populations because they make more copies of themselves and not because of the benefit of these genes to the organism.

Suppose a fly has two different genetic variants at the same position on the gene: one variant has the nucleotide G and the other has a T. What percent of this fly's offspring will inherit the variant with a G? Mendel's laws—taught in every genetics class—would say that roughly 50 percent of the offspring should inherit the G one and 50 percent should inherit the T one. Sure, there would be deviations from that exact amount because of chance, just as you would not expect to get exactly 50 heads and exactly 50 tails if you tossed a fair coin. But getting 90 offspring with G and only 10 with T would be surprise—a violation of Mendel's laws—just as getting 90 tails and only 10 heads would be a surprise and would be strong cause to think there was something strange about the coin. And yet such deviations are found in natural populations. In fact, evolutionary biologists have examined these deviations to get a better idea of why Mendelian principles work as well as they do.[69]

An allele in a heterozygote that manages to get into much more than half of the functional gametes has a considerable advantage and should quickly spread through the population. This is not just an interesting evolutionary phenomenon but also one that could be exploited to eradicate or change populations. Evolutionary biologists had long thought about using such systems for that purpose, but until fairly recently, the technology was not there.[70]

CRISPR gene editing and a shocking discovery by a graduate student at the University of California at San Diego named Valentino Gantz would change that.[71] For his thesis on wing development in *Drosophila* flies, Gantz was working with a recessive mutation—one that only was manifested in homozygotes. Trying to get sufficient quantities of flies with recessive mutations was exceedingly difficult. Gantz, in frustration, then hit upon a clever idea—frustration is often the parent of creativity—using CRISPR gene editing to convert heterozygotes for the muta-tions into homozygotes for the mutation. This process was in effect a chain reaction that used CRISPR to copy and paste the mutation. Gantz and his advisor, Ethan Bier, recognized that this process, which they called "the mutagenic chain reaction", could eventually be used for gene therapy, controlling pest populations, and other purposes.[72]

During the late 2010s, such systems, which became known as gene drives, have generated a great deal of buzz. They are promising tools for controlling and pos-sibly eradicating pest populations, including mosquito populations. There are a few concerns with them. These concerns fall into two general categories: first, will they work? And second, are they safe? Can they be controlled?

One factor limiting the effectiveness of gene drives is the targets could evolve resistance to the drives.[73] CRISPR, which was originally a bacterial defense sys-tem, works by targeting specific sequences. In such a CRISPR-generated gene drive, there is strong selection on the target to become resistant to the drive. There also can be selection for unlinked parts of the genome to resist the drive under some circumstances. When is such resistance most likely to evolve? Richard Gomulkiewicz and his colleagues have modeled different gene drive systems.[74] Their simulations suggest that resistance is more likely to evolve with a single strong gene drive than when multiple gene drives of moderate effects are used in combination.

CRISPR gene drives may share some of the same limitations as sterile male technique, such as the reduced competitiveness of the males discussed earlier. Another potential limitation of both gene drives and sterile male techniques is inbreeding. Mosquitoes in the wild often engage in inbreeding. If the wild mos-quito populations are inbreeding, they may be less likely to mate with the released mosquitoes.[75]

But what of safety? Can we safeguard against accidental or deliberate releases? As science writer Jennifer Kahn notes: "Gene drives seem almost tailor-made to tap into our worse fears: a powerful, invisible technology that spreads on its own accord, carrying out a fundamental transformation of nature."[76]

Gene drives usually are species specific. In most cases, leakage from species to species is unlikely. Nevertheless, untoward things could happen. To these ends, sev-eral containment strategies have been devised; some of these rely on separating out components of the drive system. One group of these methods is the so-called daisy-chain drive wherein one element (call it A) causes the next (B) to drive, which in turn causes C to drive. The A element does not drive and thus is limited. Mathematical modeling show that such daisy drives should not spread unlimited.[77] In addition to containment strategies, systems have been designed to allow for overwriting and removal of gene drive systems.[78]

Even with the potential safety concerns and the fact that bringing mosquitoes to Africa to target malaria probably won't happen until the late 2020s,[79] the potential benefits are huge and need to be considered. I recall a seminar some seven or so years ago by a speaker talking about modeling gene drive. The speaker, though cog-nizant of the risks of acting, asked us to think about the risks of not acting. Given the difficulty we have had controlling malaria and given the tremendous mortality,

How gene drives work

Gene drives use CRISPR to insert and spread a genetic modification through a population at higher than normal rates of inheritance. Researchers plan to use drives to eradicate malaria-carrying mosquitoes and other pests.

Once a gene drive is engineered into an animal's genome, the animal's offspring will inherit the drive on one chromosome and a normal gene from its other parent. During early development, the CRISPR portion of the drive cuts the other copy. The cut is then repaired using the drive as a template, leaving the offspring with two copies of the modification.

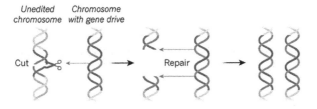

Standard inheritance

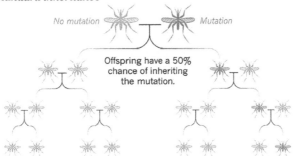

Gene-drive inheritance

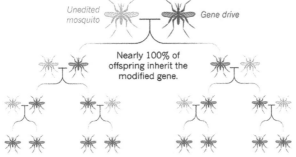

©nature

How gene drives work. Gene drives allow genetic modifications to spread through the population.

Source: From Scudellari, M. 2019. Hijacking evolution: Gene-drive technology could alter the genome of a species. *Nature* 571: 160–162.

morbidity, and reduction of quality of life that malaria and other mosquito-borne diseases bring, those costs are high.

Evolutionary principles are central to increasing the effectiveness and reducing the risk of gene drive. We will pick up discussion of gene drives in subsequent chapters.

NOTES

1. Shah (2010, p. 59).
2. Winegard (2019, p. 2).
3. After mosquitoes and other humans, the next agent of mortality is snakes at about 50,000 deaths a year. Dogs, sand flies, the tsetse fly, and the assassin bug are next on the list. See Winegard (2019, p. 1).
4. See Spinney (2017) for more detail.
5. Shah (2010, p. 69).
6. Winegard (2019).
7. The French effort was also mired with financial scandal. See McCullough (1977) for details.
8. Panama was essentially a puppet state of the United States. Prior to 1903, Panama had been part of Columbia. Using gunboat diplomacy, the United States had compelled Columbia to grant Panama independence. After the United States recognized Panama as an independent nation, it was granted permanent ownership over a 10-mile-wide strip known as the canal zone. See McCullough (1977) and Winegard (2019).
9. Winegard (2019, p. 437).
10. I thank David Fitzgerald, our US history teacher at Matoaca High School, for taking us on this trip. The experience has stuck with me, now almost 40 years later.
11. Winegard (2019, p. 269).
12. Malaria Genomic Epidemiology Network (2014).
13. Although sickle cell anemia is due to a single genetic change, its severity is modulated both by several genetic factors and environmental conditions. See Williams and Thein (2018) for details.
14. Early on when the mutation is rare, a mutation that benefits heterozygotes but hurts homozygotes has dynamics much like a mutation that has a similar benefit to heterozygotes without harming the homozygotes. This is because when a mutation is rare, it is almost always found in heterozygotes and only very rarely in homozygotes. See Johnson (2007, chapter 5) for more details.
15. Azvolinsky (2018).
16. Shriner and Rotimi (2018).
17. More information on the different *Plasmodium* species, the malarias they cause, and their evolution can be found in Shah (2010, chapter 2).
18. Vergara et al. (2008).
19. Note that the population of African Americans is a mixture of various African, European, and other populations. While common, Duffy negative is not close to being fixed in African Americans. See He et al. (2008) for more details, as well as for suggestions about possible mechanisms.
20. HIV and AIDS in East and Southern Africa. www.avert.org/professionals/hiv-around-the-world/sub-saharan-africa/overview. Accessed January 3, 2021.
21. Abu-Raddad et al. (2006).

22. The episode is called "The Red/White Blues". See Winegard (2019, pp. 34–36).
23. Fry et al. (2008).
24. Reidenbach et al. (2009).
25. This dichotomous decision tree obscures a great deal of complexity. We know that many pairs of entities that are normally considered "good species" can hybridize somewhat in nature. Evolutionary biologists of today also many other species concepts and definitions. Still, the vast majority of contemporary evolutionary biologists view reproductive incompatibility as an important feature of species. See Coyne and Orr (2004) and Hillis et al. (2020) for details.
26. See Mayr (1963, p. 35).
27. Basic information on the anatomy and neurobiology of the olfactory system of mosquitoes can be found in McBride (2016).
28. Verhulst et al. (2018).
29. Smallegange et al. (2010).
30. Crawford et al. (2017a).
31. McBride et al. (2014).
32. McBride (2016).
33. Carey et al. (2010).
34. This species was formerly considered a form (the M form) of *Anopheles gambiae*. A battery of evidence, including the existence of substantial reproductive incompatibility with other *Anopheles gambiae*, led to it being reclassified as its own species.
35. Afify et al. (2019).
36. Baik and Carlson (2021).
37. Jove' et al. (2020).
38. See Rose et al. (2020) for the study and Stensmyr (2020) for commentary.
39. See Shah (2010) and Winegard (2019) for general details on quinine.
40. In her Nobel lecture, Tu (2016) describes the background of what prompted the search for new antimalarials and provides an overview of the discovery of artemisinin.
41. Interestingly, while there are at least six species of *Artemisia*, only *A. annua* has appreciable amounts of artemisinin. As we have seen with sibling species of mosquitoes, closely related species can differ in meaningful ways.
42. Wang et al. (2019).
43. Artemisinin has also been used for a variety of other treatments, including cancer (Wang et al. 2019). It even may be useful in treating COVID-19, as clinical trials show that a combination therapy of artemisinin and piperaquine reduces the time to clear viral loads (Li et al. 2021). In areas where malaria is prevalent, the value of such uses should be considered against the likelihood that the use of artemisinin will accelerate the evolution of resistance to it.
44. Wang et al. (2019)
45. Ariey et al. (2014).
46. Nair et al. (2018).
47. Phone conversation with Tim Anderson, January 2020 and e-mail follow-up in December 2020.
48. Anderson et al. (2016).
49. The resistance to these drugs is due to changes in the chloroquine resistance transporter and dihydrofolate reductase genes, respectively.
50. Anderson et al. (2016).
51. Birnbaum et al. (2020).
52. Marapana and Cowman (2020).

53. Winegard (2019, p. 359).

54. Shah (2010, p. 208).

55. *Ibid.*

56. See Shah (2010) and Winegard (2019) for information about DDT. Although it had first been synthesized in the 1870s, DDT's ability to control insects was not fully appreciated until the work of the Swiss chemist Paul Mueller in the late 1930s.

57. Flies resist DDT. *The New York Times*. October 31, 1948. www.nytimes.com/1948/10/31/archives/flies-resist-ddt-several-strains-defy-control-efforts-here-and.html. Accessed June 7, 2021.

58. Carson (1962).

59. Winegard (2019, p. 398).

60. A short history of the screwworm program. www.aphis.usda/gov/aphis/ourfoucs/international services/sterile_fly_release_programs/screwworm/screwworm_history. Accessed June 8, 2021.

61. Knipling (1959).

62. Benedict (2021).

63. See Knipling (1955) and Benedict (2021).

64. See Werren et al. (2008) for details about the extraordinary things *Wolbachia* can do. Fortunately for us vertebrates, they generally only affect invertebrates, though Werren et al. note that *Wolbachia* infections have been found in dogs infected with heartworms. We don't yet know whether vertebrates can be intermediate hosts for this bacterium, but it is likely that such transmission, if it exists, is rare.

65. Three things are important to consider regarding the evolution of *Wolbachia*-induced incompatibility. First, although males can have *Wolbachia*, the *Wolbachia* are only transmitted through the female line. Second, eggs that have *Wolbachia* are compatible with sperm that came from a *Wolbachia*-infected male and with sperm that came from an uninfected male. Third, eggs that lack *Wolbachia* are not compatible with sperm from an infected male. Thus, all other things being equal, eggs with *Wolbachia* have a selective advantage over eggs without *Wolbachia*. Under many circumstances, *Wolbachia* can sweep through the population. If different types of Wolbachia emerge in different populations, the types can be incompatible with each other. See Werren (1997) for further details.

66. Lees et al. (2015).

67. Zheng et al. (2019).

68. E-mail exchange with Mike Wade, June 2021. Wade, who was my postdoctoral advisor, notes that like with the original sterile male technique, one of the challenges with the CRISPR editing has been getting males with CRISPR constructs that are sufficiently viable and competitive with males without the constructs.

69. See Rice (2013) for example and references.

70. See Burt (2003) and Burt and Crisanti (2018).

71. Kahn (2020).

72. Gantz and Bier (2014).

73. Scudellari (2019).

74. Gomulkiewicz et al. (2021). See also Champer et al. (2020) for an example of a gene drive that appears much more robust to the evolution of resistance by its target.

75. Zentner and Wade (2017).

76. Kahn (2020).

77. Noble et al. (2019).

78. Zentner and Wade (2017).

79. Kahn (2020).

FURTHER READING

Shah, Sonia. 2010. *The Fever: How Malaria Has Ruled Humankind for 500,000 Years.* Sarah
    Crichton Books.
Winegard, T. 2019. *The Mosquito: A Human History of Our Deadliest Predator.* Penguin
    Random House.

4 Mismatch
Are We Stuck With Our Past?

SUMMARY

One prominent reason for the prevalence of disease is that the environments we live in are not the same as those in which we evolved: this mismatch illustrates the importance of genotype by environment interaction. We begin the chapter discussing how the rise of allergies, asthma, and autoimmune diseases is partly due to the loss of contact with microbes with which our immune systems evolved. Dietary deficiencies—such as scurvy seen in 18th century sailors and others who have had prolonged lack of access to vitamin C—are also examples of such mismatch. But the idea of mismatch can be taken too far; human populations do have the capacity to evolve reasonably quickly and are not permanently stuck. Examples of such comparatively rapid evolution are adaptions to high altitudes and lactase persistence, which provides the ability to continue consuming the sugar lactose present in dairy products. Also discussed are evolutionary mutant models, the notion that some animals may have phenotypes that are adaptive given their environments but mimic human disease and disorders. Examining these phenotypes and the genes that underlie them could have clinical applications. We conclude with a discussion of the ways that humans are continuing to evolve.

"Your body is a jumble of adaptations that accrued over millions of years. An analogy for this hodgepodge effect is a palimpsest, an ancient manuscript page that was written on more than once and thus contains multiple layers of text that begin to mix up over time as the more superficial texts rub away. Like a palimpsest, a body has multiple related adaptations that sometimes conflict with each other."—Daniel Lieberman.[1]

"Being mismatched, however, is different from being stuck."—Marlene Zuk.[2]

THE CONSEQUENCES OF MISSING OLD FRIENDS

Our story begins with a large cohort of British children who were born in the same week of March 1958. These individuals were followed for the next 23 years.

In the late 1980s, David Strachan analyzed data from more than 17,000 of these individuals and found that those from smaller families were more likely to get "hay fever", also known as allergic rhinitis. This is the typical allergy with itchy eyes, running nose, and sneezing. It is one form of the so-called atopic diseases, ones caused by over-active immune system responding to commonplace allergens such as pollen. Strachan proposed an intriguing hypothesis for the rise of such ailments: children in

DOI: 10.1201/9780429503962-5

large families got infections from their older siblings, which reduced their likelihood of getting allergies. He noted:

> Over the past century declining family size, improvements in household amenities, and higher standards of personal cleanliness have reduced the opportunity for cross infection in young families. This may have resulted in more widespread clinical expression of atopic disease, emerging earlier in wealthier people, as seems to have occurred for hay fever.[3]

Could our improved hygiene be responsible for our allergies?

A decade later, Strachan followed up on what had then become known as the hygiene hypothesis. The inverse relationship between allergies like hay fever and family size had been confirmed in multiple, diverse studies. Moreover, those who had a dog in childhood had lower rates of allergies than those who did not.[4] The effect of family size was considerable: children who have three or more older siblings are about half as likely to have hay fever compared with only children. However, changes in family size between 1961 and 1991 explain only a small fraction of the increase in allergies over that 30-year span.[5] Other factors, perhaps a decline in overall infections, predominate.

Even by the end of 1990s, cracks began appearing in the hygiene hypothesis. Notably, evidence for the key assumption that childhood infections reduce allergy development was not appearing. The data did not show an inverse relationship between infection and allergies.[6] Notably, a study of a cohort of British individuals born in 1970 found no difference in those who had had measles vaccinations and those who had not.[7]

Allergies are not alone in their increase in frequency. Also rising in prevalence is Crohn's disease, a form of inflammatory bowel disease. In Crohn's, the lining of the digestive tract is inflamed. Symptoms include abdominal pain, rectal bleeding, and bouts of diarrhea and constipation. The frequency of Crohn's increased greatly during the latter 20th century, especially in rich countries. So have the frequencies of other inflammatory bowel diseases, type 1 diabetes, multiple sclerosis, and asthma. Are these trends related? If so, do they shed light on the hygiene hypothesis? What, if any, is the underlying mechanism?

Observing these trends, Graham Rook at the Windeyer Institute of Medical Sciences realized that the hygiene hypothesis as it had been originally formulated was incorrect. In the mid-2000s, he proposed a new wrinkle on it. Rook hypothesized that researchers were looking in the wrong place. It wasn't the reduction of harmful infections that was behind the increase in allergies and in inflammatory bowel diseases. Rather, it was the loss of harmless and even beneficial microbes. We were missing our old friends![8]

To see why Rook proposed the old friends hypothesis to replace the original hygiene hypothesis, we need to take a look at one of the players in the immune system: the helper T cells. These are a class of white blood cells (leukocytes) that are produced from stem cells in the bone marrow. These helper T cells mediate several immune responses including the release of cytokines, the signaling molecules we discussed in Chapter 2. Overstimulation of the helper T cells can lead to type 1 diabetes, allergies, and the other diseases and maladies that have been rapidly increasing in rich countries.

Important in the discussion about the hygiene and old friends hypotheses is that not all helper T cells are the same. In fact, helper T cells come in two different types—type 1 and type 2—that play different roles. Too much of these different types has different effects. Overstimulation of type 1 helper T cells can lead to type 1 diabetes, multiple sclerosis, and inflammatory bowel disease. In contrast, overstimulation of type 2 helper T cells can lead to allergies.

A leading mechanism for the original hygiene hypothesis was that infections depress allergies because the pathway leading to type 1 helper T cells depresses the pathway leading to type 2 helper T cells. Two lines of evidence challenge this mechanism. First, the disorders resulting from overstimulation of the type 1 helper T cells (like type 1 diabetes) were rising along with those resulting from overstimulation of the type 2 helper T cells. Rook argued that compartmentalization of biomedical research is part of the reason why it took some time to make the connection that both groups of diseases were rapidly increasing: the researchers working on each group did not sufficiently communicate with one another. Second, there was little evidence that the pathway leading to type 1 helper T cells and that leading to type 2 helper T cells repress each other. Both the epidemiological evidence and the immunological studies cast doubt on the hygiene hypothesis as it was first stated.

Instead, more likely, both the type 1 and the type 2 helper T cells are overstimulated in the absence of certain bacteria that we experienced in previous generations. Rook hypothesizes our ancestors often were around mud and fermenting vegetable material as would be found in farms and gardens where they were exposed to mycobacteria and lactobacilli. These old friends, which coevolved with us, trained and regulated the immune system of our ancestors. The lack of these old friends is from our removing ourselves from farms and gardens and not as much due to bathing and soaps.[9] The use of such domestic hygiene in rich countries generally predates the rise of the allergies, the inflammation, and the autoimmune diseases in these countries. As Graham Rook quipped in a recent perspective: "We know an awful lot about why our immune system's regulation is not in terribly good shape, and it's got absolutely nothing to do with hygiene."[10]

Another important reason for the lack of these old friends is the overuse of antibiotics. Doctors overprescribe antibiotics, treating colds and other upper respiratory infections with antibiotics. Such infections are usually viral, unaffected by antibiotics.[11] On average, an American child gets almost three courses of antibiotics in the first two years of life. That average American gets another dozen or so by age 40. While some of these courses of antibiotics are necessary, many are not. The deluge of unnecessary antibiotics not only drives selection pressure for the evolution of resistance but also depletes our natural reservoir of useful bacteria. We lose some of our old friends with each dose, and with this loss, our immune system is subtly compromised.

We don't just get antibiotics directly from doctors' prescriptions. We also get them indirectly through our food. The meat and poultry industries have long known that adding antibiotics to farm animals allows them to gain weight much faster. This is no small thing: about three-quarters of the antibiotics in the United States go to get farm animals as big as possible as fast as possible. Moreover, food sold in supermarkets often retains residues of antibiotics. We are wasting our valuable antibiotics to make meat slightly less expensive. The consequences of this misuse of antibiotics are not yet fully known, but this misuse clearly hurts us both in accelerating the evolution of resistance and in throwing our immune systems out of balance.

Let's return to Crohn's disease. Is there evidence that people with Crohn's disease have altered microbiota? Yes. Moreover, this imbalance begins early in the development of the disorder. In a study of children with previously untreated Crohn's disease, those with the disorder had reduced diversity of their microbiota overall compared with children of similar age distribution who had abdominal pain and diarrhea but did not have the inflammation that hallmarks Crohn's disease.[12] Those with Crohn's disease had more of certain groups of bacteria such as the Enterobacteriaceae, which ferment sugars to lactic acid, compared to the control. The children with Crohn's also had less of a few groups of bacteria, such as Eryipelotrichia, a class of bacteria within the Firmicutes. Children with more pronounced symptoms had a greater imbalance of bacteria. Interestingly, this imbalance also correlated with prior use of antibiotics. This result illustrates of a possible example of mismatch: the novel environment of antibiotics disrupts the normal gut microbiota, which increases the risk of Crohn's disease.

Could we use this link between bacteria and these diseases to develop treatments? Probiotics, the ingestion of live bacteria, sometimes work in restoring balance. There are multiple cases where ingesting bacteria has alleviated symptoms to pollen and other allergies, asthma, and inflammatory bowel disease.[13] For instance, taking *Lactobacillus rhamnosus*, a bacterium sometimes found in dairy products, has been shown to be very effective at de-sensitizing children with peanut allergies; double-blind controls shows that more than 80 percent of children who had the probiotic treatment became unresponsive to peanuts, while less than 4 percent in the control group did.[14]

The current COVID-19 pandemic likely has altered our microbiota still further. Our patterns of who we interact with, our hygiene habits, and how we feed ourselves changed due to the steps governments and individuals have taken to combat the pandemic.[15] While these measures were imperative to controlling the pandemic, they may have had unintended consequences. Just how they will affect the useful bacteria that surround and inhabit our bodies and how these altered microbiotas will affect our immune systems are areas of research that have barely been surfaced.

Incidentally, Darwin had an interest in pollen and allergy. In the 1870s, Darwin engaged in correspondence with Charles Harrison Blackley, a Manchester doctor who linked hay fever to pollen. Soon after publication of Blackley's book, Darwin excitedly wrote to him: "The power of pollen in exciting the skin and mucous membrane seems to me an astonishing fact,—Would it not be worth while to kill the pollen by a dry heat rather above the boiling point, and see if it retains its injurious properties?"[16] Darwin also noted that pollen that Blackley had noted as causing hay fever are generally from wind-pollinated plants as opposed to those pollinated by insects.

MISMATCH

Despite their differences in details, the original hygiene hypothesis and the old friends hypothesis have the same underlying framework: something has changed about our environment, and the resulting mismatch between the current environment and the environment in which our immune systems evolved is the source of these autoimmune and inflammatory diseases. Mismatch is an important concept in evolutionary biology: changed environments, especially rapid changes in our environments, causes some of the traits that we had evolved in previous environments to be harmful to our reproduction, survival, or our happiness.

That the effects of genes would differ in different environments—that a trait would be beneficial in one environment and deleterious in another—is an old idea in evolutionary biology. In fact, Ernst Everett Just, who was a forerunner of what would now be called evolutionary developmental biology, wrote in 1933: "Environment and organism are one; neither can be separated from the other. . . . In a certain sense we should not speak of the fitness of the environment or the fitness of the organism: rather we should regard organism and environment as one reacting system."[17]

Match and mismatch are not necessarily binary, all or none, categories; rather, they often form ends of a continuum.[18] We can speak of degrees of match and mismatch. Generally, the costs ensuing from mismatch correlate with the degree of the mismatch and the importance of the trait.

Cases of mismatch are all around. While working on this chapter, I came across an arresting sentence written by Nathan Lents in his book *Human Errors*: "Scurvy is a dystopian novel written by the human body."[19] Scurvy is an insidious condition that starts almost imperceptibly but progresses to an ever-worsening array of symptoms. It begins with lethargy and soreness in the limbs. Next to come is bleeding from the gums or even from the skin and hair loss. Inability to heal from wounds and to defend against infections follow. At the end, convulsions and even death can ensue. In addition to these physical symptoms, scurvy often led to emotional volatility. Some historians have speculated that Melville encapsulated these emotional symptoms of scurvy in the capricious and occasionally violent mood of his Captain Ahab.[20]

Regardless of whether Melville intended Ahab's volatility to be due to scurvy, the disorder is one that is most often associated with long ocean voyages of prior centuries. An estimated 2 million people who lived on ships died from it between 1500 and 1800. Far more survived but suffered in agony. We now know that scurvy arises when the body receives too little vitamin C. Fortunately for most of us, vitamin C can be found in abundant quantities in many foods, including citrus fruits, tomatoes, strawberries, broccoli and other vegetables from plants in the cabbage family. So scurvy is rare in areas where most people have reasonably good access to fruits and vegetables. But sailors on long-distance voyages in centuries past often lacked access to such foods.

Mice, dogs, cats, and most other mammals and most birds do not and cannot get scurvy. Their bodies naturally produce vitamin C. Ours do not. We and other apes and monkeys can get scurvy because we lack a functional copy of *GULO*, the gene that produces the enzyme, L-gulonolactone oxidase, needed to make vitamin C. We used to be able to make this enzyme, but somewhere, about 70 million years ago, the gene needed for the enzyme became nonfunctional.[21] The gene is still mostly there in our DNA, but because it can no longer make a product, it accumulates changes that are not visible to natural selection. Geneticists call such a gene a pseudogene.[22] Humans and the other primates lacking a functional *GULO* usually do just fine without it because we usually can get fruits and other foods rich in vitamin C. It's only in the context of the environments without these foods—such as a long ship voyage—where there is a mismatch. Such mismatches probably have been rare within the primate lineage during the tens of millions of years of years since the gene lost function.[23] Interestingly, the enzyme has been lost in a few other lineages, such as some fruit bats, as well; the animals of these lineages generally get plenty of vitamin C from their diets.[24]

Other vitamin deficiencies also reveal mismatches. Our preparation of food sometimes causes vitamin deficiencies because the preparation removes these essential nutrients. Consider beriberi, a disorder that affects both the nervous and circulatory systems. It arises from having too little vitamin B_1, also known as thiamine. Beriberi is often found in vegetarians who consume a great deal of white rice. Brown rice has substantial quantities of the vitamin; in contrast, the processed white rice does not. White rice rose in prominence across the world because is more easily stored for extended periods of time, but this processing removes vitamin B_1 from it. Because meat has considerable vitamin B_1, meat-eaters are at much less risk of acquiring beriberi.[25]

The practice of agriculture has generated considerable mismatch. Although agriculture allowed its practitioners to produce more food and to generate surpluses, it has reduced the diversity of foods that we eat. This reduced diversity increased the risk of nutritional deficiencies.[26] The reduction in diversity also sets up the prospect for catastrophic failure if something should happen to the crops we rely on. One such extreme example was the potato famine that occurred in Ireland during the 1840s.[27] We will discuss this tragedy in more detail in Chapter 8.

The urbanization that accompanied agriculture created the conditions that fostered many viral diseases. Recall from Chapter 2 that the average infected person needs to infect at least one more person for a virus to sustain itself in the population. The populations of bands of hunter–gatherers are typically far too small for such sustained transmission. Epidemics of polio, smallpox, and many other viral diseases need clusters of several thousand or more individuals; they burn out quickly in smaller groups.[28] In addition, agriculture led us to have close associations with other mammals and with birds. These associations fostered the spillover of disease into humans.

Agriculture also changed our demography. Children, even fairly young ones, could help out with chores on the farm much more readily than children of hunters. As Lieberman puts it, "a large part of the success of farming is that farmers breed their own labor force more effectively than hunter–gatherers, which pumps energy back into the system, driving up fertility rates".[29] Large families were more economically beneficially to farming families than to hunter–gatherers. But this increase in family size created a rat race, as farmers needed more food to support their large families.

Mismatch also is key in evolutionary psychiatry, an emerging field that applies evolutionary principles toward a better understanding of mental health and mental disorders.[30] Some of the distress of contemporary life arises from the mismatch between the environments in which our brains evolved and our modern life. Our ancestors did not experience the stress of our complicated multitasking and multi-taxing schedules.

They did not experience rushing to make a meeting, compounded by the frustrations of bumper-to-bumper traffic. They did not experience the challenge of meeting deadlines imposed by their boss. They did not experience the bombardment of news and other media that we face in this 24-hour-media-cycle world. They did not experience large crowds; in fact, they probably only encountered about 100 to 150 individuals.[31]

This mismatch can generate chronic stress, which can have profound and diverse physical effects. In fact, Charles Blackley, who had linked pollen to allergy, noticed in 1873 that certain temperaments were more likely to become sensitive to pollen: "If there is one temperament which, more than another, predisposes to attacks of hay fever, it is the nervous temperament".[32] Subsequent research over the last century and a half has confirmed that psychological stress can enhance the risk of acquiring allergies as well as inflammatory and autoimmune diseases. A recent review summarizes such effects: "Psychological stress can be viewed much the same as the genetic factors associated with immune disease: an evolutionary ancient condition that becomes pathological as a result of manifestation in a modern industrialized environment."[33] The changed patterns of stress combined with a greater availability of mind-altering substances also helps to explain the prevalence of substance abuse in contemporary societies.[34]

PALEOFANTASY

Although mismatch is an important concept about evolution in general and the application to evolutionary medicine, it can be overly emphasized and misapplied. In her book *Paleofantasy*,[35] Marlene Zuk, now at the University of Minnesota, argued against the excesses of the applications of mismatch to explain the ills of modern society. Zuk summed up her concern in a *Slate* interview: "What's harmful is when you misunderstand the way evolution works and end up worried because humans didn't use to X, we shouldn't do X now."[36] Specifically about the paleo diet and other suggestions that we should avoid starches and other products of agriculture, Zuk continued:

> I think everybody agrees that we evolved eating certain things and we're going to be very unhealthy if we subsist on Diet Coke and Cheetos. But it gets more complicated when you look at the details. Should we eat a lot of meat, less meat? Should we eat dairy?

Zuk's argument contains several planks. One is that humans have lived in many different types of environments over evolutionary time. There was no single environment that we have adapted to; with respect to food, our ancestors ate a wide variety of food items that differed in space and time. A related argument is that we never have been perfectly adapted to any specific environment. Finally, Zuk contends that selection can be rapid.

With respect to the admonition not to eat starches, there is substantial evidence that our ancestors have long consumed starches.[37] Moreover, these starches (along with cooking) have been important in allowing our energy-greedy brains to grow big. Over the course of our evolution, the number of copies of amylase, the enzyme that breaks down starches into simple sugars, has expanded. This expansion appears to coincide with cooking.

Zuk's interest in countering the idea that we were stuck being mismatched was sparked in part by her own research wherein she has shown that evolution could happen rapidly. Her lab has worked on crickets (*Teleogryllus oceanicus*) where males sing two distinct songs: a long-distance song used to advertise their presence to females and a short-distance song used to court females that come into range. In the Hawaiian Islands, these crickets face the threat of a fly that uses the males' own songs against them. Upon hearing the long-distance song of a male cricket, a female fly—loaded with recently hatched larvae—will pounce and try to deposit those larvae into the cricket. If successfully deposited, these larvae will then burrow inside the cricket, killing it within a few days. Nature can be cruel. Getting parasitized is also not good for the male's fitness. In the early 2000s, Zuk noticed that in some areas the singing had greatly diminished, despite an abundance of crickets. What had happened was that most of the males had stopped singing due to a wing mutation that Zuk had named "flatwing".[38]

Flatwing is an X-linked mutation; given that males of this cricket have only one X chromosome, males are either flatwing or normal. The flatwing males do not sing but can use the long-distance song of singing males to find females. Flatwing males have much lower mating success compared with normal males, but this mutation has reached high frequency because it increases male survival. Flatwing mutations have quickly (within a dozen generations) spread, independently in at least two locations.

Even though we humans have much longer generation times (and consequently slower evolution) than Zuk's flies, traits sometimes can evolve reasonably quickly in humans—certainly within the span since we invented agriculture. Recall from Chapter 3 how the Duffy receptor was lost in most African populations as a defense against *vivax* malaria. While both the flatwing mutation and Duffy are examples of adaptations involving loss of traits,[39] humans also have evolved adaptations to changed environments through gains or changes in traits.

In the next two segments, we will explore two exemplars of comparatively rapid evolutionary adaptation in humans. First, we will look at the adaptations that allow some populations to thrive at lofty elevations. Next, we will turn to dairying and the evolution of the ability to continue to digest the sugars in the milk of farm animals, such as cows, throughout one's life span.

CLIMB EVERY MOUNTAIN

Most people experience effects of reduced oxygen in the air (hypoxia) at about 2,500 meters (8,000 feet): they feel tired and sluggish and their breathing can be labored. They may also develop a headache. Although we can acclimate somewhat to this hypoxia, there are limits to how much we can. But evolution to hypoxia has occurred repeatedly: several different human populations have evolved remarkable tolerance to living at extremely high altitudes—above 4,000 meters (13,000 feet)—including in the Tibetan plateau in Asia, the Andes mountains in South America, and the Ethiopian Highlands in Africa.[40] Adaptation to the reduced oxygen (hypoxia) at high elevations is an excellent example of relatively rapid evolution in humans that counters mismatch. In each case, the adaptation involves several genetic changes. While some of the adaptations of these different populations are the same (and thus examples of convergent evolution), other changes differ between the populations both in the genetic changes and the physiological responses, as we will discuss.

Although there is some disagreement, most researchers have placed the dates of the settlements of the high elevation areas of the Tibetan plateau and the Andes to about 25,000 and 10,000 years ago, respectively.[41] These settlements were successful and well-populated: as many as 10 million people lived in the peaks of the Andes just before the Europeans came in 1492, and an estimated 6 million lived on the Tibetan plateau in the early 20th century.

The evolutionary responses in the different populations differ. Take a person unaccustomed to life at high altitudes from low elevation to high elevation and they will breathe more rapidly. But eventually they will acclimate; and with this acclimation, their rate of breathing will return close to its original rate. At high elevations, the native Andes people breathe similarly to the acclimated lowlander. Interestingly, the native Tibetans breathe more rapidly than the acclimated lowlander! The Andes and Tibetan natives also differ in how much hemoglobin they have and how many red blood cells they produce, with the Tibetans having about the same amount as lowlanders, but the Andes people having more of each when living at high elevation.

Cynthia Beall, a physical anthropologist at Case Western Reserve University, has argued that the adaptation to hypoxia in the Tibetan population is likely more advantageous than the one in the Andes because the overproduction of red blood cells in the people from the Andes predisposes them to higher incidences of strokes.[42] It also puts then at higher risk for stillbirths and premature births. These negative effects are not seen in the Tibetan individuals. Beall's lab has shown that the Tibetan population, but not the Andes population, has evolved changes at the gene *EPAS1* (endothelial PAS domain protein 1), which encodes a protein called HIF2α that affects the expression of other genes in a pathway that affects the production of red blood cells. These changes at *EPAS1* allow the Tibetan people to maintain low hemoglobin levels even in low oxygen conditions. Interestingly, recent studies suggest that the Tibetan populations acquired the changed *EPAS1* via interbreeding with the Denisovans, an archaic group of humans that lived in Asia tens of thousands of years ago.[43]

Recently, studies of ancient DNA have provided additional information about the evolutionary history and genetic adaptations of the peoples living at high elevations in the Andes. One study examined a series of ancient genomes going back as far as 7,000 years ago in both low- and high-elevation sites in the Andes, which were also compared with genomes from the Aymara, indigenous to the highlands of Bolivia, and the Huilliche-Peheueche, indigenous to the coast of Chile. The study first narrowed estimates of the date of the high-elevation settlements; the split between the high- and low-elevation genomes most likely occurred 8,750 years ago.[44] The high-elevation genomes showed evidence of selection at the *DST* gene, which encodes the dystonin protein. Dystonin has multiple functions, including ones in the cardiovascular system. The study also revealed evidence of selection acting on the *MGAM* gene, which encodes maltase-glucoamylase, an enzyme that breaks down starches. This is consistent with adaptation to agriculture: in contrast with most of the populations of the lowlands of South America, Andean highlands had developed agriculture soon after settling the high-elevation areas. Even more intriguing is the evidence of recent selection at immunity-related genes, suggesting that the indigenous people adapted to smallpox and other infectious diseases the Spanish brought with them in the 1500s.

Another fascinating recent discovery is that some of the domesticated animals of these high-elevation-dwelling people also evolved adaptations to hypoxia.[45] The Tibetan mastiff, a dog breed found in the Tibetan plateau, also evolved reduced hemoglobin in response to hypoxia. This response is also due to changes at the *EPAS1* gene. Remarkably, these changes may have arisen from interbreeding with Tibetan wolves! Pigs and goats that live on the Tibetan plateau also have evolved changes at genes that affect responses to hypoxia. Domesticated animals in other regions also show evidence of adaptation to hypoxia. In the case of llamas in the Andes, adaptation to hypoxia likely evolved prior to domestication. Evidence also points to genetic adaptation to hypoxia in Ethiopian cattle breeds.

BLESSED ARE THE CHEESEMAKERS

Cheese has been central to religious and other cultural practices for several thousand years, predating the Abrahamic religions by millennia. Archeological remains and writings show cheese played a prominent role in a Sumerian religion that was practiced in the city of Uruk about 5,000 years ago. Then, Uruk was a vibrant commercial hub of about 40,000 people.[46]

One major deity in the religion of ancient Uruk was Inanna, the Queen of Heaven and the goddess of fertility. Inanna was also the goddess of plenty; she served as the guardian of the warehouses that held the agricultural surpluses of the community. One interesting legend concerned Inanna's marriage and her two rival human suitors: Dumuzi, the shepherd, and Enkimdu, the farmer. Inanna initially favored Enkimdu. But she reluctantly agreed to listen to Dumuzi on the advice of her twin brother, Utu. In his sales pitch, Dumuzi offers Inanna rich cream, butter, and cheese. He promises to continue to supply her with these dairy products in exchange for protection and prosperity. She agrees and they marry.

Even 5,000 years ago, cheese was an integral part of several cultures across the world. In fact, the use of cheese is much older than that; the first widespread use probably dates to about 8,500 to 9,000 years ago, coinciding with advances in ceramics and pottery.

Cheese is more portable than liquid milk; it can also be stored for much longer periods of time. There was another important distinction between cheese and milk: cheese contains much less lactose! Five thousand years ago, almost no adults could tolerate large quantities of lactose, the sugar present in milk. Just as now, babies and children could break down lactose into smaller sugars. But after childhood everyone was lactose intolerant.[47]

Lactase persistence (also known as permanence) is the ability to continue to produce lactase, the enzyme that breaks down lactose, throughout one's life. It is mainly due to a single gene, *LCT*, where the allele allowing for persistence is dominant.[48] In people who lack lactase persistence, that is, those who are lactose intolerant, bacteria in the intestines break down the lactose, causing the production of methane. Thus, the consumption of sizeable quantities of lactose leads to bloating and gas. It also can lead to diarrhea.

Ancient DNA studies show that lactase persistence did not evolve until about 5,000 years ago in Europe. It would not reach even moderate frequencies for thousands more years. Thus, we were using milk products well before we evolved lactose persistence. Paul Kindstedt, a historian of cheese, noted "it was cheese and butter making that enabled dairying to gain a foothold among Neolithic populations and allowed the genetic selection for adult lactase persistence to take place".[49]

Lactase persistence has evolved independently in multiple populations, including in East Africa in addition to Europe.[50] Moreover, the region of the genome responsible for the trait shows a strong signature of positive selection acting on it both in European and East African populations.[51] The use of milk products also substantially predates the evolution of lactase persistence in East Africa. A recent study of ancient proteins found evidence for the use of milk products in the form of several individual skeletons from East Africa from as far back as 6,000 years ago. These individuals had traces of milk whey protein β-lactoglobulin, a peptide found only in milk products, on the tartar scrapped off their teeth.[52] This was a few millennia before lactase persistence reached appreciable frequencies. Whether these East African populations had developed cheesemaking or other technologies to reduce the lactose is not known.

Natural selection drove lactase persistence to high frequencies in European and some African populations due to natural selection. Exactly what was being selected is not known: was it calories, proteins, calcium? We don't know yet. Regardless of the exact reason why lactase persistence evolved, the fact that it did evolve—and evolved multiple times via natural selection—shows that humans can evolve in response to mismatch in some circumstances.

EVOLUTIONARY MUTANT MODELS

Mismatch is based on the principle that some maladaptive phenotypes are not inherently deleterious; instead, they are simply not well adapted to the current environment. Change the context, and these phenotypes would be better adapted. Related to the context dependency of phenotypes is the notion that some animals may have phenotypes that are adaptive given their environments but mimic human disease and disorders. Examining these phenotypes and the genes that underlie them could be useful in understanding human phenotypes and can even have clinical applications.

While at Syracuse University, Craig Albertson, an evolutionary developmental biologist who mainly works on fish, led a team of researchers who wrote an opinion piece for *Trends in Genetics* exploring what they called evolutionary mutant models.[53] Geneticists traditionally have looked for genetic variants by generating mutations in the lab through some kind of agent (mutagenic chemicals or radiation or mobile genetic elements). The mutations typically generated in the lab have a different distribution than what is typically found in nature. Albertson and colleagues suggested that looking at natural examples of clinical manifestations in other animals would be a fruitful, complementary approach.

Albertson and colleagues pointed to species of cichlid fish that have head shapes that resemble various craniofacial deformities in humans. They also pointed to icefish that have evolved reduced bone density to increase their buoyancy; this lower bone density resembles the precursor to osteoporosis in humans. The head shapes in cichlids and the reduced bone density in the icefish are not deleterious; indeed, these are adaptive changes. What is a disease or deformity in humans is adaptive in the other species: the context matters! Albertson and colleagues called such systems evolutionary mutant models. They hope that such evolutionary mutant models will shed light on the human disease. Whether they will or not depends on an important assumption: that there is sufficient overlap in the genes that affect the human disease and the adaptive trait of the presumptive evolutionary mutant model. This assumption is not far-fetched, as the same sets of genes often affect the same trait in rather different organisms.

Albertson is now my colleague at UMass. I asked him what motivated him to explore the evolutionary mutant model.[54] He said that his interests in it go back quite a while. During his undergraduate career, Albertson worked for his uncle, an oral and maxillofacial surgeon. Here, Albertson was introduced to craniofacial malformations and had the opportunity to watch his uncle perform corrective surgeries. During his training, Albertson got to see two different perspectives. For his doctoral research, he examined craniofacial variation in cichlids, fish that have evolved tremendous diversity in head shape over a short (relatively speaking) time. During his postdoc, Albertson worked in a zebrafish development lab. While he was impressed by the power of zebrafish as a laboratory model, Albertson also saw its shortcomings. Thus, he was motivated to look for a complementary approach.

Albertson is excited by how well the evolutionary mutant model system has worked—more and more "nontraditional" models are being investigated to better understand the connections between genotype and phenotype and how developmental processes build organisms during development. These changes are partly due to the technological advantages that have made DNA sequencing and other technologies cheaper and easier. There also has been a cultural shift in evolutionary studies, with funding agencies providing grants to more and more studies of these oddball models. The top- tier journals, including *Science* and *Nature*, are publishing more of these studies.

One promising evolutionary mutant model is the blind cavefish. The Mexican tetra, *Astyanax mexicanus*, is a small fish found in much of Mexico and in parts of the southwestern United States. It has populations that live in the surface waters as well as those that live in caves. The cave populations differ from their surface-dwelling counterparts in numerous ways.[55] Many of these have lost most of their eyes and, consequently, their vision. They often also lose most of their pigmentation. Interestingly, the genes involved in the reduced pigmentation in the cavefish are some of the same genes that cause albinism and other pigment disorders in humans. But not all the changes are regressive ones: the cavefish have larger jaws with more teeth. The cavefish also have larger olfactory organs and a better sense of smell. The cavefish also have different head morphologies than the surface populations; thus, they have been proposed as a possible candidate evolutionary model for craniofacial deformities.

But perhaps the most interesting difference between the cavefish and those from the surface is in their behavior. Cavefish behave differently in many ways compared with the surface fish.[56] Notably, the surface fish swim in schools, while cavefish don't;

instead, they swim alone. They are more active and sleep less. They also engage in repetitive behaviors. The behavioral researchers examining these behaviors saw similarity to behaviors seen in people with autism spectrum disorders. Recent research has shown that the cavefish behavior is linked to several genes that are also associated with risk for autism spectrum disorders. The cavefish differentially express several of these genes that people with autism spectrum disorders do. Moreover, when treated with Prozac (fluoxetine) and other drugs, the cavefish behave more like their surface-dwelling counterparts. The combination of these results strongly suggest that cavefish could be a useful model system for studying autism spectrum disorders.

Although he is a champion of the evolutionary mutant model principle and of the development of cavefish as one, Nicolas Rohner at the University of Kansas Medical Center provides prudent caution: "Although most key developmental and physiological pathways are conserved in vertebrates, fish are not humans."[57] Hence, we should not expect all of the findings in cavefish to be "immediately transferrable" to humans. The same caution would apply to evolutionary mutant models in general. Still, they are likely to be promising complements to traditional approaches.

There are several intriguing evolutionary mutant model systems that are being developed outside of fish. For instance, consider polar bears, which diverged from brown bears less than half a million years ago.[58] Aside from the striking loss of pigmentation, polar bears also evolved shorter and sharper claws that enable them to better navigate icy surfaces. To minimize heat loss, they evolved smaller ears and shorter tails. Polar bears also have evolved changes in their diet and now mainly feed on marine mammals. This dietary shift has greatly increased their fat intake; interestingly, polar bears have evolved to digest fat more efficiently. They also have greater fatty deposits under the skin, likely for thermal regulation. This increase in fat tissue suggests that polar bears may be a promising evolutionary mutant model for studying obesity.

Polar bears have fewer copies, on average, than brown bears of the gene *NOX4*, which is involved in the regulation of metabolism. Deletions of this gene in humans leads to a disposition toward obesity due to dietary fats being stored more readily. It seems likely that polar bears evolved reduced numbers of this gene in order to store fat more readily. How polar bears can store so much fat without apparent negative consequences is not yet known but could be a promising research avenue for reducing the negative consequences of obesity in humans.

HOW ARE WE EVOLVING?

One of the quirks of history is that the death of an American president led to a research project that, in turn, has shed insight on how humans are evolving in addition to improving the health of Americans. On April 12, 1945, President Franklin Roosevelt died at age 63 from a cerebral hemorrhage. A year before, Roosevelt had been diagnosed with dangerously high blood pressure and heart failure. Even before that, UK Prime Minister Churchill and others had noted that Roosevelt looked tired and old beyond his years.

When Roosevelt died, cardiovascular disease killed many more people than it does today. Moreover, many of these people who died of heart attacks and strokes

back then did so at comparatively young ages. Roosevelt's tragic death and the growing awareness of the national toll from cardiovascular disease spurred passage of
the National Heart Act and the creation of the National Heart Institute.[59] Among
the items in the National Heart Act was the establishment of a long-term epidemiological study of cardiovascular disease. This study would be based in Framingham,
Massachusetts, a suburb of Boston, conveniently located near Harvard Medical
School and other colleges and universities.

The first cohort of participants in the Framingham Heart Study started enrolling
in 1948 and continued enrolling for the next four years. It would include more than
5,000 residents; unlike most epidemiological studies of the time, more than half of
the recruits were women. Unfortunately, this first cohort was almost exclusively from
people of European descent, limiting the applicability of the results to other groups.
Nevertheless, the results from the Framingham Heart Study were instrumental in
determining risk factors—such as high blood pressure and high levels of certain
types of cholesterol—for acquiring cardiovascular disease. As of 2021, the study is
continuing in its eighth decade of operation, in which the children and grandchildren
of the original cohort are being studied.

The continuation of the study and the examination of the offspring and grand-off-
spring of the original cohort provides a unique opportunity to trace and even predict
evolution of physical traits. To do this, one needs traits of women and the correlations
of those traits with lifetime reproductive success (roughly the number of children). One
also needs to know the heritability of each trait being examined. Recall that heritability
is the proportion of the variance of the trait that is passed on from parent to offspring.

Sean Byars and Stephen Stearns at Yale University found that the population
of women in this study should evolve to be slightly shorter and a little heavier.[60]
Because their lifetime reproductive success is negatively correlated with total cholesterol, there would be selection for decreased cholesterol. The total cholesterol of
the Framingham cohort was 224. Over the next ten generations, it is expected to
decrease to 216.

A follow-up study used the Framingham study data to look for genes that affect
traits associated with coronary heart disease and to determine whether these genes
show the signatures for being under selection. Indeed, coronary artery disease is
linked to several genes that show the signatures of being under positive selection. But
interestingly, often the variant that increases coronary heart disease is being selected
for, not against.[61]

One such case is *FLT1*. It encodes the protein vascular endothelial growth factor 1,
a cell-surface receptor that plays several roles in development and regulating the generation of blood vessels. The variant under selection (based on Framingham study) is
the one that increases coronary artery disease. Byars and colleagues hypothesize that
this variant has some positive function early in life. This is not a case of mismatch
but rather of antagonistic pleiotropy, which, as we discussed in Chapter 1, is likely a
major cause of aging.

Long-term studies of human health, such as the Framingham Heart Study, will be
invaluable tools for the study of evolution, the study of human health, and their intersection. To get a better picture of humanity and to ameliorate health disparities, such
studies need to be done with more diverse populations than has been done to date.[62]

NOTES

1. Lieberman (2013, p. 15).
2. Zuk (2013, p. 66).
3. Strachan (1989, p. 1260).
4. Strachan (2000).
5. Wickens et al. (1999).
6. Reviewed in Strachan (2000).
7. In this 1970 cohort, about half had had the measles vaccination and half did not. With large fractions in both groups, statistical power to detect differences was maximized.
8. Rook et al. (2004).
9. See Rook and Brunet (2005), Rook (2007), and Bloomfield et al. (2016).
10.. Rook is quoted in Scudellari (2017, p. 1433).
11. For a full treatment of how we overuse antibiotics and the consequences of this misuse, see Blasser (2014).
12. Gevers et al. (2014).
13. Stiemsma et al. (2015).
14. Tang et al. (2014).
15. Finlay et al. (2021).
16. Lownes (1947). The quote is on p. 22.
17. See Just (1933, p. 23). Just (1883–1941) was one of very few African Americans in biology during his time and was among the first to get a doctoral degree (from the University of Chicago). Crow (2008, p. 1735) described Just as "one of greatest biologists of the early 20th century" who managed to publish many seminal papers and two important books during a short research career despite limited time or resources for research. Among other things, Just was the first to show cell surfaces changed during different stages of egg development in marine invertebrates. He also demonstrated that the developmental trajectory of embryos could be influenced by the environment. Byrnes and Newman (2014) provide a short biography of Just and describe his contributions to evolution and development.
18. Gluckman and Hanson (2006).
19. Lents (2018).
20. Onion (2016).
21. Nishkimi and Yagi (1991).
22. Because they are under little or no constraint from selection, pseudogenes evolve faster than most genes. See Johnson (2007, chapters 4 and 12) for details.
23. Whether the loss of the function is selectively favored is not known. Production of the enzyme imposes some costs, but these are modest; it is possible that the loss of the enzyme was effectively selectively neutral.
24. Lents (2018).
25. *Ibid.*
26. Lieberman (2013).
27. Dunn (2017) provides a detailed account of the causes and consequences of the potato famine.
28. Lieberman (2013).
29. *Ibid.*, p. 188.
30. Abed et al. (2019).
31. Dunbar (1992).
32. Blackley is quoted in Brenner et al. (2015, p. 32). This reference also provides background information on how chronic psychological stress affects the body.

33. Brenner et al. (2015, p. 38).
34. Randy Nesse, who helped found the fields of evolutionary medicine and more recently evolutionary psychiatry, explores how mismatch can create the circumstances that lead to pervasive substance abuse. Nesse (2019, p. 240) notes:

   > our minds have always been vulnerable to capture by alcohol, marijuana, tobacco, coca, and opium, but problems with them escalated as advances in chemistry, transportation, and technology have increased the diversity, purity, and availability of drugs. The mismatch was bad before; now it's getting much worse.

35. Zuk (2013).
36. Zuk is quoted in George (2013).
37. Hardy et al. (2015).
38. See Zuk et al. (2006), Tanner et al. (2019) and Heinen-Kay and Zuk (2019).
39. See Johnson et al. (2012) for other examples of how trait loss can be adaptive. We'll also see examples of the losses of genes being favored in Chapter 12.
40. Witt and Huerta-Sánchez (2019).
41. Aldenderfer (2003).
42. Beall et al. (2010).
43. Huetra-Sánchez et al. (2010). We'll discuss the Denisovans in more detail in Chapter 17.
44. The 95 percent confidence interval is 8,200 to 9,250 years. This is more recent than the estimate of more than 10,000 years ago taken from estimates of modern genomes and physical evidence. See Lindo et al. (2018) for more detail about this and other results of the study.
45. Witt and Huerta-Sa'nchez (2019).
46. Material on the history of cheese is from Kindstedt (2012).
47. Segurel and Bon (2017).
48. Lactase persistence or the lack thereof is due to regulatory regions near the *LCT* gene, not the actual gene. Moreover, other genetic factors and the microbiota also play a role. See Segurel and Bon (2017) and Anguita-Ruiz et al. (2020).
49. Kindstedt (2012, p. 14).
50. Tishkoff et al. (2007).
51. The selective benefit of lactase persistence has been estimated to be about 5 percent. This is considerably strong selective pressure. See Segurel and Bon (2017).
52. Bleasdale et al. (2021).
53. Albertson et al. (2009).
54. Personal conversation with Craig Albertson in December 2019 and follow-up e-mail exchange in February 2021.
55. Atukorala et al. (2019).
56. Yoshizawa et al. (2018).
57. Rohner (2018). The quote is on p. 356.
58. Rinker et al. (2019).
59. The National Heart Institute is now the National Heart, Lung, and Blood Institute. See Mahmood et al. (2014) for details on the National Heart Act and the development of the Framingham Heart Study.
60. Byars et al. (2010).
61. Byars et al. (2017).
62. Gurven and Lieberman (2020).

5 From Genetic Mapping to Personalized Medicine

SUMMARY

We begin with a discussion of the potential of personalized (also known as precision) medicine. We segue into the success story of cystic fibrosis: finding the genetic variants responsible for it has helped to create drugs that greatly improve the life span and the quality of life for people with this disease. Personalized medicine requires the mapping of traits to genes and genetic variants, a process that involves the principles of evolutionary genetics. We discuss quantitative trait locus (QTL) mapping and genome-wide association studies (GWAS). The evolutionary genetic concept of linkage disequilibrium is central to GWAS. We discuss a few success stories of GWAS leading to better understanding of autoimmune and inflammatory diseases. A key finding that has important implications is that several genetic variants are protective of one disease but predispose toward other diseases. GWAS has also been useful in personalizing the proper dosage of drugs based on one's genotypes. We discuss some of the privacy concerns and limitations of GWAS. One unfortunate limitation is that most GWAS investigations have been done in populations of Europeans or those of European descent; more diversity in these studies can lead to better science and potentially ameliorate heath disparities.

"Human genetics has sparked a revolution in medical science on the basis of the seemingly improbable notion that one can systematically discover the genes causing inherited diseases without any prior clue as to how they function."—Eric Lander and Nicholas Schork.[1]

"Synthesizing our growing knowledge of evolutionary history with genetic medicine, while accounting for environmental and social factors will help to achieve the promise of personalized genomics and realize the potential hidden in an individual's DNA sequence to guide clinical decisions. In short, precision medicine is fundamentally evolutionary medicine."—Mary Benton and colleagues.[2]

THE POTENTIAL OF PERSONALIZED MEDICINE

In a hospital, a week-old infant presents with a baffling array of symptoms. Doctors examine her genome to assist with their diagnosis. A couple who are from the same ethnic background and who are planning to have children undergo genetic tests to determine whether they are carriers for common disease traits (share recessive, deleterious alleles). A woman with a family history of ovarian cancer undergoes tests to see whether she is at higher risk for the disease. A man whose father had early-onset Alzheimer's disease has his genome scanned to determine his risk for getting the disease. These scenarios all illustrate examples of personalized medicine, also known

DOI: 10.1201/9780429503962-6

as precision medicine. This growing field has been fueled by dramatic advances in reducing the time and cost of DNA sequencing as well as by a deeper understanding of the biology, including the evolutionary processes, behind the traits.[3]

Personalized medicine allows us to make better-informed choices. One well-known case involved the actress Angelina Jolie. Although she did not have cancer, Jolie underwent a preventive double mastectomy in 2013 upon learning that she had a specific variant in her *BRCA1* gene.[4] Because she possessed this variant, Jolie's risks of getting breast cancer and ovarian cancer were very high (87% and 50%, respectively). With the mastectomy, her breast cancer risk was dramatically reduced. A few years earlier, the journalist Masha Gessen wrote of her own diagnosis with *BRCA1* and the processes she went through in deciding on a course of treatment, ultimately also deciding on a mastectomy.[5]

Genetic screening can take place among potential parents or around birth. The population of Ashkenazi Jews had a genetic bottleneck in the past, resulting in the accumulation of recessive deleterious mutations. Couples who are both of Ashkenazi descent routinely get screened to determine whether they share recessive variants for diseases, such as Tay-Sachs disease, a neurodegenerative disorder that strikes in infancy. This screening has led to a decrease in the incidence of Tay-Sachs.[6] Newborns are routinely screened for genetic diseases; as we will discuss, this has helped improve the life span and health outcomes for people with cystic fibrosis.

Personalized medicine is predicated on the general finding that people are differentially susceptible to diseases and respond differently to treatments. It further assumes that at least some of this variation is due to genetic factors. As we have seen earlier, variation for disease risk may arise for a multitude of reasons including mismatch, tradeoffs, heterozygote advantage, and recurrent mutations.

An essential step toward personalized medicine is the mapping of diseases to genes. This mapping has led to a better understanding of disease. It has also been useful in the development of drugs. For instance, finding that variants of *PCSK9*, a gene whose protein binds to a lipid receptor, is associated with coronary artery disease and excessive cholesterol in the blood led to a new class of cholesterol-decreasing drugs.[7] These *PCSK9* inhibitors are becoming an effective replacement for statins, which do not work in some people. Moreover, the *PCSK9* inhibitors are less likely to increase diabetes risk or damage the liver, some of the adverse side effects of statins.[8] In general, drugs that are linked to a gene are much more likely to work and to get approval by the FDA.[9] Given the expense in time and money of drug trials, these gains in efficiency are of vital importance and will likely lead to better health outcomes.

But how have biologists learned which genetic variants affect which traits? There are relatively simple cases, such as sickle cell anemia or cystic fibrosis, where the trait or syndrome is clearly defined and demarcated (individuals either have it or they don't) and is passed on from parent to offspring in a Mendelian pattern with a single gene being responsible. Characterizing such genes itself was a challenge, especially before the competition of the Human Genome Project. We will discuss how cystic fibrosis was characterized in the next section.

Unfortunately for the purposes of mapping genes to traits, most traits are not like cystic fibrosis; they are due to multiple genes and environmental input. Unraveling the genetic basis of these complex traits has been more difficult. Consider height. It is highly heritable;[10] the best predictor of the adult height a child will attain is her parents' heights. This assumes of course that she grows up in an environment not too dissimilar from her parents. To be sure, there are some large effect genetic variants that result in a person being much taller or (more likely) much shorter than average. But such variants are rare. For the bulk of the variation in height, numerous genes contribute, each with very small effect. For instance, which allele one has at *HGMA2*, one of the larger effect genes affecting height, will alter predicted values of one's height by about 3 millimeters (about a tenth of an inch). In his recent book on heredity, the American science writer Carl Zimmer noted that these 3 millimeters equate to putting on a pair of wool socks.[11] Height is not unusual. In fact, the genetic basis of traits—including disease traits—is a continuum from ones that are like cystic fibrosis to ones that are like height. We will discuss examples across the continuum in this chapter. First, let's start with cystic fibrosis.

CYSTIC FIBROSIS: A SUCCESS STORY

Back in Chapter 1, we had discussed cystic fibrosis and how people with it often acquire bacterial infections that evolve antibiotic resistance. There we saw that the respiratory symptoms of the disease arise from failures in pumping chloride ions out of cells that line the lungs and that the gene involved is the *CFTR* for the *cystic fibrosis transmembrane conductance regulator*. Here, we will explore how scientists figured out that this gene is responsible and how characterizing this gene has led to advances in treatment and better outcomes.

Researchers and clinicians have made extraordinary strides in improving the lives of people with cystic fibrosis. Over the last century, cystic fibrosis has gone from a disease where most died in early childhood to one where most people with it are adults, at least in developed countries.[12] In the United States, those born with it during the mid-2010s have a life expectancy in the mid-40s; for those born only two decades before, it was barely 30.[13] This rapid increase in life expectancy is largely due to early diagnosis, which allows for early interventions such as a high-fat diet and antibiotics. In recent years, the use of various drugs that correct the underlying defect have allowed for continued progress. The story begins with the pathologist Dorothy Andersen.

As a pathologist at the Babies Hospital at Columbia Presbyterian Hospital in New York during the 1930s, Andersen observed children dying with a variety of symptoms that affected their lungs, digestive systems, and their pancreases.[14] On her autopsy table, Andersen saw fibrous cysts covering most of their pancreases. Connecting these symptoms as part of a larger encompassing syndrome, Andersen would name the disease "cystic fibrosis" based on the lesions observed in the pancreas. Although Andersen and others would soon develop diagnostic tests for cystic fibrosis, such tests—mainly based on the saltiness of the skin—would be limited in their utility. These phenotypic tests could not be performed until children were older, as newborns don't sweat like older babies. Moreover, some children would develop symptoms only later in life. Without a genetic screen, children could not be tested at birth; as a result, interventions would be delayed while the disease destroyed more of the body. A genetic screen would require knowledge of the gene that caused cystic fibrosis.

Looking at pedigrees, medical geneticists examined the genetic basis of cystic fibrosis. The disease ran in families. It behaved like a trait due to a single gene, with the disease variant being recessive to the normal allele.[15] Each parent of a child with cystic fibrosis was a carrier: the children got one allele for the disease from each parent. Each child from a pairing of two carriers had a one-quarter risk of getting the disease.

Until the last couple of decades, tying a disease—even one that had a straightforward genetic basis like cystic fibrosis—to a specific gene was an arduous process. In the case of cystic fibrosis, finding the gene involved took the efforts of several teams both in the United States and in Canada, sometimes competing and sometimes collaborating. These teams included researchers with a diverse set of expertise: some focused on the genetic analysis of pedigrees, while others looked at the molecular genetics, the biochemistry, and the structural chemistry.

Comparison of the genes from individuals in the same family—some with cystic fibrosis and some unaffected—helped pinpoint the gene. If, for one marker along the chromosome, all of those with the disease had one pattern and all those unaffected either lacked the pattern or had one chromosome with the pattern and one without, then that marker was likely near the gene that caused cystic fibrosis. By examining overlapping fragments of the chromosome and using molecular genetics techniques, the researchers could walk across the chromosome to get to the gene and declare that it was the one that caused cystic fibrosis. This method, which relies on the position of the gene to hunt it down, was referred to as positional cloning and was state of the art in the late 1980s. Positional cloning is easier said than done: using pedigrees and this linkage analysis to find the causal gene was a Herculean struggle, especially in the days before the human genome project.

One important step in this process directly involved evolution. Once the gene was in sight, the next step was in figuring which of several nearby differences was the one that caused the disease as opposed to others that were unrelated noise. How to separate the signal from the noise? To determine whether the DNA they were examining had meaningful function, the researchers first performed zoo blots—examining the stretch of DNA in multiple species—to see whether there was conservation across these species. We would expect a functional gene to be more conserved across evolution than one without function. Even though evolution is often associated with change, an irony of biological evolution is that genes that are functionally important usually evolve more slowly than ones that are not.

These efforts paid off. Just as the 1980s ended, these teams of researchers found the *CFTR* gene.[16] They noted that most of the people with cystic fibrosis had *CFTR* genes with a deletion that caused a defective protein. Others with the disease had different defects in the *CFTR* gene. As the years went on and the researchers learned more about the *CFTR* gene and its relationship to cystic fibrosis, they would realize that the type of mutation in the *CFTR* gene often corresponded with the severity of the disease: not all cystic fibrosis was the same. These differences would matter as drugs were developed for cystic fibrosis.

Developed by funding from the nonprofit Cystic Fibrosis Foundation and approved in 2012, ivacaftor (brand name Kalydeco) became the first cystic fibrosis drug that fixed the underlying protein defect.[17] Kalydeco works exceptionally well—often essentially eliminating the symptoms of cystic fibrosis—but it only works well for about one in 20 people with the disease. It works for those people with at least one copy of the *CFTR* gene that produces proteins with a specific defect: they are fully functional but have difficulty transporting chloride through the chloride channel across the membrane. Kalydeco acts like a doorman opening and closing the channel, allowing the *CFTR* protein to push the chloride across.

But Kalydeco does not work with most people with cystic fibrosis, including those with the most common cystic fibrosis mutation. This mutation, noted as F508del, is missing one amino acid; specifically, the amino acid phenylalanine (designated as F because P was taken by the amino acid proline) in the 508th position. Because of this missing amino acid, the protein cannot fold properly. Most of the product gets degraded by the proteasome, structures in the cells that destroy damaged proteins, but some leaks through. In late 2019, the FDA approved Trikafta, a combination of ivacaftor (Kalydeco) and two other drugs, for use to treat people with the F508del mutation.[18] These other drugs help the CFTR protein fold better and stop it from being chewed up by proteasomes. Trikafta is selling briskly: in its first ten weeks, sales exceeded 400 million dollars. Drugs that work on other rarer *CFTR* mutations are still being developed.

Differences in the frequencies of different *CFTR* mutations are at least partially behind variation in the clinical manifestation of cystic fibrosis in different populations. People of Asian descent who have two defective *CFTR* alleles have less salt in their sweat than their counterparts of European descent. In fact, some fall into the normal range of salt concentrations. They are also less likely to have bacterial infections in their lungs. The reasons for this are partly due to differences in the specific mutations they have: a lower proportion of Asian individuals with cystic fibrosis have two copies of the F508 deletion.[19] But even those Asians who had two copies of the F508 deletion had less salt in their sweat compared with Europeans with two copies of that allele. Differences in other modifying genes and/or differences in the environment, such as their diet, are likely explanations.

Cystic fibrosis is reasonably common. It affects about one in every 3,000 people born of European descent, and the frequency of carriers approaches 5 percent. It is less common in Asia and the Middle East and extremely rare in sub-Saharan Africa. The frequency in Europe is high enough to suggest that heterozygous carriers, like in sickle cell anemia, have a selective advantage. But what is the selective agent? Several candidates—cholera, typhoid fever, and tuberculosis—have been proposed over the years. A study modeling historical epidemiological data supports tuberculosis as the most likely candidate.[20] Carriers being protected from tuberculosis, which was a major cause of death in Europe during the 17th through 19th centuries, can explain the geographic patterns of cystic fibrosis. In contrast, the historical prevalence of cholera and typhoid fever cannot explain the cystic fibrosis pattern. Moreover, carriers of cystic fibrosis are less likely to get tuberculous.

Tuberculous has been contained in most of Europe and in the United States. Accordingly, we would not expect continued selective pressure favoring the cystic fibrosis allele. But tuberculous is still endemic in much of the world. For instance,

Brazil has high incidence of tuberculosis. Brazil also has high proportion of people with European ancestry. Thus, we may expect to be able to observe continued selective pressure acting there. Indeed, a recent study found a strong negative correlation between tuberculosis and the frequency of cystic fibrosis is found in Brazil: cystic fibrosis is less common in regions with high rates of tuberculosis than in regions where tuberculosis is rare.[21] While other possible explanations exist, this is evidence supporting carriers being less likely to get tuberculosis.

ASSOCIATION MAPPING, LINKAGE DISEQUILIBRIUM, AND HAPLOTYPE BLOCKS

Francis Collins, recent director of the National Institutes of Health, was the head of one of the teams who characterized the cystic fibrosis gene. After cystic fibrosis, Collins, then at the University of Michigan, continued to lead efforts to track down the genetic basis of human diseases. An advocate for the positional cloning approach, Collins noted that this method had identified about 40 disease genes in the five years after cystic fibrosis. Among these were genes linked to Huntington's disease, early-onset breast and ovarian cancer, and a few kinds of muscular dystrophy. But Collins saw that such genes were low-hanging fruit: their simple genetic basis made them accessible to such analysis. But most diseases—like most traits—have a more complicated genetic basis. In 1995, Collins noted:

> The major future challenge to positional cloning (and its rapidly growing offspring—the positional candidate approach) will be the elucidation of genes responsible for predisposition to common polygenic disorders such as diabetes, asthma, hypertension, many forms of cancer and the major mental illnesses.[22]

How can these diseases involving multiple genes—the polygenic diseases—be mapped to genes? During the 1980s and especially the 1990s, an approach toward linking traits to candidate gene regions (known as quantitative trait locus or QTL mapping) became popular with geneticists. Here, controlled crosses are performed, resulting in the first generation of hybrids (F_1). These F_1 hybrids are then crossed with other F_1 individuals, resulting in the F_2 hybrids, or are backcrossed to one of the parents. While the F_1 individuals all have the same set of genes (one from each of the parents), the F_2 hybrids differ amongst themselves: recombination causes them to have different combinations of genes. The same is true for the backcross hybrids. These F_1 or backcross hybrids can then be examined both for what genes they have (by using molecular markers) and for what traits they have. If hybrids that have a specific genotype at a marker tend to differ with respect to a trait and this tendency is not simply due to chance, one can infer that something near that marker is affecting the trait.[23] Now a mainstay of genetic and evolutionary genetic research, QTL mapping is a powerful way to find candidate regions of the genome that affect the trait of interest. Depending on the question being addressed, this information itself can be useful. It is also a first step toward cloning the genes.

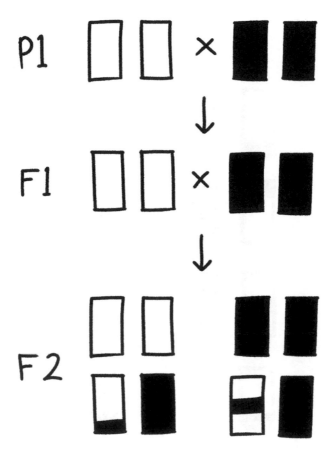

Initial crosses used for QTL mapping. The boxes represent chromosomes of individuals. In the P_1 generation, one line (light shading) is crossed to another (in dark shading) to generate the F_1 generation (one chromosome light and the other dark). The F_1 can either be crossed to itself to generate the F_2 (as shown in this diagram) or can be backcrossed to one or the other parent. In either the F_2 cross or the backcross, a diversity of genotypes can be generated.

Source: Illustrated by Michael DeGregori.

The QTL method reveals some of the limitations of human genetics faced during the 1990s: QTL analysis generally is not feasible with humans for a couple reasons. First, unlike *Drosophila* or even mice, humans don't have many offspring. Large sample sizes are needed to infer variants that have modest effect. The second limitation is that ethical considerations preclude controlled breeding. So, what to do?

Controlled breeding is not the only way to look for associations between genes and traits. If you had an idea of which gene or genes might contribute to a trait, you could examine individuals with certain genotypes to see if they differ in the trait compared to those with other genotypes. If they did, there would be an association

Trait

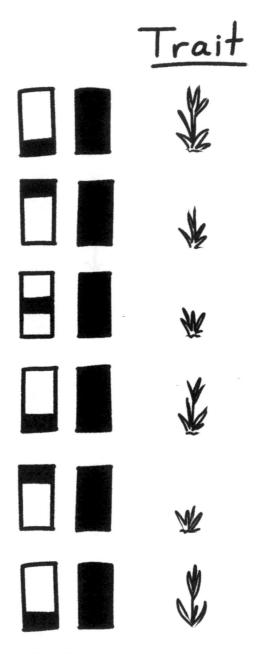

Use of markers to find regions of the genome that contribute to the height of the tree (here, the height of the plant). The boxes represent the chromosome composition of individuals. DNA markers allow the researchers to assess the origin (dark or light) of any given region of the chromosome for any individual. Here, individuals that have both chromosomes shaded dark at the lower tip of the chromosome grow taller than other individuals.

Source: Illustrated by Michael DeGregori.

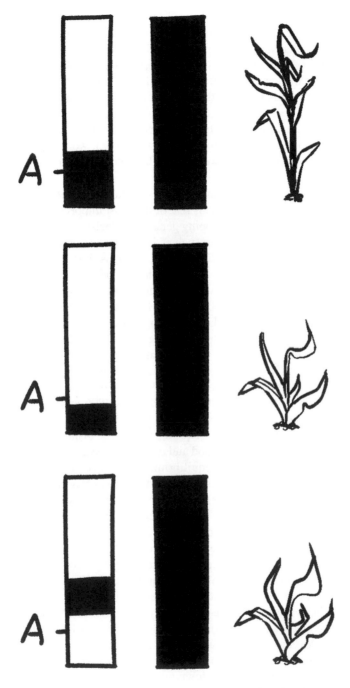

Refinement of the location of the causative genetic factor for plant height. Here, individuals that have two copies of dark shading at the position labeled with A are taller than other individuals. Thus, we can conclude that the causative factor for plant height is around the position of A.

Source: Illustrated by Michael DeGregori.

between the gene and the trait. Such association studies are feasible if you have some idea that certain genes based on their function or some other attribute would contribute to the trait of interest. But such knowledge is often hard to come by.

Would it be possible to use an association approach in the absence of prior knowledge about where the genes that affect the trait are? Could we just throw molecular markers down like darts in hopes that we find such genes? In fact, this approach is now possible. It's a technique called genome-wide association studies or GWAS— yes, biologists love acronyms. In the last decade and a half, GWAS has become a huge boon for mapping of genes. To get a sense of how it works and how evolutionary principles are essential for it, we need to delve deep into the concept of linkage disequilibrium.

Linkage disequilibrium, the nonrandom association of genetic variants at different positions along the chromosome, is one of the most important concepts in evolutionary biology and is critical to mapping genes by association. We have encountered the concept a couple of times earlier, such as in the discussion about the differences in the genetic structure in *Plasmodium* in Africa versus those in southeast Asia and how it may be responsible for differences in the evolution of resistance to artemisinin. Although the term is a bit unwieldy, the concept is relatively straightforward.[24] In the absence of linkage disequilibrium (i.e., linkage *equilibrium*), what variant an individual has at one position tells us nothing about what variant they will have at another position. In contrast, strong linkage disequilibrium means that we can predict what variant an individual will have at one position based on what variant they have at another. For instance, if individuals that are G/G at position 4896 of a gene are more likely to be A/A at position 5689 of that gene than individuals that are T/T at position 4896, then we can say that positions 4896 and 5689 are in linkage disequilibrium. Generally, positions that are close together on the chromosome are more likely to be in strong linkage disequilibrium; as the distance between points on a chromosome increases, the likelihood and extent of linkage disequilibrium is likely to decrease. Recall that linkage disequilibrium is a metric of predictive power: just as we would expect two nearby cities (New York City and Newark, New Jersey) to be more likely to have similar weather than two more distant cities (New York City and Atlanta, Georgia), we should expect variants between closely linked sites to associate nonrandomly more than those that are further apart.

Linkage disequilibrium is a central concept in evolutionary genetics in part because so many evolutionary processes affect it. Most processes—including selection and genetic drift—increase linkage disequilibrium. When a variant favored by selection increases in frequency from being exceedingly rare to being common, it drags along other variants that are on the same chromosome. Thus, these variants are associated with the selected variant: as a result, this association increases linkage disequilibrium. Variants that increase simply due to random genetic drift also drag along other variants; this association, too, increases linkage disequilibrium. In contrast, genetic recombination breaks down associations between variants at different locations; thus, recombination reduces genetic variation.

So why is linkage disequilibrium pertinent for association mapping? The answer is that without linkage disequilibrium, association mapping could not work. Association mapping looks for associations between differences in the trait of interest (for example, blood pressure) with differences in genotypes at many markers across the genome. If people with one genotype at a marker have different average blood pressures than those with a different genotype and the difference is too large to be explained by chance sampling variation, then the marker is associated with the trait. This does mean that variation at the marker itself affects blood pressure. Rather, the most likely explanation is that the marker is in linkage disequilibrium with a variant that affects blood pressure. Even though the marker is not causing the difference, it is assumed to be close enough to the causal variant that molecular methods can be used to find the causal variant.

For association mapping to work properly, the patterns of linkage disequilibrium in the genome of the population need to be known. If markers are too sparsely distributed across the genome, then the analysis may miss the causal variants because they are too far away from any marker to be in linkage disequilibrium. The density of markers required depends on how quickly linkage disequilibrium declines as the distance between two markers increases. If it falls off quickly (after a few thousand base pairs), then markers would have to be placed really close together. An extremely large number of markers would be needed to cover the 3 billion base pairs in the human genome. Conversely, if linkage disequilibrium declines more gradually—if sites located 10,000 base pairs or more away show some degree of correlation—then the markers do not have to be packed as closely together. Fewer markers would be needed.

Toward the end of the 20th century and the anticipation of completion of the human genome, there was a hope that such association mapping at a genome-wide scale could be realized.[25] This optimism seemed to be squashed by a simulation study by Leonid Kruglyak. This model made the prediction that given the approximate effective population size of humans, linkage disequilibrium would not be expected to extend more than about 3,000 base pairs (or 3 kilobases).[26] If that were the case, genome-wide association mapping would require at least half a million markers, a number that was prohibitive at the time.

A couple years after Kruglyak's more pessimistic study, the pendulum swung back to optimism with observations of the structure of linkage disequilibrium in the human genome. Contrary to the prediction that linkage disequilibrium would fade after only a couple of kilobases came reports of much larger stretches where variants stayed in strong linkage disequilibrium.[27] The linkage disequilibrium was so strong that these stretches were like a couple of blocks where certain variants at one site were always or almost always associated with variants at sites several kilobases away. These blocks would become known as haplotype blocks. The term haplotype comes from haploid chromosome. Recall that you have two sets of chromosomes—one from mom and one from dad—each of which are haploid. Because of these haplotype blocks, the structure of the genome is simplified, allowing for mapping traits to genes based on GWAS.

The likely reason that haplotype blocks exist is that recombination rates vary across the genome at a fine scale: there are recombination hotspots, where recombination shuffles the genetic deck frequently over evolutionary time, interspersed with recombination cold spots. At the time when the haplotype blocks were discovered, the mechanism for why recombination varied so much on a local scale was unknown. Now, we know that genes like *Prdm9* affect the location of the recombination hotspots: this gene encodes an enzyme that binds to DNA at specific locations, where it adds methyl-groups to proteins associated with DNA.[28] These changes affect where breaks in DNA are likely to occur. Breaks in DNA allow for recombination. Nucleotide changes at *Prdm9* affect where the protein binds and thus where the hotspots are. Interestingly, this gene also leads to hybrid incompatibility between different subspecies and species of mice: hybrids of these species are sterile in part because they have different variants of *Prdm9*.[29]

These haplotype blocks make GWAS feasible even with the technology that existed in the middle 2000s. Researchers can just throw markers down across the genome; in most cases, they should be able to recover associations with at least some of the variants that affect the trait or risk of the disease in question.

GWAS SUCCESS STORIES IN AUTOIMMUNE DISEASES

GWAS have been extraordinarily successful in many respects. As of the start of 2019, more than 50,000 statistically significant associations between traits and DNA variants had been recorded.[30] These associations have been found for traits from schizophrenia to diabetes to coronary heart disease to just about any other prominent disease or disorder. The advances from GWAS have led not just to a better understanding of a variety of diseases but have also spurred the discovery of new drugs, some of which have been successful at treating the ailments. Moreover, in some cases, the knowledge gained from GWAS has been useful in making assessments about interventions based on genetic and other risk factors.

One of many realms where GWAS has been successful in getting linking genes to disease traits is in autoimmune diseases and the related disorders of chronic inflammation.[31] We had briefly discussed how changed environments have likely led to an increase in autoimmune diseases as an example of mismatch due to changed environments back in Chapter 4. In addition to the environmental factors, variation at many genes affects one's propensity to get autoimmune diseases.

GWAS has been useful in revealing the biology behind different forms of inflammatory bowel disease, a catch-all term for conditions that cause chronic inflammation of the digestive tract. As we discussed in Chapter 4, one form of the inflammatory bowel disease is Crohn's disease. It results from the combination of genetic variants and the gut microbiota as well as diet and other environmental conditions. GWAS and other studies have enhanced our understanding of the processes that occur in cells of people who have Crohn's disease. Notably, these studies have revealed new, unexpected cellular processes that are involved in the development of inflammatory bowel diseases.

One of these surprises is the link between Crohn's disease and autophagy, the mechanism wherein the cell gets rid of misfolded proteins, damaged organelles, parts of bacteria, and various unneeded larger molecules.[32] Autophagy is the cell's way of taking out the trash. In this process, the trash is first tagged as such by a small protein called ubiquitin, called such because it is ubiquitous. Labelling what should be trash and not labelling what should not be trash is vital to the garbage disposal service. The ubiquitin-labelled material is then recognized by autophagy adaptor proteins. These adaptors take the trash to membrane-bound structures that become

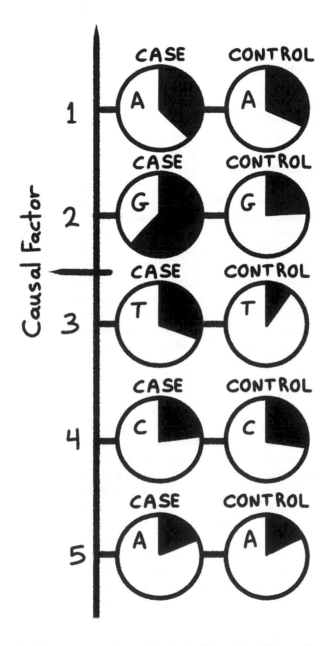

GWAS mapping. Here are proportions of a focal allele at five different (1 to 5) markers for case (individuals with the trait) and control (individuals without trait). At markers 1, 4, and 5, there is no significant difference between the frequencies of alleles in cases and controls. At markers 2 and 3, the case and control populations differ substantially in the frequencies of the alleles. Thus, we can infer that the causative factor is likely to be between marker 2 and marker 3.

Source: Illustrated by Michael DeGregori.

autophagosomes, vesicles that eventually lead trash out of the cell. You can think of the adaptor proteins as the sanitation workers that remove the trash as they go from building to building. If autophagy is working properly, the trash does not pile up. If it malfunctions, signals are sent; these signals can trigger inflammation.

A GWAS found strong associations of Crohn's disease to variants of *ATG16L1*, a gene that encodes an autophagy adaptor protein.[33] To extend the metaphor, Crohn's disease is associated with a variant of the cell's sanitation workers. Under metabolic stress, the variant of the ATG16L1 adaptor protein that is linked to increased risk of Crohn's disease is chewed up by another protein called caspase 3. Thus, under stress, the sanitation workers are hindered from doing their job. As a result, inflammatory signals are sent. Prior to the GWAS link of Crohn's disease to *ATG16L1* and other genes in the autophagy pathway, the notion that the disorder was caused in part by a breakdown in autophagy had not been considered.

Recall from Chapter 4 that Crohn's disease was associated with a reduction in the diversity and other changes in the microbiota. It's not just the microbiota and it's not just the genes; instead, it's the interaction. Some genes linked to Crohn's disease interact with the microbiota of the intestines. For instance, the gene *NOD2* was linked to inflammatory bowel disease early on, through traditional association mapping and then by GWAS.[34] The protein product of *NOD2* senses muramyl dipeptide, which is used in bacterial cell walls. Hence, this chemical is a sign of microbial activity. Failure to properly signal the hallmarks of bacterial activity could cause trouble for the microenvironment of the intestinal lining. Indeed, excessive inflammation has been associated with insufficient NOD2 signaling. Moreover, mice studies show that the experimental knockout of *NOD2* and another gene led to a condition akin to Crohn's in mice exposed to the bacterium *Mucispirillum schaedleri*, which is often found in the intestinal lining in both mice and humans.[35] Clearly, genetics and the microbiota and their interaction can affect the development of Crohn's disease.[36]

GWAS has also revealed possible treatments for Crohn's disease. For instance, a GWAS association with variants at *PRKCB* (protein kinase C beta) led to the suggestion that ellagic acid could be a treatment for Crohn's disease.[37] As a protein kinase, PRKCB adds phosphate groups to other proteins; in so doing, it influences many cellular processes. Ellagic acid, a plant product, dampens the signaling of PRKCB. It is found in cranberries, strawberries, and other berries as well as in some nuts like walnuts and pecans. Bacteria break down ellagic acid and like compounds into urolithins. More recently, a study found that urolithins also enhance tight junctions, barriers between cells that limit the exchange of fluids and other materials between cells.[38] People who have inflammatory bowel disease often have deficiencies in their tight junctions. In mice that have been engineered to have inflammatory bowel disease, the addition of urolithins improves tight junction function and reduces inflammation.

GWAS and other studies have led to a better understanding of psoriasis, a skin condition that is due in part to dysfunction of the immune system. Manifesting as red scaly patches on the skin, psoriasis is the result of excessive turnover of skin cells and chronic inflammation. It affects about 2 percent of people of European descent, with reduced prevalence in people of African and Asian descent. In addition to the skin condition, about a third of people with psoriasis also develop a form of arthritis marked with inflammation of various joints.[39]

An early GWAS study linked psoriasis risk to variants at two different but related interleukin genes: *IL12B* and *IL23R*.[40] The proteins encoded by these genes are signaling molecules that promote inflammation and tissue damage. They are produced by helper T cells and both stimulate production of interleukin-17, which has numerous targets.[41] The overexpression of interleukin-17 and its targets lead to the chronic inflammation associated with psoriasis.

Linking psoriasis to the interleukin-17 pathway has helped with the development of treatment. Because of this linkage, an antibody against interleukin-17, which eventually became named secukinumab, was developed and tested. Clinical trials showed that secukinumab successfully and quickly treated psoriasis.[42] Approval of secukinumab came in Japan in late 2014, followed by the United States and the European Union early in the following year.[43]

GWAS also has revealed the interesting finding that variants associated with autoimmune and inflammation-related diseases are often highly pleiotropic.[44] That is, the same variant affects several different diseases at once. Sometimes, this pleiotropy is synergetic: the variants act in the same direction across several diseases. Here, the same risk factor for one disease is also a risk factor for others. The more interesting case is when the pleiotropy works in opposite directions. One example of this antagonistic pleiotropy involves the CD40 protein, a receptor found in many different types of cells and with diverse immune functions. One allele of the gene that encodes this protein increases the risk for rheumatoid arthritis but decreases the risk for getting multiple sclerosis and inflammatory bowel disease. The *PTPN22* gene encodes a protein that modulates the how various T and B cell receptors respond. One allele increases the risk of type 1 diabetes and autoimmune thyroid disease while reducing the risk of Crohn's disease. Such antagonistic pleiotropy helps explain why these disorders are so common: variation can be maintained because the genetic changes that reduce risk for one disease increase it for others.

This antagonistic pleiotropy has implications for clinical trials and clinical treatment: drugs for one disease have the potential to cause the converse disease. For instance, drugs for inflammatory bowel disease may have rheumatoid arthritis as a side effect, especially if the drug is based on the CD protein and other chemicals in the pathway. Clinical trials should be on alert for such side effects.

GETTING THE DOSE RIGHT

GWAS and genetic mapping in general have been useful in pharmacogenomics not just in assisting the development of new drugs but also in the use of drugs. People vary in how they respond to medications, and some of that variable response is genetic. Medications also can have side effects when too much is taken. Thus, being able to tailor dose to one's genotype could be useful.

I recall my grandfather's bleeding hands. Because he had had an artificial heart valve, my grandfather took the blood thinner coumadin (a brand name of warfarin) to reduce clotting. He did so daily for the last decade of his life. He frequently had bleeding in his hands, a common side effect of the drug. Originally developed as a poison to control rodents, warfarin acts by blocking an enzyme that reduces the clotting ability of clotting factors. While warfarin is effective at controlling clots, it has two limitations.[45] First, the range between where it works at stopping harmful clots while not causing bleeding like my grandfather frequently had—the therapeutic window—is narrow. Second, the therapeutic window differs widely among different people: for some people, 1 milligram a day is too much; for others, ten or more milligrams a day is not enough. Hence, getting the proper dose is tricky, and people taking warfarin need to have their warfarin levels monitored regularly. Given that warfarin is the most commonly used anticoagulant across the world, increasing its efficacy and decreasing its side effects would have significant health benefits.

Prior to the use of GWAS, differences in warfarin sensitivity had been linked to variants at *CYP2C9*, a gene whose product is an enzyme that metabolizes warfarin. A study published in 2008 examined whether recommending doses based on the genotype at *CYP2C9* provided benefits.[46] In this small clinical trial, dosing based on genotype was compared with the control group where people were given the standard dose. Providing the warfarin dose based on genotype improved outcomes compared with the control group: those who received a dose based on genotype spent more time with their warfarin dose in the therapeutic window and had less bleeding.

Despite the success of the trial, the National Heart, Lung, and Blood Institute was hesitant about conducting larger trials.[47] Their concern was that other genes might also have considerable effects on the metabolism of warfarin. If relevant variants were missing, then the estimate of the best dose based on genotype would be compromised. Soon after that, a GWAS showed that in addition to *CYP2C9*, the only other genes that have considerable effects on warfarin metabolism are *VKORC1* and *CYP4F2*.[48] Larger-scale trials have since shown that genotype-guided dosing of warfarin has benefits in maintaining warfarin concentrations within the therapeutic window.[49]

WATSON'S APOE STATUS: A TALE OF GENETIC PRIVACY

Apolipoprotein E, commonly known as ApoE, is a protein that transports cholesterol and other lipids. Three different alleles are common in humans: *e2*, which has a frequency of about 10 percent, *e3* at about 70 percent, and *e4* at about 10 percent.[50] The *e4* allele is ancestral: other mammals, including the great apes, have the equivalent of *e4*. In the 1980s, the *e4* allele was shown to increase the risk of atherosclerosis and other cardiovascular disease. It preferentially binds to large lipid particles, leading to an increase in LDL (the so-called bad) cholesterol in the blood. But during the mid-1990s, increasing evidence showed that this allele had an even larger role: it is the leading genetic risk factor for Alzheimer's disease, among other neurodegenerative disorders. People with one copy of the allele have quadruple the risk of developing Alzheimer's compared to individuals with two *e3* alleles. The risk is even worse for those with two copies of the allele: *e4* homozygotes have about 14 times the risk as *e3* homozygotes. Having the *e4* allele also reduces the age of onset by several years.

The damage from the *e4* allele comes from several mechanisms. It is associated with the increase of the amyloid plaques that usually occur with Alzheimer's. Exactly how the allele leads to more plaques is unclear. In addition, the *e4* variant of the protein is linked to neurodegeneration in ways that are independent of the amyloid plaques. For instance, when neurons are damaged, they produce the ApoE protein to help transport cholesterol for membrane repair. The shape of the *e4* variant appears to sometimes be recognized by the cell as abnormal, which causes it to be targeted to the autophagy pathway, where it is chopped up. Some of the fragments from this variant are toxic to the neurons.

Given its severely deleterious effects on both neurodegenerative and cardiovascular diseases, the high frequency of *e4* allele presents an evolutionary puzzle. There are likely several explanations that are not mutually exclusive. First, the effects generally come late in life. Recall from the discussion on the evolution of aging in Chapter 1 that alleles with effects late in life have much less of an effect on reproductive success than those that have effects earlier. Moreover, some research suggests that the *e4* allele can be beneficial early in life. One study found that 2-year-old toddlers with at least one *e4* allele showed somewhat faster cognitive development than their *e3/e3* counterparts.[51] Thus, *e4* could be maintained through antagonistic pleiotropy as predicted by evolutionary theories of aging.

The frequency of the *e4* allele varies appreciably across different populations, which suggests that variation in selection pressure in different areas of the world may affect its frequency. Interestingly, the frequency of the *e4* allele is lowest at about 35 to 40 degrees from the equator: it increases both as you go toward the equator and toward the poles. The latitude of the lowest frequency of the allele corresponds to the latitude where metabolic demands are least. A reasonable hypothesis is that *e4*, which increases cholesterol, is favored in areas where there are high metabolic demands and disfavored where metabolic demands are not as great.[52]

A study of the Tsimane, forager-horticulturists in the Amazon, suggests that the *e4* allele may be protective against the effects of parasitic worms.[53] Because the worm infestation generally reduces the absorption of calories and nutrients, it can affect cognitive abilities. Given ApoE's roles in lipid metabolism and neurogenesis, it also having a role in ameliorating the effects of parasites makes sense. The average burden of worms among the Tsimane is high, but this load varies considerably. Among those with high parasite loads, older adults with *e4* alleles fared better on various cognitive tests than those without the allele. Among those with lower parasite loads, older adults with *e4* alleles did more poorly relative to *e3/e3* homozygotes.

There is a privacy dimension to the *ApoE* story as well. In the early days when a handful of people had their (more or less) complete genomes sequenced, several wished not to know their status at *ApoE*. James Watson, the co-discoverer of the structure of DNA, was among them. Noting that one of his grandmothers had died from Alzheimer's at 83, Watson said: "I don't want to worry that every lapse in memory is the start of something. I'm not afraid of the future, but I don't want to know. Of course, I could be homozygous *APOE4* and still not get Alzheimer's, so . . . it's complicated."[54] This interview was in 2007, so Watson would have been 79 at the time.

Watson had his *ApoE* gene redacted from the database containing his and other's genomes. Soon afterward, Watson was informed of the prospect that people could still figure out his status at the gene. Any guesses as to how this was possible?

If you answered linkage disequilibrium, you are correct and get a gold star. Given information about what variants Watson had at sites near *ApoE* and knowledge of the patterns of linkage disequilibrium in the region, one could infer a probabilistic expectation of Watson's *ApoE* status. Upon learning about this, Watson also redacted

2 million base pairs of DNA that were flanking *ApoE*. This degree of redaction is likely an overreaction as linkage disequilibrium falls off long before 2 million base pairs, but Watson's response is understandable.[55]

People who have had their genome sequenced or (more likely) scanned since 2007 have been more willing to learn their *ApoE* status.[56] There may be benefits to learning whether one has the *e4* allele or not.[57] People who have learned that they have the allele are more likely to use dietary supplements and take other measures to reduce their risk of cognitive decline. Learning one's *ApoE* status, however, should be one's own choice. The researchers who had warned about the ability to infer *ApoE* status from neighboring sites noted sweeping implications of this result:

> We believe the potential for such indirect estimation of genetic risk has considerable relevance to concerns about privacy, confidentiality, discriminatory and defamatory use of genetic data, and the complexities of informed consent for both research participants and their close relatives in the era of personalized genomics.[58]

What are the general implications that linkage disequilibrium could reveal one's status if you give information to an ancestry site? If a marker used in an ancestry site is in strong linkage disequilibrium with a known risk factor, then your allele status at the ancestry-site marker could reveal information about your risk for diseases. Note that the vast majority of the genome is not in linkage disequilibrium with any particular marker, but the few kilobases or tens of kilobases near the marker could well be. These are considerations that should be addressed in dealing with consumer genetics.

LIMITATIONS OF GWAS

GWAS and other genetic mapping tools have brought many successes. The genes these techniques have found are not merely useful in exploring the biology behind the trait; they have also led to practical interventions that enhance the longevity and/or improve the quality of life of individuals with various diseases. We saw how the identification of the *CTFR* gene allowed for the development of useful drugs that have greatly enhanced the life span of people with cystic fibrosis. Similarly, identification of *RPE65*, a gene that affects the degenerative eye disease retinal dystrophy, has led to the development of the gene therapy for the disorder.[59]

Despite its remarkable successes, GWAS and other gene mapping studies still have significant limitations. While some of these have feasible fixes, others are likely to be remain challenges due to the inherent complexities of the genome and biology of humans. As we will see, some of the fixes involve evolutionary biology.

One major challenge of GWAS is that many, if not most, human genetic diseases are like height: lots and lots of variants contribute to the trait, each with minute effects. Given this genetic architecture, predictive power of any given variant is likely to be extremely limited in most cases. A way out is to use polygenic risk scores to predict and assess genetic risk.[60] Here, one would add up all the effects from all the genetic variants that affect the trait in question. These scores along with environmental information can be used to determine one's risk for diseases. Such information can be useful in determining guidelines. Successful examples of the use of polygenic risk scores include deciding whether to screen for colon and breast cancers as well as for whether people should take statins to reduce cholesterol.[61]

A related challenge is the nature of these risk factors. In 2017, a provocative paper "An Expanded View of Complex Traits: From Polygenic to Omnigenic" was published in *Cell*.[62] The paper proposed the omnigenic model: that the genes affecting traits are everywhere and nowhere. It notes that if you were to break up the 3.2 billion base pairs of the human genome into about 32,000 pieces, you would find that most of those pieces contain variants that ever-so-slightly affect height. The same would be true of most traits. Some of the somewhat larger variants are related to core pathways about that trait, but most are not. The omnigenic model states that core genes account for only a small fraction of the overall genetic contribution, with the main bulk being due to numerous background variants. These background variants are likely variants at genes or regulatory regions that are three, four, five, or more steps removed from the core pathways. These third-, fourth-, and fifth-order effects collectively make such a large contribution to traits because there are so many of them: everything is connected to everything.

A third challenge is that GWAS only identifies whether a marker is close to a causal factor; it itself does not affect the trait. Further techniques are needed to pinpoint what genetic changes directly affect the trait. Recall that the human genome has long haplotype blocks that usually stretch for a few kilobases but can be longer. The ironic problem is that these long haplotype blocks that initially made GWAS feasible now limit how well we can get to the actual causal variant.[63] As we will discuss next, additional GWAS with other populations can narrow this window.

A final challenge is the lack of diversity amongst the subjects in GWAS and related studies: the vast majority of subjects are of European descent. In the GWAS performed up to the year 2016, 78 percent of the subjects were of European descent and only 2.4 percent were of African descent.[64] Although there has been greater inclusion of non-Europeans in these studies in recent years, progress in this area has been slow.

Several consequences flow from this lack of representation. One is that the genetic risk calculation for people of non-European ancestry can be incorrect if these individuals are assessed with GWAS that have been trained on data that came from people of European ancestry. Although most common variants that affect disease risk are shared across people with different ancestry from different continents, there are some differences among the groups. For instance, Michelle Kim and Joe Lachance at Georgia Tech found that genetic risk is often miscalculated for people from African countries with GWAS developed from people with European ancestry: Africans have, on average, greater genetic risk.[65] This is opposite of what would be expected; because of the genetic bottleneck that occurred in the out-of-Africa migration that occurred about 100,000 years ago, we would expect somewhat fewer deleterious alleles in African populations. These miscalculations disappear when corrected with GWAS developed with inclusion of subjects from African populations.[66] As Joe Lachance remarked to me: "what you find depends on where you look and what technology you use".[67] GWAS panels that include the populations of interest help improve the quality of the GWAS results.

One example where people of African descent have had been improperly assigned risk is in hyperthropic cardiomyopathy, an excessive thickening of the wall of the heart.[68] This condition can impede the heart's ability to pump blood but has variable symptoms. Genetic variants had been associated with this disorder from studies done with people of European descent. Unfortunately, some people of African descent have markers for the risk factors even though they don't have hyperthropic cardiomyopathy. These people were given the wrong diagnosis and suffered needless anxiety. Had more people with African ancestry been included in the study, the likelihood of misdiagnosis would have been much lower.

Another benefit of incorporating more diverse, and particularly African, populations is that African populations have greater genetic diversity. African populations also generally have less linkage disequilibrium than non-African populations.[69] The reduced linkage disequilibrium allows for better, more fine-scaled mapping, enabling GWAS to get closer to the causal genetic variants. More diversity in our GWAS participants makes the technology more powerful.

NOTES

1. Lander and Schork (1994, p. 2037).
2. Benton et al. (2021, p. 269).
3. Rehm (2017).
4. Jolie (2013).
5. Gessen (2008).
6. Rehm (2017).
7. King et al. (2019).
8. Sabatine (2019).
9. King et al. (2019) and Nelson et al. (2015).
10. Using information from pairs of siblings from contemporary populations of European descent in Australia, the Netherlands, and the United States, Visscher et al. (2007) estimate that that the heritability of height is about 0.86. This means that about 86 percent of the variance in height is due to genetic rather than environmental factors.
11. Zimmer (2018).
12. De Boeck (2020).
13. 2017 Cystic Fibrosis Foundation Patient Registry. 2017. www.cff.org/Research/Researcher-Resources/Patient-Registry/2017-Cystic-Fibrosis-Foundation-Patient-Registry-Highlights.pdf. Accessed January 15, 2021.
14. General information on cystic fibrosis and its history can be found in Trivedi (2020).
15. A dominant trait where those affected did not survive to reproduce would not continue to run in families: it could persist in the population from the influx of new mutations balancing out the loss from selection, but such diseases would be rare. In fact, the frequency of such an allele would be approximately the mutation rate.
16. Rommens et al. (1989).
17. See De Boeck (2020) for references and Trivedi (2020) for the narrative of how these mutations were deciphered and how this led to drug treatments.
18. De Weerdt (2020).
19. Bosch, B. et al. (2017).
20. Poolman and Galvani (2007).
21. Bosch, L. et al. (2017).

22. Collins (1995). The quote is on p. 349.
23. In most plants and some animals, inbred lines can be created where large numbers of the organisms can be generated that all have more or less the same genotype. This allows for more statistical power to detect the QTL. See Young (1996) for a review of QTL mapping of plant disease traits from the mid-1990s.
24. See Slatkin (2008) for a general review of linkage disequilibrium. Some population geneticists have tried to introduce different names for the concept, such as gametic phase disequilibrium. Although these alternative names are better descriptors of the concept, inertia has set in. So we are left with linkage disequilibrium.
25. See Lander and Schork (1994) and Collins (1995) for examples of some verbal arguments about taking advantage of linkage disequilibrium to map traits to genes across the genome. Risch and Merikangas (1996) provided an early model to predict how many markers would be needed to accomplish the task.
26. Kruglyak (1999).
27. Wall and Pritchard (2003).
28. Paigen and Petkov (2018).
29. Johnson (2010).
30. Tam et al. (2019). This number continues to climb rapidly.
31. Visscher et al. (2017).
32. Sharma et al. (2018).
33. Murthy et al. (2014).
34. Graham and Xavier (2020).
35. Caruso et al. (2019).
36. Genotype by environment interactions where the environment is the microbiota is likely to be an emerging feature of evolutionary medicine research, especially in conditions involving the digestive system.
37. Ellinghaus et al. (2016).
38. Singh et al. (2019).
39. Cargill et al. (2007).
40. Cargill et al. (2007) and Visscher et al. (2017).
41. Xu and Cao (2010).
42. Hueber et al. (2010).
43. Sanford and McKeage (2015).
44. See Visscher et al. (2017) for further discussion of antagonistic pleiotropy of variants. Such pleiotropy is found not just in autoimmune diseases but also in neurobiological ones. For instance, variants that affect schizophrenia and autism often display antagonistic pleiotropy for these disorders.
45. Caraco et al. (2008).
46. *Ibid.*
47. Motsinger-Reil et al. (2013).
48. Takeuchi et al. (2009).
49. Jorgensen et al. (2019).
50. See Mahley (2016) for general information on *ApoE* and its variants. Incidentally, the *e2* variant of *ApoE* also is a cardiovascular risk factor. People with it have higher cholesterol because the variant does not bind well to certain receptors. However, the *e2* variant does not lead to increased risk of neurodegenerative diseases; in fact, the *e2* variant is protective in that area.
51. The infants were given the standard Bayley scale of infant development, which examines how well they perform tasks such as following instructions and paying attention to unfamiliar objects, in addition to language and motor skills. See Wright et al. (2003).

52. Latitude predicts *e4* allele frequency better than temperature does, but perhaps latitude is better than the current temperature in predicting the temperatures that our ancestors experienced. See Eisenberg et al. (2010) for details.
53. Trumble et al. (2017).
54. See Angrist (2010) for general information on the story about Watson's *ApoE* status. Watson is quoted on p. 138.
55. See Nyholt et al. (2010) for details about how one's status at *ApoE* can be inferred from information at nearby sites. Because the researchers were careful not to disclose Watson's genotype, they used another individual's genotype, Craig Venter, who had developed some of the new sequencing tools used to sequence the human genome. Venter had disclosed his status at *ApoE*. From the information on nearby sites, the researchers were able to correctly infer Venter's *ApoE* status.
56. See Angrist (2010, p. 150). Even by 2010, clinical trials had shown that people were more positive about learning their knowledge of *ApoE* status than they had been a few years earlier.
57. Marshe et al. (2019).
58. Nyholt et al. (2010, p. 148).
59. Claussnitzer et al. (2020).
60. Chatterjee et al. (2016).
61. Shendure et al. (2019).
62. Boyle et al. (2017).
63. Tam et al. (2019).
64. Gurdasani et al. (2019).
65. Kim et al. (2018).
66. These miscalculations occur because African and European populations differ in the frequencies of alleles, and there is a bias in the alleles that have been associated with disease. These shifts in frequencies can create statistical artifacts when association studies developed with one population are used with another. The greatest statistical power to detect associations occurs when alleles are at intermediate frequencies (not close to either 0 or 1). Given that the GWAS are done with European subjects, the frequency of disease-associated alleles is more likely to be intermediate in European populations than in non-European populations. Compared with the European populations, the non-European populations are more likely to have frequencies near 0. or near 1. for the disease-associated alleles. Remember that the alleles themselves are only in association with the casual variant.

    The African and non-African populations also differ in another respect: because of the genetic bottleneck associated with the out-of-Africa migration, individual Africans are more likely to be heterozygous for alleles that changed since the common ancestor (derived alleles), while European individuals are more likely to be homozygous for these derived alleles. These differences in the genotypes can also affect inference of disease risk. See Kim et al. (2018).

67. Phone interview with Joe Lachance. January 2021.
68. Manari et al. (2016).
69. Gurdasani et al. (2019) and Batai et al. (2021).

6 Cancer
Darwin Meets the Emperor of All Maladies

SUMMARY

While cancer's incidence dramatically increases with age and is exacerbated by mismatch, it is fundamentally different from other diseases because it is a corruption of the normal cellular processes that occur within us. We explore the hallmarks of cancer and then show how these hallmarks are deviations from the "rules" of multicellularity. We also see how cancer has been found across multicellular life. In addition, we explore the apparent paradox: given that cancers share common hallmarks, why is each one so different? Another paradox (Peto's paradox) is the lack of correlation between the size of an organism and its cancer risk: big organisms with many more cellular divisions do not get cancer more often than smaller ones. We explore why elephants have low cancer risk. We then turn to other animals that have unusually high or unusually low cancer risk. These comparative oncology studies could reveal insight into lowering cancer risk in humans. Evolutionary genetics can also be useful in distinguishing between the genetic changes (drivers) that contribute to cancer and those along for the ride. We look at tradeoffs between genetic risk of cancer and neurodegenerative diseases. We also explore two evolutionary approaches to cancer: adaptative oncogenesis (as an explanation for why cancer risk increases dramatically with age) and adaptive therapy (an approach to treatment).

"Cancer is the literal embodiment of evolution. It is evolution in the flesh. We are susceptible to cancer because we are made up of a population of cells that evolves over our lifetimes. Cancer will be here for as long as multicellular life endures on our planet. The sooner we can accept that, the sooner we can use our knowledge to effectively keep cancer under control. We can't win a war against a process of evolution, a process of ecological change in our bodies, a process of cellular free-riding over multicellular cooperation. But we can shape that process so that it is less harmful to us."—Athena Aktipis.[1]

"We can predict the future, albeit imperfectly, while cancers have no foresight. Cancer has no goals or agenda but simply adapts to current microenvironmental demands. We should be able to steer its course if we could learn how to decrease the odds of resistance development and/or manipulate its microenvironment appropriately to favor less malignant phenotypes."—James DeGregori.[2]

WE HAVE MET THE EMPEROR . . .

The oncologist Siddhartha Mukherjee used a description of cancer by a nineteenth century pathologist for the title of his extraordinary biography of cancer: *The Emperor*

DOI: 10.1201/9780429503962-7

of All Maladies. Here, he likened the history of cancer and the treatment of cancer to military history—one in which the opponent is "formless, timeless, and pervasive".[3] A notable example is President Richard Nixon's metaphorical declaration of war on cancer in the National Cancer Act, which he signed into law in February 1971.[4]

Fifty years after the declaration of war, the emperor remains, though somewhat diminished. Despite considerable successes in bringing down the age-adjusted[5] cancer rate, about 600,000 Americans currently die from cancer each year. Moreover, nearly 2 million receive cancer diagnoses each year.

Why is cancer so frequent despite natural selection? Why is it still so pervasive, despite the great deal of money and effort devoted to fighting it? These related questions go to why this emperor remains so powerful, but they also suggest how we can better approach cancer prevention and cancer treatment.

Cancer is largely a disease of older people: about four of five diagnoses are in people aged 55 or above. Because it largely strikes so late in life, selective pressure against cancer-causing has been weak. Prior to the last few centuries, cancer was much less common because few people lived long enough to get it. Moreover, as we will see, genetic variants that benefit us when we are young predispose us to getting cancer when we are older. Cancer also is often a disease of mismatch: our ancestors were not subject to environments that have cancer-causing agents such as tobacco smoking. The combination of these environmental factors (which are largely avoidable) and the largely inevitable progress of aging sets us modern humans up for being more likely to face cancer.

Extended life spans and mismatch are only part of the reason why the emperor of all maladies still reigns. There is a larger reason why cancer is so pervasive, and it's one that sets cancer apart from most other diseases: cancer is the body turning on itself. It is our own cells that are battling us. It is "a caricature of the normal process of tissue renewal".[6] We have met the emperor, and it is us.

Cancer as a caricature—a corruption—of normal cellular processes makes it a fundamental evolution problem. Cancer arises when the cells dividing and evolving within us grow unchecked and, in doing so, engulf our bodies. As we will see, cancer is a consequence of being multicellular. Our predisposition to cancer is also partly due to our genes and our long evolutionary history of dealing with cancer.

Cancer is an evolutionary process, as it arises when some cells acquire differences that they pass on to their descendant cell lineages. It is also an evolutionary genetics process because these differences are heritable changes in DNA sequence. Such differences can be traced using methods developed by traditional evolutionary genetics. Finally, cancer is an evolutionary ecology process because the fate of these altered cell lineages depends on their performance within the microenvironment of their surroundings. Moreover, cancer cells alter this microenvironment analogous to how many organisms—both large and small—alter their environments.

The 2000s and especially the 2010s saw a growing trend of evolutionary biologists turning toward studying cancer, along with a similar trend of cancer researchers turning toward thinking about evolution. The new genomic tools have enabled researchers in cancer genomics new opportunities as DNA sequence data spews out like an open fire hydrant. In some ways, however, conceptual understanding of cancer and its evolution has lagged the outpouring of data. This too is changing: a more sophisticated evolutionary view of cancer is emerging. As Athena Aktipis in the epigraph notes, the martial view of dealing with cancer has its limitations: there may be benefit at least in some circumstances in viewing cancer not as an enemy to be vanquished but as a process to be controlled.

HALLMARKS AND SNOWFLAKES

Cancer is complicated. It comes in numerous types—for instance, sarcomas are cancers of the connective tissue and lymphomas are cancers that emerge from lymphocytes, which are one group of white blood cells in the immune system. Cancers strike different organs. They have different origins and different rates of progression. But cancers have deep similarities as well.

During the 1980s and 1990s, Robert Weinberg and his lab at MIT had discovered several oncogenes, genes that are highly expressed in cancer cells, that act as stimulators for cell division.[7] His lab had also found tumor suppressor genes, ones that are knocked out in cancer cells. The lack of expression of these genes sets up cells to become tumors. In 1998, Weinberg was at a conference in Hawaii.[8] There, Weinberg observed that a great deal of new information about genes and cancer was coming out but that it was not being organized into a coherent framework. Taking a break from the proceedings, Weinberg and a colleague Doug Hanahan walked by the mouth of a volcano. As they walked, they bonded, dissatisfied with the lack of clarity about key organizing principles in cancer biology. They decided to write such a paper on the hallmarks of cancer, one that has been cited tens of thousands of times and has provided clarity and inspiration to the field.[9] A decade later, they presented a follow-up noting progress that had occurred in the ensuing years: it, too, has greatly influenced the field.[10]

So what are these hallmarks? The first is that cancer cells do not require external signals to prepare for cell division. Normally, cells do not go into a state where they devote all their resources into dividing. Instead, they are carrying out the normal housekeeping and other functions of their cell type. Much like a household preparing to move, a cell planning to divide consumes resources and compromises its normal functions. Accordingly, normal cells usually are not in the process of preparing to divide unless they receive signals from outside to do so. One such type of signal is from chemicals—signaling molecules from outside—that bind to receptors on the cells. Another type of signal is the physical interaction with other cells. In any case, there is some external signal. The cancer cell, in contrast, does not need a signal from outside. It provides its own signal, usually an oncogene that mimics the effect of those external signals. Removed from the requirement of external signals to prepare for division, the cancer cell always is ready to divide. It always has the pedal to the metal. It has the need for speed.

The second hallmark is the flip side of the first: the cancer cell does not respond to outside signals telling it to stop dividing. As with the external go signals, the stop signals can be signaling molecules or physical contact. The brakes don't respond, so the cancer cells don't stop. Some of the tumor suppressor genes that Weinberg's lab had discovered are these stop signals. In the two decades since the publication of the first hallmarks paper, cancer researchers have identified numerous go and stop signals; moreover, they also have learned more about the pathways of genes involved in these signals.

Go and stop signals for cell division are not the only signals to which cells respond. When cells show signs of potentially being damaged, internal and external signals build up that promote the cell to destroy itself. This self-destruction, known as programmed cell death or apoptosis, is an important control that multicellular organisms have evolved to ward against the uncontrolled growth of damaged cells. This control is not perfect: cells can evade it. In fact, the third hallmark of cancer cells is that they stop responding to the signals to go into apoptosis. They have internalized go signals; they don't respond to stop signals; they don't respond to the self-destruct signals: this sounds like the set up for a horror movie.

Apoptosis is a controlled, coordinated demolition. Here, the cell withers away. Within an hour or two, the cell breaks down its membranes, its chromosomes, and its components. All that is left is a shrunken hull that other cells engulf and digest within a day or less. In addition to the withering of apoptosis, there is also necrosis, wherein the damaged cell undergoes a more or less controlled explosion, spewing its remains into the microenvironment. To some extent, this process is also under genetic control. As with apoptosis, the presumptive cancer cell does not generally respond to the necrosis signals either. Let's consider the failure to respond to necrosis signals as a subset of the third hallmark.

The fourth hallmark of cancer cells is that they are unconstrained by limits of cell division. At the ends of chromosomes are telomeres, from the Greek word for end, "teleos". With each cell division, part of that telomere is chewed off: hence, there is a limit to how many divisions a cell lineage can undergo. The enzyme telomerase adds repeats to the telomeres, prolonging the reproductive lives of those cells. Telomerase is usually active in only a minority of normal cells. Only some cells possess the ability to keep dividing indefinitely. This is another check our bodies have evolved to thwart cancer. The fourth hallmark of cancer cells is their effective immortality through the continued use of telomerase.

With the first four hallmarks, cancer cells can produce localized tumors. But without a food supply and without the ability to move and invade other tissues, these tumors usually are not a threat to the body. The fifth and sixth hallmarks, respectively, allow these tumors to secure their own food supply (by developing new blood vessels) and to migrate and invade. Angiogenesis, the development of new blood vessels, enables the local tumor to get oxygen and nutrients. Angiogenesis also permits the spread of cells from the original tumor to new tissues. If these migrant cells can invade the new tissues, then the cancer can metastasize, the spread of new cells that start tumors elsewhere.

In their second hallmarks paper, Hanahan and Weinberg added two enabling characteristics to the list. One of the enabling characteristics is genomic instability. In the division of normal cells, mutations are extraordinarily rare. In nearly all cases, having cells evolve changes over the course of the body's lifetime is not desirable. Thus, organisms have evolved multiple systems to make mutation as rare as possible. But these mechanisms are not perfect; moreover, they are subject to mutation themselves. When mutations knock out an anti-mutation system in a cell, that cell and its descendants evolve a higher mutation rate. With more mutations come more chances that one or more of the hallmarks will evolve. Moreover, mutations also lead to more chances that the anti-mutation systems will be knocked out, further increasing the mutation rate. And thus we have a feedback loop that leads to genomic instability. The genomes of advanced cancer cells are often littered with not only simple mutations but large-scale deletions and rearrangements of chromosomes.

A second enabling characteristic is that many tumors are rich with various immune system cells. Originally, these immune system cells were thought to be fighting the cancer cells. While some indeed are, new research reveals that often these immune cells and the inflammation associated with them are instead assisting the cancer cells by providing chemicals that promote angiogenesis as well as growth factors that encourage more cell division. These tumors have apparently co-opted parts of the immune system.

Hallahan and Weinberg also detailed a few emerging hallmarks of cancer. One is that cancer cells have different metabolic patterns than normal cells. Notably, cancer cells, unlike normal cells, engage in aerobic glycolysis. What is aerobic glycolysis? It's the breakdown of glucose into a three-carbon acid called lactic acid in the presence of oxygen. When running a temporary oxygen deficit, our bodies will engage in glycolysis. Have you ever noticed that soon after you run at an intense pace that you sometimes feel soreness in your legs? This soreness is from a buildup of lactic acid coming from glycolysis. But normal cells do not engage in much glycolysis in the presence of oxygen. It's not a very efficient form of metabolism if oxygen is plentiful.[11] Hence, they will damp down the production of glycolysis enzymes when oxygen is abundant.

Why do cancer cells engage in substantial glycolysis even with sufficient oxygen? Robert Gatenby and Robert Gillies at the Moffitt Cancer Center in Tampa, Florida, proposed an intriguing hypothesis based on the evolutionary trajectory of tumors.[12] First, they noted that cancer cells typically experience at least transient periods of limited oxygen in the early stages of tumor progression. These cells often are more limited by oxygen than they are by glucose. These periods of oxygen deprivation select for cells that continue high glycolysis activity regardless of the oxygen conditions. The tumor's continued production of lactic acid alters the microenvironment. Indeed, tumors generally are much more acidic than normal tissues. This acidity alters the selective regime in the microenvironment: cells that can tolerate the acidity are favored. Thus, the cancer cells have altered the microenvironment and have done so to favor themselves and disfavor most normal cells. This alteration of the microenvironment is not restricted to the increased acidity; in fact, it is part of a larger pattern of cancer cells changing their environments.[13] We'll discuss the role of microenvironments in more detail later in the chapter.

During the years between the two papers on "hallmarks of cancer", the trickle of cancer genomes being sequenced and published became a stream and then a deluge. The cancer genomes showed that while there were some genetic changes in common across different cancers, individual cancers even of the same cancer type often differed substantially. Cancers were like snowflakes—no two were the same!

In fact, the snowflake nature of cancers was complicating the use of precision medicine in treating cancer. In recent years, oncologists had started treating cancer based on the molecular signals they possessed and not just the location of the cancer.

For example, patients with melanomas that were due to mutations in the *BRAF* signaling gene were given *BRAF* inhibitors. But even within this class, there was heterogeneity: different cancers with the same primary signal differed in the other signals they had and in how they responded to treatment.[14] Every cancer seemed different like every snowflake was different. Because each patient had a different cancer, generalizing was limited. Although there are ways to conduct clinical trials with these limitations, the heterogeneity amongst patients does present challenges.

If cancers share common hallmarks, why is each one so different? A lesson from the natural world helps explain this apparent paradox. In Chapter 4, we discussed the cavefish, *Astynanax mexicanus*, that lost its eyes and changed in other ways as it adapted to caves. Gatenby and Gillies at the saw parallels between these cavefish and cancer.[15] Cave-adapted populations of this fish from different caves usually are not closely related: in fact, often a cave population is more closely related to a surface population than it is to other cave populations. Cave populations have become adapted to cave life, losing their eyes and gaining larger jaws and undergoing all the other changes, independently and have done so in different ways with different mutations. Indeed, crosses between two different cavefish that both have reduced eyes and are blind often result in F_1 hybrids that have normal (or nearly normal) sized eyes and sight because they evolved the loss of sight with different mutations.[16]

Cancers starting off as snowflakes but then evolving toward the same hallmarks—the same gain of control over the go signals, the same loss of responsiveness to stop signals, and so forth—is an example of convergent evolution, which most often is a sign of natural selection. This natural selection, acting within organismal bodies, channels presumptive cancers to the same suite of hallmarks, just as natural selection channels nearly all the cavefish to evolve the same suite of characteristics.

The power of natural selection to generate such complex adaptations in cancers is a testament to the extraordinarily long duration of the human body from the perspective of a cell. Although the human life span appears short to us, livers of those lives, it is immense to dividing cells. Even the duration of a tumor is a vast time span to those dividing cells. Within a tumor, there could be thousands or tens of thousands of cell divisions. Ten thousand generations, at 20 years per generation, is 200,000 years, about as long as the duration of our species.[17] This is more than enough time for numerous and complex adaptations to evolve. A group of evolutionary biologists who specialize in the evolution of cancer performed back of the envelope calculations and came to the conclusion that

> there are more reproductive events among the cells within one host individual than there have been among individuals in the entire history of the human species. Each such reproductive event is an opportunity for mutation and selective reproduction and, thus, for adaptive evolution.[18]

CANCER'S LONG HISTORY

Cancer is an extraordinarily old disease. In his biography of cancer, Mukherjee tried to determine just how old cancer is. He noted that the famed anthropologist Louis Leakey found a 2-million-year old jawbone of an early hominid that showed suggestive signs of cancer of the lymph nodes. Mukherjee said "if that finding does represent an ancient mark of malignancy, then cancer, far from being a 'modern' disease, is one of the oldest diseases ever seen in a human specimen—quite possibly the oldest".[19]

Mukherjee did not go far enough. Cancer is far, far older than that jawbone. It is hundreds of times, and possibly almost a thousand times, older! Cancer or something that has much of the appearance of cancer has evolved in most of the lineages that have evolved multicellularity.[20]

Cancer can be found across vertebrates from mammals to birds to amphibians to fish. But it is not just in vertebrates. We find cancer in insects. In fact, *Drosophila* has been used as a model system in which geneticists have discovered many genes that also influence cancer in humans. Studies in *Drosophila* have even paved the way for new cancer treatments: the knowledge gained by studying genetic pathways involved in cell–cell communication in these flies led to the identification of new drug targets that can be used to treat cancer in humans.[21] Both annelids (earthworms and their relatives) and platyhelminths (flatworms) get cancer. In fact, Planarian flatworms are frequently used in toxicology studies to test whether specific chemicals are carcinogenic. Even plants get forms of cancer such as galls, though these usually do not metastasize like animal cancers do. Cancer is found almost everywhere we find multicellular organisms.

Why would this be? Let's consider how cells should behave in a multicellular organism. You could call this the rulebook for living as a multicellular organism.[22] For multicellular organisms to persist, its cells should only divide when they are supposed to. Mechanisms would evolve for the cells as a community to send go and stop signals for cells to divide or not. Another rule of this code is that cells should self-destruct if they are if are compromised. Signals and responses to these signals would evolve. A third rule is that cells should share resources and not take more than their appropriate share. Cells should also do their assigned job and take care of the surrounding environment.

If these rules sound familiar, it's because they are opposite of the hallmarks of cancer. Cancer cells violate the rules. They grow out of control, failing to stop when stop signals are given. They fail to self-destruct when compromised. They hog resources, alter their metabolism, and damage the environment. And then they sometimes break out of their local area; and like crime syndicates, they spread to other areas and run amok there too. Cancer cells are outlaws. They are cheaters.

The rulebook for living as a multicellular organism did not come about because of conscious thought. Cells don't think. Cells don't sign agreements. Instead, the rulebook has come about through a process of natural selection: multicellular organisms whose cells "followed" the rules did better than those that did not. But no lineage of organisms has evolved a perfect rulebook: cheaters persist. Natural selection is not all-powerful. Trying to stop all the cheating of cells, like trying to stop all cheating by students at a school, will likely come at a high price. As we will see, some lineages have evolved more anti-cheater defenses than others.

Note that if a cell lineage gains a mutation that causes these cells to subvert the rules, this lineage has a selective advantage. Breaking the rules has its perks. But it harmful to the rest of the body; controlling its cells is good for the organism. Here, we have a conflict between what is good for a cell lineage and what is good for the organism.

PETO'S PARADOX AND THE ELEPHANT GENES IN THE ROOM

Given that cancer is a disease that arises from failures of controls at cell division, you might think that animals that undergo more rounds of cell division would be more prone to get cancer. In other words, bigger animals should get cancer more often. You would be correct if you looked within species. Bigger humans are more likely to get cancer than are smaller ones. Bigger dogs too. But across species, that relationship does not hold. Larger species of mammals are not more likely to get cancer than smaller ones.[23] This peculiar lack of relationship has been dubbed Peto's paradox, after Richard Peto, who observed in the 1970s that humans and mice have comparable rates of cancer despite humans being so much larger than mice and living for much longer. Because we humans undergo many more cell divisions and yet are not riddled with cancer at an early age, we must have evolved mechanisms to suppress cancer (or mice have evolved less suppression of cancer) since humans and mice diverged from a common ancestor about 100 million years ago.

Peto's paradox should also extend to animals much larger than us. Blue whales are about a thousand-fold larger than humans. If they got cancer at the same per-cell rate that humans do, we would expect more than half of whales to get colon cancer by the time they reach the age of 50 and practically all of them to get it by 80. But they don't. Thus, we would expect whales (and other elephants and other big animals) to have evolved stronger mechanisms to suppress cancer. Examining these animals ought to lead a better understand of how cancer can be suppressed. Moreover, this information could be used to develop drugs to treat or prevent cancer in humans.

In the 2010s, news came out that elephants have an unusual number of *TP53* genes. This finding intrigued biologists interested in the intersection of cancer and evolution. *TP53* is not just a cancer gene; it is arguably *the* cancer gene.[24] It acts both as a stop signal and as self-destruction signal. In some circumstances, it can even act as a go signal. Because it performs several functions related to the cancer hallmarks, *TP53* has been called the guardian of the genome, Moreover, *TP53* is one of the most frequently mutated genes in most types of human cancer.

Lisa Abegglen at the University of Utah studied Li-Fraumeni syndrome, a genetic syndrome that greatly predisposes those with it to many kinds of cancer including bone cancer, leukemias, and brain tumors.[25] People with Li-Fraumeni syndrome most often have one functional and a nonfunctional *TP53* gene. Note that *TP53* is so important that losing the function of just one copy greatly increases cancer risk. Loss of both is fatal. Abegglen collaborated with Carlo Maley, an evolutionary biologist with interests in the evolution of cancer, who had been involved in the elephant *TP53* study.

Abegglen, Maley, and their colleagues first took a more extended examination of Peto's paradox. With access to Utah's Hogle Zoo and other facilities, they were able to collect data of the incidences of cancer for a diverse group of mammals.[26] They then looked for a relationship of cancer incidence with an estimate of the number of cell divisions each mammal would undergo during its life span.[27] Consistent with

Peto's paradox, cancer incidence did not increase with the number of cell divisions. In fact, the relationship was slightly negative, though this could have been due to chance. And yes, elephants seldom got cancer.

This group also confirmed that elephants had many copies of *TP53*: each elephant had 40 copies (20 from each parent). But what were all those TP53 copies doing? Did the added copy numbers have functional significance? To address these questions, Abegglen and colleagues subjected elephant cells, typical human cells, and cells from people with Li-Fraumeni syndrome to intense radiation. Upon radiation damage, cells from people with Li-Fraumeni syndrome were less likely to self-destruct through apoptosis than typical human cells after the same dose of radiation. The loss of one *TP53* copy made the Li-Fraumeni cells less likely to self-destruct upon damage. It makes sense that these cells are more prone to become cancerous: they don't follow the rules for being multicellular as well. In contrast, elephant cells with their extra *TP53* copies were more likely than typical human cells to self-destruct when damaged.

The amplification of *TP53* genes is part of the reason why elephants don't get as much cancer as we would expect given their large size and long life span. It is not the only change responsible. Vincent Lynch at the University of Chicago and his lab found that elephants have regained the function of the leukemia inhibitory factor pseudogene (*LIF6*), which is activated by TP53 signal to prompt damaged cells to self-destruct.[28] Recall from Chapter 4 that pseudogenes are genes that have lost their function. Ordinarily, pseudogenes do not regain function, but somehow *LIF6* did. Lynch's lab also found that elephant cells were more likely to self-destruct when damaged than cells from their small close relatives, such as the East African aardvark.

The giantism of elephants and their closest extinct relatives evolved relatively recently. Elephants belong to the Afrotherians group of mammals, most of which are small. For examples, elephant shrews and aardvarks weigh only about 60 grams (2 ounces) and 170 grams (6 ounces), respectively. Lynch's group found that many of these much smaller Afrotherians have duplicated regions of the genome that contain cancer-suppressor genes.[29] To be sure, elephants and their extinct close relatives have even more of these regions. But it is interesting that the smaller Afrotherians have some. Perhaps the presence of these duplications in ancestral Afrotherians permitted the lineage that led to elephants being able to get larger without taking on a huge risk of cancer.

Whales and dolphins have also evolved tremendously large bodies. They, too, seldom get cancer. Their resistance to cancer, however, is not due to *TP53*. Humpback whales (*Megaptera novaeangliae*) do not have multiple copies of *TP53*.[30] Humpback whales and other large whales do have rather large parts of their chromosomes that have been duplicated. Some of these segmental duplications contain tumor suppressor genes as well as genes involved in sending self-destruct signals. Elephants and large marine mammals evolved the ability to suppress cancer through some similar mechanisms but with different genetic changes.

Within the evolution of cancer suppression in marine mammals, certain genetic changes have occurred independently multiple times. For instance, consider *PRDM13*,[31] a gene that damps down the expression of other genes and appears to play a role in cancer suppression. The exact same change at this gene happened in humpbacks, orcas, dolphins, and several other lineages. The same change occurred so many times—independently—that this pattern is extraordinarily unlikely to occur just by chance. As such, it strongly suggests that the change was driven by selection, likely having something to do with cancer suppression.

COMPARATIVE ONCOLOGY: BEYOND PETO'S PARADOX

These studies in elephants and whales do not just show the mechanisms behind Peto's paradox; they also provide insight into natural, evolved processes by which cancer can be thwarted. Moreover, these studies may also lead to cancer therapies. But they have generally been haphazard. Cancer research, until recently, has not taken much advantage of the full range of the evolutionary history of cancer.

Working as a postdoc with Carlo Maley and Athena Aktipis in the middle 2010s, Amy Boddy was surprised by how little we knew about cancer in other animals.[32] She told me that while we knew about Peto's paradox and that life history theory would predict faster-reproducing animals ought to have more cancer, there was not much hard data out there. This was a wide-open area. Thus, Boddy set about contacting zoos to get more data on cancer across different mammals to look for general patterns.

Boddy and her colleagues found no correlation between cancer incidence and body size, consistent with both the smaller data set in the prior study led by Lisa Abegglen and with Peto's paradox. The most striking result was that cancer risk greatly increased with litter size: species that produced many offspring at the same time were at a high risk for getting cancer.[33]

Why would cancer risk increase with litter size? Boddy and her colleagues hypothesize that mammals with large litters need to devote a great deal of their resources to their offspring both while developing within the mother and when they are feeding on the mother's milk. The young of these mammals would be selected to grow quickly. Given these demands, they would not be able to devote much to defenses against cancer. Genetic variants that increased responsiveness to go signals for cell division or those that decreased responsiveness to stop signals and self-destruct signals would promote early growth, but at the expense of increased cancer risk. Such variants would be more likely to be advantageous in a large litter. Grow fast; die young. In contrast, such variants would not be advantageous and may even be disadvantageous when there is only one hungry mouth to feed at a time.

Looking at individual species and their cancer incidence, there are a few surprises. Boddy found that elephants do get cancer at reasonably high rates. But these tumors are usually benign; moreover, the rate is still much lower than what would expect if they had the same per-cell-division risk of cancer as humans do. Interestingly, Asian elephants (*Elpephas maximus*) get cancer more often than either species of African elephants: the African forest elephant (*Loxodonta cyclotis*) and the African bush elephant (*L. africanus*). Why the African elephants have less cancer is not quite clear. We know that they have roughly equal numbers of *TP53* genes, so the answer for the difference in susceptibility probably lies elsewhere.[34]

Tasmanian devils (*Sarcophilus harrisii*) are marsupials found only in the island state of Tasmania, Australia. They face extinction in part because large numbers of them are infected with two different odd forms of facial tumors that get transferred from one individual to another. These carnivores frequently bite each other on the face; in doing so, they pass on the cancer. That the devils have not just one but two of these transmissible cancers is extraordinarily unusual, given that such transmissible cancers are exceedingly rare. The devil having two transmissible tumors is like getting struck by lightning twice! These transmissible tumors need to evade the immune systems of the animals they find themselves in. While the low genetic diversity of the Tasmanian devils is likely a factor in the ability of the cancers to be able to spread, the genetic diversity of the devils is not extraordinarily low, as the devils will reject skin grafts from other individuals. The tumors themselves appear to have evolved mechanisms to avoid the immune systems of their hosts.[35]

Tasmanian devil (*Sarcophilus harrisii*) taken at Tasmanian Devil Conservation Park in Taranna, Tasmania.

Source: Creative Commons https://commons.wikimedia.org/wiki/File:Tasdevil_large.jpg.

Another plausible reason why these cancers have arisen and why the devils are susceptible to them is that the frequent biting that the devils engage in has led to evolutionary changes in the wound-healing processes in the cellular tissues around the face. As we will discuss, wound-healing responses are associated with cancer; thus, changes in these processes could lead to increased predisposition for cancers of those tissues.

Boddy's study found that Tasmanian devils in zoos, which have not been exposed to the transmissible cancers, are also highly prone to cancer. Exactly why they are is not known, but further genetic and evolutionary genetic study is warranted.

In stark contrast to the cancer-prone Tasmanian devil, there are some animals that are extraordinarily resistant to cancer.[36] One is the subterranean-dwelling naked mole rat (*Heterocephalus glaber*). These unusual creatures are the only mammal known where nearly all the females in the colony forgo reproduction to raise the young of the queen. This type of reproduction is called eusociality. While frequent among bees (see Chapter 10) and ants, eusociality is exceedingly rare outside of those groups. Naked mole rats are also unusual in other respects: for instance, they can live for more than 20 years, which is much longer than one would expect for mammal of their size.[37]

Part of the reason why naked mole rats live so long and usually avoid cancer is that they have an unusual version of hyaluraonan, a substance that cells excrete into the spaces between cells. In these spaces, which is called the extracellular matrix, the hyaluraonan assists in the communication between cells. In humans, a buildup of hyaluraonan is associated with poor cancer prognosis. The hyaluraonan in naked mole rats is much more stretchy than that of other animals; in fact, it has been compared to a ball of rubber bands.[38] This additional stretchiness of the hyaluraonan appears responsible for enhanced elasticity of the mole rats' skin as well as in preventing tumor invasion of tissues. Continued investigation of the material properties of hyaluraonan could be useful in the development of artificial skin and in anti-cancer treatments.

Comparative oncology can also be useful in shortening the time needed to develop new cancer drugs, a process that usually is extraordinarily expensive and time-consuming.[39] From the time that basic scientists discover a potential drug to its approval averages about a decade; during this time, about a billion dollars is consumed. Moreover, at each step along the way, many drugs fail to meet standards despite all the investment placed in them. One reason for the failures, the time, and the expense is that nearly all the development of cancer drugs is done in two animals: humans and mice. Unfortunately, a wide gulf exists between the environments experienced in the mice of mice models and those experienced by people with cancer. In mice models, the typical procedure is to use a skin graft to implant a cancer cell line onto an inbred mouse that has a compromised immune system. These mice also have a much different diet than we do. Moreover, these studies usually do not consider the tumor microenvironment.

A promising avenue is the use of pet dogs in clinical trials.[40] Here, owners of dogs with cancer enroll their pets in the trials, which proceed much like clinical trials in humans. Cancer is common in pet dogs: almost half of dogs over age 10, and a quarter of dogs overall died of cancer. Cancer in dogs is much like that in humans, with some of the same genes affecting cancer in both and in similar ways. The progression of cancer in dogs, however, is usually faster than that in humans. The accelerated progression, while tragic, does allow for more rapid clinical trials.

There are some advantages in using dogs for mapping traits to genes in genome-wide association studies (GWAS). Recall from Chapter 5 that GWAS relies on linkage disequilibrium (correlations of the frequencies of genetic variants) between markers and the genetic variant that actually affects risk. The patterns of linkage disequilibrium across breeds of dogs are about the same as that in humans. But linkage disequilibrium within breeds persists for much longer stretches of the chromosome because of the genetic bottleneck associated with the formation of the breed. This within-breed/across-breed dichotomy facilitates mapping of traits (including cancer predispositions) to genes in GWAS studies. The extended linkage disequilibrium within a breed makes it easy to get associations between markers and causative variants. Using the shorter linkage disequilibrium across breeds, these associations can be more finely mapped. It's like having a map with two levels of resolution.

One success story is the drug ibrutinib, which is used to treat various lymphomas. It was first tested in dogs before it was used in human clinical trials. After successes in clinical trials, it was granted FDA approval in 2014. Even failures in dog trials can be useful: because dog trials are quicker and less expensive than those in humans, a failure in a trial in dogs quickly shows that this drug probably would not be useful in humans. These fast failures work as a rapid triage. Running clinical trials in pet dogs benefits both dogs and people: dogs get access to the most up-to-date cancer treatments, and we humans can speed up drug development.

DRIVERS AND PASSENGERS

Cancer researchers have taken exceptionally good advantage of the explosive rise in genomic sequencing capacity to produce an astonishing and still rapidly growing amount of DNA sequence data taken from tumors. This sequence information has enabled them to point to genetic changes that have arisen during the evolutionary process that cancer cells undergo within the body. Some of these changes are simple nucleotide substitutions—spelling errors—the change of an A to a T, for instance, at a particular location. Other changes are small insertions or deletions in the sequence. Perhaps in the copying, three nucleotides were deleted. Depending on where the changes take place, the effect could be large—for instance, knocking out the function of a gene—or nonexistent or anywhere in between.

Note that loss of the function of a gene could make a cell lineage more likely to take on the characteristics of cancer and to increase its cellular fitness. Knocking out the function of a tumor suppressor gene, like *TP53*, would cause cells with that change to speed through checkpoints without delay and to complete more rounds of replication in the same period of time. Such a lineage would likely outcompete lineages that have a functional *TP53*. This change is evolutionarily beneficial to the lineage, though potentially deleterious to the person bearing the cell lineage.

Changes larger than the ones discussed here, involving larger parts of the sequence, can also occur. Large deletions, removing many genes or even a whole chromosome, could happen. Duplications of a large stretch of sequence are also possible.

Sometimes, large chunks of a chromosome carrying numerous genes get moved from one chromosome to another. This is called a translocation. Sometimes, large chunks of the sequence get turned around in orientation. This is an inversion. With the possible exception of the inversions, these larger structural changes are much more likely to have phenotypic effects than the smaller "spelling error" changes.

Most of the changes, whether they are the single nucleotide spelling errors or the much larger structural changes, do not cause the nascent tumor to be more cancerous or increase the cellular fitness of the cells that contain them. The ones that do—the drivers—are of particular interest to cancer biologists because these are the changes that are most likely to affect the clinical outcomes. Identifying the driver genes also helps to inform us about the biological underpinnings of cancer, and this could lead to new therapies.

As in many fields, the major problem is distinguishing signal from noise. In fact, statistical analyst Nate Silver wrote a book called *The Signal and The Noise* that discussed the science of prediction in areas that ranged from election prognostication to weather forecasting to sports competitions to poker.[41] Inherent in the science of prediction is the need to build models based on real-world understanding and data of the problem at hand. Evolutionary biologists have much to offer regarding detecting signal in the molecular genetic data coming from cancer genomes. In fact, the idea that neutral or even deleterious alleles may come along for the ride—owing to their association with the drivers and the genomic instability often present in cancers—is a well-established evolutionary idea.[42] Evolutionary biologists have developed an ever-growing set of tools to detect selection, some of which can be applied to cancer.[43]

A logical presumption is that genetic changes that frequently appear in cancers are drivers. If, for instance, the nucleotide A changes to a T frequently in a specific gene at, say, position 564, then this change looks like a driver mutation. The rationale for this inference is that selection is likely due to lead to the same types of changes over and over again. Convergent evolution is a general signal of selection. But this presumption assumes that mutations are roughly equally likely. What if some mutations are much more likely than others?

Recent work found that some mutations are indeed substantially more likely to occur than others. One such case is those mutations induced by the RNA editing enzyme APOBEC3A. This enzyme typically works on RNA—a single-stranded molecule—but will occasionally also work on DNA. The two strands of DNA are almost always held together by bonds, but they separate briefly during DNA replication. While transiently separated, one strand can form bonds with itself, thus generating hairpin loops, named thus because they look like little pins. Here's where APOBEC3A comes in. The enzyme has an affinity for those hairpin loops; here, it will generate mutations. Because certain regions of genes are much more likely to form hairpin loops, these regions are more prone to mutagenesis from APOBEC3A.[44] Accordingly, many genetic changes that repeatedly pop up in examinations of cancer genomes are not driver mutations that increase cellular fitness but mutational hotspots that arise from the structural properties (the hairpin loops) of the DNA. Caution needs to be taken with assigning mutations to driver or passenger status.

Vincent Cannataro and Jeff Townsend at Yale University recently developed a systematic framework to determine the most important drivers of cancer progression.

They estimate the effect size of a cancer-promoting mutation while taking mutation rate into account. Given the frequency of the mutational change and the mutation rate, they can estimate how much the genetic change helps cancer cells replicates. This effect-size metric is akin to the selection coefficient evolutionary biologists have long used in predicting how the frequency of an allele will change over time.[45]

Cannataro, who is presently on the faculty at Emmanuel College in the Boston area, often begins seminar talks with a quote from Mukherjee's *The Emperor of Maladies*. In this passage, Mukherjee discusses how women of different times were treated for breast cancer. Cannataro quotes the last section where Mukherjee jumps to the future and talks about the woman of 2050, who

> will arrive at her breast oncologist's clinic with a thumb-sized flash drive containing the entire sequence of her cancer's genome, identifying every mutation in every gene. The mutations will be organized into key pathways. An algorithm might identify the pathways that are contributing to the growth and survival of her cancer. Therapies will be targeted against these pathways to prevent a relapse.[46]

The algorithms that Cannataro and others are developing, combined with the outpouring of data from cancer genomes, may make Mukherjee's prediction come true well before 2050!

Cannataro's framework also helps to address the ongoing debate as to the extent that various cancers are due to intrinsic (genetic) factors versus external (environmental) ones. Some studies have estimated that the intrinsic genetic factors contribute only modestly to lifetime cancer risk, while others reached higher estimates.[47] Resolving this debate is not just of academic interest but also informs public health, as some of the environmental factors are presumably avoidable.

If we know that one driver associated with a specific cancer is likely to be due to naturally occurring replication errors but another is mainly from a chemical mutagen, we can estimate the extent to which cancer types are caused by environmental factors by using Cannataro's method. For instance, consider thyroid adenocarcinoma.[48] More than half of the mutations that are found in this cancer come from the APOBEC process described earlier. This mutation is often caused by viral infections, which could at least in principle be prevented. But while these mutations account for more than half of the mutation weight, they only account for about 3 percent of the cancer effect. So even if all APOBEC processes were due to viral infections and all of these could be prevented, little would be done to reduce the incidence of this cancer: 97 percent of the cancer would still exist.

Lung cancer is a different story. Just looking at the mutation weight of different mutations underestimates the true impact of environmental (mostly tobacco use) causes that would be captured under the effect. Even more of lung cancer could be prevented based on Cannataro's method. Cannataro told me that he is really excited about this work because it exists at the interface of evolutionary biology, epidemiology, and public health.[49]

The early work in cancer genomics generally neglected the importance of passenger mutations, tacitly assuming that they were just noise. What if they are not just noise? What if some of the passenger mutations have some deleterious effect on the cell lineage's ability to grow? Consider the analogy of rowdy children. One rambunctious child will have little effect on the ability of the adult to drive the bus. Ten of them complicate the job of the driver. Assuming that there are no other adults to chaperone the kids, as the number of children grows, the performance of the bus declines.

A study that modeled the evolution of tumor growth allowed for driver mutations (ones that increase the fitness of the cell lineage), neutral passengers (ones that had no effect on the fitness), and deleterious mutations (ones that decreased the fitness). The deleterious mutations were allowed to come with different strengths: some had very tiny effects, some had very large effects, and some had intermediate sized effects. Those with the intermediate sized effects—about 0.1 to 1 percent selective disadvantage—were the ones that accumulated in cell lineages and had sufficiently deleterious effects. The ones with larger effects did not accumulate. The ones with much smaller effects did not, in aggregate, pose a noticeable hindrance to the cell lineage's progression to cancer.[50]

Examining these rowdy passengers can possibly inform treatment. There are two general strategies to increase the burden of the rowdy passengers. One is to increase the overall mutation rate. This would increase the overall rate at which the rowdy passengers accumulate in cell lineages. The problem with this approach is that to be effective, the increase in the mutation rate would have to be very large, on the order of a 50-fold increase. Another strategy would be to increase the deleterious effects of the preexisting rowdy passengers.

GENOTYPE BY ENVIRONMENT INTERACTION AND CANCER

Cancer arises not just from genes—whether these be genes predisposing one to cancer in the germline or sporadic changes that happen to cells—and it arises not just from the environment. It often arises from an interaction between genes and the environment.[51] Next, we will examine a few examples of such genotype by environment interactions.

Pancreatic ductal adenocarcinoma is one such cancer that illustrates this interaction.[52] This cancer, which makes up more than 90 percent of pancreatic cancers, is the fourth leading cause of cancer death in the United States. It was the cause of death of the astronaut Sally Ride, the jazz musician Count Basie, and the actors Michael Landon and Patrick Swayze. Pancreatic ductal adenocarcinoma is ruthlessly aggressive and has a low survival rate: fewer than a tenth of those diagnosed with it surviving five years. Part of the reason it is so deadly is that is resistant to most treatments. Its incidence is increasing, in part from the rise in type 2 diabetes.

To see how a gene, through environment interactions, can lead to pancreatic ductal adenocarcinoma, we need to take a glance at its basic biology and that of the pancreas. First, adenocarcinomas are cancers that arise from epithelial cells, quickly dividing cells that form the lining of various surfaces across the body. In the pancreas, these epithelial cells, which are also called ductal cells, form the lining of the duct that leads from the pancreas to the duodenum, the first part of the small intestine. In addition to the ductal cells, the pancreas has acinar and endocrine cells. The endocrine cells produce insulin and other hormones. It is the acinar cells that we are most interested in. Ordinarily, these cells produce digestive enzymes. The cells, however, are plastic. Upon certain signals such as tissue damage or inflammation, the acinar cells can become more like the ductal cells. During this transformation, which is usually reversible, they are most susceptible to turning into proto-cancer cells, especially if they have a mutation at an oncogene.

Persistent inflammation of the pancreas, known as chronic pancreatitis, increases risk for getting pancreatic cancer.[53] Chronic pancreatitis by itself is not pleasant; it often causes pain and reduces both length and quality of life. Heavy use of tobacco and of alcohol are risk factors for chronic pancreatitis. In addition to these environmental factors, chronic pancreatitis also can run in families. Moreover, some cases of this hereditary chronic pancreatitis have been pinpointed to a few genes. In addition, the risk of getting cancer from chronic pancreatitis is especially strong in the hereditary form.

A recent study provides a mechanism to explain how inflammation of the pancreas may lead to cancer in some cases but not others.[54] Mutations in the *KRAS* oncogene leads to pancreatic cancer in mice when there has been wounding. The combination of the *KRAS* mutation and the wounding causes changes in the proteins associated with chromosomes (known as the chromatin). Such chromatin changes cause changes in the expression of genes. These epigenetic[55] changes promote progression to cancer.

Another case of genotype by environment interaction occurs in mesothelioma, cancer of the mesothelium, which is a thin layer of cells that lines many internal organs. Because therapies usually are ineffective and because this cancer is aggressive, people diagnosed with mesothelioma usually have a poor prognosis: about half die within a year from diagnosis. Fortunately, it is rare in the United States, with only about 2,500 new diagnoses a year, mostly from exposure to asbestos. In Cappadocia in central Turkey, some villages have an extraordinarily high incidence of mesothelioma, with nearly half getting it.[56] In Cappadocia, the cancer runs in families. Most likely, it arises there from the interaction of exposure to erionite, a fiber akin to asbestos, and some unknown genetic factor.

Inspired by considering the role of the microenvironment plays in the evolution of cancer, James DeGregori at the University of Colorado School of Medicine proposed an intriguing and probably correct theory of cancer evolution. In DeGregori's adaptive oncogenesis theory, cancer is a disease mainly of old age because the microenvironment that cells experience deteriorates with age and tissue damage.[57] This deterioration alters the nature of selection, giving genetic variants that predispose to cancer a selective advantage. The microenvironment in young mammals, including humans, is usually stable, and the cells are well-adapted to it. As a result, most mutations in that young microenvironment are either neutral or deleterious for the cell. Thus, stabilizing selection—that is selection to maintain the same phenotype—persists during the time of that microenvironment. In contrast, when the microenvironment is destabilized either by age or by environmental insult, there is selection on the cells for new phenotypes. This directional selection favors mutations that promote rapid cell division and violation of the other hallmarks.

Some experimental evidence supports this adaptive oncogenesis theory. For instance, introducing oncogenes, such as *BCR-ABL* and *MYC*, to the blood marrow stem cells of young mice does not usually result in leukemia. But introducing the same oncogenes into those cells in older mice does. The difference is that the microenvironments are different: in the microenvironment of the old mice, the oncogene-fueled cells have a selective advantage and start expanding; in the microenvironment of the young mice, those cells do not have much of a selective advantage. It turns out that the nature of the microenvironment matters more than the age of the cells. Oncogenes introduced into old cells in a young microenvironment do not usually result in leukemia; in contrast, oncogenes introduced into young cells in an old environment do!

Inflammation is a key factor to the deteriorating microenvironment. It increases with age as well as with environmental damage. For instance, while smoking does accelerate the accumulation of mutations in cells, inflammation is likely a larger factor in the link between smoking and cancer than the increase in mutations. Cancer biologists are discovering how smoking triggers inflammatory responses and damages the microenvironment in other ways. Viral infections can also increase inflammation: chronic inflammation from our immune system fighting with viral infections could explain why immune system cells sometimes promote cancer, as Hallahan and Weinberg discussed in their second hallmark paper.

That inflammation and other damage can promote cancer and general aging suggests that we can fight these by measures that prevent, reduce, or reverse this damage. DeGregori warns that we need to consider pleiotropy—genetic variants having more than one effect—and tradeoffs when implementing such measures. Nonsteroidal anti-inflammatory drugs like ibuprofen and naproxen may have benefits in cancer prevention and treatment, but they have their own side effects. Moreover, inflammation can be useful under certain circumstances. Nonetheless, adaptive oncogenesis suggests looking at cancer from the broader perspective of the consequences of a degraded microenvironment may have practical benefits.

CANCER AND TRADEOFFS

Let's take a closer look at cancer and tradeoffs. We are predisposed for cancer in part because of tradeoffs that emerge from pleiotropy. We could have evolved a genome that was less prone to cancer, but that would come at a cost. We can see some of these costs when we look at predisposing genetic variants. *BRCA1* and *BRCA2*, which are both among the first genes linked to cancer and ones with the strongest of effects, have physiological effects outside of cancer. Mutations at these genes greatly increase the risk of breast and ovarian cancer in women and the risk of prostate cancer in men. In most studies, they are tied to increased fertility. For instance, in a large survey of women in Utah, women who were carriers for the mutation had more children than those who were not carriers.[58] In women born before 1930 and thus unlikely to have much access to modern contraception, carriers had just over six children while the control women had only about four. This effect persisted but was smaller with women born after 1930 and thus likely to have had access to contraception. The carriers continued to have children later in life and had a shorter interval between births. An intriguing mechanism for this difference is that *BRCA* mutations are associated with longer telomeres, which allows women with these mutations to continue having children longer and to be able to reduce the interval between births. Other genes such as *TP53* also have pleiotropic functions in reproduction.[59]

In Chapter 5 we discussed how genetic variants can be associated with increasing risk of some diseases like rheumatoid arthritis but decrease the risk of others like inflammatory bowel disease. Several groups of diseases show similar negative genetic associations. These diametrical diseases can reveal tradeoffs.[60] Increasing evidence shows that cancer and neurodegenerative diseases, such as Alzheimer's disease and Parkinson's disease, are diametrical diseases.

Substantial epidemiological evidence supports the diametric relationship between the neurodegenerative diseases and cancer. People who get Parkinson's and Alzheimer's are at considerably less risk of most cancers.[61] The relationship between cancer and other neurodegenerative diseases is less clear, but suggestive evidence points to cancer and Huntington's disease also being diametrical diseases.

Some of the same genes that are involved in cancer, such as *TP53*, also appear to affect neurodegenerative diseases.[62] For instance, p53 is upregulated in mice with neurogenerative diseases. Autopsy reports show also that people who die with Alzheimer's and Parkinson's disease have higher concentrations of p53 in their brains. Moreover, p53 is elevated in neurons that self-destruct through apoptosis after brain injuries. These results suggest that p53 may be overexpressed during the pathological self-destruction of neurons. Perhaps overexpression of p53 is protective of cancer but leads adult neurons to be more likely to self-destruct upon damage. Given that adult neurons usually do not divide, the losses of these neurons could lead to neurodegenerative diseases.

The diametrical nature of cancer and neurodegenerative diseases has some clinical implications.[63] Treatments for one that act on the same pathways implicated in the other may lead to complications. For instance, chemotherapy often has side effects on cognition that appear to be due to neurodegeneration. Perhaps part of the reason for these side effects has to do with this diametrical relationship.

Cancer is a disease of failure to stop replicating and the failure of damaged cells to self-destruct. Neurodegenerative diseases appear to be diseases of the inappropriate self-destruction of neurons. But even with this diametrical relationship, cancer and neurodegenerative diseases also are diseases of aging and insult. Chronic inflammation and other damage to the microenvironment, DNA damage, and other pathologies increase the risk of both cancer and the neurodegenerative diseases. These co-pathologies are partly a function of inevitable aging but also partly due to potentially avoidable environmental factors.

In conclusion about tradeoffs, it is important to remember that tradeoffs themselves are malleable and can change over time. Environmental conditions can tip the balance of these tradeoffs. For instance, when the tradeoff is based on how energy is allocated, increased resources and less competition lead to more resources geared to cancer defense.[64] Knowledge about how tradeoffs work and how they can be tweaked could be exploited to treat and manage cancer.

RESISTING CANCER DRUG RESISTANCE

Incremental advances can add up to transformative changes. In 2005, an avalanche of papers cascading through the scientific literature converged on a remarkably consistent message—the national physiognomy of cancer had subtly but fundamentally changed. The mortality for nearly every major form of cancer—lung, breast, colon, and prostate—had continually dropped for fifteen straight years. . . . The empire of cancer was still indubitably vast—more than half a million American men and women died of cancer in 2005—but it was losing power, fraying at its borders.[65]

In the decade since Murkherjee's book, cancer incidence and deaths have continued to fall.[66] Since the early 1990s peak, cancer deaths have fallen by nearly a third. This victory against the empire of cancer has had many heroes. Fewer people smoking led to a decline in lung cancer incidence. Earlier detection brought on by better screening improved survival for those with colon cancer and cervical cancer. But the increased survival for patients with several types of cancers, including leukemia and lymphoma, came from new chemical agents. This last example of progress was the outgrowth of basic biomedical research starting in the 1980s with the discovery of oncogenes and tumor suppressors.

Understanding the biology of these go and stop signals, as well as genes and pathways of the other hallmarks, paved the way for new drugs that could be used against cancer. One such game-changing drug is imatinib, sold under the brand name Gleevec. This drug, developed by the Dana Farber Cancer Institute, is effective against chronic myelogenous leukemia and a couple of other cancers. It works by blocking a specific oncogene known as a *Bcr-Abl* tyrosine kinase from signaling its go signals.[67] This leukemia usually starts with a specific mutation that is particularly sensitive to signals from this oncogene.

Unfortunately, not long after Gleevec was introduced, resistance evolved in tumors of patients taking it. This resistance limits and can even end the effectiveness of the drug. One work-around has been the development of next-generation drugs that have similar function and are more resilient against the evolution of resistant by cancer cells. While such drugs have been developed, they are not panaceas. Another strategy has been to use Gleevec along with other drugs that also target the *Bcr-Abl* tyrosine kinase, but at different parts of the molecule. Hitting the kinase at different sites makes it more difficult for it to evolve an escape.

We are engaged in an evolutionary arms race between the cultural evolution of drug and other therapy development and the cancer evolving resistance to the new therapies. Can we use the principles of evolution to tackle drug resistance in a better, more systematic and forward-thinking manner?

Both traditional chemotherapy and more targeted therapies exert selection pressure on the tumor. How do the cancers evolve in response? Some of the responses are at the target site of the agent. Others are elsewhere—upstream or downstream in the same pathway or even in a parallel pathway. We should pay more attention to these target and off-target effects, as knowledge of these can be useful in designing combination of drugs.[68]

Robert Gatenby, who we met earlier, proposed a bold and intriguing strategy for treating cancer, which he calls adaptive therapy.[69] Before we go into detail about this approach, let's start with Gatenby's concern. He was worried that nearly all therapies thus far seek to kill the greatest number of tumor cells. Gatenby argued that this maximal approach could be counterproductive if there is variation in the sensitivity to the drug within the tumor. If there are costs to being resistant to the therapy, then the resistant cells would initially be rare because they are less fit. With a maximal approach, only these resistant cells would survive. Without the more sensitive cells, the resistant cells could proliferate. If the therapy was not as severe, then there would be some sensitive cells along with the resistant ones, and the sensitive cells would be competitive with the resistant ones.

Gatenby's adaptive therapy postulates that clinicians should adjust the therapy regimen based on the status of the tumor. The goal here would be managing the tumor so as to best maintain long-term stability. This strategy, which resembles integrated pest management for controlling agricultural pests (as we will see in Chapter 9), has shown promise in some treatments of late-stage cancer.[70]

Cancer evolution studies illustrate how principles of evolution and information from traditional evolution studies of organisms can be used to guide our understanding of cancer, and in some cases, even cancer treatment. This growing field also illustrates the need to break down silos between different disciplines: the progress being made is in large part due to people from backgrounds in ecology and evolutionary biology engaging in dialogues with cancer researchers.

CODA

In April 2019, I had the privilege of attending a small conference/workshop on cancer and evolution that was funded by the Society of Molecular Biology and Evolution. Organized by Jason Somarelli, a cancer biologist at Duke University, and Jeff Townsend, an evolutionary biologist at Yale University, this conference attempted to forge such dialogues and collaborations.[71] While one of our goals has been to apply evolution to cancer, the exchange is not one-dimensional: cancer genomes and other information from cancer studies provide rich troves of data for evolutionary biologists to use as a model to better understand evolutionary processes.

In February 2021, I spoke with Jason Somarelli about how he got into cancer evolution and the new challenges he saw for the field.[72] An environmental science major at Nazareth College (outside of Rochester, New York), Somarelli had long wanted to combine his interests in ecology and evolutionary biology with something that had a medical application. He soon discovered cancer as the ideal setting to combine his interests, as it is fundamentally an ecological and evolutionary problem.

Over the next decade, Somarelli sees cancer researchers making more progress dealing with metastasis and therapy resistance by incorporating more ecology and evolution into these arenas. Echoing the adaptive therapy framework, Somarelli thinks that shifting to a view of cancer as a chronic disease that we can control rather than one that we can cure holds promise.

Somarelli adds that over the horizon is use of immunotherapy, using ideas from immunology as well as evolution to treat cancer. This, he adds, is a relatively new paradigm with growing pains. One of these obstacles is the existence of communication gaps between the cancer specialists and the immunologists. But one of the exciting areas here is that immunotherapy brings in the equivalent of predator/prey interactions to cancer therapy. A key challenge in immunotherapy is that cancer cells are evolving responses to the immune system and the immune system has the capacity to evolve in response, much like prey and predators do out in the larger arena.[73]

NOTES

1. Aktipis (2020, p. 11).
2. DeGregori (2018, p. 206).
3. Mukherjee (2010). The quote is from the Author's Note.
4. *Ibid.*, chapter 23. This declaration came in the aftermath of the success of the Apollo program moonshots. Nixon and others viewed the act's newly established National Institute of Cancer as a NASA for cancer with targeted, pragmatic goals.
5. The data are from American Cancer Society (2020). Because cancer risk is greatly influenced by age, adjusting rates for changes in the age structure in the population is the appropriate way to compare across different years. American Cancer Society. 2020. Cancer Facts and Figures 2020. www.cancer.org/research/cancer-facts-statistics/all-cancer-facts-figures/cancer-facts-figures-2020.html. Accessed February 25, 2020.
6. Pierce and Speers (1988).
7. Angier (1988).
8. Armstrong (2014).
9. Hanahan and Weinberg (2000).
10. Hanahan and Weinberg (2011).
11. Normally, glucose, a six-carbon sugar, is broken down into pyruvate (a three-carbon intermediate). That pyruvate is then is normally broken down further in presence of oxygen through the Krebs' cycle and the electron transport chain. This process generates much more (about 20-fold) useful energy per glucose molecule than glycolysis alone.
12. Gatenby and Gillies (2004).
13. Cancer cells altering their microenvironment within the organism is a process analogous to organisms modifying their environment, such as beavers making dams. The latter process is called niche construction.
14. Kurzrock and Giles (2015).
15. Gatenby and Gillies (2011).
16. The F_1 hybrids have near-normal eyes and vision because the alleles causing blindness are usually recessive to those causing normal eyes and these F_1 individuals have one functional allele at each gene that changed. Had the loss of vision been due to the same genetic change, however, the F_1 would have two nonfunctional alleles and would thus be blind. Testing in this way to see whether a change is due to the same or different genes by whether the F_1 do or do not have normal function is common practice in genetics known as the complementation test.

    Some of the next generation of hybrids (the F_2) would be blind because they have two nonfunctional alleles at one or more of the genes. Others would still have at least functional allele at all genes; they would have some blind individuals.
17. The age of our species depends on the criteria used to determine the start of our species. Most estimates place this date at around 150,000–300,000 years ago.
18. Fortunato et al. (2017). The quote is on p. 3.
19. Mukherjee (2010, p. 43).
20. See Aktipis et al. (2015) for a survey of cancer across multicellular life.
21. Mohr (2018, chapter 3) details some of the studies first performed in *Drosophila* that have led to a better awareness of cancer in humans and even some promising new treatments.
22. See Aktipis (2020, chapter 3) for more detail.
23. Caulin and Maley (2011).
24. See Armstrong (2014) for a book-length treatment on *TP53*.
25. Schiffman and Abegglen (2017).
26. Abegglen et al. (2015).

27. As a reasonable proxy for the number of cell divisions, they took the logarithm of the product of the animal size and its life span.
28. Vazquez et al. (2018).
29. Vazquez and Lynch (2021).
30. Tollis et al. (2019).
31. This gene is related to other genes that affect genetic recombination.
32. E-mail conversation with Amy Boddy. March 2021.
33. Boddy et al. (2020a).
34. Tollis et al. Preprint.
35. Stammnitz et al. (2018).
36. For a general overview of examples of animals at the extremes of cancer risk, see Boddy et al. (2020b).
37. Buffenstein (2005).
38. Kulaberoglu et al. (2019).
39. Somarelli et al. (2020a).
40. Gardner et al. (2016).
41. Silver (2012).
42. The idea traces back to theoretical treatments by Maynard Smith and Haigh (1974).
43. See Johnson (2007, chapters 3–5) and references within for descriptions of these tests.
44. See Buisson et al. (2019) and Carter (2019).
45. See Cannataro et al. (2018) and Cannataro and Townsend (2018, 2019).
46. Mukherjee (2010, p. 464–465).
47. See Wu et al. and references within (2016) for examples.
48. Cannataro et al. Preprint.
49. E-mail interview with Vincent Cannataro. February 2021.
50. McFarland et al. (2013).
51. Carbone et al. (2020).
52. Orth et al. (2019).
53. Beyer et al. (2020).
54. See Alonso-Curbelo et al. (2021) and Vassiliadis and Dawson (2021).
55. There is some confusion about the meaning of the term epigenetics, as different biologists use the term differently. Broadly speaking, it is heritable changes in the phenotype that are not from changes in the phenotype. See Jamniczky et al. (2010) for more detail.
56. Carbone et al. (2007).
57. DeGregori (2018). In her recent book, *Rebel Cell*, Kat Arney notes, "cancer cells are like the very worst neighbors you can imagine, sucking away all the oxygen and nutrients, and polluting the environment with their waste" (Arney 2020, p. 212).
58. Smith et al. (2012).
59. Kang and Rosenwaks (2018).
60. Crespi and Go (2015).
61. Driver (2014).
62. Jacobs et al. (2006).
63. Driver (2014).
64. Boddy et al. (2015).
65. Mukherjee (2010, p. 401).
66. Jones (2021).
67. Burgess and Sawyers (2006).
68. Venkatesan et al. (2017).
69. Gatenby et al. (2009).
70. Aktipis (2020, chapter 7).

71. Somarelli et al. (2020b).
72. Phone interview with Jason Somarelli. February 2021.
73. Somarelli (2021).

FURTHER READING

Aktipis, A. 2020. *The Cheating Cell: How Evolution Helps Us Understand and Treat Cancer.* Princeton University Press.

Arney, K. 2020. *Rebel Cell: Cancer, Evolution, and the New Science of Life's Oldest Betrayal.* BenBella Books, Inc.

DeGregori, J. 2018. *Adaptative Oncogenesis: A New Understanding of How Cancer Evolves Inside Us.* Harvard University Press.

Mukherjee, S. 2010. *The Emperor of All Maladies: A Biography of Cancer.* Scribner.

7 Human Life History

SUMMARY

Features of human life history—some of which are shared among placental mammals and some that are specific to apes or to just humans—affect our health. We begin with the unusual Prader–Willi syndrome, which like its counterpart, Angelman syndrome, is an exception to the rules of Mendelian genetics. In these parent-of-origin disorders, the parent from whom the mutation arose is essential. These disorders, which are failure of genomic imprinting, also illustrate the tug of war that goes on between the maternal and paternal genes. Preeclampsia and some cancers are also consequences of this tug of war. We discuss the interaction between the hormone progesterone and its receptor, a relationship that has evolved since humans diverged from chimpanzees. We also discuss new work that shows that human gestation duration is constrained not by anatomy but rather metabolic limits. We discuss the likely causes of menopause, including the grandmother hypothesis. Finally, we examine male-biased evolution, a consequence of the differences between oogenesis and spermatogenesis.

"[W]e rarely ask ourselves why gestation does not always proceed as smoothly and reliably as the lifelong beating of our hearts or filtration of blood by our kidneys. . . . A crucial contrast distinguishes obstetrics from cardiology and nephrology. The coordinated activities of heart and kidneys take place within an individual comprised of genetically largely identical cells, whereas pregnancy involves an interaction between genetically-distinct individuals whose cooperation is obviated by evolutionary conflicts of interest."—David Haig.[1]

"Humans do have a long developmental period during which we grow our enormous brains and during which we wire them up in wonderful ways, like for music and wit and other wows of humanity. But this does not deserve a uniquely human explanation. All big-brained primates take longer to develop than their smaller-brained relatives, and while they do so, they learn complex behaviors, just not as complex as our own."—Holly Dunsworth.[2]

PRADER–WILLI AND THE PREGNANCY TUG-OF-WAR

Prader–Willi syndrome is a rare disorder that affects about one in 20,000 births. It strikes males and females equally. Babies with Prader–Willi are smaller, about 15 percent smaller than average.[3] Babies with Prader–Willi tend to be listless and do not move as much. In fact, this reduced movement is even seen in fetuses. Prader–Willi babies are much less inclined to suck and often fail to thrive. If they do thrive, they still hit developmental milestones—crawling, walking, talking—much later than typical. Mild intellectual disabilities and learning difficulties are common. In addition, children with Prader–Willi often have emotional outbursts and sometimes exhibit obsessive-compulsive behavior.

DOI: 10.1201/9780429503962-8

While babies with Prader–Willi are reluctant to eat, at some point in early child-hood, a switch is flipped: from here on, individuals with Prader–Willi develop insatiable appetites. They will obsessively seek out food. Unless they are carefully monitored, most will become severely obese. In addition, individuals with Prader–Willi will usually be short with small hands and feet, partly due to deficiencies in growth hormones.

Prader–Willi is associated with a deletion of about 6 million base pairs from one of the copies of chromosome 15. Intriguingly, deletions of this region are also asso-ciated with a contrasting collection of symptoms known as Angelman syndrome.[4] People with Angelman syndrome have exceptionally small heads. They usually have severe intellectual disability and only minimal use of language. In contrast with Prader–Willi, people with Angelman frequently smile and laugh and are excitable. They also are often bursting with activity. In sharp contrast with Prader–Willi, chil-dren with Angelman's syndrome are picky eaters. This syndrome is about as rare as Prader–Willi; it, too, affects roughly equal numbers of males and females.

Deletion of the same part of the same chromosome leads to two fundamentally different syndromes. How can this be? This conundrum stumped the medical com-munity until they realized that the identity of the chromosome that has been deleted differed in the two syndromes: people with Prader–Willi syndrome had a deletion in the chromosome that came from their father, while people with Angelman syndrome had a deletion in the chromosome that came from their mother.

If you took a genetics class, the preceding paragraph might lead you to say: "Wait a minute . . . I thought the identity of the parent who transmitted an allele does not affect the phenotype." You would be right. In classical Mendelian genetics, it should not matter which parent transmitted an allele to its offspring. Most of the time, it does not. But sometimes it does. These cases when it does matter are called parent-of-origin effects. We encountered an example of such back in Chapter 3 regarding sus-ceptibility to malaria based on blood type. We don't know why the parent-of-origin effect occurs in malaria susceptibility; as we will see, we do know why it occurs in the case of Prader–Willi and Angelman syndrome.

The genes in that deletion associated with these contrasting syndromes have an unusual pattern of expression. One set is expressed only by the chromosome that came from the mother (the maternal chromosome), while another set is expressed only by the chromosome that came from the father (the paternal chromosome). When part of the paternal chromosome is missing as in Prader–Willi syndrome, only those maternal genes are expressed. In contrast, when part of the maternal chromosome is missing, as in Angelman syndrome, only the paternal genes are expressed.

During the 1980s and 1990s, biologists noticed something about many of these mammalian genes that were only expressed when on the maternal or paternal chro-mosome (but not both). These genes had methyl-groups attached to some of the nucleotides on the chromosome.[5] The biologists also knew that methyl-groups hinder expression of the genes; thus, they are a way of silencing the genes depending on which chromosome they are on. Genes that are ordinarily expressed only on the paternal chromosome are methylated when on a maternal chromosome (and vice versa). This methylation thus marks the identity of the chromosome. For this reason, it is a form of genomic imprinting.[6] Note that these methyl imprints do not affect the composition of the DNA bases. Hence, they do not affect the identity of the message (the proteins produced). But they do affect the expression of the message. These imprints can be transmitted across generations; thus, they are a form of epigenetics: they are a form of inheritance that is on top of (epi) the genetic inheritance of the DNA sequence.

Enter the Australian-American biologist David Haig.[7] As we will see, Haig would provide an explanation for why Angelman and Prader–Willi syndrome have such different effects as part of a larger view of mammalian pregnancy as an arena involving both cooperation and conflict among different genomes. Haig's ideas have revolutionized how evolutionary geneticists view pregnancy.

Originally a plant ecologist, Haig had been investigating the evolutionary genetic puzzle presented by the endosperm of flowering plants during the late 1980s.[8] This tissue acts much like a placenta in mammals: it acquires resources from the mother plant that the developing embryo will eventually use. But unlike the placenta, the endosperm is genetically distinct from both the maternal plant and the offspring. It is triploid, with two sets of chromosomes from the mother and one set from the father. Haig was interested in the implications of the maternal contribution of the endosperm's genome being twice as great as the paternal contribution. One consequence of its unusual genome is that the endosperm has a greater relatedness to other offspring of the same mother than it does to the embryo! Accordingly, interests of the endosperm and those of the embryo differ: selection would favor the endosperm to be less aggressive than the embryo at trying to get resources from the mother. Haig then proposed a mechanism for how this could happen: gene expression that is specific to one parent. Sound familiar?

Haig's investigations of these (metaphorical) struggles between different genomes in flowering plant reproduction led him to the concept that something similar was going on in mammalian pregnancy.[9] Haig was intrigued by the weird patterns seen in imprinting in Prader–Willi and Angelman syndrome. He also was aware of data from unusual mouse embryos: both those formed from two maternal sets of chromosomes and those formed from two paternal sets of chromosomes failed during development, but they failed in different ways. Those from two maternal sets had embryos that developed reasonably well, but the membranes surrounding the embryo (the so-called extraembryonic membranes) that are used to acquire resources and exchange gases were malformed. In contrast, those from two paternal sets of chromosomes had reasonably normal extraembryonic membranes but not much embryonic development. Like the two Captain Kirks after a transporter malfunction in the *Star Trek* episode *The Enemy Within*, these two sets had mirror defects, each missing something that the other had.

Haig proposed that mammalian pregnancy is like a tug of war where the interests of different sets of genes play out.[10] To think about this, we need to consider the relationship between a mammalian mother and one of her young offspring. If the offspring is a fetus inside the mother or an infant still nursing, the mother is the primary source of resources for that individual. But that individual is not her only genetic investment: she may have other children to provide for, and she may be able to become pregnant again. To the mother, these children are equally valuable, as she is equally related to all of children. In contrast, each individual offspring is more related to itself than it is to its sisters and brothers. Each individual offspring would benefit if the mother directed more resources to it, even though such behavior may hurt its siblings. In contrast, the mother's interests would be best served by allocating resources based on what is best for both her current children and ones she may have in the future.

Moreover, we need to consider the interests of the father of each child, who may or may not be the father of the child's siblings. If there is the prospect that the siblings (both current and future) may be from different fathers, the father's interests are not going to be completely aligned with those of the mother. As the likelihood is that the siblings may not be his own increases, the father's interests diverge more and more from the mother's. At the extreme, the father's interests would be served by the offspring in question getting as much from the mother as possible. In contrast, the mother's interests would be served by not providing as much for the current offspring so that she can also provide for her other offspring, both present and future.

Considering this view of evolution where individual bodies are host to different sets of alleles that sometimes act in concert and sometimes act at cross-purposes can be mind-blowing. This view of genetic conflict within our bodies is alien to our sensibilities. We like to think of ourselves and other organisms as whole. Haig presents an economic analogy: alleles are much like shareholders and organisms are like corporations. Different shareholders can invest in multiple corporations. Similarly, a corporation can have multiple investors. By having stock in multiple companies, shareholders can have conflicts of interest. Similarly, because different shareholders invest in the same corporation, there can be conflicts. Even with these conflicts, corporations usually function. For both organic and corporate bodies, there will be times when those conflicts come to a head. As we can see, pregnancy is one of those times.

Thinking about these investment decisions and how they play out during pregnancy, we need to consider three sets of genes: the mother's own genes in her own body, her genes in the child, and the father's genes in her child. Selection favors the paternal genes in fetuses to promote getting more resources from the mother. In contrast, selection favors the maternal genes in fetuses to not be as demanding. Similarly, selection favors genes in the mother to not be as responsive to the signals from the fetus. These conflicting selection pressures usually result in a compromise: the fetus obtains more resources than are in the mother's interest but less than what is in the interest of paternal genes in the fetus.

These conflicting selection pressures can lead to arms races. The selection acting on the paternal genes to demand more resources can lead to changes that lead to selection on the mother's genes (whether in her or in the fetus) to respond less to the signals. The declined responsiveness, in turn, can lead to more selection on those paternal genes in this fetus to demand still more. And next in turn, the cycle continues. Haig notes that seeing powerful demanding signals, as is often seen in maternal–fetus interactions, is a sign that conflict has led to an escalation. As Haig puts it: "If a message can be conveyed in a whisper, why shout? Raised voices are frequently a sign of conflict."[11]

The messages here are the expression of genes and manifest as the concentration of the products (usually proteins) of those genes. Recall the methyl-groups on stretches of DNA. The methyl groups dampen down gene expression. By having maternal and paternal chromosomes with different methylation patterns, genes on maternal chromosomes can be expressed differently from those on paternal chromosomes. These epigenetic markers thus modulate the expression of genes that are the basis of parent-of-origin effects that we see in mammalian pregnancy.

It is not in the interests of the paternal genes in the fetus to be so demanding that the mother's life is jeopardized. Similarly, it is not in the interests of the maternal genes in the fetus (or the mother's genes in her body) that the fetus starves. Let's say that on some arbitrary scale (from 0 to 10), that the optimal intake of nutrients from the mother to the fetus for the paternal genes is 6 and the optimal intake for the maternal genes is 4. Under most circumstances, the actual intake would be between 4 and 6 and often right near 5. But in some cases—like in the deletions seen in Prader–Willi and Angelman syndromes—one side of the tug of war is not holding up the rope. If the expression of the maternal genes is knocked out, the intake rate becomes something like 9. There are no stop signals to counteract the go signals from the paternal genes. In contrast, if the expression of the paternal genes is knocked out, then the intake rate becomes something like 1. Neither of these scenarios is good for the mother, the father, or the child; it's just the consequence of the escalation combined with one side being knocked out. Of course, milder versions of such abnormalities are also possible with expression being either 3 (with somewhat too little paternal gene expression) or a 7 (with somewhat too little maternal gene expression).

Let's consider in more detail what the optimum is for both the paternal and maternal genes in the fetus and in early childhood. The optimum for the paternal genes in the fetus is to have the placenta be highly efficient at getting resources from the mother. These genes are benefitted by the baby being heavier than average at birth. When the baby is nursing, these paternal genes are benefitted by it sucking frequently and robustly. These genes also promote rapid early growth and delayed weaning. "Greed" is good for these genes.

In contrast, maternal genes in both the fetus and in the mother have an optimum placenta efficiency that is lower than that of the paternal genes. These maternal genes are benefited by having the baby be smaller: more than large enough to survive, but not so large that their growth comes at the expense of the mother's other children. The optimum baby for these genes would be less aggressive at sucking and would wean earlier. Maternal genes also would favor reduced lean muscle mass and more fat, which would be useful for surviving periods of famine and infection. "Prudence" is good for these genes.

Interestingly, too much prudence early in life can be deleterious for the adult body. Bernie Crespi at Simon Frasier University has argued that dysfunctions from an overweighting of the prudence from the maternal genes combined with the mismatch between the easy-access of calorie- and sugar-dense foods and the environments we evolved in may increase the risk of metabolic syndrome.[12] This collection of symptoms includes high blood sugar, high blood pressure, the accumulation of fat around the middle, and high cholesterol (especially LDL cholesterol, the bad kind). The lower efficiency of the placenta leads to the lower weight at birth and the body composition tilted toward accumulating fat as opposed to lean muscle, which are predisposing risk factors for developing metabolic syndrome later in life. Low weight at birth often leads to overcompensation afterward, especially in an environment with easy access to an abundance of food. Moreover, the body composition changes also lead to a greater likelihood of developing the accumulation of fat around the middle.

A recent study found one imprinted gene—insulin-like growth factor 2 (*IGF2*)— has a large effect on birth weight.[13] The combination of genetic and epigenetic variation at this gene explained more than a third of the variation of birth weight in a French-Canadian population. *IGF2* encodes a protein that when bound to a receptor promotes growth, especially in fetuses. It and its neighbor *H19* are on chromosome 11, where imprinting is common: *IGF2* is mainly expressed from the chromosome that comes from the father, while *H19*, which encodes an RNA that is not encoded into a protein, is expressed from the maternal chromosome. DNA methylation marks these genes. Another protein called CTCF binds with the methylated DNA, which suppresses *IGF2* expression and increases expression of *H19*. Interestingly, high *IGF2* methylation on the maternal side of the placenta is associated with more IGF2 expression in the fetus and, correspondingly, larger babies at birth.

The tug of war that goes on between the paternal and maternal genes in the fetus also may affect cancer susceptibility.[14] To see this, we need to look to the placental trophoblast, which consists of specialized cells that assist with the implantation of the embryo. These cells also interact with the maternal cells through the invasion of maternal tissues. The placental trophoblast has also evolved mechanisms to evade the immune system of the mother.

Looking at the invasiveness of trophoblasts and their evasion, we can see a similarity between them and cancer cells; they share several of the hallmarks. They also both proliferate explosively and send off signals that generate new blood vessels. In fact, the trophoblast placenta has sometimes been likened to a well-behaved tumor. Moreover, would-be cancers can usurp some of the angiogenic and immune evading signals used by the placental trophoblast. Here, too, mother and fetus differ in interest of how far the placenta should intrude into maternal tissues. Paternal genes should lean toward more vulnerability to cancer, maternal genes should put on the brakes.

Recall from the previous chapter that rates of cancer increased with litter size in mammals. There, Amy Boddy and her colleagues' hypothesized explanation was that large litters required faster growth, which had the side effect of increasing cancer risk due to the growth–cancer defense tradeoff. Haig proposed an alternative: a larger litter increased the likelihood of competition between genetically different individuals for resources at the same time.[15] Such competition would lead to a greater scramble for resources. If such competition were common, there would be selection pressure for greater invasiveness of placentas, which would spillover to increased overall cancer risk. Haig noted that one could conceivably test this hypothesis by looking at groups of species that differed in litter size and in the relatedness of individuals in the litter. One such example occurs in armadillos. The nine-banded armadillo has litters of four where all four pups arise from the same egg (and are thus genetically identical). In contrast, euphractine armadillos have litters of fraternal (genetically nonidentical) twins. According to the "live fast" hypothesis suggested by Boddy, the nine-banded armadillos with their quadruplets should get cancer more than the euphractine armadillos. In contrast, the "genetic conflict" hypothesis predicts that the euphractine armadillos should have more cancer. We do not yet have these data—and ideally, we would want to get many such comparisons—but the hypotheses do have different predictions.

Preeclampsia, which occurs in about 5 percent of pregnancies, occurs when the mother's blood pressure rises during pregnancy.[16] Often accompanied by protein in her urine, preeclampsia puts both the mother and developing fetus at risk. Many cases of preeclampsia appear to be due to this genetic conflict that plays out in the interaction between mother and fetus. In some cases, it can advance to eclampsia, wherein the mother develops grand mal seizures.

Preeclampsia is seldom seen outside of humans. This may be due to the unusual demands that big human fetuses with their big energetically demanding brains present to their mothers. Preeclampsia has a rather high—55 percent—heritability; this means that genetic factors account for more than half of the variation in the risk for getting this condition. The genetic inputs come not just from the mother but also from the fetus. Moreover, preeclampsia is more common in first pregnancies, suggesting an interaction between the mother's immune system and genes in the fetus that come from the father.

In the normal course of pregnancy, the placenta generates tons and tons of blood vessels. Moreover, successful pregnancy requires remodeling of parts of the mother's circulatory system. This remodeling is particularly extensive for the spiral arteries—named for their coiled appearance—that supply the uterus. The generation of the placenta's blood vessels and the reshaping of the maternal spiral arteries both involve a shift in the balance of the signaling proteins that promote the formation of blood vessels and those that check that formation. In normal pregnancy, the risk of too little angiogenesis outweighs the risk of too much.

In preeclampsia, this balance often is not sufficiently shifted; typically, preeclampsia is associated with proteins that check the undesired growth of blood vessels and with defects in the placenta.[17] Moreover, a frequent cause of the high blood pressure of preeclampsia is that those spiral arteries feeding the uterus are too narrow. So, preeclampsia may often be the result of insufficient angiogenesis from the fetus and/or too much thwarting of angiogenesis from the mother. Interestingly, there are hints that preeclampsia is associated with a reduced risk of some cancers.[18]

Although the high heritability of preeclampsia has been well established, knowledge of the specific genes involved has lagged. A recent GWAS has established a link between preeclampsia and a gene: *FLT1*.[19] This gene encodes a receptor protein that binds factors that promote the formation of blood vessels. But this gene can also encode a variant of the protein (sFlt-1); instead of promoting blood vessel formation, this variant inhibits it. Preeclampsia indeed is associated with an excess of sFlt-1. Variants at *FLT1* that are associated with preeclampsia risk appear to shift the ratios of the variants of the protein and thus the balance between promoting and inhibiting blood vessel formation. Molecular evolution tests show that *FLT1* has been subjected to strong positive selection.[20] This is what we would expect if there have been repeated rounds of coevolution such as in a maternal–fetal conflict.

The human placenta differs from that of nonhuman animals in several respects.[21] In fact, the differences between human and mice placentas challenges the ability to make inferences about human pregnancy from mice studies. Human placentas even differ in some important ways from those of chimpanzees and other apes. Unlike other primates, humans express siglec-6 in the placenta.[22] This is one of many proteins that are expressed on the surface of immune cells; they bind to sialic acids, small, sugar-like chemicals that are often found in glycoproteins. Human placentas, unlike those of other primates, also produce the glycoproteins that typically bind siglec-6. This suggests that at some point after humans diverged from chimpanzees and other apes, the placenta gained siglec-6 expression, apparently to interact with these glycoproteins. The reason for the gain of this interaction at the placenta is not yet known, but it may have a role in slowing down the birth process in humans. Labor in humans is much longer than it is in other apes. Siglec-6 expression peaks at the end of labor. Placentas of women who give birth by Caesarean section following little or no natural labor typically have much lower concentrations of siglec-6. In contrast, women who give birth by emergency Caesarean section have placentas with substantially higher siglec-6 expression.

Siglec-6 expression in the placenta is higher in women whose preeclampsia led to a pre-term birth.[23] Is the increased expression of Siglec-6 a cause or a consequence of the underlying problems that led to the preeclampsia? We don't know, but as we saw earlier in the chapter, preeclampsia appears to be exceedingly rare in nonhuman primates. That both preeclampsia and Siglec-6 expression in the placenta both appear unique to humans suggests that the alteration in the expression pattern may have set the stage to make humans vulnerable to preeclampsia.

PROGESTERONE AND ITS RECEPTOR

Hormones such as progesterone do not act alone. Instead, they bind to receptors. In the case of progesterone, its receptor (the progesterone receptor) is a protein on the surface of cells.[24] This receptor is encoded by the *PGR* gene on chromosome 11. Expression of this gene is highest in the endometrium, a mucous membrane that lines the uterus. Thus, the endometrium is the site where you would find the greatest density of progesterone receptors. When progesterone (or a synthetic form of the hormone) binds to the receptor on the endometrium, a cascade of regulatory genes alters the endometrium, making it more hospitable for implantation of the embryo.

A variant of *PGR* that was introduced into humans from Neanderthals is found in varying frequencies (from 2 to 20 percent) in Eurasian populations. Using data from the UK Biobank, carriers of the Neanderthal variant have more sisters and report fewer miscarriages compared with those homozygous for the ancestral allele. These data suggest that the Neanderthal variant increases fertility.

There is substantial scatter among different studies regarding the effects of the Neanderthal variant on ovarian cancer risk: in some, the Neanderthal variant appears to be a risk factor. In others, it may be protective. A meta-analysis shows that the Neanderthal variant is not a risk factor overall; it is one, however, for those of people whose ancestors are from Europe and who never took oral contraceptives.[25]

Mirna Marinić and Vincent Lynch at the University of Chicago recently have shown that the function of the receptor is substantially different between humans and chimpanzees.[26] Using phylogenetic methods, they determined the sequence of the receptor for both the ancestor of humans (modern humans, Neanderthals, and

Denisovans) and the ancestor of humans and chimpanzees. They then reconstructed the protein for each and tested in transgenic mice how those reconstructed proteins affected the regulation of downstream genes. They found a functional difference between the two, demonstrating that the gene had evolved important differences since humans diverged from chimpanzees. This suggests that "animal models of progesterone signaling may not adequately recapitulate human biology".[27] They suggest as an alternative the further development of artificial organoid models.

Selection on the receptor appears to have continued during the history of our species. Analysis of sequence data finds that selection has led to increased genetic differentiation at this receptor across different populations.[28] Relatively recent positive selection has acted on the Han Chinese population, leading to reduced genetic variation there. In contrast, balancing selection appears to have been operating in Northern and Western Europe.[29] The positive selection in the East Asian population has been in changes in the expression of the receptor. These changes appear to be advantageous only in that population, suggesting substantial gene by environment interaction. Further study examining more populations and linking these changes to observable characteristics is needed.

WHY ARE HUMAN BABIES BORN SO HELPLESS?

Compared with those of other primates, human babies are born helpless. Our babies are born with the brains with the lowest proportion—about 30 percent—of their adult size. In comparison, our close relatives, chimpanzees and bonobos, give birth to babies with brains about 40 percent of their adult size. Many monkeys have babies born with brains half their adult size. Our early development is also much more protracted as well. Some have considered human infants to be like fetuses.

Why are our babies born so helpless? The textbook answer, which has seeped into popular culture, is based on two unique human traits: our big brains and our bipedalism. While brains grew increasingly large over the course of evolution, our bipedal mode of walking constrained the size of birth canals because overly wide pelvises needed to support large birth canals hindered walking. The size of human babies at birth is a compromise between these conflicting selection pressures. This hypothesis, known as the obstetric dilemma, also provides an answer to why childbirth is so difficult.[30] This hypothesis also implies that human babies are born earlier than they should be, with the first few months of infancy being like the fourth trimester. The obstetric dilemma sounds like a beautiful hypothesis that ties everything together in a sensible package with a bow ribbon. Isn't it lovely?

The problem is that the obstetric dilemma hypothesis has little or no empirical support. Indeed, as Holly Dunsworth, an evolutionary anthropologist at the University of Rhode Island, and others have shown in recent years, there is much evidence against it. As we will see, Dunsworth has not only challenged this hypothesis but also provided an alternative.

Long fascinated and puzzled by the obstetric dilemma, Dunsworth had been looking at the gestation length in different mammals, struck by the tight correlation between body size and gestation length: bigger mothers have longer-lasting pregnancy. Dunsworth was particularly struck that the same pattern held in whales and dolphins, which lack bony birth canals. Why would these marine mammals show the same pattern if their babies don't have to pass through a bony birth canal?

Dunsworth was also looking at our closest relatives. If you removed humans from the equation, chimpanzees and bonobos would be born with considerably smaller brains relative to their adult brain size as compared with other primates. Chimpanzee and bonobo infants also experience extraordinarily rapid brain growth. They are also helpless at birth, more so than other nonhuman apes. But no one tries to use the obstetric dilemma to explain their relatively small brains at birth and their helplessness. In fact, bonobos and chimpanzees do not have a tight fit between baby and the birth canal.

Although Dunsworth's discontent with the obstetric dilemma was growing, she did not have an alternative until she was inspired by the reading material she took on a paleoanthropological expedition in Kenya. After collecting fossils by day, Dunsworth spent her nights voraciously reading Peter Ellison's *On Fertile Ground: A Natural History of Human Reproduction*. In this book, Ellison argued that the end of human pregnancy is set by energetic costs rather than by anatomy: babies are born when mothers can no longer meet the needs of the growing fetus. Dunsworth was already collaborating with Herman Pontzer examining metabolism in nonhuman primates. So Dunsworth teamed up with Ellison and Pontzer to address whether energetic conditions determine when human babies are born.[31] This group first looked at gestation time across humans and our ape relatives. The average gestation length in humans is 38–40 weeks. This is slightly longer than that of gorillas (37 weeks) and much longer than that of chimpanzees (32 weeks). Larger primates have longer gestation times, and humans are on the large end of the size continuum. If we scale gestation time to what we expect based on the size of the mother, humans still take a long time to develop within the uterus. Human gestation length is more than a month longer than that expected given the size of the mother. Adjusted for the size of the mother, human babies are born comparatively heavier than most primates. Humans are also born with huge brains after this adjustment. The next largest size-adjusted newborn brains belong to gorilla newborns. Our newborns have brains half again as large as those of the gorillas. The big brains and large overall size of human newborns point to greater, not less, investment by mothers during pregnancy.

Energetic studies show that humans can indefinitely sustain a doubling or a wee bit more than their basal metabolic rate, but not much more. By the ninth month of pregnancy, mothers are pressing up against that limit. This suggests that it is these energetic challenges that limit the length of pregnancy, not anatomy. Thus, Dunsworth and colleagues presented what they would call the EGG hypothesis for "energetics of gestation and fetal growth":[32] the length of human pregnancy was set by energetic limitations, not anatomy. Changes in anatomy according to the EGG hypothesis come later: as human brains got bigger, the female pelvis responded to selection to get wider. An additional wrinkle to this hypothesis is that agriculture likely has changed our diet in the recent past, leading to larger babies. As such, there may be a mismatch that selection has not yet had time to resolve.

The central assumption of the obstetric dilemma hypothesis is that the wider pelvis of women compromises their ability to walk and run. To test this assumption, Anna Warrener, then at Harvard, and her colleagues had young adults walk and run

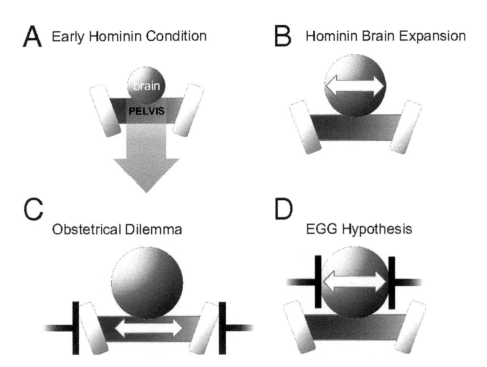

A Early Hominin Condition

B Hominin Brain Expansion

C Obstetrical Dilemma

D EGG Hypothesis

Distinguishing between the OD and EGG hypotheses. Chimpanzees and early hominins (panel A) have little difficulty during childbirth because the brain of the neonate is small relative to the size of the mother's pelvis. Brain size increased during hominin evolution (panel B). The obstetric dilemma (OD) posits that the brain of a developing fetus cannot get larger because the pelvis is constrained because a wider pelvis would compromise locomotion (panel C). The EGG hypothesis, in contrast, posits that the size of the neonate brain and the duration of the gestation period are constrained by energetic demands on the mother (panel D).

Source: From Figure 4 of Dunsworth, H. M., A. G. Warrener, T. Deacon, P. T. Eillison, and H. Pontzner. 2012. Metabolic hypothesis for human altriciality. *Proceedings of the National Academy of Sciences* 109: 15212–15216.

on treadmills, where various measurements were taken.[33] While there were statistical differences in the anatomy of the male and female subjects, these did not translate into differences in locomotion efficiency. Men and women spend equivalent energy walking and running. Selection, presumably on the ability to give birth, has led to differences in pelvic anatomy between men and women; these differences, however, do not compromise locomotion.

Until recently, most of our information about the shapes and sizes of women's pelvises came from women of European descent. We lacked information about the range of variation among different populations. A recent study by Lia Betti and Andrea Manica at the University of Roehampton starts to fill in that gap.[34] They examined variation in female pelvises in a few populations across the world, finding substantial variation within and among populations. Women in sub-Saharan populations have deeper birth canals on average, while those from Native American populations have wider canals.

The patterns of shape variation are consistent with the neutral processes of mutation and genetic drift driving changes. Pairs of populations that differ the most genetically also tend to be more different in the birth canal shape. Within populations, the variation of birth canal shape and size declines as distance from sub-Saharan Africa increases. Given that human populations further removed from sub-Saharan Africa have less genetic variation, presumably due to having undergone more genetic bottlenecks, the within-population variation data are what we expect from neutral processes.

If there were strong selection operating on the birth canal size, we would expect populations with smaller women to have relatively larger birth canals. But, in fact, smaller women do not have larger than expected birth canals. This large variation suggests that pelvic width is not tightly constrained, as had been argued by proponents of the obstetric dilemma.

But why are human babies born so helpless? The anthropologist Sarah Hrdy has argued that humans differ from chimpanzees and bonobos with respect to our extensive cooperative breeding.[35] Even in comparison with chimpanzees and bonobos, individuals take a large role in providing for and interacting with children who are not their own. These alloparents, who are often sisters, aunts, and grandmothers (but also grandfathers, brothers, uncles, and even unrelated individuals) have altered our life history. These alloparents allow for helpless infants to grow up. With the cooperative breeding they provide, predation on young declined. With decreased predation risk, selection pressures were altered: no longer needing rapid development, human childhood could last longer. This declined predation risk also reduced the selection pressure against big, energetically demanding brains. Released from this constraint, our ancestors could evolve larger brains and all the wonderful (and not so wonderful things) that we can do with them.

WHY DO HUMAN FEMALES HAVE SUCH A LONG POST-REPRODUCTIVE LIFE?

Humans are highly unusual, though as we will see not unique, in that female reproduction usually ceases in the fifth or sixth decade of life, considerably before most women in most countries die in today's world. Life expectancy at birth has dramatically increased in the last few centuries, but most of that increase has been in adding years in youth and early midlife. Centuries ago, many women survived into their 70s and 80s and occasionally beyond. This is well beyond the age where female reproduction ceases. Why menopause?

The mechanistic answer is the depletion of the precursors of eggs.[36] On the way to becoming eggs, oocytes go through a series of steps before they ovulate. Only a tiny fraction of oocytes actually get to ovulate; most are lost somewhere along on the way in a process called atresia. The typical human female can ovulate a maximum of about 400 oocytes in her lifetime. For most of the woman's life, her would-be eggs are in the stage known as primordial follicle—here they can last about 50 years. When too few follicles remain, the hormones produced by the follicles that mediate the menstrual cycle no longer can do so. Hence, she ceases to menstruate.

Why? What is the evolutionary reason explaining why women stop menstruating while some have decades left to live, in contrast to men, who can remain capable of reproduction well into their 80s? Are there other species wherein the females have an extended post-reproductive life?

The last question is the easiest to answer. We know of four species of mammals, all of which are marine mammals—orcas, short-finned pilot whales, beluga whales, and narwhals—in which females also have a considerable post-reproductive life.[37] With orcas,[38] *Orcinus orcas*, females spend nearly a third of their lives in a post-reproductive state. This is almost as much as that for hunter–gatherer humans—the comparable proportion for one such group (the Hadza of what is now Tanzania) is 44 percent. In contrast, female elephants spend only 4 percent of their lives post-reproduction; in female chimpanzees, the post-reproductive life is even less.

Studying the Hadza during the 1980s and 1990s, the anthropologist Kristen Hawks noticed that the older women were among the most productive foragers in the community.[39] She would then show that these grandmothers were beneficial to the survival of her grandchildren. These observations led Hawks to champion the grandmother hypothesis for the origin of menopause. Herein, selection favors the cessation of reproductive capacity because the older women would gain more evolutionary currency (in copies of alleles) by helping their children raise offspring than by continuing to bear their own children. This idea had been brought up in passing by some of the first investigators of the evolution of post-reproductive life like George Williams and William Hamilton, whom we met in the first chapter, but Hawkes took the first serious investigation of the idea.

Hawkes argued that grandmothers providing benefits to their grandchildren could inhibit physiological aging in women.[40] Recall that the evolutionary theory of aging posits two ways by which alleles that reduce performance and survival at old age can persist or even increase in frequency in populations: (1) genetic drift leading to the accumulation of mutations that are neutral in early life but harmful in late life and (2) selection favoring alleles that are beneficial in early life but harmful in late life. The benefits that grandmothers provide strengthens selection against the first category of mutations and changes the balance of the tradeoffs for the second category. Thus, with these benefits of grandmothering, we should expect fewer alleles that are harmful late in life.

Hawks would later examine the overall reproductive patterns of female humans from nonindustrial societies in comparison to chimpanzees. She found that human females start having babies at a later age than chimpanzee but have shorter intervals between births.[41] The help from others, including grandmothers, probably allowed our ancestors to have children more frequently. Chimpanzee and human females have their last child at about the same age. Hawks also examined the depletion of follicles in chimpanzees. The data from both the fertility trends over age and the rates at which follicles are depleted show similarities between humans and chimpanzees. These data "suggest that chimpanzees would reach menopause at about the same ages humans do—if they lived long enough".[42]

Hawkes is thus arguing that the grandmother hypothesis explains why selection did not slow the rate of follicle depletion so as to increase the female reproductive span as more and more human women lived longer. To investigate this grandmother hypothesis, we first need to see whether the presence of grandmothers did help the survival of their children's children.

A study of a population in preindustrial 18th- and 19th-century Finland examined whether grandmothers provide benefits.[43] In this study population, children often died, mainly of infectious disease, at young ages—nearly a third before their fifth birthday. Provided that they reached adulthood and had at least one child, more than half of the women lived past 50 and many lived well beyond that. Thus, post-reproductive life was common.

Having a grandmother, and especially a maternal grandmother, around benefitted their young grandchildren. Survival of grandchildren from age 2 to age 5 was 30 percent higher if they had a living maternal grandmother (aged 50 to 75) with them compared with those without living grandmothers. But this benefit did not extend when their grandmothers were either older than 75 or in their last couple of years of life. In fact, the presence of a paternal grandmother who was in the last year of life surprisingly decreased survival of the grandchildren. There was no detectable effect for the case of a maternal grandmother in the last year of life. Assuming that these results are representative of preindustrial populations, grandmothers probably do provide considerable benefits to their grandchildren, but these benefits disappear as the grandmothers age.

Another study looked at French settlers in the St. Lawrence Valley of what is now Canada during the 18th and 19th century.[44] Mothers who had a living mother had more offspring, about one additional child who survived to 15. The grandmother effect declined as the geographic distance between grandmother and grandchild increased, with no sharp breaks.

The existence of benefits from grandmothers rearing their daughters' children seems quite clear. But the existence of these benefits does not necessarily mean that the grandmother hypothesis is correct: it's one thing to show that having grandmothers around is beneficial; it's another to show that these benefits outweigh the costs of not continuing to bear children. Because grandmothers are only half as related to their grandchildren as they are to their own children, they would need to improve the survival of their children's children by roughly twice as much as they lose out by not having children.

One factor that may tilt this balance is cancer prevention. Given the benefits from grandmothering, there could be selection pressure to increase the life span of older women by shutting off their ability to reproduce. There is evidence that the cessation of reproduction in middle-aged women is protective against cancer—as repeated ovarian cycles accelerate breast, ovarian, and other cancers—and the risk of these cancers increases with age.[45] In modeling this, we need to consider what the risks would have been in environments before industrialization.

Before concluding this section, let's take a look at the post-reproductive life in marine mammals. Recent investigations show that toothed whales have evolved extended post-reproductive lives in females multiple times.[46] Beluga and narwhals are closely related; thus, the presence of the extended post-reproductive life period could have evolved once leading up to the common ancestor of these marine mammals. But the presence of post-reproductive life in short-finned pilot whales and in orcas cannot be due to a single evolutionary event, as these species are rather distantly related. As the narwhals and the belugas are distantly related to the orcas and the short-finned pilot whales, at least three independent evolutionary transitions to extended post-reproductive life must have occurred in tooth whales. Further investigation of more species may reveal still more independent cases of the evolution of extended post-reproductive lives in female whales. These repeated evolutionary transitions suggest that some feature of the ecology of some whales may drive extended post-reproductive life.

In whales, the common hypothesis for why the evolution of the end of reproductive life has remained behind the increase in life expectancy—thus allowing for a long post-reproductive life in females—is that it limits reproductive conflict between generations. Younger and older females of one social unit both producing offspring could lead to competition for resources. If such conflict occurs, evolutionary demographic theory predicts that the older females should suffer more. A recent study found support for this prediction of the so-called reproductive conflict hypothesis: in orcas, when older and younger females of the same group both are having offspring at the same time, the offspring of the older females have a substantially higher mortality rate.[47] This enhanced mortality rate is not seen in the offspring of the older females when there are no younger females in the social group having offspring.

What selective pressures are acting on the depletion of follicles? Is the limit of about 50 years that we see in both apes and marine mammals happenstance or selection-driven convergent evolution?

How recently did significant female post-reproductive life evolve in our lineage? My UMASS colleague Lynnette Sievert, who studies age at menopause and symptoms at midlife in different populations, argues that menopause probably evolved before the dispersal of modern humans from Africa because of the similarity of physiological changes across different populations.[48] But did Neanderthal and Denisovan females have a significant post-reproductive period? Sievert says that hominin longevity has most likely exceeded that 50-year threshold for at least a million years.[49] So menopause was likely a common feature of life for female Neanderthals and other archaic humans.

FATHERS PROVIDE MOST MUTATIONS

We saw earlier that the precursors to eggs in mammals are produced exceedingly early on. Although there is a maturation and a shuffling of the genetic deck as oogonia become eggs, there is only a single round of DNA replication in this process. In total, the eggs of women of any age have undergone only 23 rounds of replication. But sperm are different. Sperm are continuously produced from precursors. Thus, many more rounds of DNA replication occur in sperm production. The average 20-year-old man has sperm that has gone through about 150 rounds of replication. By the time he is 30, it will have undergone about 400 rounds. By 40, 600 rounds![50]

One consequence of the numerous rounds of DNA replication is that sperm accumulate mutations. Most mutations that are passed on from parent to offspring—the germline mutations—come from the father. Some rare diseases increase in frequency with the age of the father. One of these is achondroplasia, which causes many cases of dwarfism. Here, those with the condition have short limbs but average size torsos. It's due to a single gene, and most cases are from new mutations.

Prostate cancer, likely influenced by many genes, also increases with the age of the father but not the mother. This suggests that the input of new mutations from the father can have a detrimental effect on health.

In humans, the male mutation rate is as much as 20 times greater than the female mutation rate. Because they are found only in males, genes on the Y chromosome should accumulate mutations and evolve faster than genes on other chromosomes. Moreover, genes on the autosomes evolve faster than those on the X chromosome, all other things being equal. To think about why that is, consider that genes on the autosomes are equally in male and female parents, as every individual has exactly one mother and one father. In contrast, X-linked genes are twice as often found in female parents than they are in male parents. Because X-linked genes spend more time in female parents, their evolution should be slower.

Melissa Wilson, then a graduate student at Pennsylvania State University, looked at the factors that affect the ratio of the mutation rate in males versus that of females in 32 different mammal species.[51] A large factor was, as expected, the generation time. Species with long generation times have greater ratios of male to female mutation rates, presumably because the males of these species have sperm that undergo more rounds of replication within them. Wilson said that what drew her to this study was a fascination with how differences how and when sperm and eggs are produced could affect mutation rates across the genome.[52]

Across different populations, men tend to die earlier than women. Surely, cultural factors such as men smoking more and engaging in more risky behavior lead to the lower life spans of men as compared to women. But there may be something innately biological to the difference as well. Human males, like males of nearly all mammals, have but a single X chromosome, while females have two. Because males have only one X chromosome, any recessive and harmful alleles they have on the X chromosome will be exposed. In contrast, the two X chromosomes provide redundancy. If one X has a harmful recessive variant at a particular site, the other X is likely to have a fully functional variant. So females are largely sheltered from the harmful X-linked recessive variants.

If the nature of the sex chromosomes that each sex has affects this difference in longevity in humans, then we should expect to find such differences in other mammals. In fact, we do. Females of most mammals live longer than their male counterparts. And the findings extend beyond mammals. Recently, Zoe Xiricostas and her colleagues at the University of New South Wales in Sydney, Australia, examined a substantial (229 species) and widespread data set, confirming that the sex with two X chromosomes outlived the sex with only one X chromosome.[53] This was not only true in groups where the females have two X chromosomes but also was true in groups like birds, where the males have two X chromosomes. Interestingly, in cases where two closely related species are hybridized, the sex with two different sex chromosomes is most adversely affected. This observation, named Haldane's rule after J.B.S. Haldane, who first noted it nearly a century ago, remains a topic of study among evolutionary geneticists interested in speciation.[54] Whether the cause for the small differences in life span between males and females within species and the more pronounced differences among male and female hybrids are the same is not yet known.

NOTES

1. Haig (2019, p. 455).
2. Dunsworth (2018, p. 258).

3. Costa et al. (2019).
4. Buiting (2010).
5. Methyl-groups consist of a carbon and three hydrogens. For more on methylation and genomic imprinting, see Ferguson-Smith (2011)
6. The term genomic imprinting comes from Helen Crouse, who was studying sciarid flies (tiny flies that are sometimes called dark-winged fungus gnats). In these flies, the zygotes get two X chromosomes from their mother and one from their father. In the somatic cells—the cells that do not become the sperm and eggs—X chromosomes are eliminated: for females, one of the X chromosomes from the father is eliminated; for males, both of the paternal chromosomes are eliminated. Crouse used the term "imprinting" for the marks that identified the origin of the X chromosome. See Ferguson-Smith (2011) for details and references.
7. For biographical details on Haig, see Cromie (2003).
8. Haig and Westoby (1989).
9. Moore and Haig (1991).
10. See Haig (1993) and Haig (2000) for general information on the genetic conflict in pregnancy and the role of imprinting.
11. Haig (1993, p. 506).
12. Crespi (2020).
13. St-Pierre et al. (2012).
14. Haig (2014).
15. Haig (2020).
16. Phipps et al. (2019).
17. Ceredeira and Karumanchi (2012).
18. For instance, Wright et al. (2018) found that preeclampsia appears to be protective against breast cancer before menopause; in contrast, they found no relation for breast cancer after menopause.
19. See McGinnis et al. (2017) and Gray et al. (2018).
20. Arthur et al. (2018).
21. Schmidt et al. (2015).
22. Brinkman-Van der Linden et al. (2007).
23. Rumer et al. (2012).
24. For general background on the progesterone receptor and for the study about the Neanderthal receptor, see Zeberg et al. (2020).
25. Liu et al. (2014).
26. Marinić and Lynch (2020).
27. The quote is on p. 13.
28. Li et al. (2018).
29. The study examined a population of Americans from Utah who have ancestors from Northern and Western Europe. In both this population and the Han Chinese, the researchers examined the frequency spectrum of variants, looking for deviations from expected patterns in the absence of either positive or balancing selection. They used the Taijma's D test in this analysis.
30. Background information on the obstetric dilemma and Dunsworth's involvement in the studies criticizing it can be found in Dunsworth (2018).
31. Dunsworth et al. (2012).
32. Dunsworth (2018, p. 258) noted that she initially wanted to call it HAM and EGG hypothesis—with the HAM standing for "humans are mammals".
33. Warrener et al. (2015).
34. Betti and Manica (2018).
35. Hrdy (2009).

36. Sievert (2011, 2016).
37. Johnstone and Cant (2019).
38. These animals have sometimes been called "killer whales". In the wild, orcas are not a threat to humans.
39. Lambert (2019) and Lambert and Donihue (2020).
40. Hawkes et al. (1998).
41. Hawkes and Smith (2010).
42. *Ibid.*, p. 50.
43. Chapman et al. (2019).
44. Engelhardt et al. (2019).
45. Thomas et al. (2019). Note that early reproduction also reduces the risk of breast cancer. Delayed childbearing increases the risk of breast cancer. Likely, the increase in estradiol with each ovarian cycle increases the risk of breast and other cancers, not reproduction.
46. Ellis et al. (2018).
47. Croft et al. (2017).
48. See remarks by Sievert in Landau (2021).
49. E-mail conversation with Sievert, April 2021. See also Sievert (2016).
50. Crow (2000).
51. See Wilson Sayres and Makova (2011) and Wilson Sayres et al. (2011).
52. E-mail conversation with Melissa Wilson. March 2021.
53. Xiricostas et al. (2020).
54. See Johnson and Lachance (2012) and Delph and Demuth (2016) for information on Haldane's rule.

Section II

Food

8 Darwin at the Farm

SUMMARY

Evolutionary biology and agriculture are deeply entwined. From Darwin through today, evolutionary biologists and plant breeders have engaged in mutually beneficial exchanges of ideas and techniques; indeed, many plant breeders have been evolutionary biologists. Focusing on potatoes, tomatoes, coffee, and corn, this chapter examines some of the ways the principles of evolutionary biology have informed agricultural practices. With potatoes and coffee, we discuss how insufficient genetic variation led to disease outbreaks and how evolutionary principles can be used to combat their pests. We explore how artificial selection led to poorer-tasting commercial tomatoes and the efforts taken to improve their taste. One interesting feature of domesticated plants is their higher rate of recombination: this enhanced shuffling of genes appears to be a consequence of domestication selecting for more recombination, allowing for more efficient further selection in return. We conclude with the story of Nikolai Vavilov, his search for the centers of diversity of crop plant species, and his championing of seed banks.

"Agriculture has done more to reshape the natural world than anything else we humans do, both its landscapes and the composition of its flora and fauna. Our eating also constitutes a relationship with dozens of other species—plants, animals, and fungi—with which we have coevolved to the point where our fates are deeply entwined."—Michael Pollan.[1]

"The very success of the plant breeders' efforts was eliminating the raw material that made their work possible."—John Seabrook.[2]

LESSONS FROM POTATOES

In Andy Weir's science-fiction novel *The Martian*,[3] Mark Watney is a botanist turned astronaut, who finds himself stranded on Mars after a freak accident. Watney, played by Matt Damon in the movie, faces numerous challenges just trying to stay alive, not least of which is supplying his nutritional needs. His solution is growing potatoes from seeds left by his team. Given his predicament, Watney—fear his botany powers!—made a good choice. Thanks to their tubers, which are storage organs for developing plants, potatoes are calorie-dense and contain many of the other nutrients needed to sustain humans. The importance of these tubers is highlighted in the species name of the cultivated potato, *Solanum tuberosum*. Potatoes are also relatively easy to propagate and grow both from seeds and tubers. Look at the so-called eyes of a potato tuber. These are buds from which new plants can grow.

DOI: 10.1201/9780429503962-10

The potato originated in the Andes of South America several thousand years ago. Phylogenetic evidence supports a single origin of potatoes in what is now Peru.[4] Potatoes were a major source of food for peoples in the highlands of South America, including the Inca Empire. Potatoes were a prominent feature in Incan culture.[5] The Incas used potatoes with pigmented skin in dyeing fabrics. Rituals also reflected the importance of potatoes to the Incas. The deity most associated with the potato, Axomama, "potato mother", was a daughter of Pachamama, "Mother Earth". One ritual of burying potatoes underground to allow Pachamama to eat before the workers led to the method of roasting potatoes underground known as Pachamama.

The Incas developed long-term storage of potato products. Taking advantage of the sub-freezing nighttime temperatures in late fall at high elevations, they created a durable, freeze-dried potato product that the Spanish would call *chuño*. The process of making *chuño*, which required several nights of freezing and days of thawing, not only gave it a shelf life of at least a year but also removed much of the bitter and somewhat toxic alkaloidaloids. Processing allowed the Incas to eat a nontoxic food and to have potato plants that still had natural insecticides. The durability of *chuño* allowed it to become a commodity; Inca peasants could pay taxes with it, which the rulers could then distribute to pay for the various needs of the empire. After Francisco Pizarro and his conquistadors conquered the Inca Empire in the 1530s, they took over the administration of *chuño*.

Potatoes were introduced to Europe multiple times starting in the 1500s. Adoption of this tuber was slow, but eventually potatoes spread across the continent.[6] Note that the varieties that spread through Europe were unlike the ones filled with toxic alkaloidaloids that were common in the Andes. Interestingly, recent genetic analysis of historical samples suggests some hybridization with potatoes of Chilean ancestry as potatoes spread across Europe during the 19th century.[7] As Chile is further from the equator than Peru, the input of variants of Chilean ancestry may have assisted the adaptation of the potato to the longer summer days of Europe. Nonetheless, the potato set off a boon. Together with maize (corn) from North America, the ready and abundant supply of calories from potatoes helped to propel the population to nearly double over the century that ended in 1850. An ample supply of potatoes reduced the need for grains like wheat and corn, which often were plagued with fungal infestations.[8]

Ireland grew heavily reliant on potatoes during the first half of the 19th century.[9] A typical adult in 1840s Ireland likely ate several dozen potatoes each day! Why? The root cause was exploitation combined with extreme inequality. Wealthy landowners, of British descent, charged steep rent to the impoverished people who rented and worked the land. The landowners would then send surplus potatoes to sell to residents in cities. Survival for the sharecroppers meant getting the most bang for the buck; so, like the fictional Mark Watney, they turned to potatoes, which could supply lots of calories per unit area. For a while, it worked: the potato, especially when supplemented with milk, fulfilled their nutritional needs. In fact, it was working so well that the population was exploding due to more babies being born and fewer people dying. The downside to this good news was the growing dependence on the potato as more and more lived on the same stretch of land. Other socioeconomic causal factors also came into play; for instance, the Irish farmers were only given the poorest quality land, on which mostly just potatoes could be grown (the richer producing land was for the mostly English landlords, who grew a variety of other crops for their own tables).

Compounding the Irish dependence on the potato was the unfortunate circumstance that the potatoes grown there had minimal genetic variability: a single cultivar[10] of potato, the lumper, predominated the Irish landscape. The Irish grew the lumper mainly because it was one of few cultivars that could grow there: Ireland not only has a short growing season but also has long summer days. Moreover, the Irish did not have the institutional memory of the Andes farmers and the Inca people who had grown potatoes over the centuries.

This lack of genetic variation put the Irish potato at risk for the spread of pathogens: if a pathogen evolved adaptations to grow on one potato, it could grow on all. The propagation of potato from tuber to tuber also meant that there would be no shuffling of the genetic deck through sexual reproduction. The reliance on the monoculture of the potato meant that if the potato crop failed, the people of Ireland would face famine.

In time, they did. In September of 1845, the disease late blight hit Ireland. Unchecked, this disease could destroy just about any part of the potato plant—leaves, stems, and tubers. As the fields turned black with rot, they also emitted a foul, ominous odor. Hope of recovery during the following spring turned to despair in the fall as the blight reappeared. Over the next couple of years, the blight would destroy the Irish potato crop and, with it, the Irish food supply and economy. In his book *Never Out of Season*, the Duke University biologist Rob Dunn describes the cataclysmic tragedy:

> The scale of the horrors of the Irish potato famine is almost beyond our ability to conceive. The young died first, then the old, then everyone else. People died in the ditches where they slept for the night, en route to what they hoped might be someplace better. They died in their fields. Whole villages disappeared. More than a million people would die before it was all over—a million in Ireland, that is.[11]

Concern and then alarm about the potato blight spread quickly through Europe. One amateur potato grower responding to his former mentor noted, "I have to thank you for several printed notices about potatoes etc. etc. What a painfully interesting subject it is."[12] This potato grower was Charles Darwin; his correspondent, John Henslow, a botanist. At the time, Darwin was known as gentleman scholar and naturalist who had achieved prominence with his account of the voyage of the *H.M.S. Beagle* of a decade earlier and for studies on barnacles. Although Darwin had completed a draft of his ideas on evolution by then, only his family and a few of his closest friends knew of it.

Darwin had collected potatoes across northern South America and on islands, including in the Galapagos, during his 1830s voyage on the *Beagle*. Although he was aware of other plant diseases and had even identified disease on wheat from Argentina, Darwin did not record any sign of disease on the potatoes he saw. Some of the South American potatoes in Darwin's collection were used in early tests to find resistance: the ones tested were susceptible to the blight. Darwin also had collaborated with Miles Berkeley, whose studies would change the nature of research on the blight. Berkeley demonstrated that the blight was not due to weather or air but rather a biological agent, which would eventually be known as *Phytophthora infestans*. The genus was aptly named: the blight was a destroyer of plants.

Nearly two centuries later, *Phytophthora infestans* has not gone away. Blight remains a serious problem. The combined global cost of damage from blight and efforts to control it are almost $7 billion each year.[13] Potatoes also remain a vital part of the world's food supply. The world's production of potatoes—388 million metric tonnes or about 850 billion pounds—works out to about 110 pounds (or 50 kilograms) per person per year.[14] Only maize, wheat, and rice are consumed more than potatoes. *P. infestans* also infects tomatoes, causing periodic outbreaks in tomato plants. In tomatoes in the United States, these outbreaks usually are due to infected nursery stock sold in big store chains.

The agent of destruction, *P. infestans*, is an oomycete. Long thought to be fungi or closely related to fungi, oomycetes are evolutionarily far removed from the fungi. Recent phylogenetic studies show that oomycetes are members of the Stramenophiles, also known as heteroknots, a group of single-celled eukaryotes that includes kelp and other algae.[15] Like other Stramenophiles, oomycetes have a mobile stage with zoospores that are powered by flagellum. *Phytophthora* and other oomycetes have both sexual and asexual phases, both of which have been identified in *Phytophthora*. Oomycetes differ substantially from fungi in the chemical composition of their cell walls. This difference means that fungicides, particularly those that attack the fungal cell wall, should not be effective against *P. infestans* and other oomycetes.

Plants have immune systems, one component of which are the R genes. These are batteries of genes that provide resistance to specific pathogens. R genes encode products that bind to factors produced by the pathogens, targeting them for destruction; hence, differences in the product encoded by different R genes allow for resistance to different pathogens. Evolutionary change of the pathogens in their factors can allow the pathogen to evade detection, In this way, plants and pathogens can be in arms races, with plants evolving to match their R genes to catch pathogens and pathogens evolving to escape from the R genes.[16]

Breeders have tried to hold off the late blight with R genes; unfortunately, all too often, *P. infestans* evolves the ability to evade the R genes. Breeders have even tried breeding together several different R genes against the blight into the same cultivar, a process known as pyramiding, to try to stop it. Even these attempts usually fail; *P. infestans* is adept at escaping the R genes.[17] Given the frustrations with the R gene approach, potato breeders are also examining an alternative: selecting for partial resistance.[18] This partial resistance, known as field resistance, has been more difficult for the blight to escape. In fact, field resistance can be stable in cultivars over decades. This stability likely occurs for at least two reasons: the selection pressure is lower than in R gene resistance and multiple genes are involved in field resistance. Using various techniques, biologists can identify the genes involved in field resistance and engineer them into cultivars.

Potatoes have several unusual genetic features.[19] As previously noted, clones of potatoes can be propagated in several ways: through the tubers, by cuttings or through tissue culture. Most edible cultivars are tetraploids: instead of having two copies of each gene, they have four. In contrast, most of the indigenous varieties and species of potato are diploid. This difference in ploidy is problematic, as getting variants from a tetraploid into a diploid requires patience and a good deal of genetic trickery.

As noted in the epigraph by John Seabrook, the selection that plant breeders employ exhausts the variation that selection operates on. Hence, breeders often look to additional sources of new variation, one of which is wild species. Despite the challenge of the barriers due to differences in ploidy, potato geneticists are keen on prospecting for variation in wild species and landraces to introduce into cultivars.[20] They are motivated by the extraordinary diversity waiting in nature: Over 100 species of wild potato—found in a wide range of habitats to cloud forests—are out there. These wild species and landraces have been shown to contain useful genetic variation for several important traits, such as resistance to late blight, drought tolerance, and nutritional quality.[21]

In recent years, potato geneticists are rethinking how we breed potatoes. They are considering moving away from potatoes being propagated asexually as a tetraploid and toward it being bred through seed propagation of inbred diploid lines.[22] Genetic studies are easier to perform in a diploid than in a tetraploid. Assessment of the genetic contributors to traits of interest would be much easier with such inbred lines. Moreover, the computational tools that have been developed in other crop plants (which are mostly diploids) would be more easily used. In addition, the difficulty in placing useful variants from diploid wild relatives into cultivars would be eased. These geneticists suggest that the conventional wisdom that tetraploid potatoes produce more yield than diploid ones is not sufficiently grounded in data. They also argue that the reason cultivated potatoes are tetraploid may be an accident of history rather than having four copies of each gene instead of two resulting in an inherently better plant.

Considerable work would be needed to convert potatoes into a diploid. Potatoes show considerable inbreeding depression—the reduction of fitness resulting from unmasking deleterious recessive genetic variants. Making potatoes into a diploid crop would require investigation of the genetic basis of inbreeding depression, followed by purging of the variants that contribute to the inbreeding depression.[23] These efforts are underway. After they are created, inbred diploid lines would be crossed to generate cultivars that have desired characteristics.

Evolutionary genetic studies, combined with other tools, can inform the biology of how the pathogen damages the host and can potentially contribute to developing targets against the pathogen. One example is recent work on the gene *PSR2* (*Phytophthora* suppressor of RNA silencing) from a team at The Henrich Heine University in Dusseldorf, Germany. Potatoes and other plants use small segments of RNA to defend themselves against pathogens; these work by silencing the expression of the pathogen's genes. *PSR2* is a pathogen gene that acts to suppress the silencing. The Henrich Heine University team found an interesting evolutionary pattern in this gene.[24] Across different species of *Phytophthora*, the gene is well conserved; purifying selection appears to be weeding out changes. Within *P. infestans*, however, a different pattern emerges: the gene shows more variation at sites that change the amino acid, a sign of diversifying selection.[25] This is evidence that the gene has been involved in the arms race between the pathogen and the potato: turnover of function within species is expected if the pathogen is evolving to escape host defenses. Examining the expression of *PSR2* over the course of the infection reveals that it is most highly expressed in early stages of infections. The authors speculate that the gene may act as a master switch to alter the regulation of plant genes as the infection takes hold. Because the gene is conserved across different species of *Phytophtora*, it is an appealing target for further efforts to control the pathogen.

Lack of genetic variation made the lumper potato vulnerable to infection from a pathogen. Similar stories can be told about other varieties of crop. For example, the United Fruit Company grew lots and lots of one variety of banana—the Gros Michel—in Guatemala during the 1950s. Like the lumper potato in nineteenth century Ireland, the Gros Michel banana was propagated in 1950s Guatemala. Like the 1950s suburban tract houses that Malvina Reynolds mocked in her song "Little Boxes", the bananas were all the same. The lack of genetic variation set the Gros Michel up. During the early 1950s, it was all but wiped out by a fungus, *Fusarium oxysporum*.[26] After the collapse of the Gros Michel, United Fruit[27] switched to the Cavendish banana.

Insufficient genetic variation is not the only reason why crops are vulnerable to pathogens and pests. The mere presence of so many individuals of the same species—indeed, the same variety—presents potential pathogens with a rich target. Any potential pathogen that could establish itself in a crop would reap a large evolutionary reward. Moreover, crops are often more vulnerable to disease by virtue of being crops.[28] The goal of crop production is quantity (and sometimes, quality) of food. Genetic variants that grow faster or more economically are favored by our artificial selection. Often tradeoffs occur; efficient and fast growth can have the price of increased vulnerability to disease. Unless the breeder or farmer is aware of the specific disease, resistance to disease will take a back seat to faster and more efficient growth. Consequently, most crops are likely to be susceptible to disease.

As we saw earlier, Darwin had a keen interest in the potato and the blight; indeed, his potato collections were useful in the early work on resistance to blight. Darwin regularly exchanged correspondence and materials with other plant and animal breeders. These activities helped shape Darwin's argument from analogy of artificial selection to natural selection that was instrumental to his ideas on evolution. A decade after writing *The Origin of Species*, Darwin wrote a two-volume treatise on how animals and plants have been shaped by domestication.

Darwin is no exception; there has been a long history of exchange of ideas and methods and worldviews between evolutionary biologists and plant and animal breeders.[29] For one example, consider American evolutionary geneticist Sewall Wright, who with the British geneticists Ronald Fisher and J.B.S. Haldane, developed modern population genetic theory. Wright began his career in 1915 as a staff scientist at the USDA (United States Department of Agriculture) before joining the faculty at the University of Chicago.[30] At the USDA, Wright developed a system of calculating the extent of inbreeding of an individual from pedigree analysis and devising methods to minimize inbreeding. Interestingly, Wright, who lived to the age of 98, was himself a product of inbreeding: his parents were first cousins. Wright also began his interest in the random changes in the frequencies of alleles at the USDA. These experiences were instrumental to Wright's development of evolutionary theory.

MAKING A BETTER TOMATO

Tomatoes and potatoes are more similar than you might think. Appearances can be deceiving, but consider that we eat the tubers of potatoes and the fruits of tomatoes. Potatoes and tomatoes are in the same genus, *Solanum*. Tomatoes are also vulnerable to the late blight. Like the potato, tomatoes originated in South America.

Ana Caicedo, my colleague at UMASS, studies tomatoes and their domestication. Her lab recently discovered that the path to the domesticated tomato was more complex than once thought. The standard view for the domestication is that wild tomatoes in South America (*Solanum pimpinellifolium*) that were about the size of blueberries evolved into the cherry-sized tomato (the *cerasiforme* variant of *S. lycopersicum*) through domestication. Then, that *cerasiforme* variant was further improved as it moved north into Central America and Mexico and evolved into the large-fruited *lycopersicicum* variant.

The actual story is more interesting. Phylogenetic analysis performed by Caicedo's lab shows the cherry-sized *cerasiforme* variant diverged from the blueberry-sized *S. pimpinellifolium* occurred almost 80,000 years ago.[31] Thus, this initial divergence long predates humans being in the Americas. This variant then spread north into

The evolutionary path to domesticated tomatoes. Left to right: Mature ripe fruits of *Solanum pimpinellifolium* (tomato's wild ancestor), *S. lycopersicum* var. *cerasiforme* (a semi-domesticated intermediate), and *S. lycopersicum* var. *lycopersicum* (domesticated tomato).

Source: Courtesy of Jake Barnett.

Peru, while exchanging the occasional gene with *S. pimpinellifolium*. About 10,000 years ago, populations of the *cerasiforme* variant reached Mexico. Throughout this process, genetic variation remained high, suggesting that no significant bottleneck was involved in this process. The evolution of the large-fruited *lycopersicicum* variant occurred about 7,000 years ago through domestication. Low genetic variation and an excess of rare variants suggests that the domestication of the large-fruited variant entailed a strong genetic bottleneck.

Caicedo's analysis also reveals a fascinating common pattern of trait evolution. The path toward the domesticated traits has not been a straight line. For overall size and many other traits, the tomato became more like the domesticated variant we know during the transition from the blueberry-sized *S. pimpinellifolium* to the cherry-sized *cerasiforme* variant. Picture the fleshy wall of the tomato. This is the pericarp. During the transition to the *cerasiforme* variant, it grew thicker. These proto-tomatoes also became more watery and had less citric acid, the acid most associated with lemons. Such changes are all in the direction of that of the domesticated variant that we know; and yet these occurred long before humans were in South America. Even more interesting is what happened next when the *cerasiforme* variant moved north into Central America and then Mexico. During the journey north, the direction of trait evolution reversed. The pericarp became thinner. The tomato became less watery and had more citric acid. Only much later as the *cerasiforme* variant was domesticated into the *lycopersicicum* variant (less than 10,000 years ago) did the path of trait evolution reverse once again.

This revised evolutionary history of the tomato suggests that we should focus looking within *cerasiforme* tomatoes located in South America—not Central America or Mexico—for variation to cross into our cultivars. These southern variants should have alleles that contribute to larger fruits and the other traits we desire.

Outside of the closest relatives to domesticated tomatoes, there exists considerable variation across South America. About a dozen species of wild tomatoes have been discovered there; more species and more varieties within species likely are waiting to be discovered. These could be economically useful.

One widespread complaint about modern commercial tomato cultivars is their bland taste. Most gardeners produce tomatoes that are substantially tastier than those we get at the grocery store. Why? Work by Ann Powell and her colleagues at the University of California at Davis illustrates an object lesson as to how breeders bred taste out of the tomato.[32]

Starting in the middle of the 20th century, uniformity became one of the goals of tomatoes. Fruits that are evenly green at the time of harvest are preferred. These mid-century breeders had linked this uniformity to a genetic locus, known as the *uniform ripening* (U) locus: individuals that had two copies of the u variant (uu) had uniform light green coloring, while those with at least one copy of the U variant (UU or Uu) have patches of dark green near the stem. With twenty-first century molecular techniques, Powell's group characterized the U locus: it encodes a transcription factor—a gene that affects the expression of other genes—that affects how and where chloroplasts develop. Chloroplasts are the factories in a leaf where photosynthesis takes places, converting light energy into sugars. The uu individuals with the uniform light green that the breeders liked possessed defective transcription factors. Failure to

Tomato diversity. An array of ripe wild tomatoes showing the range of possible colors. Clockwise from the red fruit in the top right: *Solanum pimpinellifolium*, *S. arcanum*, *S. neorickii*, *S. cheesmaniae*, and *S. galapagenese*.

Source: Courtesy of Jake Barnett.

properly turn on the expression of the right genes then cause these tomatoes to have fewer chloroplasts and consequently less fruit sugar and starches. The selection for pretty fruit led to an inferior product.

Consumer preference is challenging to assess.[33] It is more expensive than measuring the weight of a tomato or even sequencing it. It does not work on an assembly line. Hence, breeders have been reluctant to breed for it; instead, they concentrate on what is easy to measure and what the producers want: yield, firmness, and resistance to disease. Another challenge is that many things factor into our perception of taste. How we perceive how tomatoes taste is partly due to how sugars, salts, and bitter compounds triggers our taste buds, but more is involved. Our olfaction sense—smell—enters the equation twice: once before we ingest the tomato as odor-emitting compounds in enter our nostrils and again when swallowing forces these compounds back into the nasal cavity from behind. This double involvement of odorous compounds helps explains why impaired olfaction can make food taste bland. Environmental factors, including our moods, also contribute to how foods taste.

Denise Tieman and her colleagues at the University of Florida are attempting to breed better-tasting tomatoes. They first examined the association between various compounds and consumer preference. Consumers generally prefer sweeter tomatoes. While sugar content is correlated with the perception of the tomato, some volatile,

More tomato diversity. Several ripe fruits from 13 species of wild tomatoes. Three varieties of cultivated tomato are shown in the inset.

Source: Courtesy of Jake Barnett.

odorous compounds also affect sweetness. In principle, we can have sweeter-tasting tomatoes without increasing sugar content. Consumers prefer tomatoes with minute amounts of 3-pentanone. Some odorous compounds, such as guaiacol and methyl salicylate, contribute to a smoky or medicinal flavor that consumers do not like.

Interestingly, the concentration of the odorous compound is not a good predictor of how the consumers will detect it. Moreover, detection is not a good predictor of preference: consumers can detect some compounds but not have strong preferences about the chemicals. There is no substitute for testing consumer preferences directly.

Once a chemical has been associated with consumer preference, the next step is to find genes that are associated with the chemical. While this work is painstaking, it is feasible, especially for crops like tomatoes that have well-developed genome resources. Recent work has shown that genetic variation associated with these desired traits does not only come from changes in the nucleotides at specific positions. In addition, duplications and deletions of genes—what are sometimes known as structural variation—are also important.[34] Using the cutting-edge technique nanopore sequencing, tomato geneticists have uncovered substantial structural variation among different genomes of tomatoes and have linked this variation to traits. One structural variant affects the expression of a gene affecting the production of guaiacol, a chemical whose presence gives tomatoes a medicinal flavor. The final step would be breeding or engineering via genome editing the desired genetic variants into tomato cultivars.

Considerable genetic variation exists within the wild tomato, *S. pimpinellifolium*, some of which is correlated with traits of interest. Further evolutionary genetic characterization of this species would assist the search for useful variants that could be crossed into cultivars. A recent study by Leonie Moyle and her colleagues at Indiana University finds that genetic differences within the wild species are more closely associated with environmental features, such as temperature and precipitation patterns, as well as soil conditions, than with geographic distance.[35] These findings suggest that natural selection imposed by the local environment is shaping this species. They also suggest, for instance, that searching for drought-resistant variants in an arid area is likely to be fruitful.

MORNINGS ARE FOR COFFEE AND CONTEMPLATION

I enjoy a good cup—or three—of coffee; indeed, numerous cups have been consumed in the writing of this book. Coffee is more than just a pleasant beverage and a mild stimulant; about 100 million people across the world make a living from it. Much of coffee production occurs in areas where conditions—poor soil, steep slope, forest—prohibit the production of other crops. Most of these areas are in relatively poor countries.

Coffee comes from species in the genus *Coffea*. Nearly all the coffee grown and consumed today comes from two species: about 60 percent comes from *C. arabica* (Arabica coffee), with the remaining 40 percent from *C. canephora* (robusta coffee). Prior to the 20th century, Arabica coffee dominated the market, but several minor species each contributed tiny but respectable shares. In the last 100 years or so, robusta has risen at the expense against Arabica coffee and the minor species.[36] Like nearly all species in the genus, *C. canephora* is a diploid. In contrast, *C. arabica* is a tetraploid that begun as a hybrid between *C. canephora* and *C. eugenioides*.[37] There is considerable variation among different populations and landraces of *C. canephora*, with populations from Uganda being genetically more distant than other populations. This species appears to have much variation there that potentially can be tapped.

The genus *Coffea* contains over 100 known species, distributed across Africa, Asia, Australia, Papua New Guinea, and various islands. Although the genus clearly evolved in the Eastern Hemisphere, phylogenetic analysis is not yet able to pinpoint its source any closer: Africa is a possibility, but so is Asia or the Arabian Peninsula. Several lineages evolved caffeine production independently. While some moderate caffeine producers can be found in Madagascar, all the most caffeine-rich coffees emerged relatively near the equator in West and Central Africa. A likely reason why caffeine-rich coffees evolved in those regions is that insect pests are more abundant in the humid tropics. Caffeine is commonly found in plant leaves as an insecticide. The genetic basis of caffeine production is relatively simple; a single gene controls production, and modifier genes modulate how much is produced.[38]

Coffee was brought to the Americas by various waves of immigrants. It grew well in the tropics and near tropics of the Americas. It has supported untold thousands, if not millions, of farmers over the centuries. But now across much of Central America, coffee plants are dying.[39] Infected with a fungus, the plants are stunted, and most of their leaves have withered away. The upper sides of the remaining leaves are marked with signs of death: yellow spots with a brown bullseye. But it is the rusty orange dust of the undersides of these leaves that give the name to the disease: rust, specifically coffee rust (*Hemileia vastatrix*). This fungus destroyed the livelihoods of nearly 2 million farmers across Central and South America between 2012 and 2017.

Rust was not supposed to be a problem in the Americas. Periodic outbreaks of the rust had plagued coffee production in the tropics of Africa and Southeast Asia for a century and a half, but the Americas were supposed to be free of it. A 1952 map from the USDA had bifurcated the world: east of the Prime Meridian was labelled "Diseased", west of the line was declared "Not Diseased". The Atlantic Ocean was supposed to protect us. Rust was not supposed to be a problem in the Americas. Or so we thought.

Slowly and then suddenly, rust came. By 2012, it was a major pest in Central America. Rust experts suggest that changing climate made conditions more favorable for the rust in the Americas. Hotter and more humid weather with more intense but less frequent rains allowed the rust to accelerate its life cycle. It is possible that the rust has evolved to better infect coffee plants in Central America, but strong evidence of that has yet to be found.

From the 150 years of rust in the Eastern Hemisphere, we do know that the rust readily can evolve to evade resistance. Over the decades, considerable effort had gone into the search for coffee plants that were resistant to the rust.[40] Several rust-resistant variants of Arabica coffee—such as Coorg coffee and Kent coffee—were found. These variants held out for a few years or a couple of decades, but eventually the rust evolved resistance. Early in the 20th century, a variety of *C. caenehora*—which became the robusta coffee—was found to be highly resistant to rust and maintained the resistance. The early robusta coffee had a harsh "grassy" flavor, which Dutch breeders were able to breed out, leaving a neutral flavor. Several coffee blends use Arabica for flavor and robusta for bulk. Instant coffee is usually mostly robusta coffee.

Coffee rust belongs to a large order of fungi known as the Puccinales.[41] These rusts have an unusually complex life cycle. Details vary from species to species, but the basic scheme of the cycle boils down to alternation of two phases, each done on a different type of host. In one phase—known as the aecial—haploid entities come together to form a dikaryon, two haploid nuclei in each cell. These dikaryons disperse to the next phase, known as the telial phase. On the telial host, the rust reproduces asexually. This asexual reproduction may continue indefinitely until environmental conditions change, at which point meiosis occurs. Ultimately, haploid spores are brought back to the aecial host. Coffee plants are the telial host of coffee rust; we do not yet know what the aecial host is.[42] Based on what we know about other rusts, Catherine Aime at Purdue University argues that the aecial host of the coffee rust is likely a gymnosperm (conifer). Determining which, if any, plant is the aecial host is key in efforts to control it.

Phylogenetic analysis of the coffee rust found a surprise: the rust may not be a single species, as had been previously thought.[43] Instead, the rust has three distinct clusters that appear to be reproductively isolated. Only one of these clusters, the C3 group, attacks the tetraploid Arabica coffee. The other two clusters can infest diploid coffee species. Interestingly, the C3 isolates are consistently unable to infect *C. canephora* (the robusta coffee). These results suggest a tradeoff: performing well on one type of coffee may impair ability to perform well on others.

In addition to the devastation of coffee species by the rust in both hemispheres, coffee plants face other burdens, including other pest species and a changing climate. The risks from these burdens compels investigation of new species to develop new species of coffee for cultivation. An especially promising species is the West African coffee *C. stenophylla*, which has been cultivated for more than 200 years and has superior taste.[44] Unfortunately, more than half of wild coffee species are at risk of extinction.[45] Moreover, we lack repositories for the genetic material (germplasm) for many of these species.

UNINTENDED CONSEQUENCES OF IMPROVING CROP YIELD

The Green Revolution led by Norman Borlaug and others helped feed many hundreds of millions of people across the world and led to greater food security. From 1960 to 2000, wheat tripled in yield, with doublings in both rice and maize.[46] With the projected continued increases in population, increases in yield are also needed, else more will go hungry.

These gains generated by the Green Revolution came from selective breeding and the use of fertilizer. One such effort was selection for crops that tolerate being planted at increasingly higher densities. For instance, varieties of dwarf rice were developed that could be grown at high density, allowing for much-increased yield. The dwarfing allele could readily be introduced into different local rice varieties. This allele was not an unmitigated success, however; varieties with it often were sensitive to drought. This is not a trivial concern, as nearly half of the areas of the world where rice is grown experiences some degree of drought in most years.

How geneticists address this challenge depends in part on the evolutionary history of the association between dwarf status and drought sensitivity. One possibility is that the allele that causes dwarfism also causes the sensitivity. Biologists call this type of physiological connection pleiotropy—from the Greek *pleion* and *tropos*, meaning many ways. Pleiotropy is common in biology. For example, chickens that have a dominant variant of the frizzle gene not only have outward-curling feathers, but also have higher metabolic rates and reduced egg laying capacity than those without it.[47] The other possibility is that the dwarf status and the drought sensitivity are due to two genetic variants that are closely linked on the same chromosome. Due to this linkage, the variant conferring dwarfness and the variant conferring drought sensitivity are usually inherited as a unit. Whether the association is due to pleiotropy or linkage is consequential: linked variants can be teased apart by crossing, but pleiotropic associations cannot be broken by breeding. In the case of rice dwarfing, the association with drought sensitivity fortunately is due to linkage. Researchers at the International Rice Research Institute in the Philippines were able to break the association through crossing.[48]

Similar increases in density have occurred in maize. The average density went from about three plants per square meter (yard) in the 1930s to about eight per square meter (yard) in the 2010s. Producing plants that can grow at such high densities is a challenge. One of the ways plant biologists have met this challenge is by altering the architecture of the plants, through selective breeding and more recently through biotechnology. Specifically, having leaves that are more upright allows for more efficient photosynthesis. With more upright leaves, the leaves shade each other less. The extent to which leaves are upright is commonly expressed with by the angle of the leaf, with more upright leaves having lower leaf angles.

In order to continue to improve the performance of maize in still greater planting densities, further decreases in leaf angle are needed. Various mutations can result in reducing the leaf angle, but these typically come with undesirable side effects. Recently, researchers at the Beijing Key Laboratory of Crop Improvement took advantage of natural variation within teosinte, the precursor to maize, to further improve maize yield at high density by reducing the leaf angle.[49]

The researchers undertook a series of crosses between maize and teosinte in order to produce what geneticists call recombinant inbred lines. Individuals in each line all have the same genotype, with some regions of the genome coming from teosinte and the rest coming from maize. Accordingly, the phenotypes of multiple individuals of the same genotype can be tested. Using these recombinant inbred lines, the researchers were able to identify a region of the genome that affects leaf angle. With additional genetic tricks, they determined the gene, which they call *UPAC2*, for *Upright Plant Architecture 2*, responsible for the leaf angle difference. *UPAC2* turns out to be a regulatory sequence that affects the expression of other genes. We need not go through all the gory details of that genetic pathway, but the bottom line is that a very small change (just two nucleotides) in this sequence has a noticeable effect on the leaf architecture. Furthermore, this regulatory allele that results in more upright leaves is not present in any known maize plants and is relatively uncommon in teosinte. Quite likely the allele was lost during the genetic bottlenecks involved in the domestication of maize.

So the teosinte allele of *UPAC2* results in more upright leaves. Will it produce the desired effect of greater yield of maize under high-density growing conditions? The answer seems to be yes. The Beijing Key Laboratory of Crop Improvement researchers show that genetically engineering this teosinte allele does improve yield of elite crop lines under high-density planting conditions.

Linkage of deleterious variants with desired traits is one of the ways deleterious traits can accumulate during selective breeding. Such accumulation also occurred during the original domestication process. This phenomenon has been called the cost of domestication.[50] One intriguing aspect of the cost of the domestication is that it is weakest in regions of the genome where recombination is most frequent.[51] Fewer deleterious changes occurred in those regions compared to those parts of the genome where recombination was rare. The shuffling of chromosomes during meiosis reduces the likelihood of deleterious alleles hitchhiking along with the ones under selection during domestication and diversification.

Evolution can be subtle and complicated; indeed, it can be seemingly counterintuitive at times: one of these occasions is the oddness that the evolutionary pressures and properties themselves—like recombination—can themselves be acted on by selection. The extent of recombination—that is how often chromosomes are shuffled—is thus both a condition that affects evolutionary outcomes and an outcome of evolution. Recombination rates can be selected on in the lab, often using *Drosophila*. A few generations of selection can yield flies with considerably more or considerably less recombination. Lab selection can even be targeted at specific parts of chromosomes, altering recombination rates between this pair of genes but not others. Although less is known about recombination in natural settings, we know that closely related species and even populations within species can have significantly different recombination milieus.

Intriguingly, recombination rates are higher in most domesticated plants and many domesticated animals (though not generally in domesticated mammals). Why? One possibility is that the process of domestication selected for more recombination in crops and animals. Another hypothesis is that the progenitors of domesticated crops and animals happened to have high rates of recombination to begin with because for some reason such organisms were more readily domesticated. Evolutionary biologists call this phenomenon preadaptation. Note that preadaptation and the selection routes to higher recombination could both be taking place, as the hypotheses are not mutually exclusive. Nevertheless, the hypotheses do make clear predictions that can be tested. If the preadaptation hypothesis is correct, then we would expect progenitors of domesticated crops to have higher rates of recombination than other plants. The selection hypothesis, on the other hand, predicts that crops would have enhanced recombination compared with their progenitors. Support for both hypotheses would consist of crops having more recombination than their wild relatives, which in turn, would have more recombination than typical plants.

Jeff Ross-Ibarra, then at the University of Georgia, set out to test these hypotheses.[52] He was interested in looking at the phenomenon across an enormous group of vascular plants. Directly measuring recombination rates in such a large group was prohibitive for at least a couple of reasons: it would entail crossing lots and lots of individuals even for a single species, and many species of plants do not have the easily scorable, heritable markers that would be required for such an analysis. Instead, Ross-Ibarra looked at published data on the extent of chiasma—the cytological manifestation of recombination—as a proxy. It's not a perfect proxy but is a good one, as the correlation between recombination frequency and chiasma rate is strong. Ross-Ibarra was able to get data from 196 species in total, 46 of which were domesticated species. The data were clear: domesticated plants had a substantially greater recombination frequency (based on chiasma) than their progenitors, but the progenitors did not have a high recombination. Accordingly, the results support the selection hypothesis but provide no support for the preadaptation hypothesis. Domesticated plants shuffle their genes more during meiosis due to selection for recombination.

VAVILOV'S LEGACY

Nikolai Vavilov had a vision.[53] Vavilov, a rising star among Russian geneticists in the 1920s, drew inspiration from Darwin's command of multitudes of facts about domesticated plants and animals. He had been mentored by William Bateson, one of the rediscoverers of Mendel's principles. He had traveled across much of Europe, Asia, Africa, and the Americas. Along the way, he sought out and collected seeds from wheat, rye, maize, potatoes, and countless other edible plants. He was in search of the origins of these domesticated plants. Here, Vavilov argued, would be the mother lodes of genetic variation—some of which would be resistant to insect pests, while others would be more tolerant of drought. Vavilov pointed to the mountainous areas as the source of these genetic treasures.

In the 1920s, Vavilov also started the first international seed bank in Leningrad of these plants. Much like the books and periodicals stored in a library, the specimens in this seed bank serve as a repository of genetic diversity. Such genetic diversity could be used to breed new plants or help save existing ones from extinction. The combination of his passion to collect lots and lots of wild seeds, his knowledge of likely sources to get the desired seeds, and his foresight to preserve these collections established Vavilov as one of the leading plant evolutionary geneticists of his time.

Vavilov had many successes and gained fame over the next two decades. He also acquired a powerful adversary—Trofim Lysenko. A decade younger than Vavilov, Lysenko was ambitious and unscrupulous. Lysenko had grown to reject Mendelian genetics and Darwinian evolution. As Lysenko's power grew, such ideas became increasingly dangerous. Vavilov was protected for a time, but eventually was imprisoned in 1941 and would die of starvation two years later.

Vavilov's seed bank succeeds him. It almost did not. In the fall of 1941, Hitler's armies surrounded Leningrad; the station, like the rest of the city, was under siege. The Nazi's strategy was to wait while the residents of Leningrad starved. Because potatoes do not store well over the long-term as seeds, the station had grown potatoes annually to keep the genetic variation—the germplasm—alive. So, during the fall and then the long winter of 1941, the station had stored a stockpile of potatoes. As the siege continued and the hunger grew, the workers at the station guarded the potatoes—and the future—rather than eat them.

Similar large seed banks exist across the world today. Just as in as in siege of Leningrad, contemporary scientists have taken great pains to protect them. During the US invasion of Iraq in 2003, the Iraqi national seed bank was destroyed.[54] Fortunately, some of the most valuable seeds were shipped in a box to Aleppo, Syria, site of the seed bank Icarda (International Center for Agricultural Research in the

Dry Lands). Although Aleppo itself would become shelled during the Syrian civil war, valiant efforts by Ali Shehadeh and others at Icarda have safeguarded multitudes of seeds.[55] As of 2017, Icarda contains seeds of 155,000 varieties. Such variation is likely to be ever more important as the planet warms.

Seed banks are being used in current research to understand the genetics and evolution of plants and to improve crops. Beginning in the 1940s, Charley Rick at the University of California–Davis made expeditions to South America to collect samples of wild tomatoes. Seeds from Rick's trips are still stored at a seed bank at the University of California-Davis, where they are available for researchers across the globe. One such researcher is Jake Barnett, a graduate student with Ana Caicedo, who we met in the tomato domestication story earlier in the chapter.[56] Barnett is using Rick's seeds to see how well these South American varieties can grow in the climate of New England and to test their ability to withstand insect pests. As of May 2021, the studies are ongoing. Stay tuned!

NOTES

1. Pollan (2004, p. 10).
2. Seabrook (2007, p. 67).
3. Weir (2014). *The Martian*. Broadway Books.
4. Spooner et al. (2005).
5. De Jong (2016).
6. In an e-mail exchange, Cathie Aime at Purdue University informed me of an interesting, though perhaps apocryphal, explanation for why potatoes may have had a difficult time getting introduced to Britain. Supposedly, Sir Walter Raleigh brought the potato to Queen Elizabeth's court. Not knowing what to do with that odd vegetable, her cooks threw away the tubers and cooked up the toxic greens, poisoning everyone.
7. Gutaker et al. (2019).
8. De Jong (2016).
9. The account of the Irish potato famine is taken mainly from Dunn (2017).
10. Cultivars are plant varieties on which selective breeding has acted to produce desired characteristics. See also the definition of landraces in note 20.
11. Dunn (2017, p. 13).
12. See Ristaino and Pfister (2016) for information on Darwin and potatoes. The quote is on p. 1035. Incidentally, in their exchange, Darwin and Henslow agreed that those of means should abstain from eating potatoes during the famine in order to allow people of less means to eat the potatoes at a price as low as possible.
13. USA Blight: A National Project on Tomato & Potato Late Blight. http:usablight.org. Accessed September 25, 2020.
14. See http:www.potatopro.com/world/potato-statistics. Accessed May 13, 2021.
15. Fry (2008).
16. Ellis et al. (2000).
17. Fry (2008).
18. Collins et al. (1999). The principles behind selection for field resistance resembles the adaptive therapy that Gatenby has advocated for in treating cancer. See Chapter 6.
19. Watanabe (2015).
20. Landraces are varieties of a domesticated plant (or animal) that have adapted to local conditions of an area. In contrast with cultivars, conscious artificial selection does not play a large part in the generation and maintenance of landraces. See also note 10.

21. Bethke et al. (2017).
22. Jansky et al. (2016).
23. See Zhang et al. (2019) and Bachem et al. (2019) for details on the genetic basis of inbreeding depression in potatoes.
24. De Vries et al. (2017).
25. The evidence that diversifying selection is operating on the gene comes from results of the McDonald-Kreitman test showing an excess of polymorphisms within species that change the amino acid. See McDonald and Kreitman (1991) and Johnson (2007).
26. Dunn (2017).
27. United Fruit has since changed their name to Chiquita Brands International.
28. Dunn (2017, p. 33).
29. Johnson (2016).
30. See Crow (2010) and Provine (1986) for detailed biographical information about Wright.
31. Razifard et al. (2020).
32. Powell et al. (2012).
33. See Klee and Tieman (2018), Tieman et al. (2012), and Tieman et al. (2017) for general studies of the genetics of tomato flavor.
34. Alonge et al. (2020).
35. Gibson and Moyle (2020).
36. Davis et al. (2020).
37. Charr et al. (2020).
38. Hamon et al. (2017).
39. McKenna (2020).
40. McCook and Vandermeer (2015).
41. Aime et al. (2018).
42. E-mail correspondence with Cathie Aime. September 2020.
43. Silva et al. (2018).
44. Davis et al. (2020).
45. Davis et al. (2019).
46. Mann (2018).
47. Vikram et al. (2015).
48. See Lobo (2008) for more information on frizzle and other examples of pleiotropy.
49. Tian et al. (2019).
50. Moyers et al. (2018).
51. Lu et al. (2006).
52. Ross-Ibarra (2004).
53. See Dunn (2017) and Pringle (2008) for information on Vavilov.
54. Seabrook (2007).
55. Sengupta (2017).
56. It's been a long journey for wild tomatoes. UMass Amherst Center for Agriculture, Food, and the Environment. https:ag.umass.edu/news-events.highlights/its-been-long-journey-for-wild-tomatoes. Accessed May 15, 2021.

FURTHER READING

Dunn, R. 2017. *Never Out of Season: How Having the Food We Want When We Want It Threatens Our Food Supply and Our Future*. Little, Brown and Company.

9 Managing Agriculture

SUMMARY

Agricultural crops are threatened by insect pests as well as by weeds; both of these natural enemies cause billions of dollars of damage and can affect food supply. We have developed multiple agents and tools to curb these pests; however, the weeds and the insect pests can evolve resistance to the agents used to control. For herbicide resistance, we focus on the use of glyphosate (trademarked as Roundup) to thwart weeds. This chemical has been genetically engineered into crop pests; it worked great for few years, but then weeds evolved resistance to it. As a case study, we look closely at the evolutionary genetics of glyphosate resistance to it in morning glories. One key question is whether there is a cost of resistance; for morning glories, the answer is "it depends". We discuss the factors that lead some insect species to have a wide diet niche and to become pests. Finally, we consider the use of biocontrol and gene drives to manage insect pests.

"Let a man profess to have discovered some new Patent Power Pimperlimpimp, a single pinch of which being thrown into each corner of a field will kill every bug throughout its whole extent, and people will listen to him with attention and respect. But tell them of any single common-sense plan, based on scientific principles, to check and keep within a reasonable bounds (sic) the insect foes of the farmer, and they will laugh you to scorn."—Benjamin Walsh.[1]

"Weeds are smart. They keep figuring out how to survive whatever we throw at them. The reason some people ended up with herbicide-resistant weeds is that they often used a really good product over and over again and the weeds weren't exposed to other control practices."—Anita Dille.[2]

BENJAMIN WALSH, DARWINISM, AND PEST CONTROL

During the summer of 1992, between my defense of my PhD dissertation and the start of my postdoctoral fellowship, I spent three weeks in Grant, Michigan, helping a colleague, Jeff Feder, with his field work on the apple maggot fly, *Rhagelotis pomonella*. About the size of a house fly, the apple maggot fly has alternating series of black and white bands on both its wings and abdomen. It is aptly named, as this fly's life cycle revolves around apples. Adult females and males meet on apples to mate. Mated females lay eggs on apples, which then hatch into larvae that feed on the apples.

We spent the days collecting flies from trees. During the evenings, we would use whiteout to mark the flies captured.[3] We would release these marked flies the next day. Capturing, marking, releasing, and recapturing the flies enabled us to track their movements as part of a larger project to investigate a possible example of speciation taking place in real time in the same place. This so-called sympatric speciation, wherein populations become separate species without geographic barriers, has long

DOI: 10.1201/9780429503962-11

been a fascinating but contentious topic in evolutionary biology.[4] Sympatric speciation is a source of controversy in part because evolutionary theory predicts that gene flow between populations that occur in the same place will wipe out any differentiation between the populations unless there is strong selection acting to diversify the populations. How likely speciation can occur in the absence of geographic barriers remains an active research topic.

The apple maggot fly has been one of the best studied cases of the first steps of sympatric speciation. You see, apples are not the original host of the fly. Before the middle of the 19th century, the fly lived exclusively on hawthorn (*Crataegus*), whose fruit resembles crabapples.[5] In the decades that preceded the Civil War, some individual *Rhagelotis* flies that had been living on hawthorn jumped to apple. Over the years, the apple type of flies begun to diverge from the original hawthorn flies—in their DNA sequences, in their behavior, and in developmental timetables. Feder and many others have tracked and characterized these differences to advance our understanding of the initial stages of speciation.

The first person to posit that these flies are an example of sympatric speciation was Benjamin Walsh, author of the first epigraph. Walsh was also an early champion of both Darwinian evolution and the need to control insect pests.[6] Born in 1808, Walsh was a classmate with Darwin at Cambridge University. Like Darwin, Walsh was fascinated by insects; indeed, Walsh had been impressed with Darwin's beetle collection. After graduating, Walsh stayed on in an academic position at Cambridge, but that did not work out. During his childhood, Walsh's father was caught embezzling funds. The father narrowly missed being executed but did spend considerable time in jail. That family scandal likely dimmed the younger Walsh's chances for promotion. Moreover, Walsh expressed growing dissatisfaction about the university's practices and the British clergy. At the age of 30, Walsh left the United Kingdom to head off to the United States, settling in Illinois, where he became a farmer and scholar. Walsh maintained interests in entomology, where he eventually became the first state entomologist of Illinois.

Upon reading a copy of *The Origin of Species*, Walsh resumed correspondence with his former classmate and became a dedicated advocate of Darwin's theory of evolution. Walsh also tirelessly championed the use of resources to control insect pests, arguing for the need for more studies of insect natural history (and what would now be called evolution) to guide their management. This passage summed up the urgency Walsh felt:

> Were a foreign army to invade our shores, our law givers would vie with one another in large expenditure and preparation to oppose such invaders. No one would think of objecting. And yet the ravages of such an army would be insignificant in comparison with an army of insects. . . . [W]here are the "army appropriations" in amount, to meet and fight this army of insect invaders?[7]

Walsh then claimed that the expenses to fight the insect invaders totaled only about $250 a year, while the price of the destruction from insect pests to crops was about $1 million. Even taking inflation into account, the costs involved today are orders of magnitude more than they were in Walsh's time. Walsh would be pleased, however, by how the management of agricultural pests has grown into a well-funded applied science. We have generated many agents, tools, and techniques to control the insect pests that nibble on plants, the weeds that compete with the crops for nutrients and space, and pathogens that sicken them.[8]

The use of these pest management measures has substantially expanded in recent decades, however, and along with expanded use has come the evolution of resistance. Unlike the temperature, precipitation patterns, and other physical features of the environment, the natural enemies of crop plants are Darwinian creatures that can and often do evolve resistance to whatever killing agents we impose.

HERBICIDE RESISTANCE

If weeds were permitted to grow unchecked, we would lose about half the yield of corn and soybeans.[9] The cost of those losses of just those two crops in the United States alone would exceed $40 billion each year. Losing these crops would likely increase food insecurity both in the United States and across the world. But controlling weeds is fraught with problems. Not only are the agents that we use to control the weeds potentially harmful to the crops or to wildlife (including pollinators) or to humans, but weeds can adapt to those herbicides. As Anita Dille from Kansas State University said in the epigraph, they are smart. Not smart in the sense of being conscious or in having a brain, but smart in the sense of being adaptable. Because they can crank out lots of progeny and because they can quickly cycle through generations, weeds can evolve quickly. They can solve evolutionary problems quickly. They are especially good at solving these problems when, as Dille noted, we keep presenting the same problem to them.

First predicted by the plant ecologist John Harper in the mid-1950s, herbicide resistance is commonplace across a wide range of plants: more than 200 species of plants widely scattered across the phylogenetic tree have been documented as having evolved resistance to one or more herbicides.[10] There are two general classes of adaptations weeds have evolved in response to herbicides: target-specific resistance and non-target-site resistance. The former are responses in the gene that the herbicide is targeting. Such responses are almost by definition single gene responses. For instance, numerous compounds in at least five different families of herbicides inhibit the enzyme acetolactate synthase. This enzyme sits at the top of the pathway for the synthesis of specific amino acids[11] Because the chemicals that target this enzyme are extremely potent in killing weeds, they can be effective at low doses. They are not very toxic to animals. Moreover, this class of herbicide selectively kills weeds over crops. For these reasons, these compounds are popular weed killers. Not surprisingly, resistance to these compounds has frequently evolved. This resistance is specific to the target site: more than 100 species have evolved resistance to the compounds by changes in the acetolactate synthase; moreover, most of these resistance mutations are due to the same changes at two different sites at the enzyme—one at position at 197 and one at position 574. In contrast with the target-site resistance mutations, the non-target-site resistance mutations are much more difficult to predict and to find. Non-target-site resistance also often involves changes at multiple genes. We know much less about the non-target responses than the target ones, but recent studies suggest that the former is more important than previously thought.

For much of the rest of this section, we'll focus on one commonly used herbicide: glyphosate. In 1974, the agricultural company Monsanto[12] initiated commercial use of its patented weed killer Roundup, a glyphosate-based compound.[13] Glyphosate works by targeting an enzyme with a very long name, 5-enolpyruvyl-3-shikimate phosphate synthase (EPSPS).[14] The enzyme's name provides clues as to its function. Synthases are enzymes that catalyze the synthesis of a product. The word "shikimate" refers to the shikimate pathway, a network of reactions used to make certain amino acids. Hence, this enzyme contributes to the production of amino acids that are used in proteins. By blocking this enzyme, glyphosate interferes with protein synthesis. Because the shikimate pathway is found only in plants and microorganisms—and not in animals—glyphosate ought not interfere with protein synthesis in animals, though it can affect the microbiota of animals.

Glyphosate works slowly and steadily. It also can travel easily in the phloem, the vascular system plants use to move nutrients and other materials down from the leaves to other parts of the plants. Because it acts slowly and because it is easily transported, glyphosate can kill all of the actively growing parts of the plant, including the roots. Hence, unlike other weed control agents, glyphosate does not kill just where it is applied. For these reasons, it is very effective at controlling perennial weeds.

Because glyphosate kills plants nonselectively, harming both crops and weeds, its initial use in crop fields was limited. It could be used to clear away plants before crops were planted—a practice known as burndown. Similarly, it could be used after crops were harvested. It had also been used on roadsides, around railroad tracks, and in other areas away from crops. However, glyphosate could not be used as a matter of course while crops were growing.

With the mid-1990s came a game-changer: genetic modification (GM) to give the crop plant a bacterial gene that enabled the crop plant to be highly resistant to glyphosate. With GM plants—now with the name Roundup Ready—the weed killer could be applied around the crops without concern for hurting the crop. The decade that followed the initial use of GM soybeans was a golden age. Within about a decade, about 90 percent of soybeans, maize, and cotton grown in the United States was genetically modified for glyphosate. The farmers loved it. Managing weeds with the simple application of glyphosate on GM crops was easier, cheaper, and more effective than the previous use of multiple weed killers. This new method was particularly beneficial for small farms.

Moreover, Roundup Ready has environmental advantages, as noted by Mark Lynas, a British journalist who had started as an anti-GM activist who subsequently became persuaded by the benefits of GM foods. In his nuanced and thoughtful book *Seeds of Science*, Lynas points out that Roundup Ready

> facilitated the wider adoption of no-till farming and conservation farming, where farmers largely stopped ploughing and drilled seeds through crop residue left on the ground. The tractors that had previously been needed for tilling and repeated weed control hoeings idled in the farmyard, leading to savings on fuel.[15]

The undisturbed soil also had more carbon—good for limiting carbon emissions—and was less susceptible to erosion.

Stephen Duke at the USDA described the middle 1990s as "the golden age of weed management". With genetically modified crops producing glyphosate, we had "simple, economical, and outstandingly effective weed management with reduced environmental impact, using a herbicide active ingredient that was considered to be virtually non-toxic to humans".[16] But even during this golden age, there was Darwinian trouble brewing. Although resistance to glyphosate had not evolved in field populations prior to the commercial use of GM resistant crops, there was concern that increased use could result in the evolution of resistance. Monsanto tried to allay those concerns. A group of the company's researchers had argued in a 1997 commentary, one year after the introduction of GM crops, that glyphosate resistance would still be difficult to evolve.[17] One line of evidence that the Monsanto researchers presented was the extreme difficulty they had in obtaining resistance in crops through selective breeding. The researchers noted that the failures to obtain resistance in crops had necessitated use of GM technology. Another line of argument they presented involved examining the possible mechanisms by which glyphosate resistance could happen and then showing that such changes were unlikely to evolve. For instance, one way to evolve resistance would be a change in the target site—the EPSPS enzyme, in this case. The Monsanto researchers maintained that resistance would be unlikely due to changes in the enzyme. They similarly dismissed other ways resistance could evolve.

The Monsanto researchers were quickly proven wrong. A year after the commentary arguing for resistance being extraordinarily difficult to evolve came a report of glyphosate resistance in rigid ryegrass (*Lolium rigidum*) populations in southern Australia.[18] Rigid ryegrass, which is a weed in some locations but is used as a crop for foraging livestock in others, is known to quickly evolve resistance to herbicides. Hence, resistance to glyphosate in it should have been a signal that resistance could happen in others. Also troubling was the finding of cross-resistance: populations of rigid ryegrass that were resistant to glyphosate also had some resistance to another herbicide, one that they had not encountered before.

In the two decades since the reports of resistance in rigid ryegrass, glyphosate resistance has been found in more and more weeds. A decade after he had found resistant rigid ryegrass—say that ten times fast!—Stephen Powles reviewed what was known about resistance to glyphosate and how it evolved. The common thread he observed was "that glyphosate-resistant weeds can evolve where there is insufficient diversity in weed management systems. Conversely, maintenance of diversity can lead to glyphosate sustainability".[19] Two end points on a continuum illustrate this theme. In Argentina, where weed management for soybeans—which are 99 percent GM—relied almost exclusively on applying glyphosate, resistance evolved readily. In Canada, multiple measures are used to manage weeds. Moreover, GM crops are rotated on each field, ensuring sufficient gaps between glyphosate use on that field. Hence, selection pressures for glyphosate should be lower in Canada. And indeed, resistance has been slower to evolve there. In the epigraph, Anita Dille had noted that weeds excel at adapting to the same control measures; they do less well, she contrasted later in the interview, when challenged with different measures. Weeds are smart Darwinian creatures, but they are not omnipotent.

We'll dig in further to glyphosate resistance in a single genus of plants—the genus *Ipomoea*. Plants in this genus are known as morning glories due to the showy but ephemeral flowering of many species in the genus.[20] Their vibrant flowers bloom early in the morning but will have wilted soon after high noon. These flowers have led several species to be commercially valued as ornamentals. Evolutionary biologists have also become interested in the morning glories. Indeed, ecological and evolutionary genetic research in these morning glories has been anything but ephemeral: over the last few decades, this genus has become a model system for studying adaptation in an ecological setting. In addition to the showy flowers, morning glories can turn out generations quickly and produce large numbers of offspring. These traits have facilitated the development of morning glories as a model system. They also have helped these species become weeds: several *Ipomoea* species are agricultural pests.

Ipomoea purpurea, the common morning glory, can be found across much of the United States. Although this tall flowering vine is a weed that competes with several important crops, including cotton, soybeans, and peanuts, it is also a prized ornamental because it has large, trumpet-shaped, vibrantly colored flowers. This plant originated in Mexico, where native populations have mostly purple flowers.[21] Populations in the southeastern United States vary greatly in flower color: in addition to purple, flowers there can be blue or white or shades in between. In contrast with the diversity of floral color, genetic diversity is low in the populations in the southeastern United States. This finding and other analysis support the hypothesis that the southeastern US populations arose from the Spanish explorers taking Mexican morning glories home, where they would be cultivated across Europe. Latter settlers would then bring the morning glories back to North America.[22]

In the southeastern United States, populations of morning glories vary considerably in the degree of resistance to the herbicide.[23] On average, the dose required to kill half the morning glories is about the same as the recommended dose for use in the field. But averages are deceiving; that recommended dose would wipe out nearly all plants of some populations and barely touch others. Interestingly, the populations do not cluster with respect to resistance. Instead, we see a mosaic of herbicide resistance: some populations that are highly resistant are near others that are very susceptible. This mosaic pattern strongly suggests that resistance evolved multiple times. Analysis of DNA data from these populations is also consistent with multiple independent cases of evolution of resistance. Moreover, the DNA data suggests that gene flow among the populations occurred before the widespread use of the herbicide. It also suggests, though not conclusively, that the ancestral populations had genetic variation for resistance before herbicide was widely used.

That multiple populations evolved resistance independently and that resistance probably came as the result of selection on existing standing genetic variation instead of on new mutations suggests that no field subject to frequent and widespread use of the herbicide is completely safe from the evolution of resistance. Consequently, surveillance for incipient resistance and for increases in resistance where the herbicide is widely used is warranted.

Plants vary along a continuum from those that exclusively or almost exclusively self-pollinate to those that mostly outcross, with many in the middle of the range. Many species even show variation in the propensity of selfing across different populations.

How does selfing affect the likelihood that a population will evolve resistance to an herbicide? It's not an easy question to answer based on first principles. On the one hand, some population genetic theory would predict that outcrossers would more easily adapt to a new environment, such as the presence of a new toxin, because outcrossing species generally have more genetic variation and higher effective population sizes than selfers do.[24] On the other hand, if the application of the herbicide kills off most of the population, selfers might have the upper hand, because outcrossers may not find a mate to pollinate them.[25] Moreover, once resistance had been established, resistant individuals would more likely thwart the influx of susceptible (and thus, maladaptive) genetic variants that come in from migration. Adam Kuester, a postdoc in Regina Baucom's lab, found that populations of morning glories that were resistant to the herbicide were more likely to self than those susceptible.[26]

Are there costs involved in the evolution of resistance in morning glories? Megan Van Essen, another postdoc in Baucom's lab, took samples from 43 different populations of the morning glory and grew them in a common garden setting.[27] She examined the plants for their resistance to glyphosate, the ability of their seeds to germinate, and the growth of the resulting plants. Germination success was strongly negatively correlated with resistance: as resistance increased, the probability of germination sharply declined. The decline was basically linear throughout the range of resistance. Resistance to the herbicide also was negatively correlated with the quality of the seeds: seeds from the more resistant populations were smaller and more likely to germinate abnormally than those from less resistant populations. Discussing, this finding, Van Etten and colleagues note: "Although most studies use seed quality as a proxy for fitness, our results highlight that reductions in progeny quality are an equally, if not more, important cost of adaptation in *I. purpurea*."[28] This is a salient point; looking at just the number of seeds might lead one to overlook the negative correlation indicative of a tradeoff. More broadly, researchers need to be cognizant of the full suite of possible fitness differences across the life cycle of the organism when assessing fitness.

The tradeoffs continued after germination. The plants from populations more resistant to the herbicide were smaller and had shorter roots. While the populations that were resistant to glyphosate all incurred costs, there were differences in how the costs were displayed: some resistant populations had poor seed quality, while others had smaller adult plants.

I spoke with Baucom more about costs, and she replied: "the interesting thing is that costs are identified in only about 50–55 percent of herbicide resistant species, so one of the themes in applied evolution has been 'why no cost?'"[29] She continued:

> That said, costs are not examined over time, and I only know of one study that looked at allele frequency of resistance alleles in a weed population, and allele frequencies did decline. So, I wonder if, in addition to looking at multiple traits that could be important, researchers should also examine costs temporally too.

Whether costs exist and how large they are will guide how well we can manage resistance. If costs are large, then we should expect resistance to fade after the herbicide is removed. If costs are nonexistent or too small, then resistance may persist even after removal. We saw in Chapter 1 similar dynamics occurring in the evolution of antibiotic resistance: loss of resistance could occur if costs were sufficiently high.

Van Essen, Baucom, and their colleagues have recently used genomic sequencing to examine the genetic basis of resistance in natural populations of morning glories. The first striking finding was the dog that did not bark—there was no difference in EPSPS, the target site of glyphosate, between the resistant and the sensitive populations.[30] Recall that the Monsanto researchers had predicted that resistance would be difficult to evolve at the target site. The problem is that several non-target-site changes yielded resistance. In all, five different regions of the genome—each containing a good number of genes—showed the signature of evolution by natural selection, presumably due to evolution of resistance.

Why are plants able to evolve resistance via genetic changes that do not involve the target site of the herbicide? Think about the environments in which plants have evolved. There, plants have had to deal with toxic compounds long before humans used herbicides—indeed, long before humans. Because plants are stuck where they are, they cannot escape such chemicals by running away. Instead, they have evolved a suite of mechanisms to defend themselves from such chemicals. These defenses fall into two general categories: (1) chemicals, usually enzymes, that break down and detoxify the toxins and (2) molecules that transport and place the toxins (or the by-products of such) in compartments.[31] An example of the first category is the P450 cytochromes, a large family of genes that encode numerous enzymes that engage in a variety of metabolic reactions. It is these enzymes that break-down suites of toxic compounds. An example of the second category is the ABC transporters; these molecules render toxins and their by-products harmless—some by excreting the toxins from the cell and others by shunting them into compartments within cells such that the toxins cannot impede cellular functions.

The genomic study of morning glories found that both types of genes were involved in the resistance to glyphosate: changes to P450 cytochromes and other detoxifying genes as well as ABC transporters contributed to the response. Because changes in P450 cytochromes often lead to cross-resistance to other herbicides,[32] the involvement of these in the resistance of morning glories suggests that the morning glories may be prone to evolve resistance to other herbicides.

The next major finding from the genomic study is that several independent resistant populations showed the same responses. The repeatability of the same resistance genes in independent populations suggests some degree of constraint to the evolutionary genetic process of resistance, that the number of genetic solutions to the evolutionary problem of becoming resistant is limited. This constraint is not too severe, as this genomic study also revealed different genetic changes in the different, independent resistant populations. The extent to which the genetic responses are constrained could inform resistance management. If the evolutionary response to resistance is relatively unconstrained (many genetic responses are available), then it may be harder to check than if the response is more constrained.

One of the five regions of the genome that differ between the herbicide-resistant and herbicide-sensitive populations of morning glories (and thus, a likely candidate for selection) shows an interesting pattern. In the sensitive populations, several variants show the signature of balancing selection; that is, selection appears to be acting to maintain genetic variation in that region. This could be because these fields are experiencing different selective pressures from year to year as farmers cycle the use of different herbicides. It suggests that the so-called sensitive populations have genetic variation for resistance, which could manifest itself under the right selective conditions.

Managing resistance to herbicides is easiest when caught early. Can we predict the emergence of the early stages of resistance and control it before it is too late? A recent study by David Comont and his colleagues examining variation in glyphosate resistance in black-grass (*Alpecurus myosuroides*) provides such a roadmap.[33]

Black-grass, also known as slender meadow foxtail, is a weed that threatens cereal crops in the United Kingdom. It is resistant to most herbicides used to control it; however, as of 2019, it is still sensitive to glyphosate. Comont and his colleagues examined the historical records for glyphosate use in different populations. They also determined LD_{50}, the amount of glyphosate needed to kill 50 percent of the weed, for each population. The greater the LD_{50}, the more resistant the population is. The researchers found that black-grass in the fields with the greatest use of glyphosate over time had the highest LD_{50}. Resistance is likely with the cumulative use of the herbicide; moreover, the findings could be used to set guidelines on how much glyphosate can be used over time. Evolution of resistance is occurring with greater exposure to glyphosate because resistance is heritable in black-grass. In fact, the researchers found heritability to be reasonably high: more than a quarter of the variation in resistance as measured by LD_{50} is attributed to the effect of genes.[34]

INSECTICIDE RESISTANCE

If he were alive today, Benjamin Walsh would be stunned by the current financial toll insect pests inflict on agriculture in the United States. A 2015 study by the Animal and Plant Health Service (APHS), a division of the USDA, pegged this figure at over $100 billion a year![35] Recall that Walsh had estimated the cost in the 1860s to be about a million. One especially damaging insect is the Asian citrus psyllid (*Diaphorina citri*), a bug in the order Hemiptera. Not only do these psyllids damage citrus plants by sucking their sap, but they also can transmit a bacterial disease called citrus greening (or huanglongbing) from citrus tree to citrus tree while sucking the saps. An early manifestation of this disease is the appearance of white and yellow blotches on the leaves. As the disease progresses, leaves fall off and roots die. Even if the tree survives, its fruits are smaller and the juice from the fruits is bitter and of poor quality.[36] The total damage from this tiny—smaller than an apple seed—insect can be in the billions each year, mostly in Florida.

Various pesticides have been used to control the psyllid. Unfortunately, several populations of psyllid in Florida have evolved resistance to various pesticides, including imidacloprid and malathion. Interestingly, the field populations that have evolved resistance to the pesticides have greater concentrations of glutathione S-transferase and other chemicals that are used to break down toxins. This suggests that the evolved resistance could be partly due to a generalized detoxification response rather than a target-site change.[37]

Experimental studies show that rotating the regimen of pesticides may thwart the evolution of resistance. Rotation is most likely to work when the psyllid evolves different resistance mechanisms to each of the different pesticides, as the different mechanisms may have tradeoffs. To the extent that the populations have evolved generalized responses, the less well we should expect rotation to work.[38] Just like the weeds in the earlier section, insect pests do better at adapting to the same challenges rather than to different challenges.

Are certain insects more likely to be tolerant of or resistant to insecticides? An old idea, dating back to the early 1960s, was that insects that had natural exposure to chemicals that were similar to synthetic insecticides would be more likely to be able to evolve resistance to the synthetic compounds. These naturally occurring compounds that resemble insecticides come from plants, which make the chemicals as an evolutionary response against insects feeding on them. The question "Which insects would most likely be preadapted to synthetic insecticides?" then becomes "Which insects would be most likely to experience the natural insecticides?"

Nate Hardy at Auburn University and his colleagues set out to test this hypothesis. They examined a large number of agricultural insect pests—more than 300 species across several insect orders—looking to see how whether their diet breadth predicted their likelihood of evolving insecticide resistance.[39] The data supported the preadaptation hypothesis: those insect pests that fed on a large variety of plants were more likely to have evolved insecticide resistance. Moreover, factoring out the number of host plants they had, those insects that fed on host plants that had diverse chemical defenses were more likely to evolve resistance. Finally, insects that fed on herbs—small plants that lack wood and branches—as opposed to trees were also more likely to evolve resistance. Knowing which features of the biology of insects predict insecticide resistance can be useful in surveillance programs. As Hardy and colleagues conclude, "Not all pests are the same; their specific evolutionary histories and biologies can have a big impact on their potential to evolve insecticide resistance."[40]

My UMASS colleague Ben Normark is one of the most creative and perceptive people I know. In our roaming conversations over the years, I am frequently stunned by the insights he reaches. As an undergraduate, Normark studied linguistics before he became a biologist studying phylogenetic relationships. There are similarities between linguistics and evolutionary biology; indeed, the methodologies of tracing word evolution and gene evolution bear some similarity. Normark studies scale insects, tiny insects in the order Hemiptera (true bugs), one group of which is the armored scale insects (Diaspidids), named thus because females are covered by waxy scales that they excrete. Yes, they are covered in their own waste! These females look like tiny blobs; in one paper, Normark once described them as having "a simplified morphology, with no distinction between head, thorax, and abdomen, no wings, no legs and only rudimentary eyes and antennae".[41] Many species of these scale insects occur in prodigious numbers and are pests, often affecting agricultural products. Some species also have extraordinarily wide diets, being able to eat plants from dozens of plant families.

Around 2008, Ben Normark started bending my ear about a syndrome he was observing in these scale insects. In addition to their immense numbers and their enormous diet breadth, these scale insects share several other characteristics. Notably, their adult females cannot fly; moreover, all but the smallest stage of the developing scale insects (the first instar larvae) are sessile. These first instar larvae disperse by wind or through the soil. Consequently, they do not choose which plants they wind up on. To survive, the tiny larvae must successfully complete development on the plant that they land on. If they fail, they die! Thus, we would expect intense selective pressure for being able to complete development on a wide group of plants. These and other characteristics are not just found in scale insects but also in root weevils (beetles), bagworm moths, and tussock moths; many species of these groups are also pests.[42]

What explains this syndrome? Normark proposed that insects with this syndrome persist in enormous numbers and are able to eat a great variety of plants due to a feedback loop between these two quantities. Because they have extraordinarily large populations, these species can exploit a wide variety of host plants. A huge population size, especially when combined with a huge effective population size, makes selection more efficient: more beneficial mutations are produced and more of these will get fixed in the population, while deleterious mutations are more likely to be weeded out. Because they have such a diversity of host plants to develop on, these species can maintain huge populations. Normark's infectious enthusiasm got me to join in a collaboration, with my role primarily being the examination of the population genetic aspects of the problem. We called this feedback loop the niche explosion.

One consequence of the niche explosion hypothesis is that if true, it provides a new explanation for the evolution of extraordinarily successful pests. Biologists generally assume that when pests are very abundant in an environment that has been altered by human activity that the alteration of the environment by human activity caused the outbreak. This is likely to be the case for many, if not most, outbreaks. But if the niche explosion hypothesis is correct, then there should be some cases where insects that eat a wide diversity of plants have huge numbers as a simple aspect of their natural history.[43]

While the niche explosion hypothesis has not been formally tested, some of its assumptions have been validated. The niche explosion hypothesis rests on the assumption that tradeoffs in performance on different host plants are rare. If such tradeoffs were common, then evolving such a large niche breadth would be exceedingly difficult, as adaptations to one plant species would make adapting to other species less likely. Daniel Peterson, a graduate student in Ben Normark's lab, tested this assumption by looking at the correlations of performance of armored scale insects across different host plants.[44] If tradeoffs were common, then we would expect to see numerous instances of negative correlations: species of armored scale insects that perform well on one host plant would perform poorly on others. In fact, contrary to the expectations if tradeoffs were common, most of the correlations were positive. This suggest that adaptations to feeding on one host plant facilitate—rather than constrain—adaptation to feeding on other host species. While these results are not direct evidence for the niche explosion, they do support its feasibility. Such a pattern of rampart positive correlations of performance across different hosts may not be general; in fact, it might be restricted to armored scale insects and a few other groups.

There is still much that we do not yet know about the evolution of niche breadth in insects.[45] Further exploration of the ecological, genetic, and other factors that differentiate generalists and specialists is sorely needed. The evolution of changes in diet breadth is not just of academic interest. Because agricultural pests often expand their diet breadth as they become invasive—though not necessarily to the extent of the niche explosion syndrome species—those applied biologists examining agricultural pests should be on the lookout for diet breadth expansion.

BIOLOGICAL CONTROL

Benjamin Walsh lambasted the blind enthusiasm people had for the pesticides of his day, what he called the "Patent Powder Pimperlimpimp", and the lack of interest in common sense plans based on scientific principles (of his time). The plan that Walsh was advocating was an early version of what would now be called integrated pest management, and it included the use of parasites to control the Hessian fly, a notorious pest of wheat and other cereal crops.[46] Walsh was thus advocating biological control.

In *Silent Spring*, Rachel Carson argued against the overuse of insecticides—and most specifically DDT—due to the harmful effects on birds and other wildlife from this glut of pesticides.[47] Carson did not oppose insecticides in general; instead, she wanted to limit their use. She too proposed use of biological control as part of integrated pest management. Carson noted that the history of biological control stretches back quite far—well before Walsh—and mentioned that Erasmus Darwin, grandfather to Charles, had been an early advocate of biological control.

How can we apply insights from evolutionary biology to biological control?[48] One reason why the agents of biological control can fail to establish themselves is that they have insufficient genetic variability due to a bottleneck. With population genetic analysis, researchers can determine the extent of genetic variability of potential agents.[49] Another consideration is whether the agents of biological control are sufficiently well adapted to the target species. Agents with a short generation time, such as small parasites and pathogens, should be more likely to be adapted to local genotypes of the target than ones with longer generation times because short generation agents should evolve more quickly. Some evidence supports that hypothesis.

One of the advantages of biological control is its durability; evolution of resistance to biocontrol agents has been rare. Several factors mitigate against such evolution. First, the biocontrol agents themselves evolve; especially if the agents have a faster generation turnover than the pests, the agents should at least be able to keep up with the pests in the evolutionary arms race. Another factor is that the pests usually must contend with more than one natural enemy—the agent is not the only entity that they are adapting to.

The Argentine stem weevil (*Listronotus bonariensis*) became a devasting pest of grassland used for cattle grazing in New Zealand, especially as livestock production increased during the 1980s. A parasitoid wasp (*Microctonus hyperodae*) was introduced to control the weevil. Although the wasp has checked the weevil, parasitism rates have declined during the 2000s and 2010s. Resistance has evolved.

Two factors of the wasp's biology limit its capacity to adapt to its target. First, it went through a severe genetic bottleneck during its introduction; thus, it has much reduced genetic variation. Second, it is unable to take advantage of genetic recombination—and thus create new combinations of genes—because it is partheno-genic. Perhaps its inability to evolve in response to the weevil is contributing to the decline in parasitism. Yet the picture is more complicated. The declining parasitism rates is most pronounced in the ryegrass *Lolium perenne* rather than in other grasses. This ryegrass is grown at high densities as part of intensive agricultural processes. One possibility is the high density of the ryegrass creates a monoculture, reducing the availability of places for the weevils to escape from the wasp (refugia).

Another concern with biocontrol is that the agents used in biocontrol can some-times become pests themselves. One such example is the harlequin ladybeetle, *Harmonia axyridis*, which was brought from its native East Asia to North America and Europe to control aphid and scale insect agricultural pests.[50] This beetle now threatens native species, both because it competes with them for aphids and because it is a gluttonous predator. In addition, by consuming fruit, the beetle is now itself an agricultural pest. Finally, because it can reach high densities overwintering in houses, the beetle is also a human nuisance. In Chapter 14, we will discuss cane toads, another example of an agent of biological control that turned into a pest.

In the final chapter of *Silent Spring*, Rachel Carson presciently mentioned a new type of biological control: a natural insecticide that comes from a bacterium, *Bacillus thuringiensis*.[51] This soil bacterium (known as *Bt*) produces toxins that kill caterpil-lars of moths and butterflies. It is generally harmless to other insects and mammals. Since the 1990s, *Bt* has been engineered into more and more crop plants, allowing for a reduction of the use of insecticides. Mark Lynas in *Seeds of Science* noted:

> My bet is that had *Bt* corn had been Monsanto's initial product launch instead of Roundup Ready soy, things might have been very different for GMOs. Genetic engi-neering could have been associated in the public mind from the outset with the reduc-tion of chemical pesticides and might have faced less widespread opposition. Some environmental groups might even have cautiously supported GMOs as part of their long-running campaign to reduce pesticides in agriculture.[52]

But even with resistance from some environmental groups, *Bt* has been extraordi-nary successful. Nearly 100 million hectares (or roughly 200 million standard foot-ball fields) of *Bt* crops were planted worldwide in 2016, ten times more than what had been planted in 1998. Virtually all the use of *Bt* is in corn, cotton, and soybeans. Farmers are getting greater yields while using less conventional pesticides, saving money and benefiting the environment. But this success comes at the price that insect pests are evolving resistance to the *Bt* crops.[53]

Delaying resistance is not futile. Indeed, there are some success stories. One is the eradication of pink bollworm (*Pectinophora gossypiella*) from the United States. This small gray moth, a major pest of cotton across the world, had been costing cot-ton farmers in the United States tens of millions of dollars due to damage from the moth and measures to control it. Fortunately, thanks to a series of containment and then eradication measures, it was declared to be eliminated from US cotton-growing areas in 2018.[54] Early measures taken to delay the evolution of *Bt* resistance in this moth in Arizona helped progress toward the elimination of this pest. Moreover, this effort highlights general practices that work well to manage resistance.

Despite heavy use of *Bt* cotton in Arizona since 1997, the evolution of resistance was thwarted by use of refugia, areas where cotton lacking *Bt* was planted.[55] The proper use of refugia impedes the evolution of resistance, especially if certain conditions are met. In the refugia, there is no selection for resistance. In fact, if there are costs to being resistant, then susceptible individuals would be favored in the refugia. Thus, we would expect nearly all the moths in the refugia to be susceptible. If resistance is due to a single gene, most of the individuals would be *SS* homozygotes. These individuals would be free to go to fields where *Bt* cotton was planted. Recall that the *Bt* toxins do not affect the adults, so these individuals would have no difficulty living in the *Bt* fields and could mate with resistant individuals in the *Bt* cotton fields. Matings between the *SS* homozygotes from the refugia and resistant moths (*RR*) that came from the *Bt* fields would result in *RS* offspring. Now finally, if resistance had a recessive genetic basis—so that *RS* heterozygotes were susceptible—the offspring would be killed in the *Bt* fields. In summary, refugia should thwart the evolution of resistance when: (1) resistance is relatively rare, (2) resistance is due to a single gene and that resistance is recessive, and (3) there are fitness costs to being resistant.[56]

In the cotton fields in Arizona, those conditions were met: the resistance was caught early and thus was rare, resistance was costly (resistant individuals had lower fitness in the absence of the toxin), and resistance was recessive.[57] That resistance was recessive was a function of the dose of the toxin. With a high dose of the toxin, heterozygotes are killed as well as the homozygote susceptible individuals. Thus, using a high dose in the *Bt* fields helps the refugia delay the evolution of resistance.[58]

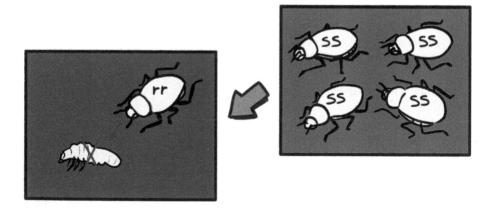

Use of refugia to slow the evolution of resistance to *Bt*. The green areas are refugia where the plants are all non-*Bt*. Here, susceptible insects can live. In the other areas (brown) *Bt* plants are grown. Here only resistant insects can survive. If we assume that resistance is recessive (heterozygotes *Rr* individuals die) and resistance is rare, then the susceptible (*rr*) insects from the refugia will outnumber the few resistant insects. Resistant insects are most likely to mate with susceptible insects. Their progeny will be *Rr* heterozygotes, which will die. Thus, the refugia should check the rise in frequency of resistant insects.

Source: Illustrated by Michael DeGregori.

Subsequentially, the genetic engineering of multiple *Bt* toxins in the same plants has helped retard the evolution of resistance. This practice, known as pyramiding, rests upon the assumption that evolution of resistance to one toxin is unlikely to lead to the evolution of resistance to the other toxin.[59] Consequently, using toxins that have different modes of action is most likely to be effective. Sterile insect technique (as we saw in Chapter 3) has also been used effectively as an alternative to or in combination with refugia.[60]

GENE DRIVES TO THWART AGRICULTURAL PESTS

We saw that in Chapter 3, gene drives can be used to either control or change mosquito vectors of disease. Similarly, gene drive can be used to control or change weeds and insect pests.

Plants have several natural selfish genetic elements that can serve as gene drives. For example, hybrids between wild rice (*Oryza meridionalis*) and Asian cultivated rice (*O. sativa*) are sterile in part because of divergence in a gene drive system.[61] The cultivated rice has evolved two genes that are not in the wild rice: *ORF2*, which encodes a poison, and *ORF3*, that encodes the antidote. Some hybrids between the two rice species that have the poison but not the antidote abort their pollen. Analysis of the DNA sequence analysis reveals that the antidote (*ORF3*) evolved first, followed by changes in *ORF2* that turned it into a toxin. The normal function of the antidote and the selective pressures that caused the evolution of it and the toxin are not yet known. With CRISPR gene editing, a drive system with these elements could be produced.

In considering the use of gene drives, plants have certain characteristics that need to be considered when using gene drives.[62] Most plants—unlike most animals—have seed banks in which seeds that were produced in prior years are stored. Because such seeds reflect the gene pool of prior generations, the presence of seed bank can alter the progress of a gene drive, usually to slow it down. Plants also routinely self, much more so than most insects and most vertebrates; this too can affect the dynamics of a gene drive.

NOTES

1. The quote is from Walsh (1866, p. 119). Walsh continued with a colorful condemnation of lawyers and Congress.

   Probably about nine-tenths of the Members of Congress and of our different State Legislatures are lawyers. . . . What do they know about Farmers, except that they have got votes? Or about Farmers' pockets, except that most of the taxes come out of them? [capitalization in original].

2. Kansas State University (2016).
3. Feder et al. (1994).
4. For a small sample of the literature on the sympatric speciation controversy, see Mayr (1963), Bush (1969), Berlocher and Feder (2002), and Coyne and Orr (2004).
5. Walsh (1864) and Walsh (1867).
6. See Sheppard (2004) for biographical information on Walsh.
7. Walsh is quoted in *The Agricultural Convention: Annual Meeting of the Illinois Natural History* (1860, p. 11). See also Sheppard (2004). Walsh's statements illustrate that the use of military analogues has a long history in invasion biology and pest management. Some

authors (e.g., Larson 2005) have criticized such use and called for a demilitarization of invasion biology. Others (reviewed in Cassini 2020) have suggested that the use of military language can motivate action.

8. Control of agricultural pests is not new. In fact, such treatments long predate Walsh. As Webster (1975) noted, the Roman historian and senator Cato the Elder recommended using amurca, the bitter residue from olive oil, to control weeds. Pliny the Elder added cypress leaves to vegetable seeds to thwart insect pests. The difference is that the scale is much larger.

9. Kansas State University (2016).

10. Baucom (2019).

11. Specifically, the pathway is for the synthesis of branched chain amino acids (isoleucine, leucine, and valine). See Garcia et al. (2017)

12. Monsanto merged with Bayer in 2018 in a deal worth $66 billion. The Monsanto name was dropped. See Fung and Dewey (2018).

13. Glyphosate, N-phosphonomethyl-glycine, was synthesized in 1970 by John Franz, a chemist employed by Monsanto. The trade name Roundup came from a "name that chemical" contest among Monsanto employees. The winner, Dottie Mills, won $50 for her entry. See Lynas (2018) for more details about Monsanto and its role in developing glyphosate.

14. For background on the historical use of glyphosate, see Duke (2017).

15. Lynas (2018, p. 94).

16. Duke (2017, p. 1029).

17. Bradshaw et al. (1997).

18. Powles et al. (1998).

19. Powles (2008). The quote is on p. 361.

20. Baucom et al. (2011).

21. Fang et al. (2013).

22. The alternative hypothesis that the morning glories grew along with maize and then spread with maize northward is conceivable but less likely based on the data.

23. Kuester et al. (2015).

24. Kreiner et al. (2018).

25. Kuester et al. (2017).

26. *Ibid*. Morning glories may be exceptions, as Kreiner et al. (2018) found that species that outcross more have more resistance mutations. There is also the possibility that species that are outcrossers are more likely to evolve resistance, but that selfing and the evolution of resistance is correlated in populations within species.

27. Van Etten et al. (2016).

28. *Ibid.*, p. 2205.

29. E-mail conversation with Regina Baucom, November 2020.

30. Van Etten et al. (2020).

31. Yuan et al. (2007).

32. Cross-resistance occurs when evolution of resistance to toxins that the plant was exposed to leads to the plant being at least somewhat resistant to other toxins. Yuan et al. (2007) document cases of cross-resistance have evolved due to P450 cytochromes.

> Here's one scenario by which P450 cytochromes could be involved in cross resistance. Suppose, upon exposure to the first toxin, there is a plant with a variant of a P450 cytochrome that is better at producing the chemicals that break down the first toxin. This variant is thus favored and will likely sweep through the population. This variant is also likely to be better able at breaking down chemicals related to the first toxin. And there we have cross-resistance.

33. The study is Comont et al. (2019). See also commentary by Baucom and Busi (2019).
34. The heritability was estimated to be 0.27. This is the narrow sense of heritability, and thus is, in fact, the proportion of the variation in resistance that arises from the additive effects of genetic variants. It was estimated via examining correlations between relatives using standard quantitative genetic methods. See Comont et al. (2019) for details.
35. Montalalvo (2015).
36. Ferrarezi et al. (2020).
37. Tiwari et al. (2011).
38. Chen and Stelinski (2017).
39. Hardy et al. (2018).
40. *Ibid.*, p. 745.
41. Morse and Normark (2006).
42. These insects also tend to be parthenogenic. For details, see Normark and Johnson (2011).
43. Normark and Johnson (2011, p. 561).
44. Peterson et al. (2015).
45. Hardy et al. (2020).
46. Sheppard (2004).
47. Carson (1962).
48. See Hufbauer and Roderick (2005) and Sethuraman et al. (2020) for discussion about how population genetic principles and studies can be used in biological control. Sethuraman et al. (2020) provide a suggested pipeline for incorporating population genomics in biological control studies.
49. See Tomasetto et al. (2017) for the study and Mills (2017) for the commentary.
50. Roy and Wajnberg (2008).
51. Carson (1962).
52. Lynas (2018, p. 98).
53. Tabashnik and Carrière (2017).
54. USDA. 2018. USDA announces pink bollworm eradication significantly saving cotton farmers in yearly control costs. October 19, 2018. www.usda.gov/media/press-releases/2018/10/19/usda-announces-pink-bollworm-eradication-significantly-saving. Accessed 26 November 2020.
55. Tabashnik et al. (2005).
56. See Tabashnik et al. (2013, 2017) and Carrière et al. (2016) for more detail.
57. Tabashnik et al. (2005).
58. Tabashnik et al. (2013, 2017) and Carrière et al. (2016).
59. Carrière et al. (2016).
60. Tabashnik et al. (2010).
61. Yu et al. (2018).
62. Barrett et al. (2019a).

10 Buccaneers of Buzz
Bees and Pollination Services

SUMMARY

Bees and flowering plants engage in a mutual exchange: bees pollinate flowers while the plants provide nectar, pollen, and other rewards to the bees. The relationship between bees and plants is not always harmonious: bees can be overly demanding and plants can be stingy with their rewards. Bee pollination is vital to agriculture: Bee-pollinated plants supply much of our nutritional needs. Moreover, bee pollination often improves the quality of crops (like strawberries) that do not strictly require pollination. We explore the evolutionary origins of bees and bee diversity. Although honey bees have an outsized role in pollination, they are unusual bees. We consider some of the genetic oddities of bees, including how their unusual mode of sex determination may have fostered eusociality in bees. We consider threats to bees, focusing on pesticides and pathogens. Finally, we consider two topics related to honey bees: the similarities between human and honey bee decision making and how to best raise bees in keeping with evolutionary principles and the natural history of honey bees.

"Bees are Black—with Gilt Surcingles
Buccaneers of Buzz—
Ride aboard in ostentation
And subsist on Fuzz."—Emily Dickinson.[1]

"The world has 20,000 species of bees spread over nine families. To toss them all together and assume that they will look the same or act the same or evolve in the same way is a bit like expecting a grizzly bear to be like a seal. I didn't pull those two animals out of nowhere. Taxonomically speaking, a grizzly bear and a seal are as closely related as two bee species from different families. We don't expect the bear and the seal to eat the same things or bed down in the same way. Why then do we expect it of bees?"

—Paige Embry[2]

I live in the center of Amherst, Massachusetts and frequently walk by the homestead of its most famous denizen, Emily Dickinson. An avid gardener, Dickinson was a keen observer of nature. References to bees, flowers, and pollination pervade her poetry.[3] In fact, I had a difficult time deciding on which Dickinson poem to choose for the epigraph of this chapter. The image of bees as loud and loudly colored pirates in the epigraph appealed to me. This poem suggests that bees are not around just to pollinate flowers; they get something out of it and are not always gentle about getting their fuzz; a rather Darwinian perspective.[4]

Fitting with the Dickinson poem, a recent book on bees described bees, from the perspective of the flower, as an overly demanding lover.[5] Some bees sometimes are

DOI: 10.1201/9780429503962-12

more trouble than they are worth for some flowers. But perhaps a more useful analogy may be that of a contractor and a homeowner. Just as the homeowner wants the best job from the contractor for the lowest possible price, the evolutionary interest of flowers is for bees to do the best job of spreading their pollen for the least cost. Likewise, just as the contractor wants the highest possible price, the evolutionary interest of bees is to get as much from the flowers as they can. We can extend the analogy still further: just as homeowners are unlikely to continue hiring contractors that do shabby jobs or are too greedy, plants will likely evolve mechanisms to prevent overly greedy or poorly performing bees from getting resources. Just as contractors are unlikely to accept offers from overly stingy homeowners, evolutionary pressure will likely cause bees to stop trying to pollinate overly stingy plants. Moreover, like contractors that specialize in specific realms, bees vary in the types of flowers they pollinate. These business arrangements between bees and flowers do not require consciousness or forethought, just coevolution of bees and flowers.

So, what do plants have that bees want? Plant rewards include nectar, pollen, and oils.[6] In addition to supplying calories, mostly from various sugars, nectar contains trace nutrients like vitamins. Plants also add various scents to help attract pollinators. A few plants even add a sprinkle of caffeine to help boost their pollinators' memory and give them the boost they need to make it to the next plant! In a sense, some nectars can be like pre-workout energy drinks. As we will see, some nectars also contain chemicals to fight against harmful microbes. Although most nectar contains some protein, pollen is the main source for protein. Plants that have animal pollinators have pollen that is much richer in protein than that from plants who lack animal pollinators; protein contents in some plants' pollen can be greater than twice that of a steak. Because pollen contains the reproductive material that plants want bees to spread and because pollen is expensive to produce, plants generally make it difficult for bees to get pollen. Flowers have evolved a variety of structures to protect their pollen. Bees that can obtain pollen from the flowers they visit usually have evolved special adaptations to get through these structures. Finally, some bees—the oil bees—pollinate plants that produce lipid-rich oils. These oils supplement the calories the bees receive from nectar and pollen.

THE BENEFITS OF BEES

We humans reap the rewards of the business arrangement plants and their bee pollinators have established. Agriculture relies on bee pollination. Honey bees[7] alone add about $6 billion in commercial value by pollinating crops in the United States.[8] The bulk of that value comes from their pollination of almonds, mainly in California. But honey bees are not the only bee that provides value. Wild bees, native to North America, provide just over $1 billion of economic value from pollinating apples and an additional half billion from pollinating other crops. Note that honey bees are temperamental fliers that are reluctant to venture outside in chilly or inclement weather, while many other bees are busy visiting flowers. Hence, crops that bloom in the spring are more dependent on native bees than those that bloom in the summer. How dependent fruits are on pollination varies from fruit to fruit and location to location. Many fruits, such as tart cherries and blueberries in Michigan, sweet cherry in Washington State, and apples from all over the United States are pollen limited, meaning that crop yield would improve with more pollinators.

In addition to being only mediocre cool-weather fliers, honey bees have another limitation: they can't pollinate tomatoes and some other plants. The pollen of tomatoes is tucked away inside the anthers instead of being on the outside. To get to the pollen, bees must shake it out. Honey bees lack the machinery to do this, but bumble bees can get to the inside pollen. While grabbing the flower with its mouth, a bumble bee uses muscles in its middle section—the thorax—to cause vibrations, which shake the pollen out of small holes in the anthers. This is called "buzz pollination" from the buzzing generated by the rapid undulations of the bumble bee's thoracic muscles.[9]

The estimates of the economic value may undervalue just how important bees are. Bee-pollinated plants supply much of our nutritional needs. The staple crops—such as rice, wheat, potatoes, and corn—are wind- or self-pollinated; these supply calories, but they are lacking in vitamins and other micronutrients. More than 90 percent of the vitamin C we consume comes from animal- (and mostly bee-) pollinated plants.[10] Animal-pollinated plants supply about three-quarters of the plant oils we use. They provide just over half the folate (also known as vitamin B$_9$), a key nutrient that assists in red blood cell development and is especially important in pregnancy. You can find lots of folate in various fruits, from oranges to strawberries, as well as in leafy vegetables like spinach.

Speaking of strawberries, pollination also improves their quality. A recent study found that bee pollination led to bigger, firmer, and fresher strawberries.[11] Here, the researchers had used enclosures to prevent bees from pollinating strawberries and found that the strawberries that did not have access to bee pollinators were substantially inferior. How does this happen? Picture a strawberry. The rough seed-like bumps on the outside of the fruit are the functional equivalents of the plant's ovaries, which botanists call achenes. When a bee pollinates an achene, it triggers a hormonal reaction that leads to bigger and redder fruits that last longer before spoiling.

Pollination is not the only service we get from bees. Honey bees partially digest and evaporate nectar, thus converting it a durable and concentrated source of sugars—honey. While nectar is about 80 percent water, honey is less than 20 percent water. Honey is so dry that it can pull moisture from the air. This property—hygroscopicity—explains why muffins and cakes made with honey can stay moist for so long. Enclosed in honeycombs and sealed off by wax, honey is resistant to fermentation; consequently, bees can use it over the long winter months, enabling them to maintain activity all year round. We humans can use this honey by taking it from wild hives or by rearing bees.[12]

Americans consumed nearly 600 million pounds (almost 300 million kilograms) of honey in 2017.[13] This breaks down to not quite two pounds (almost a kilogram) per person. Half of that comes from the direct consumption of honey, with the rest coming from honey used in cookies and bread and other foods. At about 150 million pounds (70 kilograms) per year, domestic honey production is robust but still insufficient to keep up with consumption; consequently, the United States imports most of its honey. Overall, the honey industry adds about $2 billion to the US GDP.

Use of honey and other bee products is widespread across cultures where there are bees. The indigenous group known as the Hadza have lived as hunter–gatherers for millennia in what is now northern Tanzania. During the rainy season, a Hadza hunter may obtain as much as three pounds (1.5 kilograms) of honey. The Efé of the Ituri Rainforest in the Democratic Republic of Congo outdo even the Hadza: they feast on bee materials during the rainy season, obtaining four-fifths of their calories just from honey. The first record of humans constructing beehives dates back to about 4,500 years ago, but cave paintings much older than those hives show humans collecting honey.[14]

Bees create a waxy substance—beeswax—to store honey and protect their young. Nature writer and beekeeper Sue Hubbell listed some of its manifold uses.

> Beeswax has many uses besides making fine candles. It is a base for ointments, creams and cosmetics. It is used for waterproofings and polishes. It is an ingredient in adhesives, crayons, chewing gum, inks, ski and grafting wax. But its chief value is to bee-keeping-supply companies, which remelt it and mold it into new frame foundation to sell back to beekeepers.[15]

Hubbell also suggests that bees might be the animal we humans have written the most about, other than ourselves. We began with Emily Dickinson. In addition, Hubell lists E. B. White, the literary critic and author of *Charlotte's Web* and other children's books. His witty poem "Song of the Queen" extolled bees and mocked stuffy old geneticists attempting to breed them via artificial insemination.[16] The Roman poet Virgil dedicated his fourth Georgic to bees and honey. The ancient Greeks even had a god, Aristaeus, son of Apollo and Cyrene, who had discovered beekeeping and other useful arts such cheesemaking.

Bees even factored into an incident between the United States and the Soviet Union. During the late 1970s, refugees and soldiers from Laos and other areas of southeast Asia reported cases of suspected chemical warfare. They gave investigators from the US State Department samples of plant material coated with a yellow substance. In 1981, Alexander Haig, Secretary of State under the Reagan administration, accused the Soviet Union of using chemical weapons. But analysis of the yellow substance found that it was enriched in pollen from multiple plants. This led biologists led by Tom Seeley and Matt Meselson to further investigate. That yellow substance wasn't from chemical weapons; no, it was the fecal material of Asian honey bees.[17]

This honey production is a mixed blessing, as honey bees (which were introduced to many areas, including the United States) can have negative effects on native ecosystems.[18] Thus, in some respects, honey bees could be considered to be invasive (see Chapter 14). Honey bees do compete with native bees; indeed, their ability to gather food from several kilometers away and the communication they use in foraging often make them superior competitors. How often this competition from honey bees affects populations of native bees to thrive is not clear; there is some circumstantial evidence of such population effects but little in the way of smoking guns. Some of the best evidence for these population effects comes from looking at patterns of abundance of bees in Pacific oceanic islands: where honey bees are common, the native bees are rare or absent. Given the small size (and likely limited resources) of oceanic islands, such a result may not generalize.

A CLOSER LOOK AT BEE DIVERSITY

Bees are descendants of predatory wasps that turned vegetarian.[19] Bees are a clade, meaning that all bees share a common ancestor and all descendants of that ancestor are considered bees. In contrast, wasps are not a clade: some wasps are more closely related to bees than they are other wasps. But which wasps are bees closest to?

A recent study addresses this question, and in doing so, provides a clue about how bees turned vegetarian.[20] It found that bees are most closely related to the wasp family Ammoplanidae. Some wasps in this family hunt thrips, tiny insects that have fringed wings. Here's the intriguing part: many of these thrips feed on pollen. An appealing hypothesis is that as the Ammoplanidae-like wasps evolved into bees, they went from hunting pollen-eating thrips to eating the pollen themselves. This study also estimated the date of the origin of bees to about 128 million years ago, give or take a couple tens of millions of years. This is squarely in the early Cretaceous period, around the time of the earliest flowering plants.

Honey bees get most of the attention. Bumble bees get second billing. But most bees are the unsung solitary bees. Unlike honey bees and bumble bees, these solitary bees don't have queens and they don't have workers. Most of them eke out a living by visiting flowers and providing for their young. In fact, most solitary bees can be compared with annual plants. Both come out after spring rains. Both have a resting stage that can last for years. Thus, like plants have seed banks, some bees have a bee bank. Even with these commonalities, solitary bees display remarkable variation. They differ in how and where they construct their nests: some do it underground,

The brown-belted bumble bee, *Bombus griseocollis*, is native to the United States and is abundant throughout most of the northern and eastern regions. It feeds on numerous plants including clovers, loosestrife, and milkweeds.

Source: Courtesy of Jonathan J. Giacomini; photos taken in Raleigh, North Carolina.

A closer view of the brown-belted bumble bee.

while others do it in wood. Still others use preexisting cavities. At the extreme are bees like *Colletes diviesianus* that nest in stone. Solitary bees also differ greatly in mating behavior and activity rhythms.[21]

Eusociality, social living where only a few individuals are capable of reproducing, evolved in the corbiculate (pollen basket) bees. This group contains honey bees, bumble bees, stingless bees, and orchid bees (also known as the Euglossi). These bees get their name from a structure on their hind legs known as the pollen basket or corbicula that they use to gather and carry pollen. The first three of the corbiculate bees are eusocial, while the orchid bees are not. So the phylogeny of the group can tell us about the evolutionary history of eusociality. There is a strong consensus that bumble bees and stingless bees are each other's closest relative. But what are the relationships among the other lineages? Looking at the most reliable genes, this study found that honey bees were closely related to the lineage containing bumble bees and the stingless bees, with the orchid bees as the outgroup. Hence, according to this phylogeny, eusociality only evolved once—sometime after the split between the orchid bees and the other corbiculate bees.[22]

Eusociality has evolved only a few times; even in Hymenoptera, it has been lost more often than it has been gained. But although eusociality is rare, some eusocial groups are extraordinarily successful. This success is reflected in the disproportionate accumulation of biomass in eusocial species: while only about 2 percent of known insects are eusocial, eusocial insects account for nearly half the total insect biomass.

So why do relatively few species, even relatively few bee species, become eusocial? One possible reason is the economics seldom favor its evolution. Modeling studies by Feng Fu and Martin Nowak suggest that eusociality is a high risk-high return strategy.[23] Eusociality entails producing workers that cannot reproduce themselves. Foregoing reproduction is a risk, even if the eventual payoff is good. To overcome the opportunity cost of delayed reproduction, the payoff to being eusocial must be substantially larger than ordinary reproduction. But if that benefit is sufficiently large in a species, it can evolve to be eusocial and reap huge benefits. At the extreme, honey bees are as successful as they are in large part due to complex behaviors that they evolved. For instance, individual foraging honey bees communicate the location and quality of food sources to other foragers in the colony by means of an intricate waggle dance. Eusociality fostered the evolution of such behaviors.

Evidence showing a benefit to nesting in groups (a gateway to eusociality) comes from the sweat bee *Megalopta genalis*.[24] Sweat bees are thus named because many of them are attracted by the sweat of vertebrates. This sweat bee can either nest alone or in groups; up to half the nests are constructed by multiple females (usually two to four females). A study of these bees on Panama's Barro Colorado Island showed that social nests had greater productivity than solitary ones. Mitigating predation risk may explain at least part of the benefit to working in groups; having additional adults around may help fend off predation of the young from ants.

GENETIC ODDITIES OF BEES

Bees, like wasps and ants and other members of the insect order Hymenoptera, have an unusual method of sex determination: eggs that are fertilized and thus have two sets of chromosomes (diploid) develop into females, while eggs that are not fertilized and thus have only one set of chromosomes (haploid) develop into males. This mode of sex determination entails that males do not pass on genetic material to their sons. It also creates the unusual situation that females are more genetically related to their sisters than they are to their daughters. This quirk leads to an intriguing hypothesis about why social behavior—and at the extreme, eusociality—has evolved multiple times in the Hymenoptera.

In the middle 1960s, William Hamilton, who we met in Chapter 2, proposed the kin selection for the evolution of altruism, which he defined as one individual in a species helping another at its own expense. Because relatives share variants of genes, there are circumstances wherein helping kin would be evolutionarily favored even given that cost.[25] While forgoing reproduction to help raise your sibling's offspring is an extreme cost, if the payoff is sufficiently large, it could make evolutionary sense. Hamilton argued that in these calculations of whether altruistic behavior would be favored, we need to take into account the degree of relatedness as well as the benefit to the relatives and the costs to the altruistic individual. As the relatedness between the benefiter and beneficiary increases, so does the likelihood that helping relatives is favored. So, haplodiploidy causing a female bee to be more related to her sisters than her own offspring tilts the scale toward evolving toward eusociality. On the face of it, this is an appealing hypothesis for the evolution of social behavior, including eusociality in bees and other Hymenoptera.

Hamilton's explanation for the association between haplodiploidy and the prevalence of eusociality remains a source of controversy.[26] Ed Wilson, who in addition to his work in conservation biology is one of the leading ant experts, had been one of the strongest advocates for the kin selection argument. Wilson publicly changed his mind in 2005. Among other things, Wilson argued that the causal link between relatedness and eusociality may be reversed: "individuals do not form colonies because they are closely related. They are closely related because they form colonies".[27] One intriguing alternate explanation for the haplodiploidy link is that through genetic inputs, mothers are manipulating their offspring to help her raise other offspring.[28]

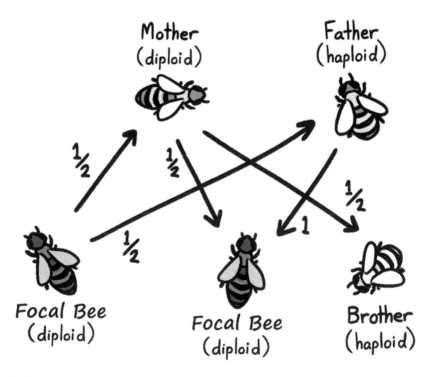

Haplodiploidy in bees makes a female bee more closely related to her full sisters than to her offspring. A female is related to her sisters through both her mother (½ times ½ or ¼) and her father (½ × 1). The total relatedness coefficient of the female to her sisters is thus ¾. The same female has a relatedness coefficient to her offspring of only ½.

Source: Illustrated by Michael DeGregori.

This manipulation hypothesis is appealing because such manipulation is more likely to evolve under haplodiploidy than when both sexes are diploid.

One interesting consequence of the mechanics of haplodiploidy in bees can send populations that are in trouble into a downward spiral. In most bees (as well as in most wasps and ants), sex is not just determined by whether the egg is fertilized. There is a genetic locus called complementary sex determination (*csd*) that adds a quirk to the sex determination: if a female mates with a male that has one of the same alleles at the *csd* locus that she has, half of her offspring (those with two copies of the same allele at *csd*) will not develop into females but instead will develop into diploid males. Such diploid males are usually sterile or effectively so; hence, they are genetic dead ends. Consequently, if too many of these sterile diploid males are produced, the population's growth rate could be hindered.

So a vicious feedback cycle can ensue.[29] Suppose genetic diversity at the *csd* locus in a species is reduced for some reason—possibly arising from a sharply decreased population size or inbreeding. With reduced genetic diversity at this locus, the likelihood of females and males sharing the same allele increases. This leads to more diploid males being produced. If enough diploid males are produced, the population growth rate could decline or even become negative. Consequently, the population size would shrink. A shrinking population leads to reduced genetic diversity all over, notably at the *csd* locus, and the cycle continues.

The Japanese bumble bee *Bombus florilegus* is found only in and around Hokkaido, Japan. During the 1990s, a European bumble bee (*Bombus terrestris*) was introduced to Japan to help pollinate greenhouse crops. Unfortunately, this bee escaped and has become a competitor of native bees, including *B. florilegus*. As a result, this native bee has declined in population. Genetic analysis has shown diploid males prevalent in colonies of *B. florilegus*.[30] We do not know just how much harm these sterile diploid males are causing for this bumble bee, but their presence bears watching. In contrast, a survey of several species of orchid bees sampled at many locations finds that diploid males are exceedingly rare.[31]

Eusocial bees tend to have a high rate of genetic recombination, the shuffling of genes on chromosomes.[32] Moreover, species with larger colonies seem to have the highest rate of recombination. While genetic recombination is a property of the genetic system of the population that affects how populations will respond to evolutionary processes, it is also subject to evolution by natural selection itself; it varies among individuals, and the likelihood of recombination is heritable. Thus, we can consider why such a link between recombination and colony size could evolve.

Why would more recombination be favored by natural selection in species that are eusocial, and especially in those with large colonies? First, consider what recombination does: Recombination makes selection more efficient; it chips away at associations between genetic variants at one locus and with the background. If a genetic variant that increases survival happens to be associated with one that decreases reproduction, it is unlikely to be favored by selection. Shuffling the genetic deck helps break that association. Next, consider the nature of the individual in eusocial colonies. Because only the queen and drones can reproduce, the genetic individual of a eusocial species is the colony. Hence, the genetic population size of such insects is likely to be low, far smaller than the number of bodies. In fact, social insects generally have low genetic population sizes, comparable to those of vertebrates.[33] So given that eusocial insects have few genetic individuals, there would be selection pressure to make selection as efficient as possible. This would be especially true in species that have large colonies. Moreover, species with larger colonies live longer and crank out fewer generations per time. Such species are also at greater risk for being infected by pathogens and parasites. Given all of these factors, a high recombination rate would be beneficial for eusocial species, especially for honey bees and others with large colonies.

A CLOSER LOOK AT HONEY BEES

Honey bees are different from other bees, even the other eusocial ones. Nearly all the other bees go dormant over the winter. Not honey bees. These buccaneers of buzz persist as homebodies, buzzing in their homes, keeping warm during the cold winter months. Colonies can persist for many years. Honey bee colonies reproduce by swarming, wherein one colony splits into two or more, each with its own queen.

Honey bees are in the genus *Apis*, from the Latin word for bee. While the bees in Europe and North America are in the species *A. melliferia* (the honey-bearing bee), there are seven other species of existing honey bees. For instance, the Asian honey bee, which was the culprit for the yellow rain, is *A. cerana*. Within *A. melliferia*, there are several subspecies; exactly how many depends on the authority.

The dark European honey bee, *A. melliferia*, covers much of Europe; its native range is bounded by three mountain ranges: the Pyrenees on the west, the Urals on the east, and the Alps on the South. Early European settlers brought this cold-hardy bee over to North America. Starting in the middle of the 19th century, immigrants from Southern Europe brought over other subspecies, notably *A. m. ligustica* (from Italy) and *A. m. carnica* (from Slovenia); these were somewhat less cold tolerant when they were first introduced to North America.[34]

Warwick Kerr (1922–2018), a Brazilian geneticist who had worked with notable evolutionary geneticists such as Sewall Wright, wanted to improve agriculture in Brazil.[35] One major impediment was poor pollination by bees: the European honey bees in Brazil were poorly adapted to the tropics, as were those imported from Portugal. Kerr knew that a subspecies of the honey bee, *A. m. scutellate*, was doing well in sub-Saharan Africa. He obtained bees from Tanzania in 1956 and started crossing them with the existing European bees in Brazil. Unfortunately, the next year, one of Kerr's hive managers accidentally let some bees escape. The escaped bees quickly spread through Brazil and then through the rest of tropical Central and South America. More aggressive than the European bees they were replacing, these so-called Africanized bees were more difficult to manage. They also sparked fear among the public as they spread.

During their spread north, some of the *A. m. scutellate* bees hybridized with the existing bees to some extent, but mainly they outcompeted them. Why have these *scutellate* bees been so successful? First, their colonies grow faster. The *scutellate* bees collect pollen more vigorously and are more efficient at converting the pollen and other floral rewards into the brood. They swarm more frequently too. The *scutellate* drones also appear to have a mating advantage; to some extent, this results from the *A. m. scutellate* bee colonies producing more drones than the bees they displaced.

Adapted to tropical climate, the *scutellate* bees, however, do not appear to fare well in colder climates; consequently, they have not spread much beyond the southernmost part of the United States.[36]

How do the *scutellate* and the European-derived bees differ in behavior? The *scutellate* bees are more prone to defend nests. They are more likely to respond to alarm pheromones and will sting much more frequently than the European bees. They will also defend nests from a greater distance than will the European bees: European bees usually do not attack intruders that are more than a few meters (yards) from the nests, whereas *A. m. scutellate* bees can start attacks of intruders that are as far away as 100 meters (300 feet). The *scutellate* bees can also pursue intruders for several kilometers (miles).[37] Why are the *scutellate* bees more aggressive? Consider that all honey bees are protecting valuable resources. They toiled hard to make the honey we and other animals take. Losing honey will likely put the hive's survival in jeopardy. The subspecies from which the *scutellate* bees are derived likely evolved in environments that had more and more aggressive predators.

There are some hints that the hyperaggressiveness of bees can be lost. For instance, *A. m. scutellate* came to Puerto Rico in 1994. Though there were some incidents of bees attacking people when the bees first appeared, these attacks markedly declined in frequency during the next few years, suggesting that the bees might be becoming less aggressive. Observation of bees confirms a decline in aggressiveness. Bees collected in the early 2000s, approximately ten years after the introduction, were not more aggressive than European bees and less so than *scutellate* bees from elsewhere.[38] These Puerto Rico bees looked like *scutellate* bees; their wing size was smaller than the European bees and comparable to those of the *A. m. scutellate* bees. Genetically, these bees had mitochondrial DNA that resembled the African subspecies, indicating that the maternal line was from African. Recall that only females transmit mitochondrial DNA. Their nuclear DNA was mixed, however, indicative of interbreeding with the European bees. Why did the bees in Puerto Rico evolve reduced aggressiveness? Perhaps aggression is not favored in the island environment. The aggressive behaviors are costly and take away resources that could be spent foraging or raising young. It will be interesting to see whether less aggressive behavior also evolves on other islands.

Wild honey bees in the United States today are a genetic composite: a mixture of multiple subspecies. Tom Seeley at Cornell University and his colleagues investigated the genetic composition of wild honey bees near Ithaca, New York (where Cornell is located). About two-thirds of the genetic makeup of those bees came from the southern European subspecies from Italy and Slovenia. Most of the rest of the genome comes from the northern European subspecies, with the rest coming from a subspecies from the Arabian Peninsula. Only a tiny sliver comes from *A. m. scutellate*. Interestingly, the genomic composition has not changed much since the 1970s: museum specimens that Seeley had collected then show a genomic profile strikingly similar to the bees collected more recently.[39] Comparable genomic studies have yet to be done on wild bees from other regions of the United States, but I suspect the composition will vary across the country.

Managed honey bees are different from most of the species we have domesticated. Humans have altered the genomes of cattle and chickens and dogs and just about all the other species that we considered domesticated. Indeed, some species like the silk-worm have been altered so much that they probably could not survive in the wild.[40] In contrast, we have not altered the genomes of honey bees much. The stuffy geneticists in E. B. White's poem have only done so much with artificial insemination. Bees differ from locality to locality, but there is nothing comparable to dog breeds or chicken breeds for managed honey bees. Instead, we have mainly just altered the environment of the bee to get them to do our bidding. For instance, beekeepers usually give bees much bigger hives than the cavities the bees themselves would use if they had a choice in order to manipulate the bees into producing more honey. Given this, Seeley and other bee experts consider bees to be semi-domesticated.[41]

THREATS TO BEES

The commercial beekeeper David Hackenberg noticed something strange about his bees in 2006: in some colonies, the adults were missing though the brood remained, untended. Other beekeepers observed the same phenomenon, which was given the name colony collapse disorder (CCD). Soon after the initial report, science journalist Myrna Watanabe wrote a report: "Colony collapse disorder: Many suspects, no smoking gun".[42] Though we know considerably more about the threats to honey bees than we did in 2008, the conclusion is still the same: many suspects, no smoking gun.

Honey bees and the honey bee industry were in dire trouble. Testifying to Congress in 2007, the University of Illinois bee expert May Berenbaum and Chair of the Committee on the Status of Pollinators of North America said, "Even before CCD came to light, our committee estimated that, if honey bee numbers continue to decline at the rates documented from 1989 to 1996, managed honey bees will cease to exist by 2035."[43] Speaking to the science writer Paige Embry a decade later, Berenbaum noted that things are no longer as bleak. Honey bees in the United States are not declining as much as they were in the 1990s. Berenbaum noted that both beekeepers and the federal government have taken steps to stem the losses.[44] Honey bees may not be out of the woods, but things are better.

It's not just honey bees that are in trouble. In fact, other bees may be suffering more. Despite losses in the United States and other locations, global honey bee stocks increased during the half-century that ended in 2015. Bumble bees have declined substantially during that same period. Other bees are probably declining as well, though we do not have as much data on them.[45]

Bees face numerous threats. Habitat destruction and change can cause their sources of pollen and nectar to disappear. The pesticides we use to control insect pests also affect bees. Bees have predators, parasites, and pathogens. Bees are also competing with introduced species. And then there's climate change. Climate change is especially challenging for bees. Dependent on flowers, bees do not just have to evolve to keep up with the changing climate but also the plants that are evolving in response to climate change. If bees and plants wind up evolving at different rates or in different ways, a mismatch between bees and the flowers they pollinate can ensue. These challenges are not acting in isolation. In fact, being burdened by one challenge can reduce bees' ability to deal with others. Dave Goulson and other bee biologists at the University of Sussex summarized how these multiple stressors may interact: "A bee or bee colony that appears to have succumbed to a pathogen may not have died if it had not been exposed to a sublethal dose of a pesticide and/or been subject to food stress.'[46] The interaction of these stressors suggests that reducing one or more of the stressors may improve bees' ability to cope with the others. We can reduce stress on bees in several ways: increasing the quantity and diversity of flowers, reducing exposure to pesticides, and limiting introductions of exotic species.

In fact, a recent study found that a high-quality diet improves survival of honey bees after viral infection.[47] Bees infected with Israeli acute paralysis virus survive better if they are fed pollen beforehand than if they are not. The bees not given pollen were not starved; they still had a supply of sugar for calories, but no protein or other nutrients. It's not just the amount but the quality as well. Infected bees given pollen that had more protein and more calcium and other trace nutrients survive better than those given standard pollen. In the absence of the virus, bee survival was not affected by the feeding treatments.

We will focus on two important stressors in the following sections: pesticide exposure and then pathogens and parasites.[48]

PESTICIDES

Pesticides do not have to kill bees to be detrimental to them; numerous sublethal effects of pesticides hinder the health of bees.[49] They slow development. They limit the life span. They reduce the number of offspring produced. In addition to those effects on the life history, pesticides can also alter behavior and learning. Given that pollinators rely on learning to locate food supplies, a bee that is deficient in learning is likely to have difficulty feeding. In eusocial bees, skills are even more important, because these bees also need to communicate information to their nestmates. Even subtle learning impairment could hinder a colony's survival and reproduction.

While humans have been using pesticides for millennia and industrial use of pesticides dates back more than a century (see Chapter 9), there is a new class of pesticides: the neonicotinoids. This class of pesticides rapidly rose in use in the early years of the 21st century—reaching a quarter of the market share by 2007.[50] Chemically related to nicotine, they target the insect nicotinic acetylcholine receptor. This receptor is a key player in neural transmission in the insect nervous system and controls how muscles contract. They also impair the learning and behavior of bees and other insects. Because vertebrates and insects have different nicotinic acetylcholine receptors, neonicotinoids are much less toxic to birds and mammals.

Neonicotinoids are also different from most other insecticides in how they are applied. In contrast with most other insecticides, which are applied in response to an actual infestation, neonicotinoids can be applied preemptively. The method of application also is different. Most traditional insecticides are applied topically or outside of the plant. The heat from the sun will break down them down and the rain will wash them off. Neonicotinoids are different; these are coated on seeds and get incorporated into the plant—all of the plant—as it grows.[51] So if a plant is treated with neonicotinoids, bees will be getting a dose through the plant's pollen and nectar. The neonicotinoids also leach into the soil—substantial exposure can be from non-crop plants whose seeds were not treated even years after treated seeds were sown.

Then a postdoctoral fellow at Harvard, James Crall devised a clever series of experiments to show the subtle effects of neonicotinoids on bumble bee behavior.[52] Crall and his colleagues set up a robotic observation platform to track the movements of bumble bee workers within the nest without disturbing. Exposing the bees to imidacioprid, a neonicotinoid pesticide, at concentrations often found in fields disrupts the behavior of bees. When exposed, workers are generally less active and nurse the brood less. Interestingly, these effects are strongest during the night. This effect of timing on the behavior is reason to monitor bees at various times. The pesticide also disrupted the social network of the bees. Bees that were exposed to the pesticide spent more time along the edges of the nest. Moreover, exposed bees also were deficient in thermoregulation. These subtle changes to behavior could have manifold and profound consequences on colony survival.

Crall and his colleagues also had developed computer models that can explain an interesting observation about the effects of neonicotinoids on different bees. A group from Sweden had shown that seeds coated with the neonicotinoid clothianidin greatly affected solitary bees.[53] These bees were unable to build brood cells in their nests, and their density was greatly reduced. Bumble bees were affected somewhat, with reduced colony growth and reproduction. Interestingly, no detectable effect of the neonicotinoid was observed in honey bees. As colony size increased, the negative consequences of the neonicotinoid diminished. What Crall and his colleagues had found was that as simulated bees engaged in social interactions in the nest, these interactions brought the behavior of neonicotinoid-impaired bees closer to that in their unimpaired state.[54] This buffering was strongest in the largest colonies. Such social buffering could explain why in the Swedish experiment, the solitary bees were most affected by the neonicotinoid and the honey bees were affected least. The large colonies of honey bees could have provided sufficient buffering, while the medium-sized colonies of bumble bees could have given them some, but not completely sufficient, buffering.[55]

I spoke with Crall at the end of August 2020. I was curious about whether the bumble bees had responded to selection from pesticides.[56] Given what we know about social buffering, I wondered if bumble bees may evolve larger colonies—large colonies would permit greater social buffering—assuming that the variation in colony size is heritable. Crall noted that while larger colonies would improve social buffering, the real bottleneck is early in the season before the first set of workers come out. Prior to this, the solo queen is most vulnerable. Accordingly, there might be selection in getting those first workers out early.

PATHOGENS AND PARASITES

Investigating dead honey bees from Japan in the early 1980s, the virology team of Bill Bailey and Brenda Ball discovered what would be the last virus of their acclaimed careers. This virus, eventually known as deformed wing virus (DWV), would become one of the most important and deadly bee viruses.[57] For a long time after Bailey and Bell's discovery, however, this virus was little studied. It was not causing much pathology, and consequently, research on it languished.

Circumstances changed as the parasitic mite *Varroa destructor* swept through populations of the European honey bee in North America and elsewhere. These calamitous mites altered the deformed wing virus, turning it from a Jekyll into a Hyde virus. Prior to the mite infestation, the virus is well-behaved, replicating slowly and having only minimal effects on its hosts. In mite-infested bees, however, the virus replicates wildly and becomes destructive. Moreover, as the mite feeds on the bee's hemolymph (the fluid of the insect that acts like blood), it injects the bee with a load of virus.[58]

During the middle of the 20th century, the *Varroa* mite jumped from its native host, the Asian honey bee, into the European species. As we will see, the mites are not as destructive to the Asian honey bee, as this bee has evolved resistance to it. But what about the virus; where did it come from? There are two reasonable scenarios: either the DWV came along with the mite during the species jump, or the DWV is native to the European honey bee and the mite lit the fuse, reawakening the virus. Phylogenetic analysis of sequence data supports the second hypothesis as it unambiguously places the European honey bee as the ancestral host of the virus.[59]

The mite not only has increased the prevalence and abundance of DWV in honey bees; it also has increased the prevalence of the virus in *Bombus* bumble bees in England and France, even though the mites do not directly infect bumble bees. Interestingly, the honey bees and bumble bees are sharing the same genetic variants of DWV: there is no evidence of differentiation. Several lines of evidence point to a recent bottleneck in the virus, followed by a population expansion. The genetic diversity is low. More rare variants are present than would be expected under neutrality. The phylogenetic pattern also suggests a recent expansion—perhaps six years before the study. All of this evidence supports the conclusion that honey bees are the ancestral and reservoir host of the virus.[60]

Several wild populations of the European honey bee apparently have evolved resistance to mites. These either lack mites or can function well in the presence of a few mites. Interestingly, the mechanisms of resistance are different in the different populations. The Asian bees have dealt with mites by grooming themselves and other bees in the colony; this hygienic behavior has evolved in European bees in Brazil and South America but not in mite-resistant populations in Sweden. Bees in upstate New York have evolved a reduced colony size, but mite-resistant bees in Sweden have not had such a reduction. Mite-resistant bees in Brazil and South America are cranking out young faster, but mite-resistant bees in Sweden have not had a change in development time.[61] Some populations have evolved more frequent swarming to control mites, because about a third of mites are exported during the swarming.[62]

Remember the study by Tom Seeley looking at the genomic composition of wild honey bees near Ithaca, New York, in both the 1970s and the 2010s? Because the bees from the 1970s were before the invasion of the mites and those from the 2010s were collected after the population had adapted to the mites, we can infer how the mites affected the population.[63] The bees from the 2010s had much lower diversity at the mitochondrial DNA, which is inherited from queen to queen. Apparently, there had been a sharp bottleneck, likely from adaptation to the mites. The nuclear DNA, which is inherited from both queens and drones, showed high diversity. A likely reason why the nuclear DNA did not show a bottleneck is that queen bees mate with multiple drones and there is little inbreeding.

Several hundred sites along the genome showed marked changes between the 1970s and the 2010s. Many of these are in genes that function during development, which may indicate selection for changes in the development of these bees. Consistent with this finding, the 2010s bees are smaller than those from those in the 1970s. There is a strong sign of selection on one notable behavioral gene—*AmDOP3*—which is involved in hygienic practices. Future study of this and other genes may provide further information about how these bees evolved to be resistant to the mites.

My colleague at UMASS, Lynn Adler, is interested in the role that plants play in the transmission of bee pathogens and parasites.[64] Bees moving from flower to flower and flowers being pollinated by multiple bees sets the stage for the transmission of pathogens. Plants also can be meeting places where bees can pick up beneficial microbes, such as bacteria that produce lactic acid. Much like we eat yogurt and other foods that have lactic acid bacteria, bees may use the bacteria found at flowers as a probiotic.

Adler has also studied phytochemicals, chemicals plants produce in nectar and pollen; some of these can act against the pathogens of both bees and plants. Bees have been documented to use these chemicals to ward off infection. The problem is that much like bacteria can evolve resistance to antibiotics, bee parasites and pathogens can evolve resistance to the antibiotic-like phytochemicals.

Then a graduate student in Adler's lab, Evan Palmer-Young was interested in whether phytochemicals in controlling trypanosome parasites in bumble bees. Trypanosomes are single-celled protozoans that infect lots of different animals. In humans, trypanosomes cause sleeping sickness and Chagas disease. The bumble bee trypanosome Palmer-Young was investigating was *Crithidia bombi*,[65] one that resides in the equivalent of the intestines (the midgut and hindgut) of the bees.

One of the phytochemicals Palmer-Young was interested in is thymol. This chemical, which is produced by thyme, has been used to control *Varroa* mites of honey bees. It also is effective against other bee parasites and pathogens. Thymol will kill the trypanosome, and it is produced by flowers that bumble bees frequently encounter.[66]

Palmer-Young and his colleagues earlier had found variation in different strains of the trypanosomes for their resistance to thymol.[67] This variation suggested that resistance could evolve. Palmer-Young and colleagues looked to see whether resistance to thymol would evolve. They also wanted to see whether resistance to a different phytochemical, eugenol, would also evolve. As expected, the trypanosome quickly evolved resistance to thymol upon repeated exposure to just thymol. Likewise, it quickly evolved resistance to eugenol upon repeated exposure to eugenol. What was a surprise was that the trypanosome also evolved resistance to both thymol and eugenol when exposed to both simultaneously.[68] In contrast to what we often see with antibiotics (see Chapter 1) and HIV treatments (see Chapter 2), multi-drug treatment did not slow the evolution of resistance to these phytochemicals. Moreover, there was a low cost of resistance to the phytochemicals; in the absence of the phytochemicals, the resistant trypanosomes did well.

What explains the rapid evolution to the cocktail of both phytochemicals? Maybe thymol and eugenol have sufficiently similar modes of action that the evolution of resistance to one confers resistance to the other. Thymol and eugenol both disrupt the function membranes of cells and membranes. The low cost of resistance probably also contributes to the ease of the adaptation to the cocktail.

Before we leave this discussion of pathogens, one final point. As we saw in Chapter 2, many of the viruses that plague humans came from nonhuman animals. Spillover ultimately from bats led to the SARS and MERS outbreaks and likely led to the COVID-19 pandemic. Such spillover is likely in bees. Earlier, we discussed how the *Varroa* mite jumped from one honey bee species to another. Working with Lyna Ngor, Quinn McFredrick, and other researchers, Evan Palmer-Young attempted to cross-infect a variety of parasites across different bee species.[69] Several of the parasites could be transferred from one bee species to another. The trypanosomes *Crinthidia bombi* and *C. mellificae* were particularly easy to cross-infect across different social bee species. More of such experimental studies are warranted to assess the likelihood of spillovers. In addition, more surveillance studies regularly monitoring social bees for the presence of novel pathogens are needed.

DARWINIAN BEEKEEPING AND THE WISDOM OF THE HIVE

Some biologists follow a single overarching question about biology and investigate it using various tools and in several different study organisms. Others are attracted to a single organism or group of organisms—a lot of them are birders—and investigate multiple questions about that organism. Tom Seeley at Cornell falls into the second category. His research organism of choice is the honey bee. Attracted to bees since he was a child in the early 1960s, Seeley has a built a career studying their behavior and evolution.

One question that Seeley spent considerable time exploring is how honey bees decide on which home to move to.[70] Bees need to find new homes when the colony splits into two—the main way colonies produce offspring colonies—or when their current hive is unsuitable. In both cases, bees travel *en masse* in what is known as a swarm. So how do bees do it?

First, much like a human organization that forms a search committee before hiring someone in a leadership position, the bee colonies send scout bees to go out exploring for suitable sites. If a scout finds a site that meets the qualifications, she—like all workers, the scouts are female—will signal her nestmates with a dance. This dance conveys two types of information: the location and the quality of the site. The site-quality signal that scout bees use is similar to the waggle dance bees returning from foraging bouts use when they signal the quality of food at a flower. Sites that are exceptional will likely elicit a vigorous dance, while those that are mediocre are apt to lead a lackluster one. At the nest, other scouts will respond to the dancing scout corresponding to the quality of the dance: more bees are likely to respond to a vigorous dance than a humdrum one. The scouts attracted by the dancing bee then go to the site to make their own assessment of the site. Like the first scout, these second scouts make dances corresponding the location and quality of the site. Multiple sites may be judged at a time; and much like a New England town meeting or jury deliberations, opinion of the site preference may change over the course of the deliberation.

Seeley has found that, based on objective criteria of site qualifications,[71] the colony as a group makes more accurate decisions than any single scout bee. The colony can make rather sophisticated decisions despite each individual bee not having much neuronal prowess and in the absence of a leader. Contrary to popular opinion, the queen has little to say over the decision of where to go. The intelligence of the hive is distributed across its members. The study of bee decision making may provide useful for studies of artificial intelligence.

Seeley also suggests that we can apply some of these findings to human decision in groups where the members share a common interest, For instance, he notes:

> One valuable lesson that we can learn from the bees is that holding an open and fair competition of ideas is a smart solution to the problem of making a decision based on a pool of information distributed across a group of individuals.[72]

Seeley also has been interested in how we can improve the lots of bees by following ecological and evolutionary principles. Using bees that are genetically adapted to the area where they will be foraging is ideal. A bee that has adapted to the relatively wet climate of New England may not do so well pollinating almonds in the arid climate of California. Likewise, the California-adapted bee may not be ideally suited for the cold winters of New England. The non-native bees probably will survive the exotic environment, but they are starting with a handicap.

Beekeepers, particularly commercial ones, usually have bees in hives that are much larger than the ones wild bees prefer. Larger hives stimulate bees to produce more honey, but they also foster the spread of disease. Moreover, smaller hives lead to more frequent swarming. This is a nuisance to the beekeeper, for sure, but swarming appears to reduce disease spread. Moreover, frequent swarming allows for greater selection pressure for strong hives.

Seeley notes that several of these suggestions may result in a trade-off of lower yield. Thus, not all these suggestions can be used by commercial beekeepers. But for the hobbyist, perhaps it is best to let bees be bees.[73] I suspect Emily Dickinson would have approved.

NOTES

1. Dickinson (2012). A surcingle is a strap fastened around a horse.
2. Embry (2018, p. 177).
3. Guthrie (2007).
4. Dickinson, who was not quite thirty, when Darwin's *Origin* was published, was exposed to Darwin's work and Darwinian ideas through *Harpers*, *The Atlantic Monthly* and other periodicals she received. See Kirkby (2010) for more details.
5. Danforth et al. (2019).
6. Chapter 7 in Danforth et al. (2019) is an excellent source for information on floral rewards. They discuss the mechanisms flowers evolved to protect their pollen and the mechanisms bees use to get around these defenses. Palmer-Young et al. (2019) discuss the chemicals plants use in defense of pollen and nectar. Palmer-Young and colleagues find that plants use both more chemicals and a greater diversity of chemicals in pollen than in nectar. This is consistent with what we would expect given the relative value of each.
7. Should it be honey bee or honeybee? Snodgrass (1956, p. vii), an entomologist from the early and middle twentieth century, addressed this question with the following passage:

> Regardless of dictionaries, we have in entomology a rule for insect common names that can be followed. It says: If the insect is what the name implies, write the two words separately;

otherwise run them together. Thus we have such names as house fly, blow fly, and robber fly contrasted with dragonfly, caddisfly, and butterfly, because the latter are not flies, just as an aphislion is not a lion and a silverfish is not a fish. The honey bee is an insect and preeminently a bee; "honeybee" is equivalent to "Johnsmith".

8. This estimate and the other information in the paragraph are from Reilly et al. (2020). Some other estimates are even higher.
9. See chapter 2 of Embry (2018).
10. Eilers et al. (2011).
11. Klatt et al. (2014).
12. Hubbell (1988)
13. Matthews et al. (2019).
14. See Chapter 3 of Seeley (2019).
15. Hubbell (1988, p. 167).
16. The poem can be found at Hubbell (1988, p. 92).
   Here is a small section:

> There's a kind of a wild and glad elation
> In the natural way of insemination;
> Who thinks that love is a handicap
> Is a fuddydud and a common sap,
> For I am a queen and I am a bee,
> I'm devil-may-care and I'm fancy free,
> The test tube doesn't appeal to me,
> Not me,
> I'm a bee.
> And I'm here to state that I'll always mate
> With whatever drone I encounter.

17. Seeley et al. (1985).
18. Goulson (2003).
19. Danforth et al. (2013).
20. Sann et al. (2018).
21. See Danforth et al. (2019) for more information about solitary bees.
22. The study is Romiguier et al. (2016). They based their findings by using only some genes. Recall that DNA has four bases: A's, which pair with T's, and C's, which pair with G's. This study found that genes that have a large proportion of G's and C's tend to be less reliable than other genes in determining phylogenetic relationships. Using only the reliable genes, this study found a clear signal that the orchid bees were the outgroup to the other corbiculate bees; hence, eusociality evolved only once in these bees.
23. Fu et al. (2015).
24. Smith et al. (2007).
25. See Hamilton (1964, 1972) for the early work tying the evolution of social behavior in Hymenoptera to haplodiploidy. Another route for the evolution of altruism and other helping behavior is reciprocal altruism, wherein individuals sometimes help others at an expense to themselves because they expect the beneficiary to return the favor. We'll discuss this in Chapter 16.
26. For instance, see Gardner et al. (2012) and Rautiala et al. (2019).
27. See Wilson (2005). The quote is on p. 164.
28. Linksvayer and Wade (2005).
29. Zayed and Packer (2005).
30. Takahashi et al. (2008).

31. Souza et al. (2010).
32. Wilfert et al. (2007).
33. The genetic population size is better known as effective population size. See also, Romiguier et al. (2014).
34. Seeley (2019).
35. Balakrishnan (2018).
36. Breed et al. (2004).
37. Schneider et al. (2004).
38. Rivera-Marchand et al. (2012).
39. Mikheyev et al. (2015).
40. Hubbell (2001).
41. See Seeley (2019) for more on how we have altered the environments of bees.
42. Watanabe (2008).
43. Embry (2018, p. 14).
44. *Ibid.*, p. 31.
45. See Goulson et al. (2015) and Vanbergen and the Insect Pollinators Initiative (2013) for further details.
46. Goulson et al. (2015, p. 6)
47. Dolezal et al. (2019).
48. I'm using parasites and pathogens somewhat interchangeably. Parasites are typically those organisms that feed on the host that are visible to the naked eye, while pathogens are microscopic.
49. Desceux et al. (2007).
50. Jeschke et al. (2010).
51. Embry (2018).
52. Crall et al. (2018).
53. Rundlof et al. (2015).
54. Crall et al. (2019).
55. Evan Palmer-Young noted that there are other possible explanations—perhaps the bees with the larger colony size have a greater foraging range and thus more opportunity to visit non-crop flowers, which are less likely to have the toxins. Related to this, perhaps the honey bees are buffered by pooling of diverse resources (across space and time) to dilute the effects of toxic food sources or time periods. E-mail exchange with Evan Palmer-Young. September 2020.
56. Phone interview with James Crall. August 2020.
57. Martin et al. (2019).
58. See Nazzi et al. (2012) and Martin and Bretell (2019).
59. Wilfert et al. (2016).
60. Manley et al. (2019).
61. Locke (2016).
62. van Alphen and Fernhout (2020).
63. Mikheyev et al. (2015).
64. See Adler et al. (2020) and McArt et al. (2014).
65. Strictly speaking, these should be considered "trypanosomatids" because they are not in the *Trypanosoma* genus.
66. Gregoric and Planinc (2005).
67. Palmer-Young, Sadd, Stevenson et al. (2016).
68. Ngor et al. (2020).

69. Palmer-Young, Sadd, and Adler (2016).
70. Seeley (2010).
71. Seeley examined how well bees survive in cavities of different characteristics. Bees' survival corresponded with their preferences.
72. Seeley (2010, p. 75).
73. Seeley (2019) and Johnson (2020).

Section III

Environment

11 The Biodiversity Crisis

SUMMARY

We face a crisis of rapidly diminishing biodiversity: largely as the result of human activity, species are going extinct at a rate comparable to that experienced in the mass extinctions that occurred in the past (about once every 100 million years). The situation is dire but not hopeless; lessons from and research in evolutionary biology can ameliorate this biodiversity crisis. Advances in evolutionary genetics have led to barcoding, a powerful tool of cataloguing and tracking biodiversity. Genetic rescue is the introduction of distant populations to rescue an endangered species. We discuss how genetic rescue was used to bring Florida panthers back from the brink of extinction. What is the minimal size of a population for it to likely persist indefinitely? Determination of such informs conservation and management decisions. We discuss the ecological and evolutionary genetic factors that help determine this minimal viable population size, including recent work on mutation accumulation in nematodes. We also discuss the factors that determine whether and how populations will respond to the new threat of climate change. We conclude with a survey on how the internal microbiota of organisms affects their ability to survive and thrive threats.

"Nature underpins all dimensions of human health and contributes to non-material aspects of quality of life—inspiration and learning, physical and psychological experiences, and supporting identities—that are central to quality of life and cultural integrity, even if their aggregated value is difficult to quantify."—Sandra Diaz and colleagues.[1]

"Finally, there is a deeper meaning and long-term importance of extinction. When these and other species disappear at our hands, we throw away part of Earth's history. We erase twigs and eventually whole branches of life's family tree. Because each species is unique, we close the book on scientific knowledge that is important to an unknown degree but is now forever lost."—E. O. Wilson.[2]

THE BIODIVERSITY CRISIS

Up to a million species face the prospect of extinction by the year 2100. This grim prediction comes from a United Nations report issued during the spring of 2019.[3] With increasing alarm over the last several decades, biologists and others have been warning about the loss of biodiversity, but the UN and other reports released in 2019 point to an even more stark reality of what we are losing.

Birds are disappearing. Three billion fewer birds inhabit North America now than in the 1970s.[4] This is a decline of about 30 percent. Old World sparrows, larks, and blackbirds were hit particularly hard. Losses were uneven across different habitats: for instance, birds living in grasslands suffered the steepest declines. These birds, which are now less than half as abundant as they were in 1970, probably are declining due to the loss of habitat. Incidentally, birds that live in wetlands are seeing rises in their numbers, likely from conservation efforts during the latter part of the last century.

It's not just birds. Insects are much less abundant and less diverse than they were a few decades ago. These declines may be driven by agricultural management; in Germany, where the most extensive survey of insect abundance and diversity was performed, the most severe declines are in landscapes near farmland.[5] Though their decline is less severe, insects in the forest were also less abundant. The loss of insects is particularly troubling because insects pollinate flowers (as we saw in Chapter 10). Insects are also important sources of food for a staggering array of organisms, including other insects. Insects also decompose dead organisms and other waste. Without insects, ecosystems are in peril.

Plants are going extinct alarmingly fast. On average, three known species have gone extinct each year since the beginning of the 20th century.[6] These are the species we know about; countless unknown species have been wiped out too. Certainly, extinctions occurred before human influence, but much more slowly. The current extinction rate is roughly 500 times greater than it was before human influence.

The current extinction rate is exceptional, but it is not completely without precedent. By most accounts, there have been five episodes since the Cambrian explosion just above half a billion years ago where large chunks of the world's biodiversity disappeared in a relatively (by geological standards) short period of time. These so-called mass extinctions thus happen about once 100 hundred million years on average, though not like clockwork; they are spaced irregularly. The last one—the one that wiped out the non-avian dinosaurs—was 66 million years ago. We now face the danger of a sixth mass extinction,[7] and unlike the other five, this one would be due to a single species—our own.

We neglect this biodiversity crisis at our own peril. As we saw in earlier chapters, our global food system greatly relies on pollination by insects and other animals. Many medicines and other products ultimately come from animals, plants, and other biological organisms. For instance, as we saw in Chapter 3, the antimalarial drug artemisinin was isolated from the plant sweet wormwood (*Artemisia annua*). Our ecosystems, both marine and terrestrial, sequester carbon out of the atmosphere; in doing so, they retard the pace of global warming. Disruption of these ecosystems could lead to a deadly circle of accelerated warming and loss of biodiversity. Along with these direct, practical benefits of biodiversity (sometimes called ecosystem services), aesthetic and moral arguments compel us to preserve biodiversity. With each species gone, our world is diminished. As Ed Wilson noted in the epigraph, when we lose species, we lose twigs (and sometimes large branches) of the tree of life.

In the past, the threats humans have posed to biodiversity have been largely from (in approximate order of importance) habitat loss and deterioration, the overexploitation of animals, pollution, and threats from invasive species. More recently, human-induced changes in climate have intensified these threats. These threats do not act in a vacuum—all too often, they reinforce each other. Species whose numbers have declined due to habitat deterioration may have reduced genetic variation, which makes them less able to respond to climate change.

Although the picture appears bleak, it is not hopeless. Conservation efforts thus far have thwarted extinction somewhat. Writing in *Half-Earth*, Wilson notes that global conservation measures have reduced the extinction rates of land-dwelling vertebrates by about a fifth.[8] Though a much greater effort is required, conservation can work. Sandra Diaz from the National University of Cordoba in Argentina and the lead author of the UN report noted, "We need to build biodiversity considerations into trade and infrastructure decisions, the way that health or human rights are built into every aspect of social and economic decision-making."[9]

We can point to some specific successes: among the most visible is the California condor. With a wingspan approaching ten feet, the California condor is the largest bird in North America. Thankfully, we can still say it is the largest North American bird because we almost lost it. During the 19th and 20th centuries, its population precipitously fell. The most likely cause for this steep decline is lead poisoning. As carrion feeders, the condors often ate animals that hunters left behind, along with lead pellets and sometimes even whole bullets. By 1982, only 22 birds were left. Captive breeding programs have brought this species back from the brink, with the 1,000th chick hatching in captivity in the spring of 2019 in Utah's Zion National Park. Although the condor remains listed as critically endangered, its prospects are much improved. As of 2019, more than 500 condors are living and more than half of those are in the wild, outside of captivity. Once again, these majestic creatures soar![10]

Evolutionary biology is of relevance to conservation efforts. In addition to threatening populations, human drivers have caused rapid evolution of these species. And as the UN report explicitly states: "Understanding and working with contemporary evolution can address important concerns surrounding pollination and dispersal, coral persistence in the face of ocean acidification, water quality, pest regulation, food production and options for the future."[11] Evolutionary biology has also made many contributions to conservation biology including developing an approach to estimate minimal viable population sizes. The concept of genetic rescue, genetic adaptation that permits a population to recover from demographic challenges, is a more recent contribution. We will discuss these contributions later in this chapter. Related to conservation biology's attempts to maintain a species that has perilously low numbers is the challenge from invasive species, wherein too many individuals of the invader pose challenges on the ecosystem. We will take up invasive species in Chapter 14.

The principles of evolution are also useful in cataloguing and identifying the biodiversity that we have.

BARCODING TO CATALOGUE BIODIVERSITY

"We are astonishingly ignorant about how many species are alive on earth today, and even more ignorant about how many we can lose yet still maintain ecosystem services that humanity ultimately depends upon."[12]

Sir Robert May, a polymath whose work spanned from epidemiology to ecology to economics,[13] wrote those words lamenting our ignorance regarding biodiversity in the abstract of a commentary of a paper that attempted to estimate the total number of eukaryotic species of life currently existing on Earth.[14] Such estimates have ranged from 5 million to nearly 100 million! Given that about 1.5 million have been described, what we do not know is more than we do know. Moreover, we are ignorant about what we do not know.

One method of assessing biodiversity sounds much like the tricorder in the *Star Trek* television shows and movies.[15] With their hand-held tricorder, the crew of the Enterprise could quickly scan for and identify lifeforms. The promise of having a real life tricorder begins with what has been dubbed the DNA barcode.[16] This method, developed by Paul Herbert (pronounced like "E-bear") and colleagues at the University of Guelph, relies on sequencing just a short piece of DNA taken from the *cytochrome oxidase I (COI)* gene in the mitochondrial (mt) genome. Because mitochondria are abundant in any given cell, harvesting mtDNA from cells is easy. From looking at the *COI* data of samples and comparing those sequences from those in databases, biologists can quickly and accurately determine the identity of a specimen to species. The same technology permits the discovery and putative classification of new species. Writing in 2005, Paul Herbert and Ryan Gregory declared:

> DNA barcoding allows a day to be envisioned when every curious mind, from professional biologists to school children, will have easy access to the names and biological attributes of any species on the planet. In addition to assigning specimens to known species, DNA barcoding will accelerate the pace of species discovery.[17]

To say that the initial response to the barcoding project was controversial would be an understatement. An early commentary described the ends of the continuum of the response as "unbridled enthusiasm, especially from ecologists" and "outright condemnation, largely from taxonomists".[18]

An early success of barcoding was a study finding much greater than expected diversity of amphipods of the genus *Hyallela*. Named after their feature of having two different (*amphi*) types of feet (*pod*), amphipods are generally small, aquatic crustaceans. The *Hyallela* amphipods live in springs and lakes across the Great Basin region of the western United States. DNA barcoding revealed strikingly diversity in these amphipods; for instance, specimens that had all been thought to belong to the species *Hyallelella azteca* actually cluster into at least 33 provisional species, some of which are very different from others in their DNA sequence. This extensive but cryptic diversity of amphipods has implications for conservation. Agricultural use of water has exacerbated water shortages and threatened aquatic habitats in the region; consequently, the springs that these amphipods use have shrunk. Given the diversity found in the springs, more conservation effort should be undertaken there.[19]

Because it is easy to use and relatively inexpensive, DNA barcoding should be particularly amenable to assessing biodiversity in the developing world. The DNA barcoding community has recently focused on studying the biodiversity of Africa, a continent that not only is the birthplace of humankind but is also home to extraordinarily isolated lineages of plants, animals, and other organisms.[20] Given its diverse climates and habitats, Africa is expected to be abundant in biodiversity, but this biodiversity is woefully understudied. Within Africa, the island nation of Madagascar is a hotspot for biodiversity. It also is home to animals and plants not found elsewhere.

Species lost here cannot be recovered from elsewhere. Unfortunately, the biodiversity in Madagascar is threatened by human disturbance.

Barcoding recently has been used in Madagascar to discover and classify the diversity of small moths called micromoths.[21] These are a catch-all category within Lepidoptera (moths and butterflies) consisting of those moths that have wingspans less than about 20 mm (0.8 inch). These tiny micromoths are exceedingly difficult to identity by morphological characters. Barcoding has shown how the patterns of diversity of these moths vary with space and time. Barcoding also has revealed a diagnostic difference that identifies a group within the micromoths called the macro-heterocera, a group that includes owlet moths, sack bearers, and lappet moths, among others. These macroheterocera share a specific nucleotide difference (one not shared by moths that are not macroheterocera), providing support that this group all share a common ancestor not shared with other moths.[22]

A recent barcoding study of New Zealand birds provides tools for conservation and management of these birds, many of which are endangered. One of the challenges these birds face is mammalian predation on their chicks and eggs. Additional DNA barcoding can be used to sample the fecal material of these mammals. Such studies, which are noninvasive and do not require direct observation, are likely to be better than traditional methods in assessing predator impact.[23]

Evolutionary principles and information have been vital in developing and applying the barcoding infrastructure. Barcoding using mtDNA can distinguish closely related animal species because mitochondrial genes tend to evolve faster than genes in the nucleus. Moreover, the gene chosen for barcoding in animals, *cytochrome oxidase I*, has different nucleotides that evolve at different rates, providing more "phylogenetic signal".[24] Barcoding also works because there is generally little variation within species at *COI* as compared with the differences between different species at this gene. Mitochondrial genes usually are not variable within species as they are between different species.[25] The relatively low genetic variation at *COI* within species is due in part to the fact that all of the genes on the mitochondria are linked and act as a unit: selection on one gene has the consequence of wiping out genetic variation across the whole mitochondrial genome.[26] Evolutionary principles also factor into the construction of evolutionary trees during barcoding.[27] The process relies on algorithms that cluster each of different specimens into hierarchal groups that are presumed to reflect evolutionary relations; much debate is focused on which algorithms are appropriate.[28]

With genomic sequencing becoming cheaper and quicker, a recent trend has been to develop extended barcodes that use multiple loci obtained by genome skimming— shallow sequencing to get at parts of the genome that are present in a large number of copies per cell, such as the mitochondria, as well as plastids in plants.[29]

GENETIC RESCUE

The southernmost franchise in the National Hockey League (the Florida Panthers) is named after the sole remaining panther living east of the Mississippi in North America. These Florida panthers have the Latin name, *Puma comcolor coryi*, but also have gone by several other names including cougar, puma, and catamounts. The historic range of this panther stretched all the way from Florida as far west as the Louisiana-Texas border and as far north as southern Tennessee. Development and loss of habitat (cypress forest as well as wetlands) has restricted its range to a mere sliver of that; now, the panthers are found in just a few pockets of Southern Florida.[30]

By the 1990s, their numbers had dwindled to about two dozen. As a likely consequence of its small numbers, the panther's genetic diversity also was depleted, indicating that it was extremely inbred. Associated with the paucity of genetic variation were the phenotypic symptoms of inbreeding: cowlicks and hooked tails were the most obvious of the morphological abnormalities. More serious but hidden were the atrial septal defects. These are literal holes in the walls between their two heart chambers. Pathogens and parasites plagued this panther. Several individuals were infected with feline immunodeficiency virus (FIV). Like its relative HIV, FIV also attacks the immune system and can cause AIDS-like symptoms, though usually not as severe as what is seen in the human disease. The prevalence of infection is also consistent with the poor genetic health of this population. Perhaps most troubling of all are the reproductive consequences of this inbreeding: panther males had reduced testosterone, undescended testes, and poor-quality sperm. Given all these challenges, the panther's prospects looked bleak. In fact, models conducted at the time predicted that this subspecies was extremely likely (95 percent probability) to go extinct, very soon—within two decades.

Given the dire circumstances, even risky proposals seemed worth doing. One such seemingly risky proposal involved bringing related populations of panthers from elsewhere to south Florida. The idea is that the introduced panthers would not only increase the population size but also bring in new alleles and combinations of alleles to the gene pool.

Pumas from Texas of the subspecies *P. c. stanleyana* had mated with the Florida panther in the past. Their prior mating is an important consideration because it suggests that genetic incompatibility, also known as outbreeding depression, was low or nonexistent between the populations. Genetic incompatibility would dampen the chances of the plan working by curtailing the ability of the alleles from the Texas population to get established in the Florida gene pool.[31]

Five of the eight introduced pumas bred. They had 15 kittens, which turned out to be sufficient to get the population growing quickly. By 2003, there were at least 95 adults, corresponding to an annual population growth rate of 14 percent. The genetic health of the population also rebounded: genetic diversity increased substantially, roughly doubling by 2007. Along with the improved genetic health, the panthers also appear healthier, with fewer cardiac deformities. The quality of the sperm and other reproductive characteristics are improving. The panthers are even more agile: more are escaping from trees during capture. This regained agility may keep the managers on their toes, but it is also one more sign of recovery.

Despite these promising signs, the panther is not completely out of danger. Its low population size and genetic variability are still reasons for concern. Too many individuals are still infected with FIV. Although their incidence has been reduced, the reproductive and cardiac abnormalities persist. I should also mention that the recovery was not solely due to the influx of migrants. The recovery of the panthers has been aided by the acquisition of over 450 square miles of panther habitat. Building highway overpasses has helped prevent cars from hitting panthers, which had been a major cause of mortality. Finally, the Endangered Species Act continues to provide the panther with legal protection.[32] These caveats aside, the Florida panther has come back from the brink of extinction, and the influx of migrants was instrumental in the success of the recovery plan.

We'll return to genetic rescue later in the chapter. But first, let's explore how the principles of evolution can inform our understanding of two important questions about conservation: first, what is the minimal size required for populations to persist? Second, can species adapt to the changing climate?

MINIMAL VIABLE POPULATION SIZE

Having too few individuals puts a population at risk of extinction. Small populations are at risk of being wiped out by accident. If there are only three females and one male in a population, it is doomed if something happens to the sole male. The long-term stability of even somewhat larger populations is fragile given that births and deaths occur in discrete units—there are no fraction births or deaths—and randomly. Ecologists often call these factors "demographic stochasticity" to contrast them to "environmental stochasticity", the fluctuations in the environment that can doom a small population: it's easier for a population of 500 birds to survive a winter that kills 90 percent of them than it is for a population of 10 birds.[33]

Then there are the evolutionary genetic consequences of small numbers. As population size decreases, the chances that individuals that are genetically closely related increases. Accordingly, in small populations, there is a reasonably large probability that individuals will inherit two copies of alleles that are recessive and deleterious. Lacking a functional allele at the locus, these homozygotes would have reduced fitness. The holes in the hearts of the Florida panthers are one severe manifestation of inbreeding. Cheetahs are another big cat that exhibits the dire effects of drastic inbreeding.[34] Although their numbers are now in the thousands, cheetahs went through a severe population bottleneck a few thousand years ago that has caused them to be essentially genetically identical, so much so that cheetahs will not reject skin grafts from other cheetahs. A likely manifestation of this inbreeding is that most male cheetahs have abnormal sperm, a malady that is hindering captive breeding.

Environments are continually changing. Even if the climate and other features of the physical environment remain the same, organisms are continually interacting with other organisms. Unlike the physical environment, the other organisms (the biological environment) are evolving. Lions preying on zebras do not face a static prey; due to predation by the lions and other factors, the zebras are evolving. Likewise, the zebras do not face a static predator.

Evolution by natural selection requires genetic variation. As Menno Schilthuizen notes: "standing genetic variation is a species' evolutionary capital. It encapsulates the ability of a species to dip into its genetic savings and immediately come up with any combination of genes that a changed environment requires".[35] Populations lacking genetic variation are at risk of not being able to evolve in response to new challenges.

Populations gain genetic variation through the input of new mutations, through genetic recombination, and by gene flow coming from other populations. They lose genetic variation through random genetic drift. Just like individuals, alleles are discrete units and can be lost through random fluctuations, especially when populations are small. Random genetic drift is most potent with small populations. Hence, the amount of genetic variation in a population is expected to correlate with population size.[36]

During the late 1970s and early 1980s, Ian Franklin and Michael Soulé put forth what has become known as the 50/500 rule.[37] To be viable over the short term, populations needed to withstand the effects of inbreeding. Soulé and Franklin estimated that to minimize inbreeding, a population would need at least 50 individuals that were capable of breeding. This estimate was based on back of the envelope calculations as well as the experience of animal breeders. Recent theoretical work suggests that the minimum number may be somewhat higher, on the order of 70 or 100.[38] But minimizing inbreeding is not enough. To be viable over longer periods of time, populations need sufficient genetic variation to be able to respond to changing environments. Soulé and Franklin estimated that populations would need at least 500 individuals to maintain that genetic variation. The 50/500 rule gained traction in conservation genetics circles during the late 20th century and became the conventional wisdom.

During this time, Michael Lynch, an evolutionary biologist first at the University of Oregon and then at Indiana University, was immersed in studying the appearance and effects of mutations across a variety of organisms. The results of his studies raised Lynch's concerns about slightly deleterious mutations accumulating in small populations. If the population size is small, natural selection is not very efficient: mutations that have very small (negative or positive) effects on its bearers are about as likely to increase in frequency and become fixed as those that have no effect. A single or even a few slightly deleterious mutations is not a problem, but the steady accumulation of them could be. Accumulation of these deleterious mutations could reduce the survival and reproduction of individuals in small populations. The reduced survival and reproduction in turn could lead to further reductions in population size. This vicious circle of low population leading to more deleterious mutations leading to lower population size is what Lynch called the mutational meltdown. Theoretical calculations led Lynch and colleagues to suggest that populations that numbered in the thousands (well above the 50/500 threshold) might still face extinction due to the mutational meltdown.[39]

Species with small population sizes have been shown to accumulate mutations that exhibit the signs of being deleterious, such as those that result in the premature termination of proteins. Indeed, ancient DNA studies show the woolly mammoth had accumulated such deleterious mutations as its population declined before going extinct. A mammoth that lived 45,000 years ago when mammoths were abundant has fewer deleterious mutations than a mammoth that lived 4,000 years ago when mammoths were in much fewer numbers.[40] The question, though, is whether the accumulation of mutations in populations of intermediate size puts populations at risk for the mutational meltdown.

Mutation accumulation experiments are an excellent method to examine the effects of new mutations that accumulate as populations are subjected to only the processes of mutation and random genetic drift, with the process of selection being nearly completely absent. The removal of selection allows the effects of the other processes to be revealed. Typically, evolutionary biologists employing mutation accumulation studies are interested in a trait related to fitness—survival or productivity, for instance—but the method has also been used to study morphological or even behavioral traits.

Mutation accumulation studies take place over many generations, taking advantage of the mutations that accumulate over these generations. Hence, they require patience and an organism with a short generation time. The mutation accumulation study begins with the researchers choosing a small subset in each replicate, called a mutation line, to start the first generation. The researchers choose the lucky ones not by any special property or characteristic but simply at random. The number chosen to start each line varies depending on the nature of the organism and the purpose of the experiment. We'll get into that in more detail later, but for now, as few as a single pair can start each line. For species with hermaphrodites, a single individual can start each line. The lucky ones then produce offspring, which then develop. Then the same small number of individuals is chosen to continue each line. Picking a small number to replicate the lines each generation is what (largely) removes the process of selection from the picture. And the process continues. Every generation of a mutation accumulation experiment is like *Groundhog Day* but with a twist—the researchers do the same thing, but the mutations accumulate!

Given that nearly all mutations are deleterious, we should expect mutation lines to deteriorate over the generations as the mutations accumulate. And that is usually what we see in mutation accumulation studies. Individuals do not survive as well. The number of offspring they produce declines. They are more likely to have deformities. Behavioral responses become more erratic. Without natural selection weeding out deleterious mutations, things fall apart; the phenotype cannot hold.

So how can mutation accumulation studies be used to test the assumptions of the mutational meltdown hypothesis? Vaishali Katju, who had been a graduate student with Michael Lynch and now runs her own lab at Texas A&M University, took advantage of the fact that the number of individuals chosen each generation—the bottleneck—sets the stringency of how much of the power of natural selection is removed in the mutation accumulation studies. If the bottleneck is wide and lots of individuals are chosen, then much of natural selection is still operating. Mutations with even relatively weak effects will be removed by selection. Conversely, if the bottleneck is narrow and only a few are chosen, then only mutations with very strong effects will be removed by selection.

Katju chose the nematode *Caenorhabditis elegans* for the study. This small worm with a big name became a model system in developmental genetics during the 1970s and 1980s and has been used with increasing frequency in evolutionary genetics in the last quarter century or so. Its popularity is reflected in it being called "the worm" as well as "*C. elegans*".

I asked Katju about the challenges of conducting such a long-term experiment. She said:

> Given that we were manually picking 620 worms to establish a new generation every four days for 4.5 years, it was indeed a considerable investment of time and labor. I think the greatest challenge was the time investment with no publishable results for almost five years until the experiment was completed in its entirety, given that ours was a newly established laboratory. There are stringent academic expectations to publish prolifically

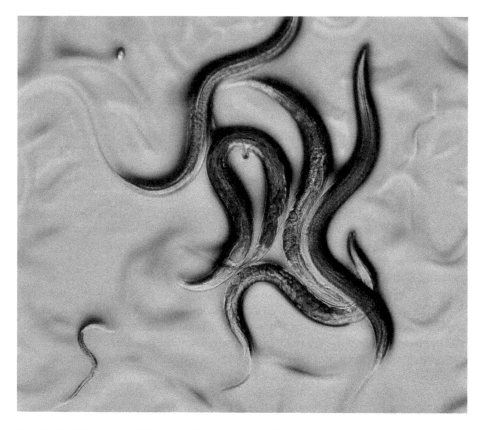

Nematodes (*Caenorhabditis elegans*) commonly used in genetic experiments.
Source: Courtesy of Austin Daigle and Vaishali Katju.

multiple papers yearly, and we were on a limb, so to speak, not knowing what to expect as there was no precedent with respect to genomic analyses of such an experiment.[41]

In Katju's experiments, lines that were maintained with a single individual starting each generation ($N = 1$) accumulated lots of deleterious mutations over the course of a few hundred generations and became sick.[42] Some of these lines went extinct. These lines, in which selection cannot purge even strongly deleterious mutations, behaved as expected. But lines maintained at $N = 10$ or $N = 100$ did not show these strong negative effects. They were not becoming sick quickly, in contrast with expectations the mutational meltdown. Similar results were obtained when the experiment was repeated using a more stressful environment.[43]

Katju's studies are just one line of research and should be repeated in other organisms to see how general they are. If similar results are obtained in other organisms, then we can dismiss mutation accumulation as a major factor in setting minimal viable population sizes. We could then return our focus on ensuring that populations avoid inbreeding,

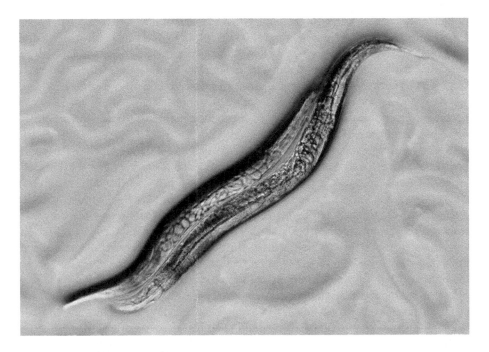

A closer view of the nematode.

have sufficient genetic variation to respond to ongoing and new challenges, and enough numbers to avoid the perils of demographic and environmental stochasticity.

Even if the mutational meltdown is not a major factor in the extinction of animals and plants, the idea behind it could be used as a weapon against viruses. Treatments that stimulate an increase of mutation could lead to something akin to a mutational meltdown in a pathogenic virus. Some evolutionary biologists have even raised the prospect of using it to treat viral infections, including coronavirus infections.[44]

THE NEW THREAT: CLIMATE CHANGE

Plants and animals are facing a new challenge in rapid climate change. Temperatures in 2019 were almost one degree C (1.7 degrees F) above the 20th century mean. Nine of the ten warmest years on record since 1880 have been in the last 15 years (2005–2019).[45] Temperatures are expected to continue climbing. How rapidly the climb over the next few decades is partly a function of the actions we have taken in recent decades, what we will do in the coming years, and what feedback loops look like.

Climate change is more than just increases of the mean temperature. It is leading to changes in climate zones.[46] For instance, the 100th meridian—100 degrees longitude— used to mark the sharp transition between the moist eastern United States and the arid west. It was, as the Tragically Hip song states, where the Great Plains begin. Or I should say where they used to begin. You see, since 1980, the transition has moved about 140 miles (200 kilometers) east to the 98th meridian. Instead of starting in the middle of Nebraska, the transition to the Great Plains is at the Nebraska/Iowa border. Hotter weather equals more evaporation and drier soils. Wind pattern shifts have also led to less rain, compounding the problem. Gardeners who have been gardening for a long time may notice that they are in a different plant hardiness zone than they once were. These zones are moving north at about 13 miles (20 kilometers) each decade.

Organisms facing climate change have a few options. They can move. They can acclimate. Over time, they can evolve with genetic changes. If they do not do one or more of these, they will go extinct.

Species are responding. Many are moving. In a now-classic study, Camille Parmesan and Gary Yohe[47] found that species were moving at an average rate of 6 kilometers (almost 4 miles) toward the poles each decade, presumably in response to climate change. Species are also moving up slope, again presumably to track climate change. Species are not just changing their ranges, they are also evolving new responses. In response to warmer weather, birds are breeding earlier than they used to. Flowers are blooming sooner too! Some of these shifts are not in synchrony—for instance, flowering plants may be responding faster than their bird pollinators are. This sets up the prospect for mismatches in timing between the plants and their pollinators. There is increasing support that these responses in timing are genetic, not just acclimation or plastic changes. For most cases, we do not know which genes contribute to these responses, but we know that the alteration in the traits have a genetic basis.[48]

But what about evolution? How well can species adapt to the climate? Related to the this, what factors determine whether and how species will respond to climate change? We will explore these questions, first with studies on plants and then with studies on insects.

The thale cress, *Arabidopsis thaliana*, is a weed that is commonly used in laboratory research (like a "green *Drosophila*") but also grows in nature, especially along the sides of roads. A recent study using this plant explored how well it is keeping up with climate change through the examination of plants from the past!

Recall the seed banks discussed in Chapter 8 that allow humans to maintain different variants of plants for agricultural and research purposes. Some plants have the natural equivalent of seed banks: with these plants, the seeds in the soil sometimes do not germinate for years after being deposited. The researchers took advantage of these natural seed banks of the thale cress. They took seeds from seed banks at different locations throughout Europe and tested how well the seeds would perform at each location. The seeds from the natural seed banks are from plants that were growing several decades ago; thus, these plants had experienced conditions from decades in the past. The seeds grown in the same region where they had been collected performed well in that home environment, showing that they were still generally adapted to that environment. However, seeds taken from further south (where temperatures are warmer) than where they were grown performed better than seeds growing in their home environment. This study thus illustrates an adaptation lag: the thale cress, an abundant plant, is not fully keeping up with the climate change.[49]

Research using the field mustard, *Brassica rapa*, in California has shown this plant can adapt to drought, but at a possible cost. The researchers took mustard seeds from before (1997) and after (2004) a drought in California. They grew the seeds at two different locations: an area that traditionally received much rainfall (the wet site) and an area that traditionally received little (the dry site). When grown under the same conditions, the seeds taken after the drought flowered earlier: two days earlier in the dry site population and almost nine days earlier in the wet site population. This change probably reflects an adaptation for a shorter growing season. The plants have evolved to grow faster in order to beat the onset of the hot, dry summer when things stop growing.[50]

Sequencing and genomic analysis of the mustard populations before and after drought reveals that the genetic changes associated with the increased drought tolerance during the recent drought are largely not those that had been associated with historical selection. This analysis also detects candidate genes responsible for the adaptation.[51] One such gene, SUPPRESSOR OF CONSTANS1 (SOC) is involved in the genetic pathway that regulates the timing of flowering.[52]

Selection for drought tolerance appears to have a negative side effect. The plants grown with seeds from after the drought are less resistant to a fungus, *Alternaria brassicae*, a common and potentially deadly pathogen to the mustard.[53] We don't know the reason why the plants are less resistant—it could be that in devoting more resources into growing quickly, they cannot devote as much resources into defense. Plants from the drought-adapted population have thinner leaves, which could make them more susceptible to the pathogen, or could be a by-product of the reduced allocation into defense, or could be unrelated to the pathogen susceptibility. Regardless of the cause, that the plants adapted to drought being more susceptible to pathogens is a tradeoff. If such tradeoffs are common, adaption to climate change may leave plants more vulnerable to other challenges. This is a theme that we will pick up again in the next chapter.

Studies with insects have addressed how well species can tolerate projected warming. Sarah Diamond and her colleagues examined this question analyzing multiple species of ants.[54] They defined warming tolerance as the difference between the warmest temperatures typically experienced in the habitat and the temperature at which the ants lose muscle coordination. So, if the highest temperature the ant species typically experiences is 37 degrees C and the ants lose coordination at 41 degrees, the warming tolerance is 4 degrees. Ants from warmer climates typically had lower tolerances, likely because the temperatures that ants can withstand without losing coordination is constrained. Interestingly, ants that live in dry habitats had greater tolerances. Some of this pattern was due to the extraordinarily high tolerances of some desert specialists, such as the *Cataglyphis* ants that run on the blisteringly hot sands of the Sahara. But the pattern is not just due to these desert specialists, as it persisted even after the desert species were removed from the analysis. They also found that the evolutionary history of a species was a strong predictor of their warming tolerance: species that have close relatives that have high warming tolerance also were likely to have high warming tolerance.

More recently, Diamond addressed how well insects could evolve in response to warming. A key metric for their ability to do so is heritability, the amount of genetic variance in a population divided by the total phenotypic variance and thus an index representing the amount of genetic variation the population has for the trait.[55] Species with higher heritability are expected to be able keep up with climate change better than those with lower heritability.[56]

Diamond's analysis found that the heritability of the upper bound of thermal tolerance differs markedly depending on the type of environment the organisms are in. Those living in climates where temperature varies considerably have nearly twice (0.4) the heritability of those that live in climates where temperatures are less variable. Because projected evolutionary responses are linearly proportional to heritability, this twofold difference in heritability corresponds to a twofold difference in the projected response. So even though the tropics are projected to experience less warming, animals that live in the tropics may be less adept at being able to keep up with the warming due to their lower heritability.

Let's put some numbers on this. Based on estimates of other parameters from *Drosophila melanogaster*, populations in the tropics (where temperature regimes are less variable) should be able to increase their upper limit by 3 degrees C in about 200 generations. Given that these flies have several generations in a year, they ought to be able to evolve at a rate more than sufficient to keep up with global warming. Species that have fewer generations per year may not be able to keep pace.

A selection experiment in *Drosophila melanogaster* suggests that the capacity for evolutionary responses may be limited. The flies did respond to selection on increasing their thermal maximum for about eight to ten generations, but after that the response plateaued. Other selection experiments with flies have found similar plateaus. The plateau suggests that these flies could only respond to about a half-degree C of further warming by evolutionary adaptation and an additional half-degree C by plastic responses.[57] The actual situation may not be that dire; the flies used in the laboratory selection experiments likely have less standing genetic variation than natural populations do. Still, this is a worrisome sign.

While sweltering at home during a hot July day in 2020, I spoke with Ted Morgan at Kansas State University, who studies the evolutionary genetics of tolerance to both cold and hot temperatures in *D. melanogaster*.[58] Morgan told me that temperature is rather complicated: the physiological and evolutionary responses to a sudden cold snap are quite different than those to gradual cooling. The same is true for responses to heat.

Morgan also stressed that the life stage of the organism must also be considered when discussing responses to heat and cold. *Drosophila* and other flies go through a series of life stages: egg, larval, pupal, and adult. Not only do the different stages look different, they also differ physiologically and express vastly different suites of genes. A recent study by Morgan's lab found that genetic variation underlying responses to heat tolerance were largely uncoupled across the different stages.[59] In other words, knowing how well a genotype does at tolerating heat as larvae provides little information about how well that genotype will do as an adult. Responses to selection on heat tolerance in larvae will be largely independent to the responses in adults.

Recently, genetic rescue has been proposed as a plausible method to assist populations that are struggling to keep up with climate change.[60] The idea is that adding individuals from populations that have adapted to warmer weather would put in more alleles that are warm adapted. For instance, suppose there is a population of salamanders in New Jersey that has insufficiently adapted to the warmer climate that New Jersey is now experiencing. Also suppose that there is another population of the same species but in Georgia. Importing individuals from the Georgia population with more of the warm-adapted alleles into the New Jersey population should help the latter population respond to the warming climate.

We will explore responses to climate change in subsequent chapters on life in the oceans and in cities. Climate change will also loom large in discussions in the chapter about invasive species their effect on endangered animals.

GENETIC RESCUE REVISITED

Genetic rescue has benefits and the benefits can be enormous, but it also poses risks—one of which is outbreeding depression.[61] This is when the different populations have different sets of alleles that do not work well together because these populations have been evolving along separate trajectories and have not been exchanging genes.[62]

Managers looking to avoid outbreeding depression can readily screen for it, as there are clear guidelines for when we expect outbreeding depression to be likely. These conditions include when populations are of different subspecies or when each population differs in its chromosomal arrangement. Populations that have not exchanged genes in at least 500 years also are likely to have outbreeding depression.

Genetic rescue of small, isolated populations that are at risk for inbreeding has worked exceedingly well. Richard Frankham of Macquarie University in Australia recently conducted a meta-analysis of cases of attempted genetic rescue where the target population likely was inbred or had low genetic diversity. Excluded from this analysis were any cases where outbreeding depression was likely, as per the guidelines listed in the previous paragraph. In over 90 percent of the examples, attempts of genetic rescue increased fitness (survival and fecundity of the populations), often dramatically so. He concludes, "the current reluctance to use outcrossing to genetically rescue isolated inbred populations is not justified on genetic grounds, given the large and consistent benefits of genetic rescue and the effectiveness of the screen for outbreeding depression revealed within".[63] He does note that other concerns, such as the financial costs and the risk of spreading disease, may argue against specific proposals of genetic rescue. Nevertheless, genetic rescue should be used much more often in the management of populations that appear to be suffering from inbreeding depression.

CONSERVATION AND THE MICROBIOTA

As we have seen in human health and in the health of bees and other pollinators, large eukaryotic organisms contain multitudes of microbes—viruses, archaea, and bacteria as well as fungi and other assorted unicellular eukaryotic organisms—in and around themselves. This microbiota functions much like an organ and affects the well-being of the host. The host and its microbiota have coevolved together. Accordingly, disruption of the microbiota could adversely affect the host. In recent years, biologists interested in conservation and management are starting to appreciate the importance of the microbiota.[64]

Changes to an animal's habitat can affect what animals eat and what animals eat affects what microbes they have. Thus, habitat change would be expected to alter microbiotas. One example of such an alteration occurred in Howler monkeys (*Alouata pigra*). Named after their bellowing calls, howler monkeys weigh up to 10 kilograms (20 pounds) and are among the largest New World monkeys. Human activity is disturbing their habitat, cutting up and fragmenting what was once continuous evergreen forest. A recent study in Mexico found that the habitat fragmentation is not only affecting the monkeys directly but also is affecting their microbes.[65] The monkeys from continuous, evergreen rainforest had much greater diversity of gut microbes than those from fragmented rainforest. Having a diverse microbiota is generally good. Moreover, the monkeys from the continuous forest had a microbiota that was healthier than those living in the fragmented forest. Notably, monkeys living in the continuous forest had more microbes that produce butyrate, a fatty acid that colon cells rely on as an energy source. Studies in mice show mice without a microbiota are deficient in energy production and exhibit signs of colon damage.[66] Adding butyrate rescues these deficiencies. Thus, it's not unreasonable to think that the reduction in butyrate-producing microbes in the howler monkeys from disturbed habitats could affect their colons and other aspects of their digestive system.

Habitat degradation is also associated with reduced diversity of the microbiota of the Udzungwa red colobus monkeys (*Procolobus gardonorum*).[67] These arboreal monkeys, which live in the Udzungwa Mountains of Tanzania, face numerous threats including habitat destruction and degradation, as well as hunting by both humans and chimpanzees. As with the howler monkeys, the red colobus monkeys that lived in the least disturbed habitats had the most diverse microbiota. The monkeys from undisturbed habitats also had more microbes that are known to break down toxins, such as ethyl benzoate, toluene, and atrazine. These toxins are typically found in the leaves that the monkeys eat. Accordingly, the monkeys from disturbed habitats may not be as adept at breaking down these toxins.

Climate change may also affect the microbiome. Increasingly, studies have linked temperature to microbiota composition in a diverse number of species.[68] For instance, a team of biologists led by Elvire Bestion at Station d'Ecologie Théorique et Expérimentale in Moulis, France, tested how a modest increase in temperature would affect the microbiota of a lizard, *Zootoca vivipara*, in a seminatural setting. They placed the lizards into enclosures that mimic natural conditions but where the temperature could be manipulated. Lizards exposed to climates only 2 degrees C warmer than the typical climate experienced a reduction in diversity of their microbiota.[69] Moreover, reduction in microbiota appears to have negative consequences: a follow-up study found that lizards with the least diverse microbiota had reduced survival in the years following the treatments. This effect persisted even after the body condition of the lizard and the treatment it experienced were factored out in their researchers' statistical model.

These studies illustrate the importance of considering microbes and the coevolutionary dance between the microbes and their animal hosts when doing conservation biology. They also highlight the need for greater collaboration between biologists who study the microbiotas and those who study conservation and management. We will further discuss the role of the microbiota in adapting to urban settings in Chapter 13.

The principles and information from evolutionary studies has helped in the development of tools to discover and track biodiversity, such as with barcoding. They also inform management and conservation of threatened species, as well as provide guidance about the ability to track climate change and other challenges. In the following chapter, we will examine how evolutionary biology provides guidance on the ability creatures in the ocean have to respond to the varied challenges they face.

NOTES

1. Diaz et al. (2019). Global Assessment Report on Biodiversity and Ecosystem Services. IPBES, p. 2.

2. Wilson (2016, p. 43).
3. Diaz et al. (2019).
4. Rosenberg et al. (2019) and Pennisi (2019).
5. Konin (2019).
6. Humphreys et al. (2019) and Ledford (2019).
7. Kohlbert (2014) chronicles the history of our understanding of extinctions throughout written history as well as current efforts to forestall and mitigate extinctions. She also discusses the previous mass extinctions.
8. Wilson (2016, p. 55).
9. Diaz is quoted in Plumer (2019a).
10. Thebault (2019).
11. Diaz et al. (2019, p. 15).
12. May (2011).
13. May died in April 2020. See Schwartz (2020) for his obituary.
14. The paper on which May was commenting on (Mora et al. 2011, p. 2011) estimates there were 8.7 million eukaryotic species, with a standard error of 1.3 million.
15. Many budding evolutionary biologists (including myself) have been inspired by the *Star Trek* franchise. Noor (2018) has written a wonderful short book on the science of evolutionary biology in *Star Trek*.
16. Herbert et al. (2003, 2004), and Savolainen et al. (2005).
17. Herbert and Gregory (2005). The quote is on p. 852.
18. Moritz and Cicero (2004). In some corners, misplaced and hyperbolic enthusiasm for barcoding remains. While the method has proven to be a useful tool for identifying and discovering species, it is not a panacea, nor a replacement for expertise in taxonomy. As Ed Wilson notes (2016, p. 163),

> Good things, however, invite overenthusiasm, and such is the case for barcoding. Some of its users see it as the solution to a shortage of expert taxonomists in the scientific world, as well as a direct route to the mapping of global biodiversity, and even a replacement of the prevailing hierarchical, name-based system of classification. However, these hopes are conspicuously in vain. The barcoding method is a technology, but it is not an advance in science or scientific knowledge.

19. Witt et al. (2006).
20. See Adamowicz et al. (2017) for a summary of barcoding research being done in Africa.
21. Lopez-Vaamonde et al. (2019).
22. Groups that all share the same common ancestor and do not exclude other descendants of that common ancestor are called monophyletic groups. Modern taxonomy strives to classify organisms into monophyletic groups.
23. Tizard et al. (2019).
24. Herbert et al. (2003). Note that while *CO1* has worked well for barcoding of animals, there is no universally accepted gene used for barcoding plants. Instead, a few different genes have been in different studies. In contrast to what is seen in animals, mitochondria DNA evolves more slowly than most nuclear genes in plants. Hence, mitochondrial genes are unlikely to be appropriate markers for estimating plant biodiversity at the species level.
25. Stoeckle and Thaler (2014). There is within-species variation at *COI* and other mitochondrial genes, but this variation is small in comparison with the divergence between species. Mitochondrial DNA has a faster mutation rate than nuclear DNA: this increased mutation rate increases both the variation within species and the divergence between species. The variation within species, however, is reduced by the mitochondria being inherited through the mother (as opposed to the case with most genes where both parents transmit copies) as

well as genetic hitchhiking (see also note 26). These factors that reduce variation within species do not affect the rate of divergence between species.

26. Because all the mitochondrial genes are linked, selection operating on one gene removes variation at the other loci. Suppose there is a variant at the hypothetical mitochondrial gene A (A1) that rapidly increases in frequency because it makes its bearers more efficient at using energy. If one variant at a site on *COI* is associated with A1, that variant will become the only variant at that site due to the selection on A1 and its association with A1. Thus, variation at that site will be lost. Evolutionary biologists call this phenomenon genetic hitchhiking.

27. See Collins and Cruickshank (2013) and Kress et al. (2015).

28. For instance, many studies rely on the bottom-up clustering method called neighbor-joining. This method has been shown to perform poorly in studies where barcoding has been used to identify specimens. Collins and Cruickshank (2013) go into further detail and discuss better approaches.

29. Coissac et al. (2016).

30. Basic information about the plight of the Florida panther can be found in Roelke et al. (1993) and from this link from the US Fish & Wildlife Service: www.fws.gov/verobeach/MSRPPDFs/FloridaPanther.pdf

31. The information on the execution and the results of the genetic rescue of the Florida panther is in Johnson et al. (2010).

32. Evolutionary biology also informs the application of the Endangered Species Act. One such example is that the choice of species concept—how we decide on what a species is and how we define species—factors into how we apply the act. See George and Mayden (2005) for an example.

33. Discussion of the early work in this area can be found in Quammen (1996).

34. O'Brien et al. (1986) describes the early work on cheetahs. O'Brien has also been a key player in the genetic recue of the Florida panther.

35. Schilthuizen (2018, p. 143).

36. Technically, evolutionary geneticists focus on a metric called the effective population size. The relative effects of genetic drift on a particular population depend not just on its census number, but also on parameters like the proportion of males and females and how steady the numbers have been over the generations. For example, genetic drift is more potent in a population that has 50 males and 450 females than one that has 250 males and 250 females. The metric of the effective population size provides a useful common frame of reference. A population that has an effective population size of 300 would experience drift like an idealized population whose numbers were constant across generations and had equal numbers of males and females. See Wright (1969) and Hartl and Clark (2007) for more information.

37. See Quammen (1996) for a discussion of the development of the 50/500 rule.

38. See Caballero et al. (2017) and Frankham et al. (2014) for recent discussions on the 50/300 rule.

39. Lynch et al. (1995).

40. Rogers and Slatkin (2017).

41. E-mail interview with Vaishali Katju, December 2019.

42. Katju et al. (2015).

43. The experiments describing the stressful conditions are described in Katju et al. (2018). The stress they imposed was high salt (sodium chloride) concentrations, a stressor that was easy to apply to the worms.

44. Jensen and Lynch (2020). Drugs like favipiravir substantially increase the mutation rate of RNA viruses. In other viruses, inducing a virus to go into mutational

meltdown has been shown to work under laboratory conditions. At sufficiently high doses, this leads to meltdown. The dose would have to be high, as the virus can adapt at low doses.

45. www.climate.gov/news-features/understanding-climate/climate-change-global-temperature Accessed July 25 2020.
46. Jones (2018).
47. Parmesan and Yohe (2003).
48. Kelly and Thapa (2016).
49. Wilczek et al. (2014).
50. Franks et al. (2007).
51. Franks et al. (2016).
52. Immink et al. (2012).
53. O'Hara et al. (2016). The changes associated with the recent drought tolerance were detected by outlier analysis. In this method, small regions of the genome are assessed based on how different the populations being compared (in this case, before and after drought) are from each other. The regions that are unusually different between the populations—outliers—are the likely targets of selection between the populations before and after the drought. To assess historical selection, a Tajima's D test is used. Here, the frequencies of different alleles are assessed across different regions of the genome. These frequencies are then compared to the theoretical expectation that assumes the absence of positive selection. If a region has many more rare variants than expected, Tajima's test is strongly negative, a sign that selection likely acted on or near that region of the genome in the historical past. The regions of the genome detected by the outlier analysis and those detected by the Tajima's D test are for the most part not the same. Hence, we can conclude that the more recent selection and historical selection are largely decoupled.
54. Diamond et al. (2011).
55. Diamond (2017).
56. Heritability is the additive genetic variance in a population divided by the total phenotypic variance. Because heritability has some limitations, some evolutionary geneticists have proposed using the related concept of evolvability either instead of or in addition to heritability in studies of evolutionary responses. Diamond notes that fewer studies in her analysis used evolvability than heritability. Moreover, in the few cases where both were used, the two metrics were highly correlated. Consequently, she used heritability in her analysis.
57. Hangartner and Hoffmann (2016). This study illustrates that heritability is a complex concept. Measures of heritability are not static across time. Selection often (but not always) leads to a decrease in heritability over time, as selection removes genetic variation.
58. Phone interview with Ted Morgan, July 2020.
59. Freda et al. (2019).
60. Aitken and Whitlock (2013).
61. A useful short review on outbreeding depression is presented in Edmands (2007). Outbreeding depression due to incompatible sets of alleles is not the only genetic risk from genetic rescue programs; see Aitken and Whitlock (2013) for more detail.
62. Johnson (2008). Populations that do not continue to have some degree of gene flow connecting them will start to diverge genetically. As they continue to diverge, genetic incompatibility begins to evolve: one population will acquire alleles that would not work well together with alleles that a different population has acquired. For an analogy, think of computer programmers at different companies updating a program independently—at some point, the changes they each make do not work well with the changes the other has made. This is what is going on in diverging populations that no longer exchange genes.

Hybrids between species often are sterile or inviable because they have combinations of alleles at different loci that do not work well together. At earlier stages of the speciation, milder forms of this incompatibility are observed as outbreeding depression.

63. See Frankham (2015). The quote is on p. 2615.
64. Trevelline et al. (2019).
65. Amato et al. (2013).
66. Donohoe et al. (2011).
67. Barelli et al. (2015).
68. Hylander and Repasky (2019).
69. Bestion et al. (2017).

FURTHER READING

Kohlbert, E. 2014. *The Sixth Extinction: An Unnatural History*. Henry Holt & Co.
Wilson, E. O. 2016. *Half Earth: Our Planet's Fight for Life*. Liveright.

12 Challenges in the Oceans

SUMMARY

Marine organisms face numerous threats. One is pollution from oil and its by-products. We begin with a discussion of the 2010 Deepwater Horizon spill and how organisms evolved responses to the oil by-products from it. One vertebrate affected by the oil spill is the Gulf killifish. We stay with this killifish to discuss its evolutionary responses to chronic pollution. In adapting to the chronic pollution, one killifish population responded through a deletion of an important gene. Interestingly, it obtained that adaptive deletion from hybridization with another killifish species. We then move to a discussion of evolutionary responses of organisms to warming oceans. Oceans are not just getting warmer; they are also becoming more acidic. We discuss the evolutionary genetic responses to this ocean acidification, focusing on the purple sea urchin. For now, the urchin can keep adapting to increased acidity, but it may soon reach its limits. We then discuss the evolutionary genetic responses of corals to various stressors. We conclude with a discussion of the factors that allow some species to adapt to environmental threats.

"Climate change is heating the oceans and altering their chemistry so dramatically that it is threatening seafood supplies, fueling cyclones and floods and posing profound risk to the hundreds of millions of people living along the coasts."—Brad Plumer.[1]

"[A]n increasing number of studies have shown that species can adapt rapidly to environmental challenges. . . . Failure to adapt means that species go locally, and possibly globally, extinct. The challenge for biologists is to explain why some species rapidly adapt when others fail."—Karin Pfennig.[2]

THE THREATS TO LIFE IN THE OCEANS

A September 2019 report released by the International Panel on Climate Change (IPCC) warned of dire consequences of continued climate change for life in the oceans and for living conditions of numerous humans across the world.[3] As Brad Plumer explained in the *New York Times*, greenhouse warming has already reduced fish populations across several regions and is putting added stress in other regions. The same increase in carbon dioxide that is leading to global warming is also causing ocean water to become more acidic. Greenhouse warming and the associated acidification are not the only threats for life in the seas. Oil spills, such as the 2010 Deepwater Horizon blowout, and pollution from industrial production also harm life in the oceans.

As Karin Pfennig, an evolutionary ecologist from the University of North Carolina at Chapel Hill noted in the epigraph, some organisms—including some marine organisms—have been able to adapt to these stressors, at least for now. But

DOI: 10.1201/9780429503962-15

not all of them survive; many have perished or are in serious danger of doing so. Which organisms are able to keep up? What physiological and genetic changes have they evolved to enable this adaptation? Can we use the principles of evolution by natural selection to see which marine organisms and under which circumstances are best equipped to evolve the necessary adaptations to make it in the new and ever-changing normal that human activity has imposed? These are the questions we will address in this chapter.

THE DEEPWATER HORIZON DISASTER

Evocative of disaster of enormous scope, Deepwater Horizon was the drilling rig embroiled in one of the worst environmental disasters and the largest marine spill ever.[4] In fact, calling it a spill does not really capture what happened. It's more like an oil spew.

In the spring of 2010, the rig, which was owned by Transocean and under lease to BP (formerly British Petroleum), was drilling an exploratory well. This well was unusually deep—almost a mile (1,500 kilometers) below sea level—and located in the Macondo Prospect, an oil-rich region located in the Gulf of Mexico about 50 miles (75 kilometers) from the coast of Louisiana. Just before 10 p.m. local time on April 20, the well exploded, a condition known as a blowout. Minutes later, Deepwater Horizon caught on fire. By the morning of the 22nd, the rig had sunk to the bottom of the gulf and 11 people were dead.

This disaster, like many, resulted from a long list of errors, negligence, and happenstance events. Failures of leadership at several companies, as well as cultures of dysfunction and cost-cutting, all contributed. For want of a properly functioning sealant; for want of compliance with appropriate safety checks; for want of any number of things, this tragedy could have been averted. But the initial blowout and loss of life was only the beginning; a second tragedy would follow, one that would severely affect the environment of the Gulf and surrounding coastal areas for years.

Blowouts happen. They are not particularly rare. In fact, in the year and a half prior to the Deepwater Horizon tragedy, four blowouts occurred in the Gulf. What made this one different and more deadly was the location. Most blowouts, like most drilling operations, occur in shallow waters. In deep water drilling, blowouts are much more consequential and much harder to contain. The struggle to contain the well stymied BP, the other companies, and multiple federal agencies. During the nearly three months before the well was capped, 5 million barrels (over 40 million gallons) of oil billowed out from below the substrate. Moreover, in response to the oil spill, the EPA approved the use of dispersants, chemicals designed to break up the oil. Unfortunately, the dispersants did more harm than good; by breaking up the oil, the dispersants made it easier for marine life to ingest it.

Animals living in the waters contaminated by the oil spill did not just have to adapt to toxins but also faced wholesale changes in their biological communities. The oil and its breakdown products brought profound changes in the microbial communities. The bacteria that were specialized to break down the most common hydrocarbons flourished until their preferred sources of energy were depleted. Then, their numbers declined; with these declines, new bacterial taxa replaced them.[5]

Where did the oil-metabolizing microbes come from? To a large extent, they are a natural part of the ecosystem. The Gulf of Mexico is extraordinarily rich in oil, methane, and other hydrocarbons that naturally seep out from numerous locations of the gulf floor. Hence, biological communities in the gulf face natural exposure to these hydrocarbons, exposure that varies considerably in both space and time. Natural seepage of oil and methane, though far lower than what was spilled from Deepwater Horizon, does maintain a collection of microbes that digest oil and its by-products. The Deepwater Horizon spill differed in being a much larger and more concentrated supply of these chemicals than natural seepage. In addition, the dispersants likely altered how microbes responded to the released hydrocarbons.[6]

Because most Gulf fish are spawning in late spring, the release of oil, oil by-products, and methane from Deepwater Horizon was particularly damaging to the marine ecosystem. Trillions—millions of millions—of fish larvae were directed killed by the spill. Moreover, the exposure also affected those that survived. These survivors were left with morphological deformities, reduced heart function, and compromised swimming performance.

So what were the evolutionary consequences of the spill? Smooth cordgrass, *Spartina alterniflora*, is an abundant perennial grass in costal saltmarshes. This cordgrass, which can grow to be over a meter in height, typically grows at the edges of marshes, and its growth there allows mussels and other invertebrates to settle. As such, the smooth cordgrass acts as an ecosystem engineer, substantially altering the ecosystem by its presence.

The cordgrass populations that were contaminated by the oil spill turned out to be genetically different from uncontaminated populations from surrounding areas. Though it is conceivable that the genetic differences are due to some other factor, that they are repeated strongly suggests that the genetic differentiation arose as an evolutionary response to the oil spill. Interestingly, the populations that were exposed to the oil are about as genetically variable as the uncontaminated populations. This suggests that the selection from the oil exposure did not substantially deplete genetic variation.[7] If so, this suggests that the microbes will be able to continue to adapt to new changes.

Copepods are small aquatic crustaceans that are often among the most abundant animals in the aquatic ecosystems. We'll focus on one copepod here: *Eurytemora affinis*, a species living primarily at the bottom of streams and estuaries in eastern North America. These tiny creatures are just over a millimeter long, or almost as long as the thin edge of a dime. They are of particular importance to coastal ecosystems because of their abundance and the roles they play in the food web. The copepods feed on bacteria, algae, small protozoans, and organic material. In turn, they are fish food. In fact, fisheries—including flounder, salmon, and herring—rely heavily on these copepods. The copepods' ability to produce large numbers of offspring in a short period of time (their generation times is only a couple weeks) make them attractive model systems for ecological and evolutionary experiments. The rapid generation time and their large population sizes make these copepods likely to respond evolutionarily to the challenge of the Deepwater Horizon spill.

Carol Lee and her colleagues at University of Wisconsin at Madison, who have been examining how *Eurytemora* copepods respond to other stressors, had a population of copepods that had been collected prior to spill.[8] They looked for copepods immediately after and in the next year (2011) after the spill and found slim pickings: very few copepods remained following the spill. By 2013, the copepod population had sufficiently recovered such that Lee's team could get samples. They then raised the two sets (pre-spill and post-spill) of copepods under the same conditions—a "common garden" in the lingo of evolutionary ecologists—to ensure that any difference they observed were genetic (and thus due to evolution) rather than acclimation. Upon exposing the copepods to high concentrations of oil, those collected after the spill survived better and developed more quickly than the pre-spill copepods. The common garden experiment thus provided strong evidence that the copepods had evolved increased tolerance to oil.

Lee's team also conducted another experiment. They tried to select for increased tolerance to oil in the pre-spill copepods. They failed. After eight generations of selection for increased tolerance to oil, the copepods actually survived high oil concentrations a little less well than they had before. That should not happen if there is genetic variation for tolerance to oil. So what gives? One possibility is that out in nature, the copepods have an egg bank. Females lay eggs in the sediment. Some of those eggs hatch soon after, while others lie dormant—sometimes for years. It's possible that the egg bank maintained the sufficient genetic variation for the copepods to evolve tolerance to the oil. The selection experiment that Lee and her colleagues conducted in the lab had no such egg bank. The egg bank also could explain how the copepods were able to bounce back after being decimated by the oil spill. Other species without an egg bank may not have been as fortunate.

The oil spill has also affected fish.[9] In addition to crude oil, fish often need to contend with the by-products of oil as it breaks down in the environment, a process known as weathering. Two common classes of organic pollutants that come from the weathering of oil are polychlorinated biphenyls (PCBs) and polycyclic aromatic hydrocarbons (PAHs). For both, their names accurately describe their structures. From "biphenyl", we know that PCBs have two six-carbon phenolic rings attached to each other. Phenol rings contain six carbon molecules linked together by a special type of bond that is neither a single bond nor a double bond but something in between. These special bonds give such chemicals increased stability, making them difficult to degrade. Many chemicals with these phenolic rings—including the simplest such chemical, benzene—have odors associated with them. Hence, they are often called aromatic molecules. Polyaromatic hydrocarbons thus have many phenol rings. In their addition to the stability, these phenolic rings also make these molecules become highly soluble in fat. Hence, they tend to accumulate in fatty tissues, including the liver. PCBs also have multiple chlorines attached to the biphenyl rings, making them even more likely to cause damage.[10]

The aromatic compounds are bad for fish, especially during their early development. Among other abnormalities, fish exposed to these toxins can have heart and other cardiac deformities. The extent of the cardiac deformities varies, depending both on how much toxin they encounter and when during development the exposure occurs. Sometimes these effects are relatively mild. Here, the fish may have elongated hearts, or the atrial and ventricular sides of their hearts do not align properly. They may also have fluid around their heart. At the more extreme end, the fish die early in development because their hearts do not develop separate chambers and instead remain as elongated tubes.[11]

One particular fish affected by the Deepwater Horizon spill is the gulf killifish (*Funduls grandis*), a small ray-finned fish. The genus of this species comes from the Latin word for bottom (fundus). And yes, killifish of this genus are aptly named given their propensity to swim at the muddy bottoms. The spill affected these killifish in several ways. Egg hatching success was reduced in those fish collected from Grand Terre, Louisiana, a site contaminated by the oil spill, compared with other sites. The fish that did hatch were developmentally delayed and showed cardiovascular abnormalities.[12]

Compared with uncontaminated sites, the Grand Terre killifish also express a different suite of genes. Among these changes is increased expression of genes in the AHR pathway, which breaks down toxic material, among other functions. The increase in the expression of genes in the AHR pathway is likely a sign of distress. Another change in the Grand Terre population is decreased expression of genes involved in egg production. This reduced expression could reflect the fish allocating fewer resources to laying eggs as they devote more resources to coping with the toxic chemicals.[13]

Clearly, killifish are affected by the toxic hydrocarbons coming from the spill. They respond with changes in the expression of genes. But this response is not necessarily an evolutionary change. Given the generation times of killifish, it's still a little too early to determine whether and how they have responded with evolutionary changes. But stay tuned.

We will stay with two species of the killifish, the Atlantic and the Gulf, in the next section on responses to chronic exposure to hydrocarbon pollutants.

CHRONIC EXPOSURE TO TOXINS

Galveston Bay is highly polluted due to production from oil and natural gas industries as well as from occasional small oil spills. Pollution concentration in the Galveston Bay area exhibits a sharp gradient: areas not too far away have much less of the pollutants than those right by the source. Populations of the gulf killifish that live in the most polluted region of the bay are much more tolerant of the pollution. That tolerance declines as the pollution does.

As a graduate student at Baylor with Dr. Cole Matson, an environmental toxicologist specializing in the genetic effects of contaminants on wildlife, Elias Oziolor worked on the resistance of the Gulf killifish to the toxins in the Houston and Galveston areas. Oziolor found that in addition to being more tolerant of PCBs and PAHs, fish from Galveston Bay were also more tolerant of the pesticide carbaryl and various chemicals that promote oxidation.[14] Oziolor continued this work as a postdoc with Andrew Whitehead at University of California–Davis. Whitehead had been studying genomic responses when killifish face environmental challenges, including the Deepwater Horizon spill.

When Oziolor, Whitehead, and their team used genomic analysis to determine the genetic changes that led to the tolerance, they found something interesting: while several genes contribute to tolerance, a large part of the change is from a deletion in the *AHR2* (acyl hydrocarbon receptor 2) gene.[15] This gene is a key player in the AHR metabolic pathway. The deletion was at high frequency in the most polluted areas, at intermediate frequencies in the moderately polluted areas, and rare or absent in areas with little or no pollution. The analysis also showed that this deletion is extraordinarily advantageous, with a selection coefficient of 0.8. Selection this strong is hardly ever seen in the wild: the fish without a functional copy of the gene are much more likely to survive in the highly polluted areas as compared with fish with the functional gene.

We've seen several other instances where environmental changes resulted in the loss of genes or gene function. For instance, deleting the *CCR5* chemokine receptor makes humans resistant to HIV and very likely had protected us from other pathogens. Increasingly, evolutionary biologists are appreciating the important role the loss of a gene or a trait can play in adapting to new environments. Writing in the late 1990s, the molecular evolutionary biologist Maynard Olson called this the "less is more" principle. In discussing this counterintuitive idea, Olson noted:

> We like to think that organisms achieve better fitness by having "better" genes, not broken ones. Over the broad sweep of evolutionary time, this principle must be true, but loss and regain of gene function may be common over shorter stretches of a species' history.[16]

Evolution by knocking out a gene is most likely when selection is intense. With milder selective pressures, adaptation usually occurs by smaller steps, usually involving several genes. But larger changes—even knocking out a gene—are more likely under intense selection, as in the highly polluted Galveston Bay. In support of this contention, experimental studies of gene knockouts in bacteria show that losses of genes are particularly likely to be beneficial when the bacteria are under severe nutritional stress.[17]

That the tolerance to the pollution came from a deletion of a gene is not the only surprise in this story. Based on their analysis of the evolutionary relationships of the different variants of the gene, the deletion did not arise within the gulf killifish. Instead, it probably arose first in the Atlantic killifish (*F. heteroclitus*) and then was brought in the gulf killifish through a rare hybridization event between the two species.[18]

Evolutionary geneticists call the phenomenon of genes being exchanged from one closely related species to another "introgression". When the introgression provides genes that are beneficial in the environment, it is an adaptive introgression. Over the last several decades, plant evolutionary biologists have championed the importance of adaptive introgression.[19] Evolutionary biologists working with animals were slower to see the importance of adaptive introgression. But since about 2005, it's been increasingly difficult to deny its importance in animal evolution, as more and more examples of adaptive introgression in animals have piled up—for example, adaptive introgression has been seen in mimicry in butterflies, in beak morphology in Darwin's finches, in the skin color of chickens.[20] Adaptive introgression has even played a part of our own evolutionary history, as we have acquired genetic variants from Neanderthals that are useful for immune functions. In the case of the Galveston Bay population of killifish, the tolerance to the toxic oil by-products appears to be largely due to adaptive introgression.

I asked Andrew Whitehead about the origin of the deletion.[21] He thinks it unlikely that his team sampled the population where the *AHR2* deletion first arose. He noted that *F. heteroclitus* has two subspecies—one northern, and one southern—with the break being somewhere south of the Jersey shore. There is a population of Atlantic killifish in the Elizabeth River of southeastern Virginia that also is resistant to the PCBs. It, too, has a deletion in *AHR2*. But this is not the population that was the source of the introgression into the Gulf killifish. We know this because the two *AHR2* deletions have different molecular signatures: the different deletions start and end at different breakpoints in the two different populations. Whitehead suspects "that the source population for introgression was another *heteroclitus* population, from somewhere south of New Jersey, [and] probably adapted to another polluted urban estuary—one that had enriched that population for this AHR deletion".

So that is where the source population was located. What about when the introgression occurred? The molecular population genetic analysis points to it happening about 34 generations before the fish were collected in 2014 and 2015. Whitehead told me that the generation time of these fish is between one and two years. That's about 50 years ago (give or take a decade), which is consistent with the peak of the PCB pollution, which started in the early 1950s and peaked in the late 1960s. Whitehead speculates that the transfer could have come through ballast water dumping, as this event was before the implementation of strict regulations regarding the dumping of ballast. But this is just speculation; we don't have a good sense how the hybridization took place.

Let's look at the Atlantic killifish and its responses to chronic hydrocarbon pollution. This species has a long history dating back at least to the Narragansett, the Indigenous people who lived in what is now Rhode Island. They noted the abundance of the Atlantic killifish, which they called the mummichog. Its Linnean name is *Fundulus heteroclitus*.

Diane Nacci,[22] then at the EPA's National Health and Environmental Effects Research Laboratory in Narragansett, Rhode Island, and colleagues surveyed populations of the Atlantic killifish for their ability to tolerate PCBs. These populations spanned the northern Atlantic coast from northeastern Massachusetts (New Bedford) to southeast Virginia (near Norfolk). This team also surveyed the concentrations of PCBs in the sediment at these localities. The tolerance of these fish varied enormously; in fact, a lethal dose could not be established for the Norfolk population. No amount of PCB that the researchers administered to these fish could kill them! In the other 23 populations in which lethal doses could be established, the most resistant population could withstand several thousand times the dose that the least tolerant fish could. Moreover, the lethal dose for each population (its tolerance) was strongly correlated with the PCB concentration in the sediment. In general, the areas with the more polluted sediment had more tolerant fish. This relationship was not perfect, but it was strong: just over half of the variation in the lethal dose could be explained by sediment PCB.[23]

Nacci and colleagues found that another *AHR2* deletion was also responsible for a large amount of the tolerance for pollution in a northern population (New Bedford) of the Atlantic killifish.[24] So deletions of this gene have been selected for repeatedly in several populations of the Atlantic killifish, illustrating that it is a favored target of selection.

I asked Whitehead why reducing AHR would be beneficial in fish exposed to lots of PAH and PCBs. He pointed to the pathway's diverse functions. It regulates the immune system. It plays roles in hormone signaling and in responses to reduced oxygen. It functions during organ development in embryos. It also activates the metabolic pathways that break down some toxins. So you would think that activating AHR in response to PAH and PCB exposure would protect the fish, but instead it's the opposite. Why? Whitehead said that the inappropriate activation of the system in early development, especially during certain critical periods of cardiovascular system development, leads to developmental defects. Moreover, the inappropriate activation of the system in embryos occurs at very low doses of the toxins, lower than those that affect the health of adults. So shutting down the pathway response seems to protect the developmental process, even though it may reduce the ability of adults to subsequently metabolize the toxins.

So what about side effects from AHR desensitization? Are there negative consequences? Whitehead said while there is some evidence that desensitization reduces fitness in some populations, the results are not consistent, and there is nothing obvious. Whitehead provides three possibilities. First, the costs are actually small or none. If so, then adaptation to the polluted waters is facilitated because costs do not get in the way. Second, there are large costs, but the researchers have not yet measured the appropriate traits. Measuring fitness components—especially in a natural setting—is painstaking work that is far from trivial. In particular, costs could be context dependent. If substantial costs do exist, then the propensity to continue adapting could be limited. A final possibility is most intriguing: ramping down the sensitivity of the AHR pathway could entail costs, but those costs have been mitigated by subsequent adaptive evolution.

This type of evolution, which is often called compensatory evolution[25], has been observed in several other situations. Recall the discussion of the costs of antibiotic resistance in *Pseudomonas*; such costs can appear low or nonexistent if there is rapid compensatory evolution. If so, then it may be difficult to regain antibiotic sensitivity simply by switching to a new antibiotic. Similarly, as we saw in Chapter 9, compensatory evolution in insect pests to insecticides (or weeds to herbicides) could thwart our ability to manage threats to agricultural crops. Here, however, compensatory evolution in the killifish may assist in their resilience in the face of chronic pollution.

Some genetic evidence supports compensatory evolutionary responses in the Atlantic killifish. Northern populations that are tolerant to the toxins share multiple copies of the *CYP1A* gene, which is a target of the AHR pathway, due to gene duplication events.[26] Given the depressed AHR signaling, having multiple copies of *CYP1A* may be beneficial. The tolerant populations also share a different variant of the estrogen receptor 2b as well as different variants of hypoxia-inducible factor 2a and genes involved in immune function.

Returning to Karin Pfennig's quote at the top of the chapter, Atlantic and Gulf killifish appear to have several attributes that enable them to readily adapt to even heavy PCB/PAH contamination. As a consequence of their large population sizes, they have a large reservoir of genetic variation, enabling them to adapt easily. Their limited home ranges allow populations to adapt to local conditions. And yet there is enough gene flow among populations to provide the requisite genetic variation. Moreover, the trickle of gene exchange from one species to the other may also benefit each species. Think of the other species as a backup reservoir. The nature of the toxin and the associated genetic response also promote easy adaptation. Substantial resistance can be generated simply by ramping down the AHR pathway! As we've seen, it can happen by knocking out the *AHR2* gene. The mutational target is thus large because many types of mutation can lead to knocking out the gene.

This reservoir is not limitless. The Galveston Bay population shows reduced genetic diversity compared with other populations. This is what we would expect after selection for increased tolerance, but the reduced genetic diversity puts the Galveston Bay population at a greater risk of going extinct if there is another challenge. It may have reduced capacity for further adaptation.

Although the killifish have adapted to PAH- and PCB-contaminated water, potential downsides still exist. The adapted killifish likely have accumulated large quantities of these toxins, which put their predators—birds and other, larger fish—at risk of PCB exposure. Because humans eat these larger fish, we, too, might face a greater risk. And while killifish have a limited home range, their fish predators may not.[27]

WARMING IN THE OCEANS

In her 1955 book *At the Edge of the Sea*, Rachel Carson[28] observed that species ranges of Atlantic coast aquatic organisms such as the green crab were changing. After pointing out that Cape Cod, southeast of Boston, had been the historical northern limit for many species, Carson noted:

> Although these basic zones are still convenient and well-founded divisions of the American coast, it became clear by about the third decade of the twentieth century that Cape Cod was not the absolute barrier it had once been for warm-water species attempting to round it from the south. Curious changes have been taking place, with many animals invading this cold-temperate zone from the south and pushing up through Maine and even into Canada.

Carson presciently attributed the range changes to climate change, observing that:

> This new distribution is, of course, related to the widespread change of climate that seems to have set in about the beginning of the century and is now well recognized—a general warming-up noticed first in arctic regions, then in subarctic, and now in the temperate areas of northern states. With warmer ocean waters north of Cape Cod, not only the adults but the critically important young stages of various southern animals have been able to survive.

In the six decades since Carson's visionary words, warming in the oceans has continued. This warming poses several challenges. In addition to the altered thermal landscape, warming also affects respiration capacity, as warmer waters generally can hold less oxygen.[29] Compounding this predicament, metabolic rates—especially in ectotherms (colloquially referred to as cold-blooded organisms)—increase with temperature.

As Carson observed, one prospect is that species will continue to move to keep up with the changing thermal regime. This may be easier said than done. One recent projection notes that to keep up with the change in the climate, some species will have to shift their ranges by as much as 1,000 km, if not more, by the year 2100.[30] The extent of these projected shifts is heavily dependent on assumptions about how quickly we can reduce emissions. If we get our collective acts together and quickly make a huge reduction in emissions, ranges are projected to change much less than under the assumption of the status quo.

Another not mutually exclusive option is evolutionary adaptation. In a review article on the prospects for marine organisms to adapt to the altered regimes that climate change will bring, Steve Palumbi, an evolutionary geneticist who specializes on marine organisms, and his colleagues noted in a recent review paper many species appear to have abundant adaptive genetic variation. This variation can be maintained when organisms have large effective population sizes and sufficient gene flow across the range of each species. They note:

> The high level of adaptive genetic variation shown here suggests an ability for marine populations to adapt in the face of climate change, but many questions remain about how fast, complete, and effective this evolution will be.[31]

Organisms are not just facing the current warming and the prospect of future warming. They also contend with other challenges. As we will discuss in detail in the next section, the oceans are not just getting warmer, but are also getting more acidic due to human activity. Organisms that live in or near estuaries also must cope with changes in salinity.

Morgan Kelly and her colleagues at Louisiana State University explored how organisms respond to selection involving multiple stressors.[32] For their study organism, they chose copepods in the genus *Tigriopus*. These copepods, which are somewhat larger than the *Eurytemora* copepods that we encountered earlier in the chapter, live in the intertidal zones of the Pacific Ocean. Thus, they typically experience variable salinity. In one trial, the copepods were selected for five generations in low salinity (1.5 percent salt) conditions, in contrast with the 3.5 percent typical of oceans. Those copepods that were selected to perform well in low salt conditions have a lower heat tolerance than the control lines that were raised in salinity conditions. The authors suggest that the reduced heat tolerance occurs because the responses to the two stressors (low salt and high temperature) each impose energetic demands. The tolerance to low salinity is due to the expression of genes related to the regulation of ions, which can be costly energetically. The heat tolerance is likely due to regulation of protein folding, which also is energetically expensive. Adapting to reduced salt concentrations results in a tradeoff of having reduced tolerance for high temperatures. Given this tradeoff, adaptation to warmer oceans may limit the copepods' ability to cope with changes in salt concentrations. For species that live in estuaries where salinity gradients can be steep, a lack of ability to respond to changes in salt concentrations may limit their range.

A change in temperature can do more than just change physiological requirements. In some cases, temperature change can even disrupt the balance of males and females in populations. In most terrestrial vertebrates, sex is determined by chromosomes.[33] But most reptiles, some amphibians, and a few fish do not play by those rules; instead, which sex an individual becomes is determined by environmental factors while they are developing. A common sex-determining factor is temperature. One example of temperature-dependent sex-determination occurs in green turtles: those that develop in cooler waters generally become males, while those that develop in warmer waters generally become females.

If temperature determines the sex, as it does in green turtles, what happens as the temperature changes? Green turtles are facing that scenario now. Turtles in the Great Barrier Reef have a skewed sex ratio, with many more females than males. This phenomenon is particularly pronounced on the northern side of the Great Barrier Reef, where it is warmer. There, about 87 percent of adult-sized turtles and over 99 percent of the juveniles are female. In the southern side (closer to the equator), the skew is not as pronounced; about two-thirds of the turtles there are female. How well the turtles will adjust depends on how quickly the threshold temperature evolves. If they can evolve a higher threshold temperature to keep up with the ocean warming, they may be able to continue producing males. If not, the population faces the prospect of extinction, as it cannot persist without at least a few males.[34]

OCEAN ACIDIFICATION

The burning of hydrocarbon fuels leads to the emission of carbon dioxide, disrupting the natural carbon cycle. Some of the additional carbon dioxide goes in the atmosphere. The oceans absorb most of the rest, a process that has positive and negative ramifications. The good news is that ocean absorption of carbon dioxide limits the increase of the concentration of carbon dioxide in the atmosphere, thus checking the speed of climate change through the greenhouse effect. The bad news is that the carbon dioxide being absorbed by the water makes the water more acidic.

Why does carbon dioxide make water more acidic? Carbon dioxide reacts with the water to yield carbonic acid. You can see this by blowing bubbles in water and checking the pH of the water with litmus paper. If you blow long enough, the water will become more acidic.

Why is ocean acidification a problem? A change in the pH of the water complicates the homeostasis organisms do every day to maintain the proper concentrations of acids and bases in different locations of their bodies. Ocean acidification is particularly challenging for the marine calcifiers—those ocean dwellers who build skeletons or shells or spines from calcium carbonate animals who live in the oceans.[35] These marine calcifiers include crustaceans and mollusks and echinoderms and reef-forming corals. Even some algae are marine calcifiers. Building these shells and keeping them from dissolving is made more difficult when the waters become more acidic because the concentration of available carbonate declines along with pH.

Some marine animals can withstand much more acidic waters. One example is the vent mussel *Bathymodiolus brevior*, which can be found in considerable numbers in the underwater Eifuka volcano.[36] Drop a line south from the east coast of Japan and a line west from the Hawaiian Islands—where they meet is not far from this volcano. Hydrogen sulfide and carbon dioxide (in liquid form at the pressures this far deep) coming from the volcano result in very acidic waters. And yet mussels can be found in water with pH as low as 5.36! But even though the mussels are in the low pH waters does not mean that the acidity is without consequence. The mussels collected from Eifuku had much thinner shells than those of the same species collected where the pH was higher. In fact, the Eifuku mussel shells were so thin that the researchers examining them could read text through those shells. The Eifuku mussels also grew more slowly, as determined by examination of growth rings. Like trees, these mussels have patterns of growth that allow one to calculate their age and growth rates. Interestingly, the mussels on Eifuku showed no sign of damage from crab predation, unlike those from elsewhere. Perhaps Eifuku mussels can live there because it is a refuge from crabs.

The exceptional species like the Eifuku mussels aside, calcifiers do appear to face a substantial challenge from ocean acidification. A meta-analysis (an analysis of multiple studies) examining the studies of acidification up to 2010 demonstrated that marine animals showed negative effects from lower pH, but that the effects were variable. The study found that calcifiers were most at risk, and calcification was the most sensitive process.[37]

Ocean acidification is not just an academic concern; it also has direct economic consequences, as ocean acidification hinders the growth of commercial marine invertebrates that calcify.[38] The oysters, clams, scallops, and other mollusks that build shells are particularly sensitive to ocean acidification. These mollusks are an important component of certain regions, such as southern Massachusetts, where the industry brings in $300 million a year. One especially vulnerable city in this region is New Bedford. Once known for its importance in the whaling industry, this city's livelihood is now reliant on a single species—the scallop—which accounts for about four-fifths of its fishing revenue. Ocean acidification also has affected oysters in Washington State, eventually leading to then Governor Christine Gregoire to form a Blue-Ribbon Panel on Ocean Acidification in 2012.[39] Other states affected include Maryland, Maine, and Washington.

We'll focus on one species—the purple sea urchin, *Strongylocentrotus purpuratus*— for the remainder of the discussion of responses to ocean acidification. Yes, this sea urchin is purple. It lives in intertidal and near-shore zones off the edge of the Pacific Ocean from British Columbia, Canada, down to the Baja California peninsula of Mexico, some 2,500 kilometers. These regions are often called kelp forests, named after the brown algae seaweeds that dominate the vegetation. Kelp forests are among the most productive marine ecosystems. Urchins feed on kelp and, in turn, are preyed on by bigger animals—lobsters, various fishes, and even otters and other marine mammals. While urchins can be part of healthy kelp forests, they also can undergo unchecked population growth and overexploit the kelp resources, leading to what marine ecologists call "urchin barrens".[40]

Sea urchins have been used as a model system in the study of embryology and developmental biology for well over a century. Urchins are easy to keep in the lab at relatively little expense. Like many marine organisms, urchins spawn by shooting both sperm and eggs into the water. Sperm then fertilize eggs. Having eggs that are relatively large made urchins an attractive model for the study of how sperm fertilize eggs. That urchin eggs and early embryos are large and mostly transparent enabled biologists of the late 19th century and early 20th century to quickly learn much about how cells divide and how the fates of cells are determined during early development.[41] Later in the 20th century, the purple sea urchin was one of the first

Purple sea urchin (*Strongylocentrotus purpuratus*).

Source: Courtesy of Melissa Pespini.

species wherein developmental geneticists figured out the wiring diagrams of gene regulatory networks during development. In other words, they showed which genes turned on and off the expression of other genes and when and where in the developing embryo that these switches took place.[42]

As we consider how sea urchins (and other organisms) adapt to more acidic ocenas, we should think about the parallels to responses to warming temperatures. While there are similarities between these two responses, there is also a salient difference.[43] Both in the waters and even more so on land, temperatures vary greatly, and that variation usually follows a steep gradient. It's much warmer in waters near the equator than it is in waters near either pole. The change in temperature due to global warming is expected to be small relative to the existing gradient. Accordingly, a genetic variant that is well adapted to current local conditions would not have to go very far to find a suitable environment under global warming. In contrast, no such global gradient exists for acidity. While there is variation in pH, the variation generally does not follow a continuous gradient in the same way that temperature does. So both movement and adaptation in response to ocean acidification is likely to be limited. A more likely scenario is that regions with already low pH are likely to face local extinctions.

Before she had her own lab at LSU, Morgan Kelly was a postdoc with Gretchen Hoffman at UC Santa Barbara. There, she and her colleagues performed a series of experiments with the purple sea urchin to examine the effect of ocean acidification and carbon dioxide on size and to estimate the genetic variation of that trait.[44] Measuring the effect of the ocean acidification is straightforward—expose developing urchins to different treatments. Here, the treatments were water with a carbon dioxide and pH typical of the oceans of the 2010s and water with higher carbon dioxide and lower pH (about 7.6), corresponding to conditions expected by the middle of the century. Obtaining estimates of the genetic variation is more complicated. To do this, Kelly and her colleagues used a complex breeding design enabling them to determine both the additive genetic variation (the variation that selection can work on) as well as variation that arises from differences in the environment of their mothers (maternal effects).[45]

On average, the urchins that experienced the high-carbon dioxide, low pH conditions were almost 10 percent smaller than those that experienced the carbon dioxide and pH conditions typical of the oceans of the 2010s. So we would expect urchins to be about 10 percent smaller by midcentury. Considerable genetic variation (both additive genetic variation and maternal effects) was found at both the low and the high carbon dioxide treatments. The additive genetic variance was higher in the high carbon dioxide treatment. Hence, we would expect sea urchins to evolve in response to ocean acidification.

But how much? To get a sense of whether the urchins could evolve sufficiently fast to keep up with the increasing acidity of the oceans, Kelly and colleagues also incorporated a theoretical model. Here, they assumed that in the absence of ocean acidification, body size was under stabilizing selection: individuals that were near average size would be selectively favored over those much bigger or much smaller than average. Such an assumption is common in evolutionary modeling.[46] With ocean acidification projected to lead to smaller sea urchins, the distribution of sizes of urchins would shift toward most urchins being smaller than optimal. Given this distribution shift, there would be selection to ameliorate the reduction in size caused by the increased acidity. Based upon the extent of genetic variation existing in the urchins and realistic assumptions about the strength of the current stabilizing selection, Kelly's model projects that adaptation will partially (but not completely) counterbalance the reduction in size caused by the acidification.

Given its ecological relevance and its long history as a model system for experimental biology, the purple sea urchin was a natural focus for genomic researchers. So genomic resources for the urchin were developed relatively early on; in fact, a version of its genome was published in 2006, around the same time that genomes for the chimpanzee and the dog were published. The early adoption of genomic resources permitted the sea urchin community to get a jump on genomic studies of adaptation to environmental stressors, including ocean acidification.[47]

These genomic tools attracted young scientists who were interested in using them to address environmental problems. One of these was Melissa Pespeni, who begun working as a graduate student in Steve Palumbi's lab at Stanford as the sea urchin genome was coming to completion. As Pespeni told me, the genomic tools were one factor that attracted her to studying the sea urchin.[48] The population genetic features of this organism were another. Because it is broadly distributed and has both high rates of dispersal and high population sizes, Pespeni thought the sea urchin would be a robust system for detecting local adaptation. In other organisms, demographic features—such as changes in population size—can make it difficult to distinguish the signal of adaptation from noise.

Pespeni and her colleagues performed a common garden experiment, exposing different populations of sea urchins to the same environment for three years.[49] They then cut off part of the spine from some individuals from each of the populations they had. As sea urchins in the wild face lobsters and other predators, regrowth from damage is a prospect that many would likely face. The ability to regrow could be compromised with further acidification. Pespeni's group found that the southern population (California) grew back their spines faster than those from the Oregon population. Interestingly, the California population also had increased expression of certain biomineralization genes, suggesting that products of these genes might be involved in the regrowth.

Pespeni and her colleagues were also interested in whether there were genes that showed signs of being locally adapted to the pH conditions.[50] Such a sign would be having allele frequencies that varied depending on the conditions. For instance, a gene would show the signature if it had an allele that was consistently high in frequency in populations that often experienced low pH was common but consistently low in populations where low pH was rare. The reverse pattern would also be such a signature. The important thing is that the allele frequency associated with the acidity regime. They then sampled sea urchins from six sites from the Pacific (two off the coast of Oregon and four off the coast of California) and took measurements of the acidity of the waters, assessing how often the pH read below 7.8. Acidity generally fell along a north/south gradient, with the Oregon samples being most prone to spend time with pH below 7.8.[51]

With the ranked environments and with genomic data from urchins from the six populations, Pespeni was able to identify genes that had allele frequencies correlating with the environments. A couple hundred genes—more than what is expected simply from chance—showed such a correlation. These tended to play roles in cell adhesion and biomineralization. Pespeni also looked at the genes that showed changes in frequency in response to experiments where the urchins had been exposed to substantially more acidic water. Of the genes that showed both signatures of local adaptation and changed during acidification experiments, seven stood out as the top candidates. Several of the top candidates appear to function in lipid metabolism.[52]

A follow-up study confirmed and extended these findings: urchins that experience different acid conditions also express different sets of genes in response to acute acid stress. Urchins from the same six populations across the Pacific acidity severity gradient were exposed to high carbon dioxide concentrations. Here, the researchers focused on genes that were turned on the most in populations from the highly acidic populations, but not the ones that experience little acid stress (and vice versa). Those from areas that often experience low pH differentially ramp up expression of genes in oxidative metabolism pathways, while those that do not experience increase the expression of genes involved in protein synthesis.[53] The urchins that experience acid stress often appear to be increasing expression of genes that will permit them metabolizing fatty acids and carbohydrates in order to maintain growth and calcification. Moreover, even before the acute stress, these metabolism genes are already expressed more in the urchins that encounter the acid stress more often. This phenomenon, known as preloading, appears to be an adaptive response, as it allows the urchins to get a head start when they experience acid stress.

The urchins that experience acid stress pay a price in having their metabolism ramped up high, and not just when they are actually experiencing the stress. Consequently, they will need more resources just to carry out housekeeping functions. This difference could explain why the sea urchins from the northern population that frequently encounter acidic conditions are slower to regrow spines in the common garden experiment compared with those from the southern population, which do not. What I have not told you until now is that the areas where acidity is most frequent are in upwelling zones. Upwelling, a frequent occurrence near the west coast of the United States, arises when winds cause mixing between surface and subsurface waters. This mixing makes the waters more acidic but also releases nutrients. Perhaps in the wild, these additional resources allow the sea urchins to cope with the enhanced metabolic demands.

The combination of a large population and prior adaptation to local environments appears to provide the sea urchin with adequate reservoir of standing genetic reservoir to cope with rapid ocean acidification and other evolutionary challenges. But is it enough?

I asked Pespeni, who now runs her own lab at the University of Vermont, what she thought about the prospects of the sea urchin. She said that they seem to be coping well for now. But, she continued, can they handle multiple stressors? As we have seen, nature is complicated, and organisms often need to handle more than one stressor. Recall the experiments Morgan Kelly did showing that copepods had a difficult time adapting to both temperature and salinity. The adaptation to acidic waters may make the urchins more vulnerable to other stressors. If alleles that convey adaptation for increased temperature and alleles that convey adaptation for more acidic waters come together—either by being genetically linked or by pleiotropy—adaptation to both would be relatively easy. If not, adaptation to both could be more challenging.

Pespeni's group has some results that suggest the genetic reservoir of sea urchins has limits. Looking at the urchins in experimental treatment groups, they found a considerable loss of genetic variation in both the low and especially the high carbon dioxide groups. In a single episode of selection, there was a 19 percent reduction of the nucleotide diversity in the high carbon dioxide selection experiment versus only 10 percent reduction in the low carbon dioxide one.[54] This loss of genetic variation is similar to what we observed in the killifish of Galveston Bay: both the urchins and the killifish appear to be coping with the environmental stressors through evolutionary responses, but at a cost of their reservoirs of genetic variation.

Taking advantage of the extraordinary fecundity of the sea urchins, Pespeni's lab took 400,000 eggs (!) from each of 11 females: half of the eggs were placed in mildly acidic conditions (pH 8.0), while the other half were placed in more extreme acidic conditions (pH 7.5).[55] Remember that the pH scale is not linear but instead is logarithmic, such that small changes in the number can correspond to large differences in the degree of acidity. The eggs were fertilized by sperm from all the males pooled together. After the larvae had developed, the researchers sequenced the genomes of those that survived. These responses to extremely low pH came from alleles that were at relatively low frequency. Responses to both the extremely low pH and the moderately low pH involved alleles that were at a higher starting frequency. These results suggest that the extreme fecundity and the large population sizes of sea urchins may be what may enable it to continue to adapt to ever-more-acidic waters. Other marine organisms may not be as fortunate.

Does the change required to track ocean acidification need to be evolutionary adaptation? As we noted earlier in the book, responses to environmental change can be acclimation (changes in the organism) or adaptation (changes in the genetic composition in the organism). Such is the case with the responses to ocean acidification. Moreover, as we saw earlier, the relationship between acclimation and adaptation can be complex.[56,57] Buffering provided by acclimation to ocean acidification can give populations time to catch up to the altered environment. Having some individuals persist and reproduce due to acclimation responses could keep the population going until the population responds with genetic changes to the insult. But acclimation can have a dark side. It could hinder genetic adaptation by making the fitness gradient shallower without altering the genetic composition of the population. This happens because acclimation pushes the average phenotype closer to the optimum. Acclimation itself could have costs.

CORALS

The science writer Elizabeth Kohlbert once described coral reefs as "organic paradoxes" that are "obdurate, ship-destroying ramparts constructed by tiny gelatinous creatures. They are part animal, part vegetable, and part mineral, at once teeming with life, and at the same time, mostly dead".[58]

Corals are ecosystem engineers par excellence. As testimony to how well corals build communities, some of the most productive and speciose areas of the world are coral reefs. Coral reefs also provide what are called ecosystem services. They add not only beauty but commercial value via tourism. Their presence also reduces damage from storms; a recent estimate of their benefit to coastlines runs to several billion dollars each year.[59] Remarkably, about half of federally managed fisheries in the United States rely on the habitats created by coral reefs. The commercial value of the benefit of coral reefs to US fisheries is an estimated $100 million.[60]

Coral reefs are even sources of material that could benefit biotechnology. For example, venomous cone snails live in the coral reefs around New Caledonia. Analysis of the venom of these snails has helped to guide the development of ziconotide, a highly efficacious pain reliever that is hundreds of times more powerful than morphine. It is primarily used for pain management and sedation of cancer patients that are terminally ill. Another example of application from coral reef life from the same region is the development of the potent antimicrobial agent arsenicin A, which was developed from extracts of compounds from the sponge *Echinochalina bargibanti*.[61]

Corals face numerous threats, including temperature rises, ocean acidification, and environmental catastrophes like the Deepwater Horizon. Because they are sessile, they can't simply move to a new environment. Many corals also grow very slowly and have long generation times; hence, their potential for rapid evolution may be limited.[62] One example of these long-lived, slow-growing corals is *Paramuricea biscaya*. This species, which can live for several centuries, grows at an estimated rate of 0.3 to 14 microns per year.[63] A micron is a thousandth of a millimeter or a millionth of a meter. To put this in perspective, North America and Europe are moving apart at about an inch (25 mm or 2500 microns) per year. These corals are growing at a rate that is hundreds, if not thousands, times slower than continental drift!

Corals have a symbiosis with an algae (zooxanthellae). These algae are responsible for the vibrant colors and assist the coral with the ecosystem engineering. When corals are stressed, they expel the algae in a phenomenon known as bleaching, During the heat wave of 1998, about one sixth of reef-building corals died.[64] Some models predict that even 2 degrees C warming above preindustrial times will be catastrophic for most corals.

Experimental studies show that both higher temperatures and increased acidification will reduce calcification in coral species. The negative response varies considerably among the different species. Interestingly, a tradeoff may exist, as the species that calcify the most in the absence of thermal and acidity stress have the greatest declines when exposed to these stressors.[65]

"The dire projections of coral reef futures are rooted in the assumption that corals bleach when water temperature exceeds a threshold level, and critically, that this threshold remains constant over time."[66] So wrote Thomas DeCarlo at the Australian Research Council for Coral Reef Studies and his colleagues. They wanted to examine that assumption and did so by examining the historical record—namely, skeletal cores of the coral of the genus *Porites*. Like with trees and the Eifuku mussels we encountered earlier in the chapter, the patterns of growth of corals can be accessed by looking at their skeletons. High temperatures and other stresses are recorded as stress bands. DeCarlo and colleagues' finding was that temperatures that once resulted in stress bands no longer are. This implies that the corals are adapting to higher temperatures.

Addressing whether there is enough genetic variation for corals to adapt is exceedingly challenging because corals are not easily reared in the lab and have long generation times. However, biologists are able to look at different populations within the same species and are increasingly able to do genomic analysis to look for genes that may affect the responses.

Morgan Kelly's group studied responses to low pH in *Balanophyllia elegans*, a solitary coral that lives off the Pacific coast. They examined two different populations, one from Northern California, where upwelling is common, and one from Southern California, where upwelling is much less common. Although caution must be taken given that only two populations were observed, the results are in keeping with expectations and with the studies with sea urchins: Corals from northern California, which experience low pH more frequently, are better able to maintain their respiration rates and stores of protein and lipids in comparison to those from Southern California. Notably, the Southern Californian corals appear to be depleting their energy reserves after a month's exposure to low pH. Moreover, upon low pH exposure, corals from the high upwelling site will increase the expression of genes that are involved in calcification, such as those that bind and transport calcium ions. The authors suggest that the corals from the low upwelling site are unable to regulate the expression of these genes upon low pH conditions, presumably because they are energetically limited. Corals from the low upwelling site showed more differentially expressed genes—similar pattern to what in seen in other corals, copepods, and snails.[67]

Along similar lines, a pair of biologists at Temple University, Erik Cordes (a coral expert) and Rob Kulathinal (a genomics specialist), have collaborated to investigate genomic responses to oil exposure in corals. With their graduate student Danielle DeLeo, they first found that three different species of coral showed more severe effects upon exposure to the dispersant or to a mixture of oil and dispersant than they did upon exposure to just oil. So this is further evidence that adding dispersant to clean up the Deepwater Horizon oil made things worse.[68]

They then focused on *Paramuricea biscaya*, the very slowly growing coral we met earlier, looking for genes that were expressed differently in the contaminated versus reference sites. Consistent with tissue sloughing and necrosis seen in the affected corals, a suite of wound healing genes was overexpressed in the affected corals. Genes that contribute to the inflammatory response were also expressed higher in the corals at the contaminated site. Consistent with the contamination imposing oxidative stress, genes involved in oxidoreductase activity were overexpressed. Genes associated with DNA replication were under-expressed in the contaminated corals, suggesting that growth would be curtailed.[69]

Cordes and Kulathinal are currently engaged in searching for corals that are very resistant to ocean acidification. Their work is featured in the film *Acid Horizon*.[70]

THE BIG PICTURE

At the start of the chapter, we saw Karin Pfennig challenge evolutionary biologists to assess which species are most likely to adapt to changing conditions and environmental insults. In general, we can say that odds favor those populations and species with numerous, fecund rapid-developing organisms. Being numerous and fecund allows for the maintenance of genetic variation required for continued evolution by natural selection. The short generation times of fast-developing organisms will reduce the rate of environmental change per generation. The animals we saw adapting— the *Tigriopus* copepods, purple urchins, and killifish (for a vertebrate)—all fit that description. The factors that make them well able to adapt to these environmental changes are also ones that make them suitable as model systems. As such, evolutionary biologists studying adaption to changed environments should keep in mind that the performance of their study animals may not reflect how well other species are doing.

But even these numerous and fecund, rapid-developing animals may not be able to keep up with continued and accelerating environmental insults. The Gulf killifish in Galveston Bay appear fortunate in that they acquired an adaptive deletion from a closely related species. In a similar way, because different sea urchin populations have adapted to local conditions, these sea urchins may globally have more of the requisite genetic variation to adapt to changes occurring with climate change.

Costs of adaptation also are likely to be important. The killifish may be fortunate regarding the costs of the deletion that allowed them to tolerate the extremely toxic environments of Galveston Bay. Although costs of this deletion could still be found, the more likely case is that they are either mild or were ameliorated through compensatory evolution. If so, then this killifish could continue adapting to new challenges even after adapting to the extremely toxic Galveston Bay. In contrast, costs to adaptation appear greater in the sea urchin and in the copepods. These high costs may hinder the ability of the copepod and the urchin to keep up as oceans warm and become more acidic. These results illustrate the need to study costs of adaptation (and selection to multiple stressors) in assessing which populations are likely to thrive in the ever-changing seas.

NOTES

1. Plumer (2019b).
2. Pfennig (2019).
3. Special report on the ocean and cryosphere in a changing climate. IPCC. www.ipcc.ch/srocc. Accessed May 24, 2021.
4. Background information on the Deepwater Horizon spill and its consequences is primarily from Juhasz (2011).
5. King et al. (2015) and Head et al. (2006).
6. Juhasz (2011) summarizes the immediate biological effects of the spill. Joye et al. (2016) and Beyer et al. (2016) provide updated accounts.
7. Robertson et al. (2017).
8. Lee et al. (2017).
9. Joye et al. (2016).
10. Instead of being attached by ionic bonds as in sodium chloride (table salt), these chlorides are attached to the phenolic rings by covalent bonds. These covalent chlorides are highly reactive, making PCBs even more likely to cause damage.
11. Clark et al. (2010).
12. Dubinsky et al. (2013).
13. Whitehead et al. (2012).
14. Oziolor et al. (2016).
15. Oziolor et al. (2019).
16. Olson (1999, p. 21).
17. Hottes et al. (2013).
18. Oziolor et al. (2019).
19. Arnold (1997).
20. Hedrick (2013).
21. E-mail interview with Andrew Whitehead, May 2019.
22. Nacci et al. (2010).
23. One striking exception is in Jamaica Bay, New York, which abuts the southern edge of Long Island and is where Kennedy Airport was constructed. Jamaica Bay plays a role in numerous ecosystem services, including protection from storm surges such as that seen with Superstorm Sandy in 2012. While Jamaica Bay had one of the highest PCB concentrations, the fish there were only moderately tolerant. Nacci and colleagues suggest that the fish may not be experiencing as high of a concentration of PCBs as the readings would imply.
24. Nacci et al. (2016).
25. Whitehead et al. (2017).
26. Reid et al. (2016).
27. Whitehead et al. (2017).
28. Carson (1955). The quotes are from p. 22.

29. Doney et al. (2012).
30. Morley et al. (2018).
31. Palumbi et al. (2019). The quote is on p. 71.
32. Kelly et al. (2016).
33. Nearly all mammals, including humans, have chromosomal sex determination: with a few exceptions, XX individuals are female, while XY individuals are male. In birds and in butterflies, it is reversed: ZZ individuals are male, while ZW ones are female.
34. Jensen et al. (2018).
35. Fabry (2008).
36. Tunnicliffe et al. (2009).
37. Kroeker et al. (2010).
38. Srinivas (2015).
39. Brown (2017).
40. In fact, populations of the purple sea urchin are currently devouring kelp forests off the coast of California. See Katz (2019).
41. Ernst (1997).
42. Davidson (2001).
43. Kelly and Hoffman (2013).
44. Kelly et al. (2013).
45. To obtain estimates of genetic variation, Kelly and her colleagues used a breeding design where males were crossed to several females and the offspring of these crosses were divided up into groups that each were subjected to different treatments. Within a treatment, variation among offspring from the same father can be used to calculate the additive genetic variance—this is the type of variation that natural selection works on in the short-run. Information from the full siblings—having both the same father and the same mother—can be used to estimate maternal effects. Because they share the same mother, these full sibling individuals likely share effects from sharing the same mother. Perhaps one mother had exceptionally good nutrition; her offspring are likely to benefit. Perhaps one mother had exceptionally poor nutrition; her offspring are likely to be stunted. The breeding design can capture the extent of these maternal effects.
46. I asked Morgan Kelly in a phone conversation (June 2019) about the assumption of stabilizing selection. She said that while there was no clear evidence that larger individuals are being selected against, assuming that there is such selection is reasonable. She added that the other possibility for the maintenance of the body size distribution in the absence of ocean acidification is mutation-selection balance, where larger individuals are favored but mutations reduce size such that most individuals would be smaller than average. In this scenario, there would be similar selection against the smallest individuals. Hence, we should expect similar dynamics with ocean acidification with the mutation-selection balance model that we saw with the stabilizing selection model.
47. Evans et al. (2015).
48. Phone conversation with Melissa Pespeni, August 2019.
49. Pespeni, Barney et al. (2013).
50. Pespeni, Chan et al. (2013).
51. Fogarty Creek in Oregon was ranked as the most severe because pH most often fell below 7.8. The Alegria site in southern California experienced sub 7.8 pH the least; thus, it was ranked as the least severe site.
52. Pespeni, Chan et al. (2013, pp. 864–865). Some of the genes that responded are "phosphoribosyltransferase, a catalytic and regulatory protein involved in nucleotide synthesis and salvage; lin-10, a gene that functions in development; a sterol-carrier protein; a nonspecific lipid-transfer protein."

53. Evans et al. (2017).
54. Lloyd et al. (2016).
55. Brennan et al. (2019).
56. Sunday et al. (2014).
57. DeBiasse and Kelly (2015).
58. Kohlbert (2014, p. 129).
59. Woodhead et al. (2019).
60. www.oceanservice.noaa.gov/facts/coral_economy.html. Accessed June 21, 2020.
61. Motuhi et al. (2016).
62. Freler et al. (2013).
63. DeLeo et al. (2018).
64. Freler et al. 2013.
65. Bove et al. (2019).
66. DeCarlo et al. (2019, p. 2).
67. Griffiths et al. (2019).
68. DeLeo et al. (2016).
69. DeLeo et al. (2018).
70. www.acidhorizon.com/sciemce. Accessed June 22, 2020.

13 Challenges in the City

SUMMARY

Far from being barren, urban areas often are teeming with life, though the composition of species in cities is often different from those found elsewhere. Moreover, natural selection often has changed species that live in urban areas. Here, we consider how organisms have evolved in urban areas. One of the classic examples of evolution by natural selection—the peppered moth—is also an example of evolution in cities. We look at how living in cities has altered the faces of foxes and caused plants to restrict the dispersal of their seeds. Cities are often warmer than the surrounding areas; this urban heat island has affected the evolution of animals and plants. We also look at the behavioral and neurological changes that have occurred in some animals that thrive in cities. We look at how roads and other features of cities affect animals and their evolution. In some cases, the initial stages of speciation have happened in cities. We look at how the microbiota influences the ability to life in cities. We conclude with a discussion of rats and how the principles of evolutionary biology have been applied in managing these urban pests.

"[W]hen we talk about ecology and evolution, about ecosystems and nature, we are stubbornly factoring out humans, myopically focusing our attention on that diminishing fraction of habitats where human influence is still negligible. . . . Such an attitude can no longer be maintained. It's time to own up to the fact that human actions are the world's single most influential ecological force."—Menno Schilthuizen.[1]

CITIES ARE FULL OF LIFE

Human influence is greatest in our cities. More people live in urban areas than outside of them; as of 2018, urban areas account for 55 percent of the world's population. By 2050, more than two-thirds of the world will be in cities, with an additional two and a half fold more urban dwellers.[2] Here, in the cities is where most people encounter most nonhuman life. Far from being sterile, cities are teeming with life, though this life is altered through sorting—not all organisms thrive in the city—and evolution. Because this is where people are most likely to contact biology, studies of urban ecology and evolution can be a gateway for city inhabitants to learn and appreciate biology. The exchange is not unidirectional. Indeed, city residents who learn about the biology of the life around them are more likely to engage in community science.[3] Accordingly, as the authors of a recent book on urban evolution note, engaging city residents in the ecology and evolution of the life that surrounds them in cities can generate "an unprecedented task force to combine data collection and science education".[4]

DOI: 10.1201/9780429503962-16

The pace of life is faster in cities. So is the pace of evolution: cities typically foster rapid evolution, including diversification (and even the formation of new species). Cases of urban evolution usually are ones of rapid evolution, as urbanization usually occurs on the scale of decades. Evolutionary biologists today, as opposed to those of prior generation, are more inclined to see rapid evolution as an important aspect of the process in general.[5] Even as recently as the 1990s, rapid evolution was largely seen as the exception. Today, rapid evolution—evolution that we can see in the span of a few years to a lifetime—has become viewed more as being commonplace.

THE PEPPERED MOTH

The peppered moth, *Biston betularia*, and the change in frequency of its color morphs is a textbook case of evolution by natural selection.[6] It is also an excellent case of urban evolution, as it was the pollution of the cities that drove the evolution.[7] Prior to the middle of the 19th century, nearly all the peppered moths in England had wings that were mostly white, with darker specks that resembled pepper; hence the name. This is the *typica* morph. A much darker morph (known as *carbonaria*) started to be noticed around 1848 in the industrial cities, such as Manchester in the northwest of England. The frequency of the *carbonaria* morph rose quickly till, by the end of the 19th century, nearly all the peppered moths of Manchester were the dark *carbonaria* morph. The dark morph also climbed to high frequencies in other urban, industrial areas of late 19th- and early 20th-century Britain. In the countryside, far removed from the pollution of the urban centers, the light morph was more common. Going from the urban centers to the rural areas, a cline of decreasing frequency of the dark morph was apparent.

Why did the dark morph increase in the industrial urban centers? One hypothesis assumed that better camouflaged moths would have a selective advantage because such moths were less visible to bird predators. On light birch trees that had been common in preindustrial times, the *typica* morph was favored over the darker *carbonaria* morph. But the industrial revolution brought soot that darkened the trees and sulfur dioxide that killed off the light-colored lichens that grew on the barks of the trees. On the soot-darkened trees, the *carbonaria* morph would blend better and should have a selective advantage over the typica morph.

The camouflage hypothesis remained untested until the 1950s with the seminal work of Bernard Kettlewell, who conducted experiments examining the role of bird predation. Kettlewell showed that birds indeed would preferentially pick out the lighter *typica* moths and spare the darker *carbonaria* moths when they were placed on the soot-darkened, lichen-depleted trees of the industrial centers. Kettlewell also had the presence of mind to conduct the reciprocal experiment: looking at bird predation on the lighter trees of the countryside. There, the birds preferred to prey on the darker moths. Hence, moths that resembled the birch trees they encountered had a selective advantage. Kettlewell and others also collected data on the frequencies of the different morphs in different areas. Consistent with the camouflage hypothesis, there was a strong trend of the dark *carbonaria* morph being more common in more polluted areas.

Melanic (dark) and typica (light) morphs of *Biston betularia cognataria*, the American sub-species of the peppered moth.

Source: Courtesy of Bruce Grant.

Although I did not directly work on the peppered moth, I have a personal history with it. As an undergraduate at the College of William and Mary, I took an evolution class with Bruce Grant in 1986. At this time, Grant had been working on aspects of the peppered moth for a couple of years and passionately described the work that he was doing during class lectures. Grant's enthusiasm was infectious. The puzzle-solving aspect of his approach to doing science appealed to me. I joined his lab the following semester and did an undergraduate honors thesis there.

Although my thesis project was on evolutionary behavior of a tiny parasitoid wasp and not on peppered moths,[8] I grew up as a scientist watching the *Biston* project unfold. Grant, Karen Galloway (Grant's other undergraduate student at the time), and I had many conversations about the moth work. Moreover, Rory Howlett then a graduate student from the United Kingdom who was also working with *Biston*, visited the lab as part of his collaboration with Grant. Their collaboration was about whether moths could choose the color of tree to rest on. If trees of different colors were present, a light moth would have an advantage if it preferred to rest on a light tree. Similarly, a dark moth would have an advantage if it preferred to rest on a dark tree. The preference for trees that matched one's color would have been a wonderful evolutionary adaptation; but alas, the link between preference and color did not exist. Some individual moths did show consistent preferences, but the dark moths did not prefer dark backgrounds more than lighter moths did.[9] Seeing up close that organisms are not perfectly adapted and that biology sometimes—often?—does not give you the answer you expect was an important lesson for me as a young scientist back then. Nevertheless, the research post-Kettlewell has only strengthened the case for natural selection driving changes in the frequency of color morphs in the peppered moth: the evidence is now indisputable. Moreover, the evidence strongly implicates bird predation as the agent of selection.[10]

In the years since the 1980s, we know more about the genetics of this color polymorphism: it is due to a single gene. Moreover, population genetic analysis demonstrates that there was a single mutation that rose to high frequency in the urban areas; there were not multiple, independent selective sweeps.[11]

We also have a parallel case of natural selection driving changes in the frequency of color in the American subspecies of the peppered moth, *Biston betularia cognataria*. The color polymorphism in North America is due to the same genetic locus as the more famous case across the pond.[12] Moreover, a similar story of the rise and fall of the darker morph played out in North America. Once again, the respective causes were pollution brought on by the Industrial Revolution and abetted by Clean Air legislation. We do not know as much about the rise in frequency of the darker morph of the American moth as we did with the British moth, but its decline has been investigated by Grant and his colleague, Larry Wiseman. They conclude:

> Both American and British peppered moth populations are converging on monomorphism for their respective typical forms, correlated with reduced levels of atmospheric pollution on both sides of the Atlantic. While a correlation alone does not establish a causal relationship, common correlations suggest a common cause.[13]

The peppered moth teaches us that evolution can happen quickly and can be influenced—often unwittingly—by human activity. It also shows that at least sometimes the evolutionary changes that were the consequence of human activity can be reversed. Finally, this example shows that evolution can and should be studied in cities, as well as in areas less travelled by humans.

URBANIZATION AND THE GREAT SORT

Living in the city entails challenges. Unlike in the countryside, the presence of artificial light means that it is never truly dark in the city. Cities are also noisy. Organisms that do not tolerate noise or light at night are unlikely to make it in the city. The human-made structures in the city are different than natural ones. Human structures, like houses and office buildings, tend to be less structurally complex than natural structures, like trees. Human-made structures are also more likely to have smooth surfaces.[14] Certain types of animals and plants possess traits that make them better equipped for urban life than others. It is these organisms who are most likely to be found in urban areas. Other animals and plants who lack these traits will be sorted out.

Urban plants tend to be dispersed by wind, as animal pollinators are often less reliable in the city. They also are more likely to better tolerate smog and other air pollution. They generally prefer alkaline soils with high nitrogen content. Plants that are early colonizers of disturbed areas are more likely to thrive in urban environments given the extensive disturbance—think of construction sites and road paving, for instance—that occurs in cities. Such plants, which are called ruderal species, tend to put down roots quickly.[15] Urban plants often are not native to the area. For example, a survey of plants in London and the surrounding area found the plants that were most represented in the urban city were overwhelmingly not native to Britain.[16]

Next, we turn to *Anolis* lizards. These lizards, which are typically 4–8 cm (a couple of inches) long and come in a variety of colors, are extraordinarily diverse.[17] Not only are there nearly 400 species (about 150 of which are found in the Caribbean), but they occupy an array of different niches. For example, some live near the ground and use broad tree trunks, while others live high on the narrow twigs of trees. The near-ground lizards race quickly to catch prey, while the twig-lovers slowly creep using their large adhesive toepads. The evolution of these diverse ecological roles is a textbook case of convergent evolution, as the same types of lizards have evolved independently on the many different Caribbean islands.

Anolis lizards also differ in how well they tolerate urban areas. Which ecological and morphological factors are associated with whether a lizard species would spend time in urban areas? This is the question that Kristin Winchell at Washington University in St. Louis and her colleagues examined looking at lizards in the Caribbean. They first found that lizards that were closely related typically have similar tolerances for being in cities.[18] This is potentially troubling, as it suggests that further urbanization may endanger large branches of lizard biodiversity. Species with large native ranges are more likely to tolerate city life. This suggests that species that have had to adapt to a variety of environments may be tolerant of urban areas. Traits related to locomotion also affected the propensity to tolerate urban areas. For instance, species that have many specialized scales, known as lamellae, on their hindlimbs are more likely to be found in cities. These lamellae are likely to assist the lizards moving across the smooth surfaces that predominate in cities.

One of the particularly attractive features of the study by Winchell and colleagues is that it allows researchers to make predictions about which specific lizards will be able to tolerate urban life. For instance, they consider two species (*A. desechensis* and *A. monenesis*) that live on uninhabited islands (Desecheo and Mona, respectively). Their model predicts that *A. desechensis* would better tolerate development of Desecheo than *A. monenesis* would tolerate development of Mona. This makes sense because *A. desechensis* is closely related to *A. cristatellus*, a species that exploits urban environments, while *A. monenesis* is closely related to *A. cooki*, a species that lives in tropical forests and does not tolerate urbanization.

What about other vertebrates, such as birds? A recent study by researchers at UCLA found nesting behavior to be associated with tolerating urban life in Southern California; birds that build nests on human-built structures were more likely to be in the Los Angeles area than birds that prefer to build nests in tree cavities.[19] Other aspects of behavior also affected the sorting process: species that persist in Los Angeles were also more likely to be bold and seek novelty than those species did not. This study was unusual in that it included observational data from both the mid-1990s and the mid-2010s. For the most part, the results were consistent between time periods. But there were a few changes. For instance, dark-eyed juncos appear to be tolerating urbanization now. Whether this change is due to evolution is not clear.

As for mammals, studies point to two different routes being able to make it in the big city: mammals that can tolerate urban areas tend to have big brains and/or large litters.[20] As we have seen for birds, big brains likely enable greater behavioral flexibility in mammals, which is likely to be useful in confronting the novel challenges of the city. But even in the absence of big brains, mammals with large litters are more likely to persist in the city; being prolific allows populations to quickly rebound if the population is reduced to a few or even one.

Even after the sorting, species living in urban environments often evolve genetic differences compared with rural populations of the same species. Next, we consider a few examples of such evolution.

URBAN FOXES EVOLVE SHORT SNOUTS AND URBAN PLANTS EVOLVE TO STOP DISPERSING

Red foxes (*Vulpes vulpes*), frequently found in the cities of Western Europe, provide another example of substantial evolutionary change upon urbanization. Already present a century ago, foxes became more widespread in England, particularly in the south, during the period between the two world wars as suburbs grew.[21]

A recent morphological analysis led by Kevin Parsons at the University of Glasgow showed that urbanization profoundly altered the evolution of these foxes.[22] Comparing foxes found in and near London, they found several changes. Notably, urban foxes have evolved shorter and wider snouts. This morphological change affects how the snouts work and thus how the different foxes bite. The snouts act

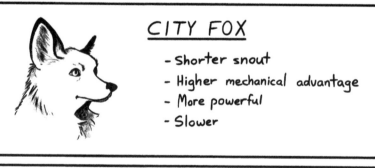

CITY FOX

- Shorter snout
- Higher mechanical advantage
- More powerful
- Slower

COUNTRY FOX

- Longer snout
- Lower mechanical advantage
- Less powerful
- Faster

Differences between the city fox and the country fox.

Source: Illustrated by Michael DeGregori.

like a lever: like any lever system, there is a tradeoff between power and speed. Compared with those of their rural counterparts, the stout snouts of the urban foxes give them the ability to generate more force. This property—what is called greater mechanical advantage—has a downside: the snouts of the urban foxes do not operate as quickly as those of the rural foxes.

I asked Andy Conith, a coauthor of the study who is a postdoctoral fellow at UMass Amherst, about the ecological context of this difference in mechanical advantage. He told me:

> In the paper, we argue that the urban foxes are typically scavenging on a wide variety of food items (that can have a range of mechanical properties). Many of these food items are discarded bits of human food waste. These discarded items aren't going to run away, so the urban foxes can optimize power over speed. Rural foxes are active predators. They must hunt and chase down their prey, but their prey is typically squishy rabbits or small rodents that don't require much force to process. However, these will run away so having a fast jaw closing mechanism is likely advantageous.[23]

The change in the urban foxes was not limited to the change in the snouts. In rural foxes, males are considerably bigger than the females; in the urban foxes, this sexual dimorphism is much reduced. The reduced sexual dimorphism and other morphological changes are consistent with patterns seen in the domestication of dogs. Moreover, they are consistent with the study where Russian scientists selected for tame foxes. Starting in 1959, Dmitri Belyaev and Lyudmila Trut at the Institute of Cytology and Genetics in Novosibirsk, Siberia, chose the most tame foxes to start the next generation. Over time, this artificial selection led to foxes that were as tame as dogs. But tameness was not the only trait that appeared in these selected foxes. Among other changes, these tame foxes had shorter snouts, much like the urban foxes observed here.[24] Short snouts might be a common feature associated with tame behavior.

The urban environment does not just affect animals. Plants too have evolved in response to challenges posed by the city. One such challenge is habitat fragmentation: because of the way landscapes are fragmented in the city, suitable habitat for a plant may be just a sliver of a city block that is far away from other such patches. In such patchy settings, dispersal of seeds may be costly, because seeds that are dispersed may not find suitable sites. Hawksbeard (*Crepis sancta*), a common weed found in much of Europe, has two types of fruits: small fruits whose seeds will readily disperse and large fruits whose seeds do not readily disperse.[25] A study in Montpellier in the south of France found that selection has been operating to reduce dispersing in the city.[26] First, more of the non-dispersing seeds are found in urban patches than in the countryside. Moreover, experiments revealed a cost to dispersal in the urban areas: dispersing seeds are less than half as likely to settle on a patch suitable for growth than are the non-dispersing seeds. Modeling suggests that the change in frequency of dispersing seeds could evolve in a few generations. The reduced frequency of non-dispersing seeds here is more likely a direct consequence of fragmentation, indirectly brought on by urbanization.[27]

White clover (*Trifolium repens*) is an important source of nectar for pollinators in North America. Some clover plants can produce hydrogen cyanide, a potent poison used to defend against herbivores. In clovers, the concentrations are sufficient to kill invertebrates and small mammals. Other plants can produce enough hydrogen cyanine to kill or at least cause severe bloating in large mammals. In fact, the production of hydrogen by plants was discovered when farmers and ranchers saw their livestock being poisoned.[28]

Hydrogen cyanide would harm the clover if it were released haphazardly, but the clover evolved a clever way of ensuring that only damaged tissue releases hydrogen cyanide. Thus, in the absence of damage, the plant is protected from its poison. The plant has an enzyme that makes a precursor molecule that could be converted into hydrogen cyanide. This precursor is not harmful unless triggered, much like a bomb.[29] This precursor is stored in special compartments of cells. The plant has another enzyme that converts the precursor to hydrogen cyanide. This enzyme is like the detonator. Ordinarily, no hydrogen cyanide is produced because the precursor molecules are stored in the compartments and the detonator enzyme is not in those compartments. But when an animal takes a bite out the plant, the compartments are breached, letting the precursor (the bomb) meet with the enzyme (the detonator). Boom! Instead of a delicious leaf, the hungry animal gets a nasty dose of hydrogen cyanide. One of the consequences of this clever hydrogen cyanide delivery system is that for it to work, an individual plant requires having active alleles at two different genetic loci—the one that encodes the enzyme for the precursor and the one that encodes the detonator enzyme. Either one without the other is useless.

Marc Johnson and his lab at the University of Toronto found that clover in northern cities often have a higher percentage of plants that cannot produce hydrogen cyanide because they lack one or more of the functional enzymes. In fact, a repeated pattern is that in several regions around cities in the northeastern United States and in Canada, the proportion of clover plants capable of producing hydrogen cyanide decreased as one approached the city.[30] Population genetics call such a pattern of a consistent change in the frequency of an allele or a trait a cline. The likely reason for these clines is that areas near cities have less snow cover; the reduced snow cover makes the plants more susceptible to freezing, which can break apart cells and cause the release of hydrogen cyanide. Releasing of hydrogen cyanide in the absence of herbivores would be costly to the plant. The fact that the clines are repeated across multiple regions strongly suggests that the loss of the trait has come about from selection.[31]

Below, we focus on three more aspects of how selection has shaped animals and plants living in cities: climate differences, cognition and behavior, and roads.

HOT TIME—SUMMER IN THE CITY!

If you watch a local television weather report, you might notice that the temperature readings for the cities are usually a bit higher than those of the surrounding less urbanized areas. You might have even noticed the change if you go from city to countryside, or vice versa. The effect is real; cities are indeed a little warmer than rural areas. This phenomenon even has a name—the urban heat island. In the previous section, we saw a likely urban heat island affecting white clover's use of cyanide production: reduced snow cover near cities led to fewer clover plants producing cyanide as a means of self-protection.

Urban heat islands arise from several factors. First, heat is generated by the densely packed numerous respiring human bodies and their use of energy to perform daily activities, including controlling the microenvironments of their buildings. Second, look at all the dark asphalt and concrete in the city. Think about walking barefoot on a concrete parking lot where the sun has been shining. Asphalt and concrete absorb and retain substantial heat, which helps keep cities warm. Note that the effects of the urban heat island are often greatest at night; the countryside cools off more quickly than the city.[32]

Urban warming is a window into the future. The extent of the typical urban heat island circa 2020 is comparable to the projected warming by 2100; hence, we can look at responses by organisms to urban warming to see what is likely to occur as warming continues. Is urban warming the same across different climates? In other words, how similar is the additional urban warming seen in climates that are already warm to that seen in colder climates? Do animals respond similarly?

A species of acorn ants (*Temnothorax curvispinosus*) found in the eastern United States suggests that urban populations in the north respond similarly to rural populations.[33] Sarah Diamond, whose work on ants we discussed in Chapter 11, and colleagues took ants from three cities—Cleveland, Ohio (42 degrees north); Cincinnati (39 degrees north); Knoxville, Tennessee (36 degrees north)—as well as ants from rural populations from the same locations and tested their tolerance to heat and cold in a common garden. For each locale, the urban populations had higher heat tolerance and reduced cold temperature. Because the experiments were conducted in a common garden setting, the differences are likely to be genetic. They conclude: "rapid evolutionary change in thermal tolerances of acorn ants across rural-to-urban gradients can recapitulate evolved differences among species in thermal tolerance across latitude".[34] They also note that because this acorn ant already has a high heat tolerance, relative to other ants, it may be difficult to continue to evolve increases in heat tolerance.

In a follow-up study, urban and rural ants were crossed in the same environment for two generations. The differences in heat and cold tolerance remained, which definitively shows that the differences are evolved genetic ones and not simply plastic changes that persist across a generation.[35] These crosses also show that the genes involved in cold tolerance and warm tolerance likely are different: the hybrids between urban and rural populations tolerated heat about as well as their urban parents, while these hybrids tolerated cold about as well as their rural parents.

Recent key studies on thermal adaptation and the urban heat island have come from a rather unassuming animal: *Daphnia*, sometimes known as the water flea. These tiny crustaceans live in a diverse array of freshwater environments, from small pools of water to large lakes. They are commonly used in ecological, evolution, and toxicology studies, as well as for classroom demonstrations. *Daphnia* also are fish food, both for fish pets and in nature; in fact, they are among the most important prey items in aquatic ecosystems.

Aurora Geerts, then a graduate student with Luc De Meester at the University of Leuven in Belgium, and colleagues took a two-pronged approach to demonstrate that *Daphnia* could sufficiently rapidly evolve increased heat tolerance to keep up with the expected warming.[36] The first line of research took advantage of the fact that many of the eggs that *Daphnia* lay can remain in the sediment for many decades. If conditions are right, these eggs can hatch. Hence, *Daphnia* have an egg bank that is analogous to the seed bank that many plants have. Just as we saw in the last chapter with seed banks, these egg banks are a window into the past. Geerts and her colleagues found that *Daphnia* taken from eggs laid soon before their study had substantially higher heat tolerance than those from eggs laid at the same sites 40 years previously. The researchers also used a selection experiment to demonstrate that *Daphnia* could evolve this increased heat tolerance even more quickly. Note that *Daphnia* are abundant, have high fecundity, and produce several generations per year—all factors that would suit them for being able to keep up with climate change.

Kristien Brans, Luc De Meester, and colleagues followed up on those experiments, showing that urban *Daphnia* could tolerate heat better than their rural counterparts.[37] Because they had conducted common garden experiments, they could be reasonably confident that the differences were evolved differences and not just plastic ones. Interestingly, the more heat-tolerant animals tended to be smaller. The urban animals developed faster but at a smaller size. Being smaller also benefits those living in the warmer cities because warmer water usually contains less oxygen. In addition to being smaller, the urban *Daphnia* develop more quickly and release progeny earlier. They have evolved a faster pace of life.[38] Yet there is a likely tradeoff because smaller *Daphnia* are less competitive against their larger brethren and do not graze for food as well. In general, populations of animals that live in colder climates are larger than those that live in warmer ones.[39] How well this pattern applies in the case of the urban heat island is still an open question.

The *Daphnia* from the warmer urban areas also evolved different physiologies. Compared to their rural counterparts, they accumulate more protein, fat, and sugars.[40] They also have evolved a more efficient oxidative stress system to take care of the consequences that come along with their ramped-up metabolism. Life in the fast lane exacts several prices—among these is the increased production of free radicals, atoms or molecules that are highly likely to engage in chemical reactions because they have unpaired electrons. The urban *Daphnia* have evolved changed in enzymes such as superoxide dismutase and catalase to more efficiently deal with the free radicals.

Recall the study about the factors that predicted which species of *Anolis* lizards would be found in urban areas in the Caribbean? In addition to those morphological and behavioral characteristics mentioned earlier, the climate of the lizards' native habitat also predicts whether they are suited for city living.[41] Lizards from hot and dry native habitats are more likely to be found in urban environments. Moreover, lizards that have high body temperatures in the field also tolerate urban settings better.

It's not just sorting that has shaped the lizards living in Caribbean cities: urban lizards also are adapting to the heat.[42] *Anolis cristatellus* from multiple urban areas in Puerto Rico have higher body temperatures and tolerate heat better in a common garden setting. Analysis of genomic sequences showed that the urban and rural population differed substantially in the gene that encodes arginyl-transfer RNA synthetase, the enzyme that attaches the amino acid arginine onto the appropriate transfer RNA during synthesis of proteins. We do not yet know how changes in this gene have led to increased heat tolerance, but a clue is that enzymes involved in attaching amino acids onto transfer RNAs are important in responses to various stresses. This gene is not the only one involved in the response, but it does appear to be a major contributor. Interestingly, the same change at this gene appears responsible for the heat tolerance of the lizards in several of the Puerto Rico populations. Selection seems to have worked on the same genetic change multiple times.

Urban lizards, like urban *Daphnia* and urban acorn ants, seem preadapted to handle the challenges that global warming will bring. The unknown question is whether will they be able to continue to evolve as cities grow still hotter.

BIG BRAINS IN THE BIG CITY?

Urban mountain chickadees living in Reno, Nevada, are quicker to explore a novel environment than those that live away from humans.[43] The urban chickadees are better problem-solvers and are better at retrieving information from their long-term memories. The telencephalon, the functional equivalent of the cerebral cortex in birds, is also larger in the urban birds.

The urban chickadees being more adept at problem solving is not an isolated example. In fact, there is a strong trend for urban animals, especially birds, to be better at cognitive tasks than rural members of the same species.[44] We will take a look at a few more examples.

The Barbados bullfinch (*Loxigilla barbadensis*) is not actually a bullfinch. Bullfinches are a genus of birds (*Pyrrhula*) that group together with other finches. The Barbados bullfinch is instead a seedeater, which do indeed eat seeds. And yes, the Barbados bullfinch can be found on Barbados, the small Caribbean island-nation located at almost the western tip of the Lesser Antilles. In fact, Barbados is the only place where you find the Barbados bullfinch in the wild; it is endemic to the island. The bird is an opportunistic species and can be found both in the urban cities as well as in the surrounding forests.

Situated in Barbados is a field station run by McGill University in Montreal, Canada. One McGill researcher, Jean-Nicholas Audet, found that urban birds were more skilled at problem solving.[45] Urban birds also were bolder.[46] The urban birds are also better able to mount a robust immune response when challenged with an antigen. This finding suggests that they are better able to handle the onslaught of parasites that often come with urban life. A follow-up study looking at multiple species in the Barbados showed that urban populations of most species were bolder and more tolerant of novelty than those from rural areas.[47]

Another approach to demonstrate that natural selection is behind differences between the urban and rural populations is to look for repeated differences in specific genes in birds sampled in urban and rural settings. This approach was performed on European blackbirds in which urban birds are bolder. Members of the same German group looked at pairs of urban and nearby rural settings at 12 different locales across Europe.[48] They genotyped the birds (792 in all) at seven different candidate genes that all were known to affect behavior, including circadian rhythms, migration patterns, exploratory tendencies, aggression, and anxiety. With the focus on just these seven genes, the likelihood of spurious associations is much reduced. The strongest signal came from *SERT*, which is also known as

5-HTT, a gene that affects anxiety and aggression not just in birds but also in humans and other mammals. The paired urban and rural populations showed strikingly consistent differences in allele frequencies of the gene: in 10 of the 12 paired populations, the allele with the highest frequency was rarer in the urban setting. We do not know exactly how changes in *SERT* leads to changes in behavior in the urban birds. Nevertheless, the combination of these approaches provides strong evidence that at least some of the differences between urban and rural blackbirds evolved by natural selection.

Birds are not alone in showing differences related to cognition. Emilie Snell-Rood and Naomi Wick at the University of Minnesota examined museum specimens of ten species of mammals from urban and rural areas in Minnesota. Overall, the urban specimens tend to have larger brains than the rural ones of the same species.[49] This trend is particularly pronounced in species that have higher reproductive rates. High reproductive rates may facilitate evolution—having higher fecundity and larger population sizes (which often go with having high fecundity) makes selection more efficient. Snell-Rood and Wick had access to specimens from different periods of time, which gave them the ability to track whether and how brain sizes were changing over time. They found no trend over time for increasing cranial capacity in urban populations. In fact, for a couple of species, the brains of urban specimens may have shrunk slightly. For some species in rural areas, there may be a trend to larger brains. Could this be because humans are using rural areas more?

ROADS AND EVOLUTION

[T]hough evolutionary perspectives have gained traction in many fields of conservation, road ecology is not among them. This is surprising because roads are pervasive landscape features that generate intense natural selection.[50]

Conservation biologists are just now coming to appreciate how roads can affect evolution in urban areas. For instance, think of the small mammals and birds that we see as roadkill. Charles Brown and Mary Bomberger Brown find that selection has decreased roadkill in cliff swallows (*Petrochelidon pyrrhonota*). The Browns, who have been studying the social behavior of these birds, have traveled across southwestern Nebraska collecting data on the swallows at roadside nesting sites. One pattern they observed was a steep and steady decline of the number of road-killed birds.[51] In the 1980s, nearly 20 birds were killed each year. In any given year since 2008, fewer than five birds were killed. In contrast to the large decline in the number of birds killed, road traffic had been steady, if not increasing. Apparently, the birds are getting better at avoiding the traffic.

Could the fewer birds killed be due to the birds learning how to avoid the traffic? That is possible. But it does not appear to be the whole story. Selection on the shape of the wings also is occurring. The birds that are found as roadkill have longer wings on average than the birds in the population as a whole. Other research has shown that birds with shorter and more rounded wings are able make vertical take offs than the long-winged birds. Hence, birds with short wings may be more maneuverable and thus better able to avoid the rushing cars and trucks!

Roads are not just influencing how urban animals evolve; they also have affected the evolution of urban plants.[52] Early studies showed that liverworts that grew along the side of roads had higher lead tolerance than individuals collected away from roads. Individuals from each population had been reared in a common garden experiment. Thus, the differences among populations was genetic and not simply due to the environments in which they were grown. Accordingly, we can infer that the difference in lead tolerance evolved by natural selection. Similar patterns of increased lead tolerance in populations near roads have also been found in other plant species, such as the ribort plantain (*Plantago ianceolata*).

Roads can also influence how animals move in a city. Recall the urban red foxes that had evolved different snouts. A recent study by Sophia Kimmig and her colleagues[53] took a landscape genetic approach[54] to infer how red foxes moved in the Berlin metropolitan area. The patterns of DNA sequence data they collected enabled Kimmig and colleagues to reach the conclusion that the foxes did not move freely across the city. Foxes found in the most urban areas were genetically distinct from foxes found on the outskirts of the urban area. The urban foxes also had less genetic variation than their counterparts in the countryside. Their results also indicated that the foxes were moving via the major roads and railways. This is interesting from the perspective of what animal behavior biologists call the "ecology of fear".[55] Like humans, nonhuman animals perceive fear; sometimes their perception may not be reality. Because fox hunting is banned within the city, being inside the urban center poses little actual risk to the foxes. In contrast, the fast-moving cars on the major roads and the trains on the railways most likely pose a far greater risk to the foxes. And yet the foxes avoid the urban center and move along the roads and railways because they likely perceive the city to be more dangerous than the roads and railways. Evolution by natural selection has not yet had enough time to select against traveling down the busy highway.

Kimmig told me that because the foxes are quite abundant, there is not much effort toward their protection.[56] She noted, though, that other species use these abandoned lands, including the much rarer European brown hare. Kimmig is now placing radio collars on foxes in Berlin to study movement patterns across the day.

SPECIATION IN THE CITY

Urban life sometimes does not just lead to genetic changes in the species; it can also create new species. A classic example of such happened in the London underground. During World War II, Londoners took cover in the Underground to escape bombardment from the Germans only to find an aggressively biting mosquito that was related to *Culex pipiens*, a mosquito that is commonly found in both Europe and North America. The mosquitoes look just about identical, but they are not the same mosquito. The aboveground mosquito prefers to bite birds, while the Underground mosquito aggressively bites mammals (humans and rodents). Mosquitoes from above ground go into a dormant state known as diapause during the London winters; with the warmth down below, the Underground mosquito stays active all year round.

What to call this subterranean dweller? The Underground mosquito has been called both *Culex pipiens molestus* (denoting it as subspecies of *C. pipiens*) or *C.*

molestus (denoting it as a species separate from *C. pipiens*). Based on the commonly used biological species concept, the two should be considered separate species; in addition to behavioral and genetic differences, the two mosquitoes are reproductively incompatible, as the Underground mosquito does not produce hybrid offspring with aboveground *C. pipiens*. Different populations of the Underground mosquitoes will produce hybrids with other Underground mosquitoes.[57]

Aside from the London Underground mosquitoes, well-established cases of speciation linked to adaptation to urban life are rare. But, as Ken Thompson and two of his colleagues at University of British Columbia argue, the dearth of such studies may be more due to lack of effort than lack of cases.[58]

URBANIZATION AND THE MICROBIOTA

What about the microbiota and urbanization? The results are mixed. In some species, urbanization appears to negatively affect the microbiome. For instance, house sparrows (*Passer domesticus*) in urban areas Belgium had less diversity in their gut microbiome compared with those from nearby rural areas.[59] This difference was particularly large during the autumn; rural birds show greater seasonal fluctuation in diversity, with a large decline as autumn becomes winter, while the city birds maintained a low diversity year-round. City birds do have a higher abundance of microbes whose sequence suggest that they break down exotic compounds more efficiently. The gut microbiomes of the city and country birds seem to reflect their diet.[60] City birds eat human leftovers like bread and peanuts as well as sunflower seeds, millet, and other commercial birdseeds. Country birds have a different and more varied diet. They eat wheat, corn, oats, and other cereal grains. They also feast on a wider variety of insects. Gut microbiome composition was not related to condition—fatter birds and skinny birds did not have distinct microbiomes. More recently, birds were fed mimics of the urban and rural diets in a common setting—the birds fed the urban diet had microbiotas like those of urban birds, while the birds fed the rural diet had ones that were like those of rural birds. Is the lower diversity harmful to city sparrows? Having a high diversity of gut microbiome seems protective against pathogens. Moreover, there are reports of declines in the numbers of sparrows in the cities.

And yet not all urban dwellers have shown declines in the diversity of their microbiota. For instance, the eastern water dragons (*Intellagama lesueurii*), semi-aquatic lizards living in the city parks of Queensland, Australia, have a more diverse microbiota than their rural counterparts. This difference may also reflect differences in diet, as the urban diet is more diverse and appears to have more animal protein.[61] Systematic systems of how urbanization affects the microbiota have yet to be performed; these may be instructive as the effects of urbanization on the microbiota of given species may provide information on how well the species is coping with urbanization.

RATS

In 2001 and 2002, Robert Sullivan spent nearly every night—interrupted briefly by the September 11 terrorist attacks—observing rats from an alley near his home in New York City. He sought out rats in order to discover his city. Rats, Sullivan noted, are living in our "parallel universe, surviving on the effluvia of human society".[62] In plainer English, they are eating our garbage. Rats are not just carriers of disease but also can be destructive; they cause more than a quarter of all electric cable breaks and about one-sixth, 18 percent, of all phone-cable disruptions. Rats have extraordinary reproductive potential—a pair of rats could have 15,000 descendants in a year.

The rat most often encountered in cities today is the Norway rat, also known as the brown rat (*Rattus norvegicus*). It displaced the black rat (*Rattus rattus*) in the last few centuries in most cities. Population genetic studies show that brown rats came to Europe in the 1500s, largely after the black plague.[63] They would come to North America and displace the black rat in the 18th century. The same population genomics data demonstrates that rats are not moving a good deal between cities; we don't see the admixture patterns that we would see if rats from different cities were interbreeding. From this, we can infer that transport of rats by humans across cities is not a big contributor of the spread of rats and is unlikely to be contributing to the spread of pathogens.

Brown rats continued evolving during the 20th century. Museum specimens from the 1890s (when Theodore Roosevelt was police commissioner of New York) and 2010s reveal that rats have evolved longer noses and shorter upper molar teeth.[64] The pace of evolution over these 120 years or about 370 generations is too rapid to be due to genetic drift; thus, we can implicate directional selection as the likely cause for the change. These morphological changes could be adaptations to the colder city of New York and/or the softer food sources.

Population genetic studies can inform city planning and pest management. As Kaylee Byers and her colleagues note: "it is becoming increasingly important to identify both common features among cities which influence rat movement as well as local features, which can be used by city planners to target and predict areas prone to rat infestation and re-infestation".[65] They also note that such efforts are most pressing in marginalized communities where residents are at greater risk of exposure to rats.

Eliminating rats from too small of an area can lead to related rats coming back to recolonize the area, because related rats usually share resources. By looking at patterns of relatedness among rats, population genetic studies can inform pest control efforts. For example, in Vancouver, Canada, highly related clusters of rats can occupy a city block or more.[66] These studies suggest that pest management should consider the entire block as unit and try to eradicate rats from the whole block instead of treating individual buildings. Population genetic data can also identify migration barriers. For example, midtown Manhattan, which has fewer residents and better sanitation than areas north and south, acts as a barrier to rat movement.[67]

The city of Salvador, Brazil, has experienced rapid population growth as well as crowding and substantial poverty. Here rats carry leptospirosis, a disease caused by bacteria that are related to those that cause Lyme disease and syphilis. This disease causes a variety of symptoms including fever, chills, and muscle soreness. Most troubling about this disease is that many people with it may recover for a bit and then worsen. Rats spread the disease through their urine. Population genetic analysis shows that some different valleys within the city have essentially different populations of rats.[68] Indeed, rats from two of the valleys that are separated by fewer than 50 meters are genetically distinct. The area between the valleys has an upgrade and a road with much traffic—these presumably impede rat movement. The genetic data suggests that each of these valleys should be considered as eradication units. Resources should be allocated to individual valleys rather than across the city.

Life in the city is unusual. A peculiar subset of organisms gets through the sorting process; some of those even thrive in the city. Natural selection has shaped this life further. The evolution of this city life is rapid, often acting at the same time scale of ecological processes.[69] Consequently, the ecosystems of cities are unusual: the interactions among species may be fundamentally different. Managers need to be cognizant of these altered species interactions.[70] Moreover, some species in urban areas may be doing too well; the success of these too successful species can hurt other species. We will take up these invasive species in the next chapter.

NOTES

1. Schilthuizen (2018, p. 4).
2. Meredith (2018).
3. Also known as citizen science.
4. Szukin et al. (2020, p. 4).
5. Resnick et al. (2019).
6. Majerus (2009).
7. Background information on the story of the peppered moth can be found from Cook (2003), Majerus (2009), and Grant (2020).
8. Johnson (1987).
9. Grant and Howlett (1988). In comments to me after reading this section, Grant added, "We do see geographic variation indicating adaptation to local, stable conditions . . . but these are largely monomorphic populations has operated. There was no behavioral polymorphism that temporally matched . . . the transient melanic polymorphism." E-mail correspondence with Bruce Grant, September 2020.
10. Grant (1999).
11. Cook et al. (2012).
12. van't Hof et al. (2011).
13. Grant (2004).
14. Grant and Wiseman (2002). The quote is on p. 89.
15. Winchell, Battles et al. (2020).
16. McKinney (2002).
17. Hill et al. (2002).
18. Losos and Schneider (2009).
19. Winchell, Schliep et al. (2020).
20. Cooper et al. (2020).
21. For birds and mammals, see Sayol et al. (2020) and Santini et al. (2019), respectively.
22. Harris and Rayner (1986).
23. Parsons et al. (2020).
24. E-mail interview with Andrew Conith. June 2020.
25. For the fox selection story, see Dugatkin (2018) and Dugatkin and Trut (2017).
26. Cheptou and Lambrecht (2020).
27. Cheptou et al. (2008).
28. Dubois and Cheptou (2016).
29. E-mail interview with Marc Johnson. June 2020.

30. Dan Ruskin first used the bomb-detonator analogy in describing the hydrogen cyanide delivery system in his wonderful book about the darker side of nature. See Riskin (2014).
31. Thompson et al. (2016).
32. Marc Johnson presented another possibility to get such clines. Given the unusual genetic basis of the trait (having at least one functional allele at two different loci is required to produce hydrogen cyanide), it is possible to get clines by genetic drift if the populations toward the center of the city have lower effective population sizes than those further away from the city. In fact, effective population sizes appear to be similar in rural as compared to urban populations. Hence, this possibility is unlikely in this case. See Johnson et al. (2018) for details.
33. General information on urban heat islands and the response of organisms to them can be found in Diamond and Martin (2020).
34. Diamond et al. (2018).
35. *Ibid.*, p. 6.
36. Martin et al. (2019).
37. Geerts et al. (2015).
38. Brans et al. (2017).
39. Brans and De Meester (2018).
40. The pattern of body size reduction with increasing temperature is sometimes referred to as Bergmann's rule, after observations that the German biologist Carl Bergmann made in the middle of the 19th century. Bergmann had intended to apply this rule to endothermic (warm-blooded) animals such as birds and mammals. Some authors have also used Bergmann's rule to apply to exothermic (cold-blooded) animals, though Salewski and Watt (2016) argue that reference to Bergmann's rule should only be made when dealing with endothermic animals.
41. Brans et al. (2018).
42. Winchell, Schliep et al. (2020).
43. Campbell-Staton et al. (2020).
44. Kozlovsky et al. (2017).
45. Sol et al. (2020).
46. Audent et al. (2015).
47. They tested birds for boldness by presenting birds with an open dish full of delicious seeds (delicious for the birds, at least) while the researchers hid and recorded the behavior of the birds. Those that quickly went for and gobbled up the seeds were scored as bold. Related to boldness is the quality of neophobia—fear of new things. The researchers measured this by repeating the assay for boldness several times and then placing a novel object—for instance, a foot-long yellow stake—next to the dish full of seeds. The urban birds were, as expected, bolder than the country birds. Those birds that lived in the urban areas were quicker to approach seeds after the researchers hid. But—to the researchers' surprise—the urban birds were more neophobic than their less urban counterparts; the novel object made the urban birds reluctant to approach the dish full of seeds.
48. Ducatez et al. (2017).
49. Mueller et al. (2013).
50. Snell-Rood and Wick (2013).
51. Brady and Richardson (2017, p. 91).
52. Brown and Bomberger-Brown (2013).
53. Atkins et al. (1982).
54. Kimmig et al. (2020).
55. Landscape genetics is a relatively new field—dating to the turn of this century—that uses patterns of genetic and genomic information about individuals at different locations to make inferences about how these organisms (usually animals) move across landscapes. A

merger of landscape ecology and population genetics, landscape genetics is being increasingly applied to management and conservation biology activities.

56. Laundre' et al. (2010).
57. E-mail interview with Sophia Kimmig. May 2020. Kimmig also provided some information about the changing landscape of Berlin and how that may affect foxes and other urban animals. Berlin as a city has undergone substantial remodeling over the last 70 years. From the end of World War II to the late 1980s, Berlin was a divided city, with West Berlin enclosed within East Germany. During this time, its city structure was quite heterogenous, with many wasteland areas scattered around both halves of the divided city. The reunification of Germany in 1990 also united the city. However, in the immediate years after reunification, many businesses (especially in the former East Berlin) failed. There was a bright side to these economic calamities; the shuttering of businesses led to many "greenish" spaces that were substantially isolated from human contact. More recently, however, these greenish spaces are disappearing as the economy booms and the housing market tightens.
58. Byrne and Nichols (1999).
59. Thompson et al. (2018).
60. Teyssier et al. (2018).
61. Teyssier et al. (2020).
62. Littleford-Colquhoun et al. (2019).
63. Background information on rats comes from Sullivan (2008). The quote is on p. 2.
64. Puckett et al. (2016).
65. Puckett et al. (2020).
66. Byers et al. (2019).
67. Combs et al. (2019).
68. Combs et al. (2018).
69. Richardson et al. (2017).
70. Hendry (2017).

FURTHER READING

Schilthuizen, M. 2018. *Darwin Comes to Town: How the Urban Jungle Drives Evolution*. Picador.
Szukin, M., J. Munshi-South, and A. Charmantier (eds.). 2020. *Urban Evolutionary Biology*. Oxford University Press.

14 Challenge From Invasive Species

SUMMARY

Some introduced species thrive a little too well in their new habitat, where they become destructive to the ecosystem. These invasive species can threaten the existence of native species. We begin with two case examples: first, cane toads that are destructive in Australia and in Florida, and then large mice that are threatening birds on Gough Island and other islands in the South Atlantic. We then turn to the general features of invasive species and the threats they pose. Can we predict which introduced species are likely to become invasive? Progress in addressing this question has been largely elusive; however, a few common suggestive themes have emerged. For instance, invasive plants tend to have small genomes and rapid growth. Native species can respond to threats from invasive species: for instance, soapberry bugs have evolved larger beaks in response to an invasive weed outcompeting the native plants. We conclude with a discussion of an invasive fungus that has been a major factor in the decline of amphibians.

"Invasions of alien species present ecologists and evolutionary biologists with an interesting paradox: why are exotic organisms, which come from distant locations and have had no opportunity to adapt to the local environment, able to become established and sometimes to displace native species, which have had a long period of history in which to adapt to local conditions?"—Dov Sax and James Brown.[1]

"Ecologists and evolutionary biologists have a love-hate relationship with invasive species. . . . Although we dislike the harm they cause to the economy and environment, we appreciate their attributes as study organisms. They are easy to propagate and often have short generation times and small genomes (at least in plants)."—Dan Bock and colleagues.[2]

THE CANE TOAD INVASION

Cane toads (*Rhinella marina*),[3] also known as marine toads, are hard to miss. First, they are toads of unusual size: they are perhaps the largest toad and often weigh considerably more than a kilogram (2 pounds).[4] Their broad heads have distinctive brow ridges on the side. These toads also have several qualities that make them pests. They are superlative gluttons; while these voracious predators will eat almost anything, they prefer beetles, ants, and other arthropods. The toads also take the Biblical adage "be fruitful and multiply" to heart: females can lay tens of thousands of eggs. These toads also have a powerful anti-predator defense, with nearly all life stages of the toad being toxic to potential predators. Adults have poison glands by their shoulders that secrete toxins that cause muscle spasms, rapid heartbeat, and

DOI: 10.1201/9780429503962-17

possibly paralysis. Ingesting an adult toad could faces can be deadly to mammals (including humans), birds, crocodiles, and any other animal naïve or foolish enough to try.

Native to Central and tropical South America, cane toads have been introduced to several areas—mostly tropical—across the globe. The toads' prodigious appetite for insects led many to consider them as attractive agents to control insect pests such as the cane beetle, which can be destructive to sugarcane crops. The problem is that the toads' fecundity, combined with their toxicity and voracious appetite, can lead them to become pests themselves. Most notable among the places where the toads have become pests is Australia. Introduced in northeastern Australia in 1935 to control the cane beetle (also known as the cane grub there), the toad has spread rapidly westward and covers more than half of the northern section of the country.

Although there are ample examples of native vertebrates dying because they ingested toads, documenting whether such deaths are affecting the abundance and distribution of these animals is more challenging. It requires having good records of population numbers before and after toad invasions. It also requires ruling out that the declines were due to other factors. One toad-associated decline for which we have strong documentation is the steep drop in numbers of the freshwater crocodile (*Crocodylus johnstoni*) in the Daly river of northern Australia.[5] Toads had established populations in the river by 1999, and there had been sightings of them a year or so before then. Between 1997 and 2013, the population of the freshwater crocodile declined by about 70 percent across the whole river, with some areas seeing even larger drops. Crocodiles of intermediate size were especially hard hit. Because these crocodiles are top predators that feed on a variety of organisms from insects and crustaceans to turtles to mammals, their decline is likely to have ripple effects through the Daly river ecosystem. The most recent assessment of this species by the IUCN (2017) still lists the crocodile as being of "least concern".[6] But this assessment is based on the species as a whole; the IUCN assessment warns that subpopulations could be at risk.

During the early 2000s, the western expansion of the toads in Australia accelerated to about 55 kilometers (35 miles) per year.[7] The rapid and accelerating expansion is likely due to evolution of the toads. Cane toads near the expansion front move more frequently and move further than those elsewhere. Long-term tracking data also shows increasing mobility with time.[8] The toads have also evolved morphologically; they are larger and have disproportionately larger limbs. Bigger toads with longer limbs can move further and thus spread faster. This adaptation, however, comes with a cost: the larger toads at the invasion front are prone to arthritis of the spine.[9] In addition, the toads at the invasion front also have larger poison glands than those in long-established areas.[10] Moreover, the toads at the front show different immune responses and may also be more prone to infections.

Cane toads at the southern edge of the range face colder temperatures as they move further away from the equator as well as uphill. The cold temperature may explain why the southern front is moving much slower than other parts of the expansion front. Study finds that toads near the southern edge are acclimated to cold temperature.[11] The tolerance to cold temperature appears to be just acclimation—and not genetic adaptation—as toads from different elevations that are raised in the same laboratory conditions respond similarly. The lack of a genetic difference could be due to the toads from different elevations being part of the same population. More study is needed here: perhaps toads away from the front are less able to acclimate to the cold. If so, that would be evidence that the toads have evolved the ability to acclimate as they moved south. Alternatively, the toads away from the front could also be adept at acclimation. This would suggest that toads were preadapted to cold temperature.

Australia is not the only place where the toads are abundant and spreading. Writing for *Outside* magazine, Ian Frazier described the surfeit of toads in south Florida:

Step out of your house in the morning and cane toads are squatting on the front walk. They are in the garage in the coils of garden hose. They climb up the screen on your lakeside cabin's front to get closer to the outdoor light. They are in your window wells and by the hot tub on the patio and next to the swimming pool filter motor. They sit and look at you as if you owe them money; a creepy shiver excites your shoulder blades.[12]

In Florida, cane toads are outcompeting the native southern toad, probably because they lay tenfold more eggs. They are also a nuisance because pet dogs in the area are being poisoned. These poisonings are especially worrisome for smaller dogs. Larger dogs, like larger crocodiles in Australia, seem to be at less risk.

Cane toad (*Rhinella marina*).

Source: Courtesy of Cinnamon Mittan.

In contrast to the case in Australia, there is good evidence that cane toads in Florida have evolved cold tolerance during their invasion. As with Australia, cane toads were introduced intentionally for beetle control in Okeechobee, Florida, during the 1930s. These introductions failed, presumably because the weather was too cold. Then, in 1955, cane toads accidentally escaped from a pet trade shipment and established a population in Miami, about 100 miles (160 kilometers) south of Okeechobee. In the ensuing decades, the toads moved further north, well beyond Okeechobee.[13]

Cinnamon Mittan, then a PhD student at Cornell University, had been interested in cane toads in Florida as a possible example of rapid evolution. She had initially planned to study the evolution of pesticide resistance but was inspired to study cold adaptation after finding an early article that suggested that the toads would never spread beyond south Florida because the planned introductions in Central Florida had failed.[14] Mittan collected toads from Tampa, near the northern edge of the toads' range, and from Miami. Freezes occur in most winters in Tampa but are rare in Miami. After the toads were kept in a common environment, their thermal tolerance was assessed. Toads from Tampa were able to stay active at notably lower temperatures than those from Miami, showing evidence for thermal adaptation. Mittan also found evidence for acclimation in addition to the adaptation.

Shane Knuth, a politician in Queensland, Australia, was concerned about the cane toad killing the native wildlife and taking over the habitat.[15] He had promoted paying people to collect toads. Although his proposal of a standing bounty was rejected, Knuth persuaded his government to hold what would be called "the Toad Day out" in March 2009. On this day, the residents were instructed to locate toads, bag them, and then bring them—unharmed—where they would be weighed and then disposed of (either by freezing or gassing). Prizes were given out for the largest toads and the greatest overall total weight.

Events like Knuth's Toad day are not unusual. Most of the measures to control toads are low tech and usually involve humans reducing their population size.[16] The toads are readily caught by hand and can be trapped. Treating water where toads are found with chemicals to suppress larval development is another tried-and-true method. Such methods are most useful when the population is not too large and when migration from other areas can be controlled. Manual removal is most effective at getting rid of toads from smaller, satellite populations as well as those in restricted areas like islands. One plausible surveillance method to track areas that could be prone to toad invasions would be to periodically check such areas for the presence of toad DNA.

Cane toads are spreading toward Pilbara, a region in northwestern Australia, at an alarming rate. Much effort is being spent attempting to stop the toads from entering Pilbara because this region is rich in biodiversity. Notably, Pilbara has a large population of an endangered species, the northern quoll (*Dasyurus hallucatus*), a cat-sized shrew whose numbers and range have dwindled due to toad predation. One proposed idea to prevent the toads from entering Pilbara is to construct an 80 kilometer- (50 mile-) wide waterless barrier.

In connection with the barrier, some researchers have proposed a radical idea that directly applies evolutionary genetics: a genetic backburn.[17] Much like firefighters sometimes will light backfires to starve the main fire, this proposal would involve bringing more toads to the front. Recall that toads at the front move much more quickly than those from established areas. Bringing toads from those established areas to the front would be expected to slow the spread. Computer simulations suggest that this could work, especially if there is a strong tradeoff between dispersal and fitness, as some evidence suggests.

GOUGH ISLAND AND ITS RODENTS OF UNUSUAL SIZE

Suppose you were to fly from Johannesburg, South Africa, to Buenos Aries, Argentina. About a third of the way during your flight, you might would fly near the Tristan da Cunha, an archipelago of small volcanic islands. One island in this archipelago is Gough Island, the focus of our attention. At 91 square kilometers (35 square miles), it is about one and a half the size of Manhattan island. Unlike Manhattan, this island is mountainous, with steep cliffs and a rolling plateau. Even at sea level, it is chilly, with blustery winds. Further uphill, it can snow during the winter months. Except for a small weather station and the occasional researcher, this island is not inhabited by humans.

The island is home, however, to multitudes of birds, several of which are endemic. One of these is the Tristan albatross, *Diomedea dabbenena*. Although this albatross will forage across much of the Southern hemisphere from South America to southern Africa to the southwest coast of Australia, nearly all the breeding population is found on Gough Island. It primarily eats fish and cephalopods, though it likely also feeds off food waste dumped by ships. This albatross is currently listed as critically endangered according to the ICUN's Red Book.[18] Its main threat is predation from a new arrival to Gough Island: giant mice.

The mice on Gough Island weigh about 30–35 grams (just over an ounce), as opposed those from the mainland, which weigh slightly less than 20 grams. These mice are not just big but are also aggressive and persistent predators. Often working in small packs, they can take down hatchling birds that are hundreds of times their size. These mice probably first established on Gough Island from seal hunters that stayed on the island during the winter of 1810–11. In 1888, the seal hunter George Comer recorded in his diary: "we are troubled considerably with mice, as there are so many here".[19] The mice mostly eat plants during the summer. During the winter, when the plants are unavailable, the mice prey on the chicks. The predatory behavior and the abundance of seabirds allows the Gough mice to escape the ordinary ecological constraints limiting body size.[20]

Mice are thick on the ground on Gough Island. Based on data from the numbers of mice trapped in areas of dense cover, researchers estimated the density to 224 mice per hectare or about 90 per acre.[21] The researchers who trapped the mice suggest that this figure may even be an underestimate of the true density because they had insufficient traps. Assuming that the estimate is correct and is representative of the island as a whole, a simple extrapolation would suggest that about 2 million mice would live on the island. A likely reason why the mice have reached such large numbers is that they are very fecund: of the female mice that were found pregnant, they carried an average of nine fetuses.

The Tristan albatross had lived on other islands in the archipelago, but human activity had essentially wiped them off those islands in the middle of the 20th century. Although some albatross adults are killed or injured by longline fishing, the greatest source of concern to the albatross is from mice harassing, injuring, and kill-ing its young. We have firm evidence that due to the mice that the albatross has been having low breeding success since at least the late 1970s.[22] Circumstantial evidence suggest that the mice have been significant predators for much longer, perhaps more than a century. The albatross has an unusual life history: it lives a long time and pro-duces few young at any given time, and these young take many years to reach sexual maturity. Given these features, the albatross population is buffered from the mice preying on its young for some time. But at some point, the low fecundity and long maturation work against the prospects for the albatross.

The Tristan albatross is not the only bird that the mice are preying on. In fact, most of the 22 seabird species on Gough Island appear threatened by the mice. A recent study examined the breeding success of some of the seabirds on Gough Island and compared these results with the breeding successes of analogues, similar birds that breed on islands without mammalian predators.[23] In total, the results suggest that if there were no mice on Gough Island, the birds could raise an additional 2 million chicks each year, about twice the total of the actual total production of about a mil-lion chicks. Note that the missing 2 million chicks is not entirely from direct preda-tion. Some of this reduction may be due to indirect effects such as harassment from the mice. Petrel species that burrow and breed during the winter were hit especially hard. Species that breed during the winter are more likely to be targets because the mice have fewer other sources of food then.

Where are the Gough Island mice from? Melissa Gray, Bret Payseur, and other University of Wisconsin researchers have used genetic evidence to determine that the mice are from the Western Europe *domesticus* subspecies of *Mus musculus*.[24] The genetic data, however, were unable pinpoint the mice to a specific country of origin. One plausible (though yet unproven) explanation for the failure to find a specific source population is that the source population was an admixture. No population structuring was found within Gough Island; the mice there are one population. The DNA data are consistent with a colonization time of a couple centuries ago.[25]

The University of Wisconsin researchers also investigated the genetic basis of the growth patterns of the Gough mice. After establishing a breeding colony of Gough Island mice, they looked at the growth trajectories of these mice as compared with the smaller mainland mice. The giant Gough Island mice retain their large size in the laboratory environment, demonstrating that the differences between these mice and their smaller counterparts are genetic.[26] These mice are larger primarily because they grow much more quickly than the mainland mice in the first few weeks after birth. Crosses between the Gough Island and the mainland mice reveal that the genetic variants responsible for larger size are mostly dominant: the F1 mice have a growth trajectory that closely resembles the Gough Island parent, especially for the first three weeks, although it falls off somewhat as the mice continue to grow.

These researchers then conducted a quantitative trait locus (QTL) analysis (see Chapter 5) by obtaining F2 mice and then looking for associations between DNA markers and weight. Their results were contrary to the expectation that the dif-ferences may be due to a handful of genes. Indeed, many genetic regions, each of small effect, contribute to the increased size of the Gough mice. In the cases where a QTL was statistically significant, having the Gough Island genetic variant always increased weight. That all the detected QTL were in the same direction suggests that natural selection and not just random genetic drift drove the evolution of larger size. Some of the QTL found in the Gough mice correspond with those found in lines of mice artificially selected for larger size, but others do not. The QTL in common may be due to common pathways. The ones that differ, however, suggest that the evolution for larger size in Gough Island acted on different genes than the selection in the lab.

Payseur's lab followed up on this QTL study by examining differences in the patterns of gene expression between the Gough Island and the mainland mice.[27] A strikingly large proportion of genes—between a fifth and a half, depending on the tissue examined—differed in expression, and they often did so by a large extent. Many of these differences likely are not due to changes in the genes themselves but rather genes that regulate them. Moreover, many of the expression differences may not have anything to do with the differences in size of the mice. In a more targeted search for genes related to the size differences, Payseur's lab examined the genes with different expression that were within the QTLs. They found 180 such genes; of these 180, 113 differed in DNA sequence between the Gough Island and mainland mice. It is these 113 candidate genes that are most likely to be involved in the size differences.

One of these genes is *LRP1*, the low-density lipoprotein receptor-related protein. This gene encodes a receptor on cell membranes that plays numerous roles, including the processing lipoproteins, the proteins that transport fat molecules. It also is involved in the breakdown of protease, the enzyme that chops up other proteins; hence, it plays a role in protein metabolism. Moreover, the product of *LRP1* functions in insulin signaling, cell growth, cell migration, and programmed cell death, as well as a host of other cellular functions. Given its multifunctionality and given that many of these functions are associated with growth, it's not surprising to see *LRP1* in the list of candidate genes. Another candidate gene is *ARID5B*, AT-rich interactive domain-containing protein 5B, a gene associated with susceptibility to obesity and differences in the types of fat deposited. Going through testing to see which genes on the list of candidates involved in the size of differences will be a challenge, given their number and their individual small effect.

Payseur's lab also looked at the genes that show signatures of being selected during the occupation and invasion of Gough Island.[28] The most striking pattern was the abundance of genes that affect neurological functions showing the signatures of selection. One is *Ston2*, a gene that encodes a protein that assists in the recognition of proteins across the synapses in neurons. Knocking out this gene in mice leads to bolder, more impulsive mice. In humans, variants of this gene are associated with schizophrenia risk.[29] In laboratory settings, mice with the genetic makeup of those from Gough Island mice are more prone to explore their surroundings compared with the mainland mice.[30] They will spend more time in the center of an area rather than hid in the corners. These mice also are more tolerant of the light than mainland mice. Payseur's lab is currently looking at the genetic basis of this behavioral difference. The Gough Island mice being more exploratory is consistent with the fact that selection has operated on neurological genes.

Curiously, there was an absence of genes affecting body size in the screen for selection. Payseur's group suggests that the dearth of such genes may be because the selection for larger size may have acted on variation that existed before the selection (standing variation) rather than new mutations. The action of selection acting on standing variation is more challenging to detect than selection on new variants through the usual scans of selection. If the founding population of Gough Island were an admixture, as we had discussed earlier, then it is more likely that it would contain the standing variation for size.[31]

The larger size of mice on Gough Island is consistent with the island rule, the trend that rodents and other small mammals get larger on islands. What about the predatory behavior? Until recently, this behavior was thought to be restricted to Gough Island. In the last few years, reports of mice scalping albatross hatchlings on Marion Island suggest the phenomenon may be more common.[32] Marion Island, which is about 2,000 kilometers (1,200 miles) southeast of South Africa, has about 22 percent of the breeding population of the wandering albatross (*Diomedea exulans*) and lesser numbers of other albatross species. As with Gough Island, mice likely came to Marion Island during the 19th century. Cats were deliberately introduced to the island in order to control the mice. Unfortunately, the cats went feral and became pests themselves, preying on native bird species, especially the burrowing petrel. The extermination of cats in 1991 left mice as the only introduced mammal on the island. Twelve years later, the first recording of mice attacking birds was recorded. Over the years, the attacks have become more frequent and more severe. By 2015, albatross young were found with fresh nape wounds. Videotape recordings show prolonged and vicious attacks by mice on the chicks.[33] Moreover, numerous chick carcasses show the mice are a significant cause of mortality in the young albatrosses.

The predatory mice on Marion Island differ in two fundamental ways from their counterparts on Gough Island. In contrast to the much larger Gough mice, the Marion mice are no bigger than typical mice. The mice also differ in where they strike the birds: Gough mice primarily attack the wings of their prey, while the Marion mice mainly go for the heads and necks.

THE THREAT FROM INVASIVE SPECIES

The toads in Australia and Florida and the predatory mice of Gough Island are excellent examples of invasive species. Whether they were introduced accidentally or intentionally, they established a foothold and spread. Both the toads and the mice are fecund. They are not just multiplying quickly but also are also threatening native species.

Invasive species pose numerous threats.[34] In addition to killing native species through predation (as in the case of the Gough Island mice) and toxicity (as in the cane toads), they can outcompete native species, as many invasive grasses have done. These activities can reduce the abundance and genetic diversity of native species. In doing so, invasive species raise the risk that native species will go extinct, either locally or globally. But their mode of destruction is not limited to simple one-on-one species interactions. Invasive species can modify the ecosystem and its functioning. Through these modifications, they create conditions that make them more likely to thrive in the environment and hurt the ability of native species to do so. They can alter nutrient cycles and microbial communities.

Various human activities promote the introduction and spread of invasive species. Transport can occur by many means that are often dependent on biological properties of the organism. Birds, bats, and many insects can reach new locations actively through flight. That is why birds are often found on remote oceanic islands, but mammals are not, unless they are brought by humans. Many mammals can swim, enabling them to reach islands that are not as remote. Plants can be transported through their pollen; although pollination is often also by the wind, animals (particularly insects and some birds), or by their seeds. Seeds, especially if they are encased in appealing fruits, could be transported long distances by mammals. Small fish and marine invertebrates can be moved great distances by ships—through several mechanisms including exchange of ballast water or by attaching and dislodging to the sea chests, recesses at the hulls of ships. In addition to the threat they pose on native species, invasive species also can hurt human livelihoods and well-being. As we have seen in prior chapters, invasive species can threaten agricultural crops, such as the late blight threatening potatoes and coffee rust threatening coffee plants. Invasive species can also cause and/or transmit disease. In this chapter, we will focus on the threat to native species.

Among researchers, there is considerable debate about what constitutes an invasive species. In 1999, President Clinton signed Executive Order 13112 on "invasive species", defining them as "alien species whose introduction does or is likely to cause economic or environmental harm or harm to human health".[35] Debate often revolves around the focus on non-native species. Recent studies, such as those by Dan Simberloff, conclude that native species seldom cause harm, and in the relatively rare circumstances when they do, it is usually because of disturbance.[36] Examination of the causes of known extinctions supports this contention and illustrates that invasions from non-native species are a major contributor to extinction. Invasive non-native species were listed as one of the drivers of the extinctions for a third of the animal species and a quarter of the plant species.[37] Native species were listed as a driver much less frequently, accounting for slightly less than 3 percent of animal species and about 5 percent of plant extinction. When native species were listed as a driver, there were always other drivers associated with them; in no case did native species act alone to cause extinction. Non-native invasive species were listed as the cause of all the extinctions of spiders and other arachnids as well as for most of the cases of birds and fern extinctions.

Another point of contention is whether we should include impact as part of the definition of invasive or simply consider an invasive species to be one that has become extraordinarily successful quickly after an introduction.[38] Many researchers follow the former definition, considering invasiveness to require a significant negative impact. Under this definition, which we will follow for the rest of the chapter, unless specifically indicated, whether a species is invasive can depend on where they are. A case in point is earthworms. These are abundant and are not native to much of the United States. Gardeners usually welcome these worms, as their churning of the soil allows for a better environment for most plants grown in American gardens. Here, they would not be invasive. But earthworms can be harmful to some forests. Earthworms accelerate decomposition, which can cause soil to be less rich in nutrients. These changes in decomposition also alter microbial communities, which can have wide-reaching effects on the ecosystem. These changes are especially concerning for the boreal forests of Canada. The leaf litter in the boreal forests in Alberta is typically about six inches deep; with earthworms, it can be less than an inch.[39]

Earthworms in the forest would be invasive. Invasiveness is thus context-dependent and not a fundamental property of a species.

HERBERT BAKER AND THE IDEAL WEED

Along the coast of California abutting Monterrey Bay sits Asilomar Conference Center, a resplendent and rustic suite of buildings. Designed by Julia Morgan,[40] who also designed nearby Hearst Castle, Asilomar's buildings were set to follow the natural contours of the hilly landscape. Asilomar has been the setting for some of the most important scientific conferences, including ones discussing the ethical ramifications of recombinant DNA during the middle 1970s[41] and similar ethical concerns about CRISPR genome editing during the 2010s.[42]

Asilomar was also the site for a conference in 1964 about colonizing—what would be known today as invasive—species. The meeting brought together geneticists, ecologists, and taxonomists. Included among this group were Dick Lewontin and Ed Wilson, two prominent evolutionary biologists who were rising stars in their respective fields at the time. But the diversity of expertise was still broader; in addition to the academic scientists, there were wildlife biologists, researchers working with weed control researchers, and other applied scientists. They came together to discuss how species might evolve to colonize and perhaps become invasive when introduced to new areas.[43]

The leaders of this meeting were the British evolutionary ecologist Herbert Baker and the American evolutionary geneticist George Ledyard Stebbins, both of whom were botanists. Baker and Stebbins worked well together as a team due to their common interests and complementary expertise and personalities. Then in his mid-40s, Baker had been thinking about invasion biology for 20 years. Contrary to most biologists of his time, Baker saw ecology and evolutionary biology as disciplines that were entangled: evolutionary thinking was essential for understanding ecological processes, and the ecological milieu was fundamental to understanding how evolution would proceed. Stebbins, one of the architects of the modern synthesis that brought together Mendelian genetics and Darwinian evolutionary biology, excelled at synthesizing ideas and information about the evolutionary genetics of plants. Stebbins, at nearly 60 at the time of conference, was a fountain of tireless energy. Though Stebbins could be mercurial, the more diplomatic Baker could help settle any feathers Stebbins had ruffled. Baker and Stebbins were able to produce a volume with 27 contributions from the Asilomar: *The Genetics of Colonizing Species*.[44]

At the conference and in his contribution to the volume, Baker had focused on the features that make a plant likely to become a weed. He suggested characteristics—14 in all—that weeds tended to share. Among the most prominent, weeds have short life spans and flower quickly. Weeds tend to have more phenotypic plasticity than nonweeds: being better able to respond to different environments would allow prospective weeds to take advantage of the new environments they found themselves in after introductions. Debate over how well weeds fit these characteristics continues to the present day, but Baker provided a useful framework in establishing the question and putting forth some plausible answers.

In the ensuing years, a new possible feature of weeds has been proposed: genome size. As noted in the epigraph by Bock and colleagues,[45] invasive species and common laboratory model organisms not only share the ability to grow rapidly but also have small genomes. In fact, one of the main model plant species, the thale cress (*Arabidopsis thaliana*), is a weed that has a small genome (only about 135 million base pairs). A small genome is a useful feature for genomic and evolutionary genomic research. Is it also a trait that makes a plant more likely to become a weed? If genome size were a good predictor of whether an introduced plant would become an invasive weed, such information would be useful in surveillance programs: special attention could be paid to species that have small genomes.

Why should we expect invasive species to have small genomes? First, consider that even closely related species can have noticeably different genome sizes. The differences in genome size are usually due to large genomes being overloaded with transposable elements and other repetitive DNA that has no function other than its own replication. These larger genomes laden with so-called junk DNA can be detrimental. The extra DNA and machinery associated with it take up space; as a result, species with larger genomes tend to have larger cells. Due to this larger cell size and the processing needed to replicate DNA, large-genome organisms generally take longer to go through cell division. This slower cellular division may impede their ability to grow and develop quickly. The large cell size that comes along with having a large genome also tends to make seeds larger. Because small seeds are easier to produce than large ones, the reproductive output of species with bloated genomes may be limited.[46]

Genome size appears predictive of invasive potential, at least at first glance. Animals and plants with small genome size may be more prone to be invasive. Certainly, ones with large genomes are unlikely to do so. But the case is not yet airtight. Most studies examining the association between genome size and invasive potential have looked at correlations using different species as data points. The problem with this is that other factors not considered in the analyses may be at play, obscuring the causal link. A better way to approach this relationship would be to look at comparisons of populations within species. As Bock and his colleagues noted, such studies are rare.[47]

I followed up with Dan Bock in October 2020 about the current status about this link between genome size and invasiveness. Bock, an evolutionary ecologist now at Washington University in St. Louis who works on invasive species, told me that "the jury is still out. We need more experimental studies, as evidence available so far is equivocal".[48] Bock pointed to an intriguing study on common reed, *Phragmites australis*. This grass is commonly found in wetlands and can grow up to a couple meters (about 6 feet) in height. The European subspecies was introduced to North America, where it has become invasive and outcompetes the native subspecies. The invasive European subspecies has a smaller genome than the native North American subspecies, consistent with the prediction.[49] Yet in a competition experiment held in a common garden setting, clones with smaller genomes do not have superior competitive ability. The authors of this study urged prudence in the interpretation of this study: "It needs to be kept in mind that the results we report in this paper are based on a common-garden experiment conducted in a single temperate garden, which somewhat limits generalization of our findings, as with many ecological experiments."[50] Clearly, much more needs to be done.

Are there properties of animals that may make them prone to be invasive? Carol Lee, whose copepod studies we had discussed in Chapter 12 with respect to adaptation to climate change in the oceans, also works on biological invasions and has a possible answer to this question. The copepods she studies, *Eurytemora affinis*, are native to brackish and marine ecosystems but have made several invasions into freshwater systems in the Atlantic and Gulf of Mexico. In a perspective paper from 2008, Lee hypothesized that environmental disturbance would select for traits that make species more suited to be able to thrive in new environments and potentially become invasive.[51] If disturbance is relatively frequent and sufficiently pronounced, selection would favor rapid population growth and the ability to tolerate a broader range of environmental conditions. Species with such traits would be more prone to quickly ramp up to large numbers upon introduction to a new locale. Frequent disturbance also can lead to fluctuating selection, wherein selection acts in different ways at different times or in different locals. For instance, suppose that most of the time, the copepods live in relatively salty water, but occasionally there is a disturbance wherein a burst of freshwater floods the estuary. In this situation, selection could normally favor genetic variants that do well in salt water, but during the disturbance, selection would favor variants that do well in fresh water. Such fluctuating selection would promote the persistence of a diversity of genetic variants. Selection pressures fluctuating over time is not the only way that genetic variation can be maintained by selection; selection favoring rare variants or selection that favors heterozygotes also will keep genetic diversity in the population. Any form of selection that maintains genetic diversity is known by the umbrella term of balancing selection.

Quantifying the extent of disturbance in an area is challenging, if not impossible, in practice. This challenge limits the ability to test the broader hypothesis that species that become invasive when introduced are more likely to have had experienced disturbance in their native range. A related, narrower hypothesis—that invasive species experienced more balancing selection in their native range—is more amenable to testing. Balancing selection leaves genetic signatures that can be detected in population genetic tests.

Recently, Lee's group found numerous genetic variants in the copepods that showed signatures of balancing selection within their native habitat, consistent with the hypothesis.[52] This balancing selection, whatever form it took, permitted the species to harbor genetic variation that could be used when it found itself in a new area. Were this the case, we would expect directional selection—the type of selection that makes advantageous variants more common—to be operating on some of those variants in the new habitat. In fact, that is just what Lee's group found: many of the genetic variants that showed signs of being under balancing selection in the native habitat also showed signatures of directional selection in the new freshwater habitat. Moreover, some of the same variants showing the signatures directional selection are shared across different invasions. That these signatures are occurring in parallel is further evidence of the generality of the results.

Several factors predispose this system to variation being maintained by balancing selection. There are tradeoffs between functioning in fresh water versus in salt water. The species also has overlapping generations—and an egg bank. The parents who produced the eggs in the egg bank likely experienced different selection pressures than the young that emerged from the eggs. These factors all help to maintain variation that selection can act on.

Ion transporter genes predominate the types of genes that show signs of selection. This makes sense, given that a major environmental fluctuation involves changes in salinity. One gene in particular—NHA—the sodium-hydrogen antiporter—is a membrane protein that brings sodium ions into the cell and kicks of hydrogen ions. In addition to ion transport genes, genes that play roles in DNA regulation, stress response, and energy production also were abundant in the lists of genes showing signs of selection.

Intriguingly, lines of copepods taken from marine and freshwater areas also differed in their expression of many of the ion transporter genes when raised in the same environment.[53] Moreover, raising copepods in fresh water alters their expression of these genes in the same direction of the evolutionary shift that copepods took when they invaded freshwater areas. Thus, acclimation and evolutionary adaptation are working in the same direction. Having the ability to acclimate likely helps the copepods coming from marine environments be able to survive the freshwater environments during the time before the population can evolve genetic differences that further their ability to survive the freshwater environment.

BECOMING INVASIVE

Species do not automatically become invasive. In fact, becoming invasive is a process that can take considerable time and contains several stages. Along the way, evolutionary changes can occur that either enhance or hinder the progress of the prospective invasive organism.

Evolutionary changes could make introductions more or less likely. As we saw with the cane toads, evolution of larger legs and an enhanced disposition to move is making the toads better able at spreading into new areas. Plants could also evolve greater dispersal. For instance, changes in fruit or seed or pollen could enhance the dispersal capacity of a plant.

After reaching an area, the potential invasive organism must be able to thrive under the area's environmental conditions. Acclimation, as is likely with the copepods invading freshwater systems, may allow the potentially invasive species time to evolve genetic adaptations.

As we saw with tracking evolutionary responses to living in urban environments, resurrection studies can be used to study how invasive species evolved such adaptations. Tufted knotweed (*Persicaria cespitosa*) is an annual plant that was introduced from southeast Asia into the northeastern United States. In its native range, the knotweed grows along stream banks and forest paths. For most of its existence in the United States, it also only grew in such habitats. During the 1990s, the weed spread into open areas with intense light and has become invasive there.

A common garden study examined three populations of this species taken from New England, with samples taken both in 1994 (before the spread into open sites) and in 2005 (after the spread).[54] The plants were grown under both full sun and shaded conditions. Based on the comparison of the before and after plants, the knotweed evolved increased reproductive output between 1994 and 2005. This increase in reproductive output was most pronounced in the high light environment, consistent with the knotweed being better able to thrive in the open areas that are exposed to the intense rays of the sun. No signs of a tradeoff were apparent: the 2005 plants did at least as well under shaded conditions as the 1994 ones. Under high light, the 2005 plants invest more in roots than in leaves, allowing them to maintain a supply of water. This suggests that the plants had evolved a plastic response allowing them to better tolerate the full sun. Unlike with the copepods mentioned earlier, this plastic response of the plants was not present in the source population; it evolved along the way.

Another case of adaptation by an invasive species is the Pyrenean rocket (*Sisymbrium austriacum*).[55] This small herb, known for its vibrant yellow flowers, has a native range in the Pyrenees and the surrounding hilly regions of northern Spain and southern France. In 1824, it was found in much lower ground in Belgium, 1,200 kilometers (750 miles) north of its original location. The most likely culprit for the introduction is that it was accidentally transported with a shipment of wool. Over the years, it has spread further north, mainly along the Meuse River, and is now in the Netherlands. Several of the genes that show large differences between the native range of the herb and samples from further north play a role in determining when flowers bloom. One such gene, *GIGANTEA*, sets the flowering time in plants that live at high latitudes where the daylength is long during the summer months. Plants that had evolved under a less-extreme light cycle at the border of Spain and France needed to adapt to a different regime further north in Belgium and the Netherlands. A time series from museum specimens confirms these findings: there are steady changes in frequencies in these flowering time genes with the passage of time. Notably, these changes in the flowering time genes and other genes that showed signs of adaptive change occurred before the rocket became invasive. These changes likely helped enable the herb to persist in the environment before it was able to become invasive.

Examining the process of how potentially invasive species become invasive helps resolve a paradox that Jason Fridley and Dov Sax proposed in an epigraph of this chapter: why are exotic organisms able to displace native species, even though they did not evolve there?[56] Well, exotic organisms do not always take over when introduced to a new area. In fact, most introductions fail. Even those species that ultimately have had successful introductions usually had multiple failures before they succeeded. Recall that the deliberate introduction of cane toads in Florida failed before they succeeded later. Part of the reason that most introductions fail goes back to the demographic and environmental stochasticity that we discussed in the first chapter on conservation genetics. Populations with small numbers—and most introductions are ones with small numbers—are in peril until they can build up their numbers.

But why can non-native species do better than the native species that evolved in the area? One factor is one that we discussed earlier with the copepods that Carol Lee studies: invasive species often evolve in disturbed areas. These species presumably have genetic variation that enables them to have a reservoir on which to draw on when introduced in new areas.

Another hypothesis, not mutually exclusive with the disturbance hypothesis, for why organisms become invasive after introduction is that they escape their natural enemies. White campion, *Silene latifolia*, is an exemplar of this. In the northern United

States, it is a weed that grows in disturbed habitats such as along roads and around the edges of agricultural plots. In Europe, in contrast, white campion is sold by seed catalog companies; there its fragrant flowers fetch customer attention. The difference between weed and valued commodity arises because white campion is held in check in Europe, where it faces greater challenges from multiple natural enemies. For one, aphids feed on the plants' phloem, the part of the stem that transports the sugars and other compounds the leaves generated during photosynthesis. In addition, herbivores eat the flowers. The plants also contend with fungal pathogens. Release from these natural enemies can allow the plant to dominate where it cannot in its native range.

Lorne Wolfe at Georgia Southern University quantified this extent of damage from the natural enemies of the white campion in the two continents. After sampling dozens of populations from both Europe and the United States, looking for the extent of damage from these attacks, Wolfe found striking differences: more than 80 percent of the European populations had damage from at least one of the enemy classes, while only 40 percent of the North American populations did. Moreover, European plants were 17 times more likely to be damaged than the North American ones.[57]

That natural enemies can check invasive species raises the prospect of using biological control to thwart them. As we saw in the chapter on managing agricultural pests and in the case of the cane toads here, biological control itself is not without risk. If biological control is used, the control agents should be from the same area as the invasive species, as populations from each are more likely to have adapted to each other. Knowledge of the origins of the invasive populations is thus useful to biological control measures.[58]

A final consideration in assessing whether an introduced species is likely to become invasive is the composition of the native community that the introduced species finds itself with. Areas that have a diverse native community are expected to be better able to withstand invasions because the diversity of the community limits the niche space that the invaders can occupy. This idea, which is known as the biotic resistance hypothesis, has become increasingly better supported, thanks in part to the collection and analysis of large data sets. Looking at nearly 25,000 field surveys from the National Park Service, my UMass colleagues Eve Beaury and Bethany Bradley found a strong negative between the plant species richness in the native communities and the probability that invading species occupied.[59] This relationship held across a wide array of different types of communities, including eastern temperate forests, great plains, and deserts. Furthermore, Fridley and Sax found that all other things being equal, species that come from regions where phylogenetic diversity is high are more likely to become invasive than those that arise from areas where phylogenetic diversity is low.[60] Given that the biotic resistance hypothesis is true, we might expect a dangerous positive feedback loop: biological invasions can lead to declines in the diversity of native species, which in turn leads to more invasions.

NATIVE SPECIES' RESPONSES TO INVASIVE SPECIES

I must admit a fondness for scientific papers with catchy titles. One such paper by Scott Carroll and his colleagues begins with a pun of an old aphorism, "And the beak shall inherit".[61] The beak here is that of soapberry bugs, seed-eating true bugs (order

Hemiptera), where one Australian species (*Leptocoris tagalicus*) has evolved larger beaks in response to the introduction of an invasive host plant, the balloon vine.

During the 1960s, the balloon vine, *Cardiospermum grandiflorum*, spread through Australia. It is damaging to the native plant community because it smothers other plants. The soapberry bug, which has exploited the balloon vine in Australia, traditionally feeds on seeds of the wooly rambutan, a tree whose species' name (*Alectryon tormentosus*) reflects its hairy leaves. The bug uses its beak, a modified mouthpart, for feeding. This beak pierces the fruit walls and then the seed coats of the plant, enabling the bug to acquire seeds. Because the balloon vine has much larger fruits than the native rambutan, possessing larger beaks would be expected to be advantageous for bugs feeding on the invasive balloon vine.

Carroll and his colleagues took two approaches to examine whether the introduction of the vine led to evolution of larger beaks in the soapberry bugs. First, they examined whether bugs living on the vines had larger beaks than those living on the native hosts. The vine dwellers indeed had larger beaks, ones about 5 percent larger than those of the bugs collected on the native hosts. These differences persisted after reciprocal transplantation (raising vine dwellers on the native hosts and the native host bugs on balloon vines). Hence, the difference in beak shape was due to genetic differences and not plasticity or maternal effects. The researchers also took advantage of museum collections to provide further confirmation and to estimate the time of the change of the beak. They found that those bugs collected after 1965 had larger beaks than those before 1965, not long after the vine had invaded.

There is also evidence that some snakes have evolved in response to the introduced cane toads in Australia. If a snake swallows a toad, it is more likely to survive the toxins if the snake is large than if it is small. Thus, we would expect selection to favor larger snakes after toads are introduced; and all other things being equal, the snakes should increase in size, provided body size is heritable. Ben Phillips and Richard Shine from the University of Sydney in Australia examined museum specimens to test this hypothesis. They looked at the red-bellied black snake (*Pseuechis porphyriacus*) and the green tree snake (*Dendrelaphis punctulatus*), two species whose numbers have declined after toads have invaded their habitats. In the time since the toads invaded the habitats of the snakes, the snakes have increased in size,[62] and relative head size has decreased over time. This decline in head size would be consistent with selection for smaller heads, as individuals with relatively smaller heads are less likely to swallow adult toads. Although selection is a likely reason for the changes, it is possible that they arose due to altered developmental trajectories or phenotypic plasticity in the snakes.

Exposure to the toad has not just led to the red-bellied black snake increasing in size; it has also led to the snake becoming more tolerant of the toxin over time.[63] Phillips and Shine collected snakes from populations that have been in contact with toads for different lengths of time. They then fed some of the snakes a non-lethal dose of the toxin and tested the running speed of snakes fed the toxin with snakes that had not had any toxin. The longer the snakes had been in contact with the toads, the less their running speed was affected by the toxin. This is strong evidence that increased tolerance to the toxin had evolved as a response to selection. In addition to this evolved physiological tolerance, snakes had also evolved to avoid eating toads. Knowing which animals are likely to evolve adaptions that promote coexistence with the invasive species will be useful in setting priorities in conservation efforts.

INVASIVE PATHOGENS: THE CASE OF THE AMPHIBIAN-KILLING FUNGUS

Amphibians are at particularly high risk of extinction. Biologists started noticing declines of amphibian species, especially of frogs, during the 1970s and 1980s. Up through the 1990s, these biologists debated the extent of these declines, with some arguing that the declines appeared to be localized. By the turn of the millennium, however, it was obvious that these declines were staggering and widespread. Observing that more than a third of amphibian species are at risk of extinction, David Wake and Vance Vredenburg noted that amphibians "may be the only major group [of animals] currently at risk".[64] Salamanders that live at high elevation in the tropics are at greatest risk, but few groups of amphibians are safe. There is an irony here, as Wake and Vredenberg observed: while amphibians are faring poorly now during this likely sixth major extinction, they persisted quite well during the previous four major extinctions of the 300 million years before humans.

While many factors (some of which are acting in synergy) are responsible for the decline of amphibians, we will focus on one here: invasive pathogens that have been identified as chytrid fungi of the genus *Batrachochytrium*. Chytrid fungi, which reproduce asexually and sexually, are typically found in aquatic habitats. That they are causing widespread disease in amphibians came as a shock because chytrid fungi do not usually cause disease, nor do they usually infect vertebrates. And yet one chytrid, *Batrachochytrium dendrobatidis* (*Bd*), is a major threat to amphibians across much of the world.[65] At least 500 species have seen declines in their numbers, and about 90 have gone extinct. The fungus has particularly hurt those amphibians that are large, are aquatic, and are restricted to a narrow elevation zone.

Bd infection causes the disease chytridiomycosis, which begins by attacking the outer keratin layer of skin. It leads to both thickening of the keratin layer and the premature loss of the layer. It hinders the amphibian's ability to regulate the proper balance of electrolytes like sodium and potassium, which can lead to cardiac arrest. Dehydration and difficulty breathing are also likely upon severe infection.[66]

Phylogenetic analysis places the likely source of the pathogen to be South Korea, most likely during the turn of the 20th century.[67] Multiple pathogenic lineages of *Bd* are found in South Korea. Although the pet trade is a likely source for the spread of the infection, the ways by which these pathogens were transmitted to so many amphibians remain unknown; more phylogenetic analysis would be useful in tracing these routes. Moreover, there are likely many unknown pathogenic fungal lineages in East Asia. Surveillance, combined with phylogenetic analysis of genetic samples, is warranted to prevent future outbreaks.

In addition to *Bd*, a related fungus (*B. salamandrivorans*, known as *Bsal*) is also infecting amphibians. Co-infection by both *Bd* and *Bsal* is particularly troubling: laboratory experiments showed that newts infected by both pathogens had high mortality, much higher than that for those infected with just one.[68] Co-infection disrupts the regulation of gene expression in the skin. Upon co-infection, genes in the complement pathway show reduced expression, and those that produce keratin show increases in expression.[69]

The stark declines of amphibians due in part by the pathogenic fungi do not come in a vacuum. The loss of amphibians can lead to declines in other taxa. For instance, a *Bd* outbreak in 2004 wiped out more than three-quarters of the amphibian abundance and resulted in the loss of more than 30 species in a site in central Panama. Soon after, the diversity of snakes declined. A likely reason for the decline in snake biodiversity is that snakes feed on amphibians.[70]

Some good news is that the skin microbiota appears to help protect at least some species from the pathogenic fungi. For example, the eastern red-backed salamander (*Plethodon cincereus*) shows some resistance to *Bd*. Its skin contains a bacterium, *Janthinobacterium lividum*, which acts as an antifungal against *Bd*, most likely by the secretion of violacein and other compounds. Some of these antifungal bacteria have been administered to other amphibians to act as probiotics. While the bacteria do not always work and the results are species specific, some of these antifungal bacteria have been promising as defense against *Bd* in some amphibians.[71]

In addition to the fungi, amphibians are also threatened by climate change. How the interaction of two looming threats will play out is still a matter of debate. A reasonable argument could be made that climate change could intensify the threat from invasive pathogens. First, the changed climate is likely to expose species to more new pathogens as ranges of both pathogens and hosts shift with the changing climate. Second, climate change also adds an additional stressor to the hosts that are already dealing with invasive pathogens. Third, pathogens would likely evolve faster than their hosts, owing to their large numbers and rapid generational turnover. But there is not much evidence supporting this idea. Moreover, some evidence suggests that parasites and pathogens may be more vulnerable to extinction threats than non-parasites.[72]

Jeremy Cohen and his colleagues have argued that hosts will be especially vulnerable to pathogens when hosts that have adapted to one regime of temperatures are confronted with a substantially altered temperature regime. They posit that the pathogens, due to their faster generation time and larger population sizes, should be able to adapt more quickly to the new regime than the hosts. Hence, as temperatures increase, hosts that have adapted to cooler climates should face a greater challenge from pathogens than those that have adapted to warmer ones. Cohen calls this the thermal mismatch hypothesis.[73]

Some evidence in support of this hypothesis from toads of the genus *Atelopus*. Known as harlequin toads, these toads live in Central and South America and face threats from *Bd*. These toads, which live at moderate elevations (up to 1,000 meters or 3,300 feet) in the wild, prefer cool temperatures. Experiments by Cohen and his colleagues show that the fugus is most deadly at high temperatures. While this is only one species in one situation, this result is consistent with the thermal mismatch hypothesis.[74]

Much like with tracking pathogens that could potentially cause disease in humans, surveillance is important in tracking pathogens that could threaten populations of animals and plants. Such surveillance efforts have been used in tracking the presence and absence of *Bd* in various locales; for instance, a DNA-based surveillance array showed that *Bd* does not appear to be in or around Papua New Guinea.[75] Barcoding has been instrumental in surveillance of plant and animal pathogens.[76] As discussed in Chapter 11, evolutionary principles and information are important in considerations of the validity and applicability of such barcoding.

A necessary attribute of invasive species is that they are numerous. Controlling invasive species thus is in part the flipside of trying to conserve species whose numbers are too small. But having large numbers is not sufficient to being invasive; the fundamental distinction is whether the species is destructive. Invasive animals, plants, and microbes are a threat to their new ecosystem and the species that live in it. Thus, they bear some similarity to metastatic tumor cells. Interactions between evolutionary ecologists who study invasive species and those who study metastatic cancer could be mutually beneficial.[77]

NOTES

1. Sax and Brown (2000, p. 363).
2. Bock et al. (2015, p. 2277).
3. The toad was originally named *Bufo marinus*. Current systematics assigns it to the genus *Rhinella*.
4. General information on the Cane toad is from Zug and Zug (1979) and Shine (2010).
5. Fukuda et al. (2015).
6. Isberg et al. (2017).
7. Phillips et al. (2007).
8. Alford et al. (2009).
9. Brown et al. (2007).
10. Rollins et al. (2015).
11. McCann et al. (2014).
12. Frazier (2017).
13. Mittan and Zamudio (2019).
14. E-mail interview with Cinnamon Mittan. October 2020. Mittan's inspiration to investigate the evolution of cold tolerance illustrates how the older literature is still relevant.
15. Sturcke (2009).
16. Tingley et al. (2017).
17. Phillips et al. (2016).
18. Angel and Cooper (2006).
19. BirdLife International (2018).
20. Cuthbert et al. (2016).
21. Rowe-Rowe and Crafford (1992).
22. Wanless et al. (2016).
23. The analogues in the study used birds found on islands that did not have mammalian predators. These analogues were chosen to match as best as possible the morphology, life history, and ecology to those of each of the focal birds found on Gough Island. See Caravaggi et al. (2019) for further details.
24. Gray et al. (2014).
25. According to analysis of the genetic data, the time that mice most likely colonized Gough was 110 years ago. However, the confidence interval for this estimate is large, and 200 years (or longer) falls squarely in the middle of the confidence interval.
26. Gray et al. (2015).

27. Nolte et al. (2020).
28. Payseur and Jing (2020).
29. Luan et al. (2011).
30. E-mail interview with Bret Payseur. October 2020 and Stratton et al. (2021).
31. Payseur and Jing (2021).
32. Dilley et al. (2016).
33. Dilley et al. (2016, p. 76) provide accounts of these attacks. In one case, after mice were seen climbing on the heads of two wounded grey-headed albatross chicks, they write: "The chicks initially appeared to try to shake off the mice, but after a while the chicks sat while the mice fed on their heads."
34. See Pyšek et al. (2020) for general information on the threats from invasive species.
35. Invasive Species. 64 Federal Register 6183. February 3, 1999.
36. Simberloff et al. (2012).
37. Blackburn et al. (2019).
38. Cassini (2020).
39. See Chroback (2019) for more details about the invasiveness of earthworms. Climate change is likely to accelerate the northward spread of earthworms and their ensuing effects on the ecosystem.
40. Leddy (2007).
41. Lynas (2018).
42. Doudna and Sternberg (2017).
43. Barrett (2015).
44. Baker and Stebbins (1965).
45. See Bock et al. (2015).
46. Knight et al. (2005).
47. Bock et al. (2015).
48. E-mail conversation with Dan Bock. October 2020.
49. Pyšek et al. (2018).
50. Pyšet al. (2019).
51. Lee and Gelembiuk (2008).
52. Stern and Lee (2020).
53. Posvai et al. (2020).
54. Sultan et al. 2013.
55. See Vandepitte et al. (2014) for the study and Frank and Munshi-South (2014) for a commentary.
56. Fridley and Sax (2014).
57. Wolfe (2002).
58. Estroup and Guillemaud (2010).
59. Beaury et al. (2020).
60. Fridley and Sax (2014).
61. Caroll et al. (2005).
62. Phillips and Shine (2004).
63. Phillips and Shine (2006).
64. Wake and Vredenburg (2008). The quote is on p. 11466.
65. Schlee et al. (2019). Amphibian fungal panzootic causes catastrophic and ongoing loss of biodiversity. Science 363: 1459–1463.
66. Whittaker and Vredenburg (2011).
67. O'Hanlon et al. (2018).
68. Longo et al. (2019).
69. McDonald et al. (2020).

70. Zipkin et al. (2020).
71. Reboliar et al. (2020).
72. Rohr et al. (2011).
73. Cohen et al. (2017).
74. Cohen et al. (2019).
75. Bower et al. (2020).
76. Choudhary et al. (2021).
77. Somarelli (2021).

Section IV

Society

15 The Sequence on the Stand

SUMMARY

This chapter examines the forensic and legal applications of evolutionary genetics. We begin with the example of how information in personal ancestry databases broke the Golden State Killer case. The perpetrator in this case was identified because he shared DNA information with distant relatives who were in such databases. We discuss how patterns of genetic relationships can be used to make these inferences and the genetic privacy ramifications of this and similar cases. Next, we examine how population genetics played a role in debates during the early days of DNA fingerprinting. Phylogenetic analysis of DNA sequence information can be used to infer patterns of pathogen transmission. Such inferences have been used in court cases where people have been accused of willingly transmitting pathogens to others. Analysis of DNA information can be used to determine the composition of microbes in a dead body; in turn, this information can be used to estimate time of death.

"I've written about hundreds of unsolved crimes, from chloroform murderers to killer priests. The Golden State Killer, though, has consumed me the most. In addition to fifty sexual assaults in Northern California, he was responsible for ten sadistic murders in Southern California."—Michelle McNamara (2018).[1]

"[O]ne interesting thing about genomic data is that one person in a multi-thousand-person family tree can expose the information of every other person without their consent or even knowledge because how many of us know our third, fourth cousin." Erin Murphy (2019).[2]

THE SEARCH FOR THE GOLDEN STATE KILLER

Michelle McNamara hunted criminals with a fervent passion and a penchant for detail.[3] She was neither a prosecutor nor a police detective. Instead, McNamara was a freelance journalist specializing in cold cases, crimes that have remained unsolved after the end of an active investigation. McNamara's self-described obsession with cold cases started during her early teen years after a neighbor was killed and his murderer was never found. Born in 1970, McNamara was in her twenties when the then-new Internet fostered her ability to track these cold cases. As stated in the epigraph, McNamara was particularly fascinated by one killer, whom she dubbed the Golden State Killer.

What was most remarkable about the Golden State Killer was the sheer volume of gruesome crimes he had committed and the scope of these crimes in both geographic

DOI: 10.1201/9780429503962-19

distance and time. Like the Internet, DNA evidence changed the tracking of crimes during the 1990s. And this DNA evidence would be instrumental in discovering the Golden State Killer: starting in 1996, DNA evidence connected more and more long-cold crimes to a single individual who committed several unrelated sprees of murder and rape. His first spree was in and around Sacramento, the state capital, located in the northern part of the Central Valley of California. Then known as the East Area Rapist, he sowed panic across the region with brutal rapes during 1976 and 1977. He then moved his heinous crimes further west, to Contra Costa county, just east of Oakland. A few years later, he committed murders much further south, closer to Los Angeles. Here, he was called the Night Stalker, and later the Original Night Stalker.[4]

Although DNA evidence had linked these numerous and disparate crimes to a single perpetrator, it alone could not establish who he was. Law enforcement has access to the CODIS (Combined DNA Index Sequence) database, which has DNA from more than 13 million people, roughly 5 percent of American adults. But the Golden State Killer was not in that database. So long as the perpetrator's DNA information remained out of CODIS or other databases available to law enforcement, his identity would remain a secret.

McNamara did not live to see the publication of her book *I'll Be Gone in the Dark*, as she died in 2016 leaving an incomplete manuscript, over 3,000 computer files, notebooks, police records, and other documents. With help from McNamara's widower (the actor and comedian Patton Oswalt), McNamara's chief researcher, Paul Haynes, and the investigative reporter Billy Jensen completed the book. In the part of the book written after McNamara's death, Haynes and Jensen noted that McNamara frequently thought that a breakthrough in the case could come from genetic databases such as 23andme.com and ancestry.com, where people submit their DNA looking to see where their ancestors came from as well as searching for distant relatives. These two companies do not work with law enforcement out of concern for the privacy of their clients. Haynes and Jensen said that McNamara considered the prospects:

> If we could just submit the killer's actual genetic material—as opposed to only select markers—to one of the databases, the odds were great that we would find a second or third cousin and that person would lead investigators to the killer's identity.[5]

McNamara's book came out in February 2018. Two months later, police arrested Joseph James DeAngelo, then age 72, for numerous crimes connected to the Golden State Killer. The break came in the way that McNamara and her colleagues had predicted: finding relatives from the use of publicly available databases. Paul Holes, who had recently retired as an investigator from the Contra Costa County District, claimed he took DNA from one of the crime scenes and used that in constructing a profile on the ancestry site GEDmatch.[6] The site returned more than 100 profiles that were likely distant relatives—third or fourth cousins—to the one Holes had used. From these profiles, the police developed leads that eventually brought them to DeAngelo, who eventually confessed to a series of crimes.[7]

Linking DeAngelo to the Golden State Killer was not the first time that public genetic genealogy databases were used to solve cold cases. Genealogist Colleen Fitzpatrick used such databases to pinpoint the prime suspect of a series of murders that occurred in the early 1990s.[8] But until very recently, such cases were exceeding rare. The high-profile nature of the Golden State Killer case may cause investigators and others working with law enforcement to rely more on public genetic genealogy databases.

The Golden State Killer is an outlier. The horrific nature and the volume of his crimes make this case an exceptional one. We can be pleased that he has been brought to justice while maintaining concern about how law enforcement may use such public databases in other circumstances. Moreover, searching public databases for information about the genetic nature of individuals is not restricted to people working with law enforcement. Others may do it for prurient or nefarious purposes.

After the arrest of DeAngelo and the reports that the investigators had used GEDmatch to link him to the Golden State Killer crimes, GEDmatch explicitly stated that law enforcement could no longer use their site for such purposes. They also changed their terms of service. One of the founders at GEDmatch, Curtis Rogers, noted:

> While the database was created for genealogical research, it is important that GEDmatch participants understand the possible uses of their DNA, including identi-fication of relatives that have crimes or were victims of crimes. If you are concerned about non-genealogical uses of your DNA, you should not upload your DNA to the database and/or you should remove DNA that has already been uploaded.

While it is true that that those who entered information into the database can remove it, this information may already be in the hands of people wanting to exploit it. Moreover, while those who entered the genetic material into the databases gave con-sent, their relatives—including distant relatives they may or may not know of—did not. In addition, your DNA is forever; you can't get a replacement for it like you can with a stolen credit card or house keys.

But how likely is it that investigators or others can find the identity of someone based on information gathered from one of their (potentially distant) relatives who put their DNA in a genetic ancestry database? Did the investigators in the Golden State Killer case just get lucky? Michael "Doc" Edge and Graham Coop, evolution-ary geneticists at the University of California at Davis, addressed these questions with some modeling. They first posted this as an entry to their lab blog and then later as a preprint.[9]

Recall that the investigators made an account on GEDmatch, the fourth largest private ancestry database, using the DNA profile left at the crime scene.[10] They did not find an exact match—they would not expect that—but they found several likely distant relatives, probably about as related as third or fourth cousins. How likely it would be for the investigators to find matches depends on the structure of the human genome, notably how much recombination occurs and where. Based on these prop-erties, evolutionary geneticists like Edge and Coop can estimate the likelihood that cousins will share stretches of their DNA from their common ancestors, grandpar-ents in the case of first cousins and great-grandparents in the case of second cousins. These identical-by-descent (or IBD) analyses can also be used to estimate how long these stretches will be: first cousins will tend to share much larger stretches than more distantly related cousins, as recombination chews away at these IBD stretches with each generation.

As we go from close to more-distantly related relatives, two processes are at work, and they operate at cross-purposes. First, the number of relatives increases as we move from near to more distant relatives. Most people have more cousins than siblings. Second, the genetic signal degrades as the degree of relation increases. With ever more distant relatives, the IBD stretches get smaller. In fact, at higher order relations, many "cousins" will not share any DNA segments inherited from the common genealogical ancestor. People can be genealogical cousins without sharing genetic material through a common ancestor! Even at third cousins, there is a small (roughly 2 percent) chance of that happening. Thus, there is a tradeoff with the optimal range for finding useful matches being about third and fourth cousin: most people have plenty of these, even if they don't know the ones they have, and the signal should still be sufficiently strong.

To sum up their findings, Edge and Coop found that the investigators in the Golden State Killer case were not particularly lucky. They conclude that the results there "are close to what we expect given the size of the GEDmatch database".[11] Most people would be matched under similar circumstances. Uploading your personal genetic information into a database like GEDmatch potentially affects a few hundred people. Another research group led by Yaniv Ehrlich came to similar conclusions: more than

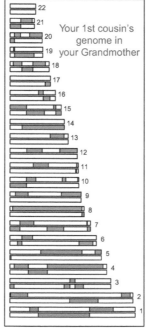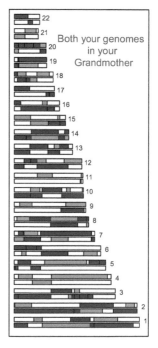

Matching of genomic segments in first cousins. First cousins share a common grandmother and will likely share numerous large chromosomal regions. Sharing through common grandfather not shown. Ancestry tests will very easily pick up first cousins.

Source: Courtesy of Graham Coop and Doc Edge.

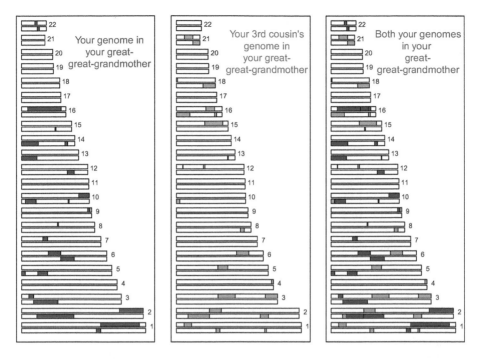

Matching of genomic segments in third cousins. Third cousins share a common great-great-grandmother and will likely share a few smaller chromosomal regions. Sharing through common great-great-grandfather not shown. Ancestry tests often pick up third cousin matches but much less easily than they do first cousin matches.

Source: Courtesy of Graham Coop and Doc Edge.

half the people of European descent could be matched to at least a third-cousin or more similar relative.[12]

There is more to this story. Doc Edge, who now has a position at the University of Southern California, told me recently that the investigators who broke the Golden State Killer case left out some pertinent information.[13] Based on accounts from the investigators, Edge and Coop had assumed that the key matches for finding the Golden State Killer were those from GEDmatch. But that was not exactly the case. Edge led me to a 2020 *L.A. Times* investigative report showing that the genealogists working with law enforcement searched two other private databases (FTDNA and MyHeritage) in addition to GEDmatch and that the key match was found with MyHeritage.[14] Edge notes that "MyHeritage was not aware they had been searched in this way until the *L.A. Times* report came out, and it's a type of search they explicitly disallow". That the investigators also used other databases has little impact on the findings by the evolutionary genetic researchers—people are relatively easy to track down by finding their fairly distant relatives in databases.

A decade prior to the capture of DeAngelo, Elizabeth Joh, a legal expert specializing in the intersection of privacy and technology, expressed concern about people's

DNA being taken without their consent.[15] She noted numerous reasons why an individual may be targeted. For instance, there is the DNA trophy hunting of a celebrity. When he was president, Bill Clinton and his bodyguards took the prospect of DNA trophy hunting so seriously that the bodyguards saved a glass that Clinton had drunk from while at a British pub. Current or potential members of a professional sports team may also be targeted, especially if the interest thought it could make a profit from knowledge about the target's genetic status. Busybody neighbors, blackmailers, and others with abject motives may also target individuals. With the databases, these adversaries can potentially gain the genetic information without physical samples.

Edge and Coop examined the ways someone armed with a bit of knowledge about evolutionary genealogy and databases can violate the genetic privacy of a targeted individual.[16] One of these is tiling, by which a target's nemesis can get knowledge about most of the target's genome. If you know the target shares a region of the genome with a random person in the database and you know the genotype of that random person, then you can make reasonably good inferences about the target's genotype in that IBS region. By uploading lots of genotypes into the database, the nemesis has many opportunities to make comparisons. Edge and Coop found that tiling is usually effective; it can recover a large part of the genome for most individuals of European descent. Tiling usually is more effective at obtaining insight into the target's genome from the numerous but smaller matching segments from the more distant relatives. The databases can thwart the nemesis by increasing the window size of the shared region needed to report matches; thus, more distantly related matches (third and fourth cousins) would not be reported.

As noted in the epigraph, Erin Murphy, a law professor at New York University, argued that the data in genetic databases are fundamentally different from other such databases because one person can link to many others, potentially exposing their genetic information. Expertise in evolutionary genetics would be valuable in informing governments about how such databases should be regulated and in informing individuals about the risks and benefits of participating in genealogical and other genetic databases.

DNA FINGERPRINTING

Forensic geneticists long had hoped to find a genetic marker that, like fingerprints, can be used to distinguish and identify specific individuals. During the middle 1980s, the British geneticist Alec Jeffreys realized this dream with a series of markers that he called DNA fingerprinting. In an early application of these DNA fingerprinting, Jeffreys use this technique to resolve an immigration dispute. British immigration authorities had flagged one of the sons of a family returning from Gabon because they thought his passport was fraudulent and that he was not related to the rest of the family. Jeffreys' DNA fingerprinting demonstrated that this child was indeed the son even though the father was not available.[17] Soon, these methods were used to settle paternity suits and to solve crimes.

These methods, especially early on, did not involve direct sequencing of DNA. Instead, they relied on looking at patterns of DNA fragments separated out by size on gels. One method used enzymes that cut DNA at specific points; differences in sequence at the sites where these enzymes cut would be reflected in the size of the fragments.[18] Another method examined regions of the genome where short sequences were repeated several times in a row. This method could distinguish individuals because humans differ in the numbers of the various repeats that they have.

Starting in the late 1980s, some evolutionary geneticists became increasingly involved in debates about DNA fingerprinting, some even testifying as expert witnesses in court. On a visit to Harvard University in the fall of 1998, I inadvertently was a witness to this involvement.

I had gone to Harvard to give a talk to Dick Lewontin's lab group about a new project I was working on. Lewontin had been a dominating figure in evolutionary genetics. He was among the first to use molecular tools to assess genetic variation in populations. In the year I was born (1966), he used protein electrophoresis—tracking the movement of enzyme variants on gels—to show that natural populations of *Drosophila* harbor much more variation than had been expected, a discovery that would alter the field. Though first used in flies, this method could be used to systematically estimate the extent of genetic variation in just about any organism.[19] A few years later, Lewontin published an extraordinary book, *The Genetic Basis of Evolutionary Change*; in elegant but clear prose, it placed the discoveries of extravagant genetic variation in context and established an intellectual framework that is still relevant today.[20] As we will see in the chapter on human genetic variation and race, Lewontin also did some early seminal work showing that most of the genetic variation in humans was within rather than among populations. I was eager and slightly apprehensive about my audience with Lewontin and his lab group.

When I reached Lewontin's lab, there was no one there save for a nondescript middle-aged man who was quietly reading a newspaper with his feet propped up on the desk. I introduced myself. He said, "I'm a friend of Dick's," quickly returning to his paper. Lewontin's graduate student Adriana Briscoe and his undergraduate student Carlos Bustamante came in, and I chatted with them. Lewontin came and talked with the guy with the newspaper about new genetic databases in Massachusetts and how they could be used. The newspaper guy left. I gave my talk.[21] After we chatted about my work for a bit, Lewontin said that the guy with the paper was Peter Neufeld, one of the founders of The Innocence Project, an organization that uses DNA and other evidence to try to exonerate wrongly accused people.[22] As of early 2021, The Innocence Project has exonerated 375 people with the use of DNA evidence; a majority of these people are African American.[23]

Even years before my visit to his lab, Lewontin and his colleague Dan Hartl had been critical of some of the practices law enforcement had made in the early days of DNA fingerprinting.[24] In addition to the general criticism about the handling of the evidence and interpretation of the results, some of these criticisms directly involved aspects of evolutionary genetics. Recall that the hope of DNA fingerprinting was that the pattern would uniquely identify individuals. When there is a match between the suspect's pattern and the DNA pattern left at the crime scene, could it come by chance? Is it possible that the suspect is not the actual culprit but merely shares a DNA profile with the culprit? Given the presumption of innocence a defendant in a criminal case is given, the prosecution must show that the defendant did it beyond a reasonable doubt. If there is a match between the DNA fingerprint of the defendant and that at the crime scene, then how common (or rare) that pattern is becomes important to the question of reasonable doubt.

How should we calculate the probability that a match between the defendant and the crime scene could arise by chance? The typical fingerprint involved several different genetic loci that generally were highly variable among the population. Let's say that there are four loci: A, B, C, and D. Now assume that the frequencies of the defendant's type at A, B, C, D are respectively 0.04, 0.01, 0.05, and 0.02. Assuming that these loci were independent—that the likelihood of being a genotype at one locus is not predictive of the genotypes at other loci—then we simply need to multiply all these numbers to get the probability of a match occurring by chance. In this hypothetical example, that would be 0.04 times 0.01 times 0.05 times 0.02 or 0.0000004 or one in 2.5 million. Far smaller probabilities—one in billions or even trillions—can be reached with having a few more loci. Judges and juries hearing such figures as something million or billion to one think that these are extraordinarily unlikely to be due to chance and that information can persuade them to convict.

Lewontin and Hartl expressed concern about the validity of that multiplication procedure. Recall that it rests on the assumption that the loci are acting independently and that the probability of being a genotype at one locus was unaffected by the genotypes at other loci. But what if the loci are not independent? We've seen in other chapters where the nonindependence of loci mattered—we called it linkage disequilibrium. And by the way, Lewontin was the one who named the concept such back in 1960, initially studying patterns of the frequencies of chromosomal types in grasshoppers.[25] With sufficient linkage disequilibrium, these astronomical odds can shrink. In addition to linkage disequilibrium, differences of frequencies of variants among populations can also affect these probabilities.

Lewontin and Hartl wrote up their concerns and provided guidance in a paper that was eventually published in *Science* in 1991. I had read the paper when it came out while I was in grad school, where we had discussed it in a journal club. I had vaguely recalled there being some controversy with letters to the editor following the publication. In researching this matter for the book, I met up with Dan Hartl at his Harvard office, where he told me that there was indeed more to the story.[26]

Hartl told me that he and Lewontin had been pressured to withdraw the paper by a prosecutor named James Wooley at the Justice Department's Strike Force in Cleveland. Hartl's comments led to search for confirmation, which I found in a *New York Times* story from late 1991 that quotes Hartl saying that Wooley "felt that publication of the paper would be a disservice to the system of justice in the United States".[27] Here, Lewontin had commented: "When someone who is an official in the Department of Justice Criminal Division Strike Force telephones a private citizen to request an action the citizen would not ordinarily take, then a form of intimidation has been used."

Hartl also told me that after they had sent back the galleys of their *Science* paper, the journal's editor in chief, Daniel Koshland, had pressured them to soften the conclusions. Such a practice is highly unusual. Hartl told me that he was more willing to make concessions, but that Lewontin refused. According to the *New York Times* story, Lewontin did try to mollify the editor: "We finally did make some changes, against my better judgment."[28] Koshland also took the unprecedented step of holding up publication while a specially invited rebuttal was being written allowing for simultaneous publication of the paper and rebuttal.[29]

In the ensuing years, the controversy died down. The National Research Commission released a report regarding best practices for the use of DNA fingerprinting in court cases.[30] Among their recommendations was the use of what became known as the "ceiling principle" for calculating probabilities of getting a match. Instead of multiplying by the frequency of each matched allele in the total population, probabilities should be calculated by multiplying by what the frequency of the matched allele was in the subpopulation where the frequency was highest. In other words, if the frequency of a specific matched allele is particularly high in one group—say, Japanese Americans—then the frequency of the allele in Japanese Americans should be used in the calculations.

THE SEQUENCE ON THE STAND

Richard Schmidt was a gastroenterologist practicing in Lafayette, Louisiana. In the summer of 1995, the police arrested him and raided his home after receiving a complaint from Janice Trahan, a nurse who had worked for his practice.[31] Trahan had alleged that Schmidt had injected her with HIV the previous year. How did this happen? Trahan and Schmidt had had a long affair but had since broken up. On the night of August 4, 1994, soon after the breakup, Schmidt had come to Trahan's house, waking her up. He persisted in offering to give her a shot of what he claimed was vitamin B12, a substance he had given her during the affair when she felt tired. Eventually, she gave in and accepted the shot. Some months after receiving this shot, Trahan came down with an illness and was subsequently found to be HIV positive. She claimed that Schmidt had deliberately infected her due to his jealousy. After a jury trial, Schmidt was found guilty of attempted murder in the second degree on October 23, 1998. What distinguished this trial was that it was the first to use phylogenetic analysis of DNA sequences as evidence.

Recall from Chapter 2 that evolutionary biologists can in some cases infer transmission of viruses from person to person and even the direction of the transmission (who infected whom) by examining the patterns of relationships—technically, the phylogenetic relationships—of viral sequences. This is what enabled biologists to determine that the early SARS-CoV-2 outbreak in the northeast United States mainly came from Europe and not China. Making inferences based on phylogenetic relationships is straightforward in the abstract, though it can be challenging in practice. The soundness of these inferences depends on proper application of evolutionary principles as well as a large body of empirical studies.

So how do biologists figure out the phylogeny? First, they figure out what samples need to be included in the analysis. Typically, the analysts obtain DNA sequences from viral samples. In this case, they would have samples of HIV from Trahan (the victim), samples from patients of Schmidt (the accused), and samples from people in the community. With these samples, computational algorithms are run to figure out the phylogeny. The task is to determine the best phylogeny (relationship patterns) from all of the possible phylogenies. This process, which is called reconstructing the phylogeny, is not simple. Many different types of algorithms can be used to reconstruct phylogenies, and considerable debate still occurs among evolutionary biologists about which methods are most appropriate in which circumstances. In practice, most methods usually, though not always, converge on similar answers. Before we discuss the results, let's consider a couple standard methods.[32]

One class of method—parsimony—derives from an idea that stretches back to the 14th-century insight of the English friar and philosopher, William of Occam, whose razor (as is frequently interpreted) states that the explanations that explain the data with the fewest causes are likely to be correct. Although parsimony sometimes fails us, scientists usually seek to find the simplest explanation—those with the fewest number of causes—that adequately account for the data, with the assumption that these parsimonious explanations are more likely to be correct. The parsimony method of phylogenetic reconstruction seeks to minimize the number of evolutionary steps involved in the evolutionary history of all of the sequences. If one phylogeny requires 34 steps and another requires 38, then the one with 34 is deemed more parsimonious, and thus superior, by the parsimony method. The parsimony method has a long history in phylogenetic reconstruction and is still commonly used.

The parsimony method does not consider whether some changes are more likely than others.[33] There are approaches that account for such knowledge of differences in the frequencies of different types of changes. One commonly used such method is the maximum likelihood. This approach begins with an explicit model of the evolutionary process of the sequences being studied that considers which types of evolutionary changes are most common. Compared to the parsimony methods, maximum likelihood methods are much more computationally demanding. The extraordinarily rapid increases in computational power have made maximum likelihood methods much more feasible now than even in the 1990s. But even outside the computational challenges, the maximum likelihood methods require a model about how the molecules and the organisms evolve, and the method only works well when that model sufficiently resembles reality. Unfortunately, models do not always come close to fitting reality, and often we do not know the extent to which they do not.

How do we know that the methods used in phylogenetic reconstruction are reliable? Do they provide accurate results? These questions are really important, because if the methods were unreliable, then we would be in danger of using shoddy science to support recommendations of guilt or exoneration of criminal defendants, a proposition that should unnerve us all. The general result of different methods usually, though not always, giving similar answers provided some comfort but can still leave one uneasy. Is there a way to independently verify the reliability of the phylogenetic reconstruction methods?

During the early 1990s, a team of biologists at the University of Texas at Austin performed experimental studies with the bacteriophage T7 to test these methods. Led by David Hillis and Jim Bull, this team split samples of phage after a designated number of generations to establish new lineages. They then sequenced the phage from each of the lineages. Next, they analyzed the data with the previously mentioned and other methods. From this, they found that most of the time, the methods (both parsimony and maximum likelihood) returned the correct phylogeny. Thus, we can generally be confident that these methods are reliable.[34]

One of the leaders of the phage studies, David Hillis, was one of the expert witnesses at the Schmidt case. He got involved after a call from the district attorney prosecuting the case, who asked Hillis to reanalyze the sequence data (originally collected by Michael Metzker at Baylor College).[35] The DA also asked Hillis to respond to a defense brief with the bizarre argument that phylogenetics was not informative to the case because HIV evolves! Eventually, the prosecution team had Hillis come to Louisiana to testify in pretrial hearings about the admissibility of phylogenetic evidence and analysis. Hillis and others argued that phylogenetic analysis was a reliable way of analyzing the evolution of DNA sequences and of testing evolutionary hypotheses. Then the defense team had moved to questions about contamination. Hillis advised the prosecution to collect new samples from the patient and victim and send them to another lab to get independent sequence data.

Hillis then analyzed the sequence data the independent lab had sent him. This analysis and others found that the victim had HIV sequences that were extremely similar to those of one of the doctor's patients, much more so than sequences from other individuals in the community who had HIV.[36] Moreover, analysis showed that the victim's sequences were nested within that of the patient: some of the patient's HIV sequences were more closely related to the sequences to those of the victim than they were to other sequences of the patient. This pattern provides support for inferring the direction of the transmission: it is much more likely that the patient's blood infected the victim than the other way around. These results, along with others, led to the conviction of Richard Schmidt. In the following years, such evidence has been used to convict and exonerate many more criminal defendants.[37]

THE MICROBES OF DEATH

A key piece of information of any homicide is the time of death. As any viewer of crime and medical mystery TV dramas knows, the time of death can be used to rule out or implicate suspects. Establishing this interval requires making inferences from multiple types of evidence. An emerging source of information used in setting this interval comes from microbes associated with the body in the days and weeks after death. This thantomicrobiome[38]—the microbes of death—changes over time in patterns that resemble the ecological succession of plants and animals that occur after a disturbance.

During the first few hours after death, the release of calcium and other ions causes the muscles to stiffen in a process known as rigor mortis.[39] Similarly, blood pools cause black and blue patches across the body. The body temperature drops at a rate dependent on the ambient temperature. From a combination of the stages of each of these processes, forensic pathologists can estimate time of death during the first couple days after death. Around this time, blowflies and other insects are attracted to the body, where they lay eggs. Often the by-products of certain bacteria, such as *Proteus mirabilis*, attract the insects to the body.[40]

Following those early markers, microbes continue to ferment carbohydrates in the body. This fermentation releases gases, which can cause bloating. Fermentation and the microbial break down of proteins also can cause changes in skin color. Similarly, blowfly and other insect eggs hatch and develop into larvae and then pupae. The bacterial blooms also attract nematodes, which then feast on those blooms. Some fungi are specialized in feeding on dead material, where they breakdown diverse organic chemicals. These saprotrophic fungi are especially abundant in the late stages of decay. Examination of these stages can be used to estimate time of death after the first couple of days through the next few weeks.

Jessica Metcalf and her team at the University of Colorado at Boulder have collected sequence information of these microbes over a time series.[41] With such data and new machine learning tools, Metcalf's team has developed a clock of sorts that accurately measures time since death based on the composition of the microbial community. The average error of their microbial death clock was about three days, considerably better than other methods.

Metcalf's team has found that early in this successional pattern, the bacterial Firmicutes and Bacteroidetes that can thrive in the absence of oxygen thrive as they ferment the carbohydrates in the body. This activity leads to bloating of the abdominal cavity. Around day 9, the bloating causes the abdominal cavity to burst. The burst releases oxygen, allowing aerobic bacteria and eukaryotic microbes to flourish. These include Rhizobiales and Brucaellaceae. In contrast, the anaerobic Firmicutes and Bacteroidetes decline in frequency. Ruptured cavities also release ammonia and nutrients into soils, causing pH to rise substantially; with this change comes declines in Acidobacteria.

This method has some advantage over the conventional use of looking at changes in insects to estimate the time of death.[42] For instance, if it is too cold, insects won't come, but microbes are still there. These microbial dating methods are still new, and knowledge gaps persist. For instance, we know that temperature is an important variable to be used in estimating time since death, but what about other environmental factors, such as humidity? Are they also important, and in what ways?

NOTES

1. McNamara (2018, p, 15).
2. NPR (2019).
3. For background on the Golden State Killer case, see McNamara (2018).
4. There would turn out to be several different killers all given the Night Stalker nickname. To avoid confusion, the one that McNamara became interested in was called the Original Night Stalker.
5. McNamara (2018, p. 271).
6. Allen et al. (2018).
7. Dowd (2020).
8. Fitzpatrick used genetic material from the Y chromosome. Because Y chromosomes and last names are passed through males, this method allowed her to zero in on a last name (Miller). Combined with other evidence, this led police to Bryan Miller, who they arrested for the murders. See Phillips (2018) and sources within for details.
9. Edge and Coop (2019).
10. At the time, GEDmatch had nearly a million profiles. As we will see, the investigators also used the other databases.
11. Edge and Coop (2019, p. 7).
12. Erlich et al. (2018).
13. E-mail conversation with Doc Edge. April 2021.
14. St. John (2020).
15. Joh (2011).

16. Edge and Coop (2020).
17. Zagorski (2006).
18. The first type of marker is called an RFLP for restriction fragment length polymorphism. The second is called an AFLP for amplified fragment length polymorphism. See Jordan and Mills (2021) for more detail.
19. Lewontin and Hubby (1966).
20. Lewontin (1974).
21. Lewontin's comments were very helpful in the development of this project. Our first paper on it was Johnson and Porter (2000). Carlos Bustamante, who would go to be Dan Hartl's student, and Adriana Briscoe have each gone on to have very successful careers in evolutionary genetics.
22. Span (1994).
23. The Innocence Project. DNA exonerations in the United States. http://innocenceproject.com/dna-exonerations-in-the-united-states. Accessed April 14, 2021.
24. Lewontin and Hartl (1991).
25. Lewontin and Kojima (1960).
26. Meeting with Dan Hartl, June 2019, and e-mail follow-up in April 2021.
27. Kolata (1991).
28. Kolata (1991) also documents other cases of alleged harassment of scientists involved in the early days of DNA fingerprinting. One example was FBI lawyers making references to the status of one scientist's visa.
29. The rebuttal was by Chakraborty and Kidd (1991).
30. See, for instance, Lander and Budowie (1994).
31. Details of the case can be found in Wapner (2019).
32. See chapter 20 of Hillis et al. (2020) for more detail on phylogenetic methods.
33. That the parsimony method assumes nothing the evolutionary processes involved is both a feature and a bug. It's a feature in that the simple nature of the parsimony method makes it comparatively easy to understand and employ. It's a limitation in that sometimes we actually know that certain evolutionary changes are more likely or less likely than others. If certain changes are much more common than others, we may be led astray if we do not account for the differences in relative frequency.

    The maximum likelihood method, in contrast, assumes an explicit model. Recall that DNA has four different bases: adenine (A), cytosine (C), guanine (G), and thymine (T). Due to the chemical properties of these bases, mutations from C to T are much more common than those of any other change. Hence, C to T changes are more likely to occur in lineages. Some other changes, like A to T, are much less frequent in most situations. Under a maximum likelihood method, the less frequent (more surprising) changes count for more; among those phylogenies requiring the same total number of changes, those that require fewer of the rare types of evolutionary changes will be favored over those that require more of these rare changes. Maximum likelihood methods also allow for the possibility that different lineages may evolve at different rates.
34. Hillis et al. (1992).
35. E-mail correspondence with David Hillis. April 2021.
36. Metzker et al. (2002).
37. See Scaduto et al. (2010) for two examples.
38. Roy et al. (2021).
39. Metcalf (2019).
40. Metcalf et al. (2016).
41. Metcalf et al. (2013).
42. Metcalf (2019).

16 Darwinian Security

SUMMARY

The principles of ecology and evolution can be applied to security issues, including national security. Game theory is important in evolutionary biology, behavioral science, and in national security. We discuss the prisoner's dilemma in detail and show how it has been applied in animal behavior studies, where the concept of evolutionary stable strategies comes into play, as well as in conflicts between human individuals and nations. Arms races are not just found between nations but also can occur in nonhuman animals. We consider arms races leading to extreme weapons (horns) in beetles and what lessons we can apply from these observations to human conflict. We then turn to what we can learn about living with threats from natural systems. We conclude with a discussion of asymmetric warfare, as seen in insurgencies, and the parallels seen in the natural world. A key lesson throughout the chapter is that systems that evolved through natural selection tend to be economical and adaptable.

"Evolution provides a lens to look at security broadly and across a number of different gradients. . . . Evolution helps us think about the relative role of cooperation vs. conflict. And a functional and historical approach can help us think about ways to better prevent and respond to security breaches."—Rafe Sagarin and colleagues (2016).[1]

"Nature is the ultimate economizer, relentlessly culling individuals who allocate resources poorly."—Douglas Emlen (2014).[2]

RAFE SAGARIN'S QUEST

Gary Hart, the former US senator and presidential candidate, presciently warned about the threat of large-scale terrorist attacks in the years immediately preceding the September 11, 2001, attacks. As co-chair of the Commission on National Security with former Senator Warren Rudman, Hart was among the persistent voices calling for preparedness for such new threats.[3] The first Hart–Rudman report, released in 1999, warned, "America will be attacked by terrorists using weapons of mass destruction, and Americans will lose their lives on American soil, possibly in large numbers."[4] Hart's commission recommended shoring up our defenses against terrorism as well as more effort in promoting democracy abroad. On September 4, 2001— one week before the attacks—Hart reiterated the threat in a speech in Montreal, though he did not specify the nature of the threat.[5]

After the September 11 attacks, Hart continued as an advocate for threat preparedness. In 2012, Hart wrote a forward to a book by Rafe Sagarin, a young academic at the University of Arizona, who had a fresh perspective on responding to threats to national security.[6] Sagarin was neither a military historian nor a foreign policy expert. In fact, Sagarin was an ecologist who studied how species interact

DOI: 10.1201/9780429503962-20

in intertidal ponds. The basic message from Sagarin's book, *Learning from the Octopus*, is revealed in its subtitle: *How Secrets from Nature Can Help Us Fight Terrorist Attacks, Natural Disasters, and Disease.* In his foreword, Hart stressed a key message of Sagarin's book: the pragmaticism and the adaptability of evolution the products of biological evolution. Hart noted:

> Sagarin shows that biological science demonstrates how fauna and some flora adapt to changing conditions. . . . Creatures change not necessarily to make themselves more beautiful or exquisite. They do so because it helps them survive in a constantly changing environment.[7]

Hart saw Sagarin's book as a guide of how to apply the approaches organisms take to make themselves secure from threats toward making our security systems more adaptable and responsive.

Sagarin was an inquisitive and passionate thinker. Just after completing his last final exam as an undergraduate at Stanford University, Sagarin wrote a letter to the *New York Times*, decrying the pernicious culture of grade-grubbing there, noting

> what I find unfortunate, and what most disappointed me about my time at Stanford, is that our narrow focus on letter grades has allowed us to forget that a university should be about learning and knowledge, rather than class rank and one-upmanship.[8]

For his graduate work at the University of Santa Barbara, Sagarin documented how marine life off the coast of California had changed—presumably due to climate change—from the 1930s to the 1990s.

After completion of his PhD, Sagarin took a path that is quite unusual for an ecologist: he joined the staff of congressional representative Hilda Solis of California, who had an interest in the environment. During his time there, Sagarin started thinking about security. Ever observant, Sagarin watched his fellow staffers as they were screened upon entrance into the Capitol. Sagarin

> noticed that they'd just put their hand over their keys in their pockets so they didn't have to waste 30 seconds putting it on the conveyer belt through the security screening—and that didn't set off the alarm when they did that.[9]

This observation made him realize that "adaptable organisms are going to figure a way to get around" security measures. An adaptable security system would be a flexible one that can keep up with adaptable organisms. Soon after, Sagarin organized a working group at the National Center for Ecological Analysis, a think tank located in Santa Barbara, California, bringing together ecologists, evolutionary biologists, anthropologists, psychologists, security experts, bio-warfare experts, and others to explore the connections between how biological organisms and societies deal with threats to their security.

On the afternoon of May 28, 2015, Sagarin had a phone call with his colleague Dan Blumstein[10] about a draft of an encyclopedia article. Sagarin then jumped on his bicycle. Minutes later, he was hit by truck driven by an inebriated driver and died. Sagarin was 43.[11]

Before we delve into the work that Sagarin and others engaged in during this century, let's go back to the insights from the last century showing that biological organisms and organizations (such as nations) engage in games, much like humans do.

GAMES ANIMALS, PEOPLE, AND NATIONS PLAY

Two individuals—let's call them Amy and Brad—are captured by the authorities, who suspect that they robbed a bank. There is enough evidence to charge Amy and Brad with trespassing and other minor charges, but not for the bank robbery itself. Amy and Brad, who don't know each other well, are placed in separate rooms; each is presented with the opportunity to give up the other and walk away without any charges. If Amy betrays Brad but Brad stays mum, Brad will probably face a sentence of five years. Likewise, Amy will probably get five years if Brad gives her up. The prosecutors also say if each betrays the other, they will likely get three years. Finally, if Amy and Brad both cooperate with each other and remain silent, they will face a minimum sentence of less than a year.

Sounds like a common situation in a legal or police drama? It's actually the premise of a basic game called the prisoner's dilemma, first formulated by researchers at the RAND corporation in 1950.[12] The dilemma is that both prisoners would fare well if they both cooperate with each other and stay quiet, but each prisoner would do better if they betray the other. Regardless of what Amy does, Brad would do better if he betrays her. Suppose Amy keeps quiet. Brad fares better if he betrays her (he goes free) than if he also keeps quiet (he still faces the lesser charges). Suppose she betrays him. Then Brad still does better if he betrays her (a sentence of three years versus five years). The same calculus is true of Amy's decision. Betraying the other serves the best interests of both Amy and Brad. Accordingly, we should not expect cooperation between prisoners facing these conditions.

But there is a way to get cooperation out of this prisoner's dilemma. In the preceding paragraphs, we assumed that Amy and Brad only meet once and never confront each other again. We also tacitly assumed that there was no connection, such as family ties or group membership, shared between them. What if Amy and Brad were likely to meet up again? What if they frequently encountered each other in similar situations? Under these conditions, their mutual cooperation would serve Amy and Brad each well.

In fact, researchers have tested such repeated games of prisoner's dilemma. They did not look at prisoner's facing possible sentences but at an analogous structure of payoffs. If the two participants (call them Carl and Diane) both cooperate with each other, they each receive a sum of money—say $20. If Carl cooperates but Diane defects, Carl gets nothing, but Diane gets $30. The reverse is the case if Carl defects but Diane cooperates: he gets $30 and she gets nothing. If they both defect, they each get the paltry sum of $5. Again, it is to the advantage of each participant to defect if they were never going to meet each other again. But if they knew that they were likely to engage in multiple rounds of this game, Carl and Diane might be better off playing more cooperative strategies. In fact, a strategy that works well is TIT FOR TAT. A player employing this strategy will cooperate in the first game and in each subsequent round unless the other player defects. If the other player then cooperates in the next round, an individual playing TIT FOR TAT would go back to cooperating; it would "forgive" the other player's defection.

The success of different strategies in playing prisoner's dilemma and other similar games can be evaluated. Starting in 1979, the University of Michigan political scientist Robert Axelrod has been conducting tournaments wherein people submit computer programs to compete in multiple rounds of the prisoner's dilemma. These computer programs, each with a predetermined strategy, face off against different players. Overly trusting programs that always cooperate do not perform well, but neither do programs that always defect. In fact, TIT FOR TAT and similar programs perform well.[13] Summing up these results, Axelrod noted:

> What accounts for TIT FOR TAT's robust success is its combination of being nice, retaliatory, and clear. Its niceness prevents it from getting into unnecessary trouble. Its retaliation discourages the other side from persisting whenever defection is tried. Its forgiveness helps restore mutual cooperation. And its clarity makes it intelligible to the other player, thereby eliciting long long-term cooperation.[14]

The prisoner's dilemma is one of numerous contests developed by researchers in the field known as game theory, the science of understanding how different entities perform and react when their actions affect the circumstances of other entities. Such games can involve more than two players; however, such games, by necessity, involve at least two. If the consequences of a decision do not involve other participants, that choice is not in the realm of game theory.

Game theory has also been applied to the behavior of nonhuman animals.[15] For instance, in laboratory settings, Norway rats employ a strategy like TIT FOR TAT when engaging in multiple games of prisoner's dilemma in laboratory settings.[16] In realms of human behavior, classic game theory assume that each participant was guided by reason to pick out the best solution knowing that their opponent(s) were also reasoning beings. In the realm of animal behavior, researchers assume that the strategies have evolved; assuming that the strategies are heritable (offspring are more likely to use strategies their parents have used), the strategies that make their bearers more likely to survive and reproduce are likely to be the ones that would be more common over time. Organisms do not require a brain to employ different strategies in different contexts. Even bacteria can play games: they can respond differently according to environmental cues, and their different responses can affect the fitness of other bacteria. In fact, one such game involves bacteria that act much like suicide bombers.[17]

Similarly, game theory can be applied to nations and other organizations. In fact, the RAND corporation, originally part of the US Air Force that was spun off into a private company, focused on developing game theory as a response to the development of nuclear weapons in the 1940s. Such games are currently being used to investigate a variety of international situations. For instance, Bellal Bhuiyan at the University of Dhaka recently presented the counterterrorism policies of India and Bangladesh as an example of the prisoner's dilemma.[18] Both countries would benefit if both took active—as opposed to reactive—counterterrorism measures. If only one country (India) took active steps, then Bangladesh would get the benefits of the measures without taking on the considerable costs entailed. Likewise, India would benefit if only Bangladesh took active measures.

Individual humans, animals (and even microbes), and multi-personal organizations (including nations) play similar games where both players can "win", that is do better, if they cooperate. Such games are sometimes called "non zero-sum games" because the sum of the gains by the players is not fixed. Some authors, such as Robert Wright, have argued that performance in such games may have led to progress in biological evolution and human civilization.[19]

One important application of game theory is investigating whether individuals engaged in conflict will escalate. In the 1970s, George Price[20] noted that disputes between individual animals of the same species seldom are all-out fights to the death. Indeed, conflicts are often ritualized: some displays are more like trash talking rather than an actual fight. To explore how such behavior could evolve, Price, along with the eminent British biologist John Maynard Smith,[21] considered a contest between two animals with multiple rounds; during each round, each individual can engage in a conventional tactic such as a display, or it can escalate to a dangerous tactic, or it can retreat. Maynard Smith and Price devised several possible strategies that individuals can perform. For instance, individuals that employ the hawk strategy always escalate to the dangerous tactic on first encounter and will continue fighting until their opponent retreats or is seriously injured. In contrast, an individual employing the dove strategy will never escalate. If its opponent escalates, it will retreat. Hence, doves will always lose in contests against hawks. On the other hand, two doves encountering each other will just trash talk before settling their differences. Encounters between two hawks are likely to lead to serious injury to one or both participants; in contrast, encounters between two doves will never do so. An individual employing the retaliator strategy will never strike the first blow, but it will fight back if its opponent escalates.

How can we determine which strategy is likely to evolve? Maynard Smith and Price introduced the concept of the evolutionarily stable strategy (ESS): a strategy is evolutionarily stable if it cannot be invaded by a rare strategy. Dove is not an evolutionarily stable strategy; although doves in a population of all doves do very well, a population of all doves will get beaten up by the introduction of a few hawks. But under some circumstances, hawks may not be an ESS either. If costs of fighting to the bitter end are sufficiently large relative to the reward, a few doves might do well in a population of hawks. Under many circumstances, retaliator is an evolutionarily stable strategy; rare hawks or rare doves will not do better than the retaliators in a population full of retaliators. Although the real world has additional complications, such as some individuals being in better condition to fight or having more to defend, these games can be modified to incorporate these factors: simple games can explain how animals have evolved to escalate or resolve their conflicts.[22]

ARMS RACES AND EXTREME WEAPONS

Sagarin and others looking for lessons from nature that can be applied to conflict between nations and similar organizations often appeal to analogies. While not infallible, due to the differences between biological evolution and human culture, such an approach can be useful at examining the conditions wherein certain features occur. One such area is in the realm of arms races and the development of extreme weapons.

Arms races are not just a feature of conflict between rival nations: they are a frequent occurrence in the biological world. Predators evolve better tricks to catch prey, and their prey evolve counter-tricks to evade their would-be predators. Parasites evolve mechanisms to invade hosts, and their hosts evolve countermeasures to resist those parasites. Arms races do not even require different species: in fact, some of the most spectacular arms races occur within species when males are competing amongst each other for mates. The selection here—sexual selection—is analogous to natural selection; in fact, many evolutionary biologists consider sexual selection to be a subset of natural selection.

Extreme weapons are found in an extraordinarily diverse collection of animals. The horns of bighorn sheep and the antlers of deer are obvious examples, but extreme weapons can be found in bees, wasps, frogs, isopods, flies, and beetles. In the vast majority where extreme weapons are evolved, they are only on males (or are much larger in males) and have been used to defend resources that are limiting and related (directly or indirectly) to mating success.[23]

We will focus on horns in beetles, some of which are extraordinarily large. Keep in mind size relationships; a beetle 3 inches (7.5 cm) long dragging around a horn that is an inch long is like a 6-foot- (180 cm) tall person dragging a horn that extends 2 feet (45 cm). What are the ecological factors that affect whether beetles evolved horns?

Doug Emeln at the University of Montana has addressed this question with dung beetles—yes, beetles that live on dung from cattle and other large mammals. Dung is a rich source of nitrogen, but it is fleeting. Dung beetles have evolved two main ways of dealing with the ephemeral nature of dung. Some beetles roll the dung into big balls and use these for food and/or for breeding chambers. Others hide the dung in tunnels under dirt. We'll focus on the tunnelers. Here, female beetles dig the tunnels while the males guard the females. Note that the males are guarding both a food source and their progeny. With males fighting in an enclosed space, horns become an effective weapon—the resident male can use its horns to block or dislodge intruders. Likewise, would-be usurpers can use their horns against the resident males. Experimental studies show that time and time again, males with larger horns will consistently win the lion's share of contests and thus gain access to females and dung. Accordingly, we would expect tunneling to be associated with the evolution of horns. And that is what Emlen found; a phylogenetic analysis involving 45 different genera of dung beetles found horns were gained eight different times independently. In all eight of these cases, the lineage had beetles that tunneled. One of the lineages that lost horns had switched from tunneling to rolling. So tunneling has a strong influence on whether horns evolved. In the tunnels, fights between beetles are one on one. In contrast, contests in beetles that roll dung often involve multiple beetles at once.[24]

And yet tunneling is not the only factor. Consider the genus *Onthophagus*, the largest genus of beetles, with more than 2,000 described species. While just about all *Onthophagus* species are tunnelers, only about half the species have horns. Moreover, in the species that have horns, the horns differ: some species have horns extending out from the front of the head, while others have them coming from the middle or the base of the head. Still other beetles have horns coming out from one of several parts of the thorax.[25]

Emlen found that lineages of beetles that had experienced increases in population density during their evolution tended to gain horns and to get longer horns. Competition over resources likely is more intense in these lineages. Hence, sexual selection for larger weapons would also be more intense. What explains where the horns grow? Horns are expensive and entail tradeoffs: beetles with large horns may have stunted wings or malformed wings or undeveloped eyes. Emlen suggests that the nature of costs of growing horns at a specific site combined with the ecological context may affect where horns grow. For example, big horns on the thorax reduce the ability of males to fly. For populations where the density of both beetles and dung is high, flight is not as important as resources, and other individuals are nearby. Hence, the costs to evolve thoracic horns may be lower under these circumstances.[26] As noted in the epigraph from Emlen, nature favors economy whenever possible.

Emeln's work on extreme weapons in beetles has led him to search for parallels in human military history. One such parallel is that the nature of combat affects whether arms races will occur. Military historians have long noted that battering rams on galleys, machine guns on early planes, and other weapons that made one-on-one combat more likely typically spurred on arms races.[27] An assumption in both the animal studies and the military history research is that putting resources into superior weaponry is more likely to be advantageous in one-on-one contests than in ones involving more than two contestants. In one-on-one bouts, the contestant with superior weapons usually wins, but this is less likely with those involving multiple contestants. A study of artificial intelligence (AI) programs playing the wargame Starcraft II supports this assumption. In games with two individuals, the player with superior power won an overwhelming majority of the contests. In games involving more than two players, superior power was less of an asset.[28]

If nature is a guide, arms races between nations are favored when competition for resources and the nature of the battles is one-on-one combat where superiority in weapons is likely to be favored, whether that is from battering rams, planes with machine guns, or nuclear weapons.

LIVING WITH THREATS

A key insight from Sagarin's work is that the systems that have evolved in animals to deal with threats are often decentralized and distributed. Consider the immune system. First, it contains several layers of defense including the skin, antibodies, and T cells. Second, it mounts responses without involvement from the centralized nervous system. This decentralized and distributed system, Sagarin argued, is more likely to be adaptable.[29] Hence, we should look to making our defense systems more decentralized—for instance, giving local officials more flexibility—in order to make them more responsive to threats.

Sagarin also suggested that we should become better at using uncertainty to our advantage. A country facing the threat of terrorism is a bit like an animal dealing with the threat of a predator. In both situations, both participants face uncertainty. Sagarin noted, "organisms seek to reduce uncertainty for themselves and increase uncertainty for their adversaries".[30] For example, consider periodic cicadas. Every 13 or 17 years—depending on brood and species—these insects come out in huge numbers, far too many for predators to eat. The timing is not an accident; by appearing in cycles based on large prime numbers, the cicadas thwart the ability of their predators to time their banner years with the cicada appearance. Sagarin noted that the color-coded security threat level used in the 2000s and the general procedures used by the TSA are examples of what not to do; they mitigated the uncertainty of terrorists while unnecessarily keeping citizens at a perpetual elevated threat.

Sagarin's colleague Dan Blumstein at UCLA studies the behavior of marmots—think of them as being like squirrels of unusual size. From his studies of antipredator behavior, Blumstein has also made suggestions about how we can improve security.[31] Like Emlen, Blumstein stresses the economy of nature, noting that natural defense systems evolved from preexisting ones. Defenses against one predator are often co-opted as defenses against novel ones. Blumstein presented the case that we should take this lesson in economy to heart, avoiding overly costly defenses:

> Whenever possible, we should seek to adapt defenses from preexisting resources. For instance, rather than creating a novel and unique "Department of Bioattack Division", health care systems can be better developed and communication among hospitals and governmental agencies improved, so that biological attacks could be quickly detected.[32]

I spoke with Blumstein during the summer of 2020 as much of the United States was seeing surges in coronavirus cases.[33] He noted the challenge involved in figuring out structuring ways of thinking to incorporate Darwinian ideas into policy about how to better respond to and cope with threats. One important challenge will be how to set up systems such that they can be "preadapted" to deal with novel threats. Blumstein saw the state of Darwinian security being like the early days (late 1990s) of evolutionary medicine (see Chapter 1). At this stage, we are not so much looking for specific solutions for particular problems but rather attempting to develop a frame of reference for thinking. Given that our bodies have actually evolved through natural selection and other Darwinian processes and that national security for the most part has not, I suspect progress in Darwinian security will be slower than progress in Darwinian medicine has been.

DIRECT APPLICATION TO ASYMMETRIC WARFARE

For the most part, work in Darwinian security has rested on looking to nature for insights that come from imperfect analogies. But Dominic Johnson at the University of Edinburgh has directly applied the principles of evolution to counterinsurgency and counterterrorism.[34]

The United States has been mired in conflicts that involve insurgencies for most of the first two decades of the 21st century. The insurgencies following the invasions of Iraq and Afghanistan have resulted in several thousand American deaths and tens of thousands wounded. Hundreds of thousands more have been deployed in these war zones, interrupting their lives and placing them in perilous conditions. Why were such insurgencies able to persist for so long, given the overwhelming strength of the American military? How can we mitigate future insurgencies?

The answers to these questions can be addressed by considering the principles of evolution. These principles are not limited to biological organizations. Recall from the Introduction how Frances Arnold and other chemists have used evolutionary principles to create more useful and effective enzymes. Similarly, engineers have used the principles of evolution in the design of airplanes and other machines. Likewise, processes using these principles and thus analogous to evolution occur in cultural contexts. Evolutionary adaptation requires variation, selection, and faithful replication.

Johnson's argument is that the conventionally "weaker" side—the insurgency—typically has the upper hand with respect to the rate of evolution because that side has more variation, faces more intense selection pressure, and has more faithful replication. The insurgency is more varied in both participants and tactics than the US military. The stronger US military put more selection on the insurgents than vice versa. The weaker insurgency also fought on the same territory day in, day out for years, while the stronger US military had new people rotating in and out. These differences made it likely that the insurgency would outpace the military in the evolution of tactics. As Johnson noted, "initial attacks following the invasion were scattered and unorganized, but quickly morphed into car bombs at checkpoints, ramming convoys, and IEDs".[35] Although the military adjusted its tactics, the insurgents changed tactics again.

The challenge to militaries facing insurgencies is how can they improve their own adaptation and limit how quickly the insurgency adapts. Sure, the insurgents usually adapt faster, but large organizations can also adapt. In the early days of the insurgency of Iraq, we saw adaptation by the US military. The adaptation was not top-down—it did not come from the Pentagon or even by senior field officers—instead, it came from the bottom up. It came from quick-thinking soldiers who armored their Humvees with scrap metal they found.[36] This "hillbilly armor", as the soldiers called it, was a practical makeshift solution to a problem that the Pentagon, with all its resources, failed to anticipate. The adaptation seen in this hillbilly armor is more like the adaptation we observe in the biological world.

NOTES

1. Sagarin et al. (2016, p. 10).
2. Emlen (2014, p. 73).
3. Homeland defense: exploring the Hart-Rudman report. U.S. Senate, Subcommittee on technology, terrorism, and government information, Committee on the Judiciary. April 3, 2001.
4. Ellis (2004).
5. Bauch (2001).
6. Sagarin (2012).
7. Foreword by Senator Gary Hart to Sagarin (2012).
8. Sagarin (1994).
9. Eilperin (2015).
10. Phone conversation with Dan Blumstein. August 2020.
11. Eilperin (2015).
12. See Axelrod (1984) for more on the prisoner's dilemma.

13. Just how well TIT FOR TAT does depends upon the pool of competitors as well as on the structure of the tournament. See Rapoport (2015) for details.
14. Axelrod (1984, p. 54).
15. Axelrod and Hamilton (1981).
16. Schweinfurth and Taborsky (2020).
17. See Axelrod and Hamilton (1981) and Lenski and Velicer (2000).
18. Bhuiyan (2016).
19. Wright (2000).
20. Despite formal training in biology, George Price had made seminal contributions to evolutionary biology in a few short years (1967–1974) in areas ranging from the theory of natural selection to the problem of altruism to game theory. Originally a chemist who had worked on detecting traces of radioactive isotopes as part of the Manhattan Project, Price jumped from field to field in the 1950s and 1960s. After a botched surgery to remove thyroid tumor, Price left the United States for the United Kingdom and started working on topics in animal behavior and evolutionary biology. Price fell into a depression during the winter of 1974, possibly in part as a complication of the thyroid operation. Then he died from suicide in early 1975. See Frank (1995) for a summary of his contributions to evolutionary biology and Harman (2010) for biographic details.
21. Maynard Smith and Price (1973).
22. Maynard Smith (1976).
23. Emlen (2008).
24. Emlen and Phillips (2006).
25. Emlen et al. (2005).
26. *Ibid.*
27. Emlen (2008, 2014).
28. Fea et al. (2020).
29. Sagarin (2012).
30. *Ibid.*, p. 134.
31. Blumstein (2008).
32. *Ibid.*, p. 152.
33. Phone conversation with Dan Blumstein. August 2020.
34. Johnson (2009).
35. *Ibid.*, p. 92.
36. Sagarin (2012).

17 Human Genetic Diversity and the Non-Existence of Biological Races

SUMMARY

A century ago, the then-new sciences of genetics and evolutionary biology often were misused to propagate racism, anti-miscegenation laws, and anti-immigration. We need to be careful about repeating such mistakes. Current evolutionary genetic research affirms ideas from mid-20th century anthropologists that there are no races, as commonly perceived; instead, we see that most variation can be found within populations and that there are clines in the frequency of genetic variants. We look at studies of ancient DNA that show modern humans exchanged a trickle of genes with not only Neanderthals but also with Denisovans and other archaic humans. Ancient DNA studies show that there is no such thing as a pure human population; we have always been mixing. Challenging common perceptions about race, the ancient DNA studies revealed that Cheddar Man, who lived 10,000 years ago in what is now England, had light-colored eyes but dark skin. We also examine the evolutionary genetics behind skin color. Finally, we turn to a book written by Nicholas Wade, a science journalist, that has misapplied findings of evolutionary genomics that incorrectly propagate ideas about the reality of races and racial stereotypes.

"Skin color is a terrible racial classifier. There really are no good biological classifiers for race."—Sarah Tishkoff.[1]

"There are no races, there are only clines."—Frank Livingstone.[2]

"We now know that nearly every group living today is the product of repeated population mixtures that have occurred over thousands and tens of thousands of years. Mixing is in human nature, and no one population is—or could be—'pure.'"— David Reich.[3]

CHEDDAR MAN

Cheddar Gorge, near Somerset in southwest England, is known for its eponymous cheese. It is also known for "Cheddar Man". Discovered in 1903, Cheddar Man was the first skeleton of a pre-written-history human in the British Isles.[4] He lived during the Mesolithic, the middle Stone Age, just under 10,000 years ago. Cheddar Man most likely practiced hunter–gathering, as agriculture did not spread to the British Isles until considerably later.

DOI: 10.1201/9780429503962-21

What does Cheddar Man look like? We can't tell directly from his skeleton what kind of hair or skin or eyes he had; the ravages of time have wiped all these features away. All we have is his skeleton. No one saw what he looked like. But that has not stopped people from depicting him. One such depiction has him with dark hair and mustache with rosy cheeks and, yes, light skin like White Europeans of today.[5]

Thanks to ancient DNA studies, we now know that the Cheddar Man does not look like the depiction in the previous paragraph. Selina Brace at London's Natural History Museum and several teams of scientists sequenced DNA from Cheddar Man. With knowledge about which alleles he had at known genes that affect skin color, Brace and the other researchers were able to infer that Cheddar Man had dark to black skin. Similarly, they inferred that his hair was dark brown or black and that his eyes were blue or green.[6]

Similarly, a 7,000-year-old skeleton found in northern Spain had ancestral alleles of skin-color genes suggesting that he, too, had dark skin.[7] In addition, like Cheddar Man, he also likely had blue eyes.[8] Dark-skinned individuals with lightly colored eyes living in northern Spain and in the British Isles several millennia ago is surprising. As the evolutionary geneticist David Reich puts it, "Britons in the past did not look like Britons today, and were genetically very unlike Britons today."[9]

With his dark skin, Cheddar Man does not look like a "White" West European of today. But he also does not look like a "Black" African. His likely appearance and his genes confound most layperson's ideas about race. As we will see, there is no good biological criterion to place humans into races, as we can for chimpanzees and some other animals. There is only one human race.

RACE IN AMERICA

I write this chapter in April 2021 just days after Derek Chauvin was found guilty of second-degree murder and other charges related to the death of George Floyd,[10] a 46-year-old security guard who happened to be Black.[11] The previous May, Chauvin had been a Minneapolis, Minnesota, police officer who arrested Floyd because he suspected Floyd was using a counterfeit $20 bill. During the arrest, Chauvin knelt on Floyd's neck for nearly ten minutes as Floyd repeatedly cried out that he could not breathe. Chauvin dismissed these concerns. Long after Floyd ceased breathing, Chauvin kept him pinned to the ground, facedown, ignoring comments from the other officers about Floyd's condition. Floyd's needless death and the cruelty displayed by Chauvin prompted rallies and protests across America during the summer of 2020—even in the midst of a pandemic—and sparked considerable soul searching about how we value Black lives in this country.

The murder of George Floyd is far from the only news event in recent years to spark attention involving "race". The list is too long to mention. One key event was the killing of Trayvon Martin in 2012.[12] After stopping at a 7-Eleven convenience store, the 17-year-old African American Martin was walking to his father's house in Sanford, Florida. George Zimmerman, a resident in the area, saw Martin and called the police. Taking matters into his own hands, Zimmerman went after Martin. In the struggle, Zimmerman killed Martin. Although Zimmerman would eventually be tried (and acquitted) for Martin's murder, no charges were filed for six weeks. The delay in charging Zimmerman led to protests that eventually would result in the

Black Lives Matter movement. This movement has expanded beyond standing against brutality against Black people to defeating the systemic devaluation of Black people.

Martin Luther King's dream remains not fully realized: all too often, individuals are judged on the color of their skin and not by the content of their character. We also have not grappled with the pervasive influence of slavery and the state-supported suppression of Black people that occurred in much of the United States for a century after the official end of slavery. As the *Miami Herald* columnist Leonard Pitts noted: "As a nation, we have never quite dealt with our African-American history—the unremitting terrorism, the ongoing violations of human rights, the maiming of human spirit."[13] Nodding to the historian Ray Arsenault, Pitts added that Americans have hung onto "mythic conceptions of what they think happened" rather than the whole truth of these experiences.

The use of race as a concept for classifying other humans is not all that old. To be sure, people living a thousand years ago or even 500 years ago recognized that other humans from different geographic regions appeared different from them, but these differences were not reified into racial categories like we see them today. The modern conception of race arose in the 17th and 18th centuries. It appeared in part to justify slavery and colonialism: it is easier to treat someone as less than human if they are considered a different race.[14]

The story of the use of race to justify mistreatment and exclusion is not restricted to what was inflicted upon Americans with African ancestry. In his recent book *The Guarded Gate*, the writer and editor Daniel Okrent chronicled the use of race—and the use of the science of the time—to restrict immigration. Even other Europeans were sometimes viewed as distinct (and inferior) races by Americans of European descent during the 19th century. Prescott Hall, a lawyer who was one of the most vigorous advocates for restricting immigration, wrote in *The Boston Herald* in 1894 about the rise of immigration of people from Italy and countries from eastern Europe: "Shall we permit these inferior races to dilute the thrifty, capable Yankee blood?"[15]

Immigration from Asia would be essentially eliminated by the 1880s.[16] Restriction of Eastern Europeans would follow a generation later, and it would be done aided by calls to science. Writing in the 1910s toward the end of his long career as a historian, author, and politician, Senator Henry Cabot Lodge tied the new science of genetics and evolution to restricted immigration: "Darwin and Galton have lived and written, Mendel has been discovered and revived, and the modern biologists have supervened."[17]

These sciences were also used to appeal to support for anti-miscegenation laws. In his book *The Passing of a Great Race*, Madison Grant, one of the founders of the Bronx Zoo and close friend of President Theodore Roosevelt, argued: "Whether we like to admit it or not, the result of the mixture of two races, in the long run, gives us a race reverting to the more ancient, generalized and lower type."[18] Although many geneticists of the time objected to the use of their science to promote eugenics and racism,[19] the use of these sciences to justify these ills was not restricted to a few cranks; indeed, it was pervasive.

NO RACES, ONLY CLINES

Although other factors were involved, the shock following the release of the information about the full extent of the atrocities of the Nazis made eugenics into a dirty word.[20] These horrors, especially the Holocaust, led to much soul searching among human geneticists, evolutionary biology, and anthropologists of the era.

Many biological anthropologists of the third quarter of the 20th century expressed skepticism that races in humans existed. Some went further and outright denied the idea of separate races. We saw in the epigraph that the American anthropologist Frank Livingstone declared that that there were "no races, only clines". Livingstone, who was among the first to investigate the genetic basis of sickle cell anemia, certainly was aware that people differed in physical appearance and in the frequency of alleles. He thought, however, that these physical and genetic differences were not discrete categories but rather gradations.[21] He reminded his critics that sickle cell anemia was not just found in Africa but also in Greek and Italian populations, albeit at lower frequencies. Loring Brace, one of the leading paleoanthropologists of the time, agreed with Livingstone's conclusions.[22] In his rejection of races, Brace noticed variation within populations and graded clines across geography in skull measurements.

Are the statements by Brace and Livingstone on the absence of discrete races based on the cherry picking of a few outlier observations and some wishful thinking, or are they supported by quantifiable analysis? Dick Lewontin would address this question in the early 1970s with analysis of allele frequencies of a set of genes across different populations. We met Lewontin in Chapter 15 in the discussion of the use of DNA fingerprinting in court cases. In the 1960s, Lewontin had become famous among evolutionary geneticists for his breakthrough work on accessing genetic variation within population by how variants of enzymes migrated on gels. In the early 1970s, he was busy working on his now-classic book *The Genetic Basis of Evolutionary Change*, which provided the intellectual framework for evolutionary genetics for several decades.[23]

Lewontin argued that we typically cue in those differences—skin color, eye shape, hair type—that typically are used to make racial differentiation.[24] We are species that heavily rely on visual information, and we like to place things (and people) into categories. To get an unbiased sense of the patterns of variation within and among different populations and within and among the entities that we typically call races, we would need a reasonably unbiased data set. Such a data set would be a sufficiently large set of genes that were chosen more or less at random, without any preconceptions. For example, genes that affect skin color would not be representative of the true patterns of variation. For his data set, Lewontin used the frequency of alleles from 17 different genes, including some enzymes, a few lipoproteins, and some of the minor blood groups.[25] A few dozen populations from seven different "racial groups" were analyzed.[26]

Lewontin found that the vast majority—over 85 percent—of the genetic variation was within populations. The variation that was among populations, but within each of the seven groups, was 8.3 percent of the total. Finally, the variation among the seven major groups corresponding to "races" was only 6.3 percent! Only about 6 percent of the total genetic variation is explained by clustering individuals into "races". We are much more alike than we are different.

Lewontin also noted the genetic diversity in the African populations. On the other end of the continuum, there was considerably less genetic variation within indigenous Americans and Oceanians. This suggests that modern humans originated in

Africa and then migrated outward. With each migration, there was a loss of genetic variation.

From these data, Lewontin concluded that there was no justification for placing humans into separate races. Randomly chosen genes show that what we call races "are remarkably similar to each other, with the largest part of human variation being accounted for by the differences between individuals".[27] In other words, there are no races, only clines.

Three decades after Lewontin's analysis, British statistical geneticist Anthony Edwards claimed Lewontin had committed a fallacy when he claimed that races could not be distinguished by examining genetic differences.[28] Edwards argued that due to the modest differences in allele frequency between groups traditionally recognized as races, it would be impossible to assign individuals to such groups based on data from a single gene or even from a handful of genes. But with more and more genes—especially if variation at different genes were correlated (linkage disequilibrium)—such assignment would become increasingly possible. Edwards' jab at Lewontin could be summed up as: I'm right and you're wrong because with enough data, I can reliably classify individuals into groups traditionally recognized as races. The problem is that Lewontin never committed this fallacy. Lewontin did not say that such assignment was impossible. He only pointed to the data showing that only a modest amount of variation was explained by differences among the groups traditionally viewed as races.[29]

Lewontin's 1972 survey was of just 17 genes, some of which are likely to have been under strong selection. Moreover, the determination of the allelic states of these genes was crude, based on the appearance of the protein instead of DNA sequence. Now we have extraordinarily more information, and we have it from a wider scope of genes. Do Lewontin's results still hold? The answer is yes: more recent surveys using much larger and DNA-based samples come to same conclusions. The vast majority of the genetic variation is within populations. In fact, these surveys show only 3 to 5 percent of the variation is among the groups commonly thought of as races.[30]

Is 5 percent a lot of the variation? I have thus far have not given context or comparisons to that figure. Let's consider two such comparisons: dog breeds and subspecies. Dog breeds have been subjected to hundreds of generations of artificial selection to generate specific traits and to maintain breed purity. The proportion of genetic variation that is among breeds of dogs is around 30 percent of the total, much higher than that among groups of humans commonly thought of as races. Human races are not like dog breeds.[31]

Species of nonhuman animals are sometimes divided into subspecies for taxonomic purposes. Subspecies are also recognized to protect variation with species where some populations are threatened. The typical cutoff for designating subspecies of animals occurs when the proportion of variation among different groups is 0.25 or greater.[32] As we have seen, the value for this statistic in the groups we typically consider races of humans is about 0.05 or less, which is well below the cutoff. In contrast, based on this criterion, we can designate subspecies or races for the common chimpanzee, *Pan troglodytes*. One such race would be those living in upper Guinea. Another would be the population around the Gulf of Guinea. All other chimpanzee populations would fall into the third race. These three groups are sufficiently differentiated from each other. Chimpanzees have separate biological races; humans do not. There are no races.

Also, consistent with what Livingstone had claimed nearly six decades ago, most of the human genetic variation that exists among different populations can be explained by clines: populations that are near each other are genetically similar.[33] Some studies have found clusters of populations that cluster according to continents, but much of those clusters could arise from the sampling. If the sample used does not fully represent all the variation in the continent, then clusters associating with continental boundaries could arise as an artifact even if the populations do indeed integrate. There might be some slight clustering associated with continental boundaries, but it is minute.

The DNA data also confirm the decline in variation with distance from central Africa, the pattern that Lewontin noticed in the early 1970s. One explanation for this pattern is a series of bottlenecks of reduced genetic variation; with greater geographic distance from central Africa, there were more bottlenecks. Recently, Joseph Pickrell and David Reich noted that admixtures with various populations (some of which may be extinct) can generate the same pattern.[34] As we will see, human populations have repeatedly mixed among one another. Moreover, a combination of admixture and bottlenecks is also plausible. We will return to the clustering of human populations in discussion of a recent book, Nicholas Wade's *A Troubled Inheritance*, toward the end of this chapter.

ANCIENT DNA AND NEANDERTHALS

Genome sequencing has extraordinarily exploded over the last few decades. During the 1990s, we started getting the occasional morsels of information like the drips of a water faucet left slightly open. By the end of that decade, the trickle was approaching a steady stream. Over the next decade, it continued expanding, becoming like the flow of a garden hose. In the 2010s, this outpouring has been more like an open fire hydrant. We are awash in genomic data.

The advances of genomic sequencing during the last decade have been especially visible in ancient DNA. David Reich, one of the leaders in this field, has noted that studies of ancient DNA double at a rate about equal to the delay between researchers submitting manuscripts to journal and their publication.[35] This is faster than the famed Moore's law of computing. Things are happening fast and accelerating. Being able to obtain (nearly) complete genomes from multiple ancient humans allows evolutionary geneticists to do ground truthing of the changes that occurred during our history rather than to rely just on statistical inference from patterns seen in contemporary populations. One inherent assumption of past studies that relied on patterns of DNA in contemporary populations is that populations in the past did not move—that the people in an area 10,000 years ago are the same people that were there 4,000 or 1,000 years ago. Ancient DNA allows for direct testing of that assumption.[36] As we will see, this assumption turns out not to be a realistic one: human populations move around and often exchange genes with other human populations.

Neanderthals disappeared from Europe about 39,000 years ago.[37] Before their disappearance, Neanderthals met modern humans and likely exchanged both genes and tools. Humans and Neanderthals likely met in the Near East in around the location of modern-day Israel. The myth that Neanderthals were brutish, unsophisticated creatures does not hold up to the facts. Neanderthals made stone tools that were as advanced as those that the modern humans of that era used. There is also abundant evidence that they tended to their sick and elderly. They were as humans as our ancestors were.

We know much more about Neanderthals from the DNA we have obtained and sequenced from them. Getting DNA from Neanderthals or any skeleton more than a few hundred years old is extraordinarily challenging. Elaborate precautions to avoid contamination—both from bacteria and other microbes as well as from the researchers' own DNA—are required. In the early days of Neanderthal DNA studies, we only had mitochondrial DNA (which is much more abundant, but only provides information about matrilineal lines of descent). The mitochondrial DNA evidence from Neanderthals ruled out extensive exchange of DNA between modern humans and Neanderthals. This DNA evidence was limited in its power; it could neither support nor contradict more limited genetic exchanges between Neanderthals and modern humans.

The ground-shaking news that Neanderthals and modern humans did indeed exchange genes came with the nearly complete sequence of Neanderthal DNA. Even with the complete genome sequence, some researchers remained skeptical. For instance, Graham Coop, whom we met in Chapter 15 in the case of ancestry databases and the Golden State Killer, raised the possibility that while there likely was interbreeding between modern humans and an archaic population, the population may not be Neanderthals but one related to Neanderthals. Getting DNA from more specimens allowed researchers to home in on just who was mating with the modern humans. With more evidence, confidence grew. Modern humans and Neanderthals mated and exchanged genes at various times in our collective pasts.[38]

People from most populations outside of sub-Saharan Africa have a trickle of Neanderthal ancestry, with people from East Asia generally (but not always) having more Neanderthal DNA than those from Europe. On average, about 2 percent of DNA in a modern human came from Neanderthals, though the range is from less than 1 percent up to 4 percent. Remember that Neanderthals and modern humans are already closely related, so the overall proportion of genetic differences that we obtained from Neanderthals is quite low.

Roughly 2 percent of the modern human genome is from Neanderthals, but the extent of Neanderthal DNA varies considerably across the genome: some regions have much more than others.[39] Some parts of the genome are completely or almost completely devoid of Neanderthal DNA. These are referred to as Neanderthal deserts. Such deserts more often occur in gene-dense regions than gene-poor ones. This suggests that natural selection has been operating to remove Neanderthal DNA over time, as genomic regions that have a lot of genes are more likely to be subject to selection.[40]

Two further observations suggest that the selection against Neanderthals has come out because some Neanderthal DNA is incompatible with human DNA: the X chromosome has much less Neanderthal DNA than the autosomes (the non-sex chromosomes), and the Neanderthal contribution is reduced in regions that include male fertility genes. In crosses between species and subspecies of other mammals, the fertility of male hybrids is usually more affected than the viability of hybrids or the fertility of female hybrids.[41] Moreover, the X chromosome typically plays a disproportionately large role in this reduction of hybrid male fertility.[42] In addition, ancient DNA samples of modern humans from Europe show that the proportion of Neanderthal DNA has declined over time.[43] Whether we assign Neanderthals to a distinct species from humans or a subspecies, there clearly is some degree of post-mating reproductive isolation between modern humans and Neanderthals. In contrast, there is no indication of such reproductive isolation between different groups of contemporary modern humans.[44] There are no biological races in contemporary humans.

Until recently, researchers had assumed that sub-Saharan African populations had practically no Neanderthal ancestry, as they were unlikely to have had contact with the Neanderthals. These populations had been used as a reference to look at the Neanderthal contribution in other populations. A new method that does not rely on a reference suggests that African populations have considerably more Neanderthal DNA than had been assumed, but still less than the non-African populations.[45]

ANCIENT DNA: DENISOVANS AND GHOSTS

In the 1700s, a Russian hermit named Denis lived in a cave in the Altai Mountains of southern Siberia.[46] We know much more about the cave, which is about as spacious as a four-bedroom house, than we do about Denis. This cave, the Denisova cave, has become famous because of the discoveries of archaic humans from it. Over the last few hundred thousand years, it has been home to several groups of archaic humans, including Neanderthals and their namesake, newly discovered archaic humans called Denisovans. These Denisovans, which existed until about 50,000 years ago, mated with both Neanderthals and modern humans. Indeed, some of their DNA lives on in the genomes of some individuals.

The first Denisovan bone found was just part of a pinky finger from a child.[47] The researchers knew it was a child's because the growth plates had not fused as they would have in an older individual. DNA from the finger revealed this individual to be belonging not to modern humans or Neanderthals but a distinct group that had diverged from both modern humans and from Neanderthals several hundred thousand years ago. Analysis of the DNA from the Denisovan and from contemporary human populations across the world led to the conclusion that Denisovans and humans had exchanged genes, leading to some Denisovan DNA being present in some contemporary human populations. Denisovan DNA is most abundant in the genomes of people from New Guinea—an estimated 3 to 6 percent of their DNA came from Denisovans! But it is much less common in most other populations; it is found at 0.2 percent in the genomes of most East Asian individuals and only in trace amounts elsewhere. Biologists studying Denisovans are not sure where and when Denisovans and modern humans mated and exchanged genes. David Reich thinks the most likely place is in or near southern China.

Most of the Denisovan DNA, like most of the Neanderthal DNA, found in contemporary human genomes is likely either neutral or slightly deleterious. A few exceptional genes that came from Denisovans have been adaptive. One notable example, as we had discussed in Chapter 4, is the change in red blood cells that enables Tibetan individuals to live under the low oxygen regimes found at high elevations.

Modern humans exchanged genes with two archaic humans, Neanderthals and Denisovans. From genomic analysis, we have strong signals that modern humans have exchanged genes with other archaic humans. For instance, when he was a postdoc working with Sarah Tishkoff at the University of Pennsylvania in the early 2010s, Joe Lachance sequenced and analyzed whole genomes of 15 individuals from populations of hunter–gatherers in Africa.[48] In some regions of these genomes, there were clusters of variants that were in high linkage disequilibrium that looked different in comparison to other genomes. Such a pattern resembles what is seen with the regions associated with Neanderthal and Denisovan introgressions. It looks like there had been an exchange of genes with some archaic human population in the past. What's different here is that we do not have direct evidence of the archaic human population; we only have the signature it left.

David Reich calls such populations "ghosts".[49] His team, led by Pontus Skoglund, found further evidence for such a ghost population in Africa by looking at both ancient DNA and contemporary populations. The San, who consist of several populations of hunter–gatherers in southernmost South Africa, are perhaps the oldest cultures and oldest peoples known. Most of their genomes come from the lineage that split from the lineage that led to all other contemporary human populations.[50] One would predict, then, that the San should be equally distant from hunter–gatherers in West Africa and those further east. But that is not the case; the San share fewer mutations with populations from West Africa than those from Central or Eastern Africa.[51] One explanation is that the Western African populations exchanged genes with an ancestral ghost population, thus reducing their apparent similarity with the San.

Skoglund's analysis also found a north–south cline showing a gradation of genetic variants across geography; going south, the resemblance of ancient individuals to people currently living in southern Africa increased. Because the analysis included seven different individuals who lived in Malawi over a span of several thousand years, the researchers could check to see whether temporal changes occurred. They found no consistent change over time in Malawi between individuals that lived there 8,000 years ago to 2,500 years ago. Interestingly, the current population of Malawi does not resemble this ancestral one. It was replaced during the Bantu expansion that we discussed in Chapter 3.

Such ghosts keep popping up: sometimes, these are ancient ghosts of archaic human populations like the one Lachance and Tishkoff found in the hunter–gatherers in sub-Saharan Africa; sometimes, the ghosts are more recent modern humans. Either way, genomic inference has picked up these signals, and these signals strongly suggest that human populations have been exchanging genes with other populations—both those that are similar to them and those that are more different—for the last million years, if not more. In addition, what populations were around several thousands of years ago may not correspond to the populations there now. Ten thousand years ago, West Eurasia (from Europe to approximately Iran) was inhabited by at least four major populations.[52] The two easternmost groups farmed; one was in Iran and the other was in the so-called Fertile Crescent that stretches from western Iraq to northern Egypt. The two western groups were hunter–gatherers. Cheddar Man belonged to the westernmost one of these, whose boundaries extended across western Europe. The other group was in central and eastern Europe. As David Reich notes, these groups were about as similar—or distinct—as East Asians and Europeans of today, but none of them exist today. They've all either disappeared or have been mixed up. The same story plays out elsewhere across the world.

Populations appear, mix with one another, and disappear. We have always been mixing; there are no pure populations.

ONLY SKIN DEEP

Why do different populations have different shades of skin color? Natural selection has played the predominant role in shaping this variation.[53] Sexual selection and other cultural practices likely have been contributing factors in some regions.[54]

Most of the selection on human skin color is from two competing pressures: too much ultraviolet radiation is harmful, but some ultraviolet radiation[55] is needed to spur on the chemical reactions to make vitamin D. Although excessive ultraviolet radiation can cause cancer, that probably is not the main reason why too much of it is bad for fitness. Excessive ultraviolet radiation also depletes the body's supply of folate, one of the B vitamins, as the body uses it during repair of cellular damage. Folate is particularly important in rapidly dividing cells as well as during pregnancy; in fact, several major birth defects arise from folate deficiency. Folate deficiency also can affect male fertility, as it has been linked to poor sperm quality and quantity. Hence, dark skin would be advantageous in areas where solar radiation is extremely high; by blocking the ultraviolet radiation, folate levels could be maintained.

Vitamin D is important not just for bone development and maintenance but also for the immune system. Lighter skin thus would be advantageous in areas where ultraviolet radiation was limited. Moreover, the ability to obtain darker skin when ultraviolet radiation is most intense (natural tanning during the summer months in temperate latitudes) would be adaptive. Note that tanning is an acclimation, while the ability to tan is an adaptation.

Given these two selection pressures, one can see that large-scale migration— either voluntary or forced (as in the subjugation of numerous people from Africa into slavery)—can create mismatches. For instance, people with light skin in areas with high solar radiation are more prone to skin cancer. They may also be at risk for folate deficiency. In contrast, people with dark skin in areas with low solar radiation are at greater risk for vitamin D deficiency.

How is skin color produced?[56] Our skin has numerous specialized cells called melanocytes which produce melanin, the main class of pigment that determines skin color. Melanin is contained in organelles called melanosomes. Melanocytes transfer melanocytes to keratinocytes, the main cells that line the epidermis, the outer layer of our skin. Compared with lighter skin, darker skin has more and larger melano-cytes. The melanocytes are also spread more evenly in dark skin. Other factors that determine the color of skin include the rate at which melanin is produced and how long melanocytes remain in keratinocytes before they are degraded. In addition, there are two types of melanin—eumelanin, the dark pigment, and pheomelanin, the reddish-yellow pigment. With more eumelanin and less pheomelanin, the skin is darker. With more pheomelanin and less eumelanin, the skin is lighter.

In the last two decades, more and more genes have been linked to changing skin color. The first gene linked to variation in skin color in humans was the melanocortin-1 receptor (*Mc1R*) gene, which encodes a receptor protein found on melanocyte cell surfaces.[57] The nature of the Mc1R protein affects the production and the relative concentrations of eumelanin and pheomelanin. Variation at this gene also often affects the color of other mammals. In fact, just one change in the Mc1r protein leads to the lighter color in beach mice, lighter variants of *Peromyscus polionotus*, that blend in the sand near beaches.[58]

For most of the time since the lineage that led to our species' split from chimpanzees, our skin color was likely intermediate in color. Our thick hair protected us from the sun's strong ultraviolet rays. Then about a million years ago, our ancestors lost that thick hair across most of their bodies. This shift was likely due to changes in the way we thermoregulate; because we had shifted to sweating more, the thick hair we had that was previously adaptive now became a burden. As that hair was lost, our exposed skin became more vulnerable to the ultraviolet rays. Hence, darker skin became advantageous. Corresponding with this shift, we have evidence that selection at *Mc1r* led to the fixation of a new variant along with decreased variation near that site.[59] This change is found in all populations.

Another important gene affecting human skin color was identified by looking at a mutation in a model system, zebrafish (*Danio rerio*).[60] In zebrafish, the *golden* mutation has reduced melanin and looks lighter. Examination of the golden zebrafish under electron microscopes revealed that they have fewer melanosomes than typical zebrafish. The melanosomes that the *golden* fish did have were smaller and less dense. The characteristics of these melanosomes match those of people with lighter skin. What was the genetic change that led to the *golden* mutation? Does the same gene that affects melanosomes and color in golden fish also affect melanosomes and skin color in humans?

Researchers found that the *golden* mutation was due to a change in the solute carrier family 24 member 5 protein—whew, that is a mouthful!—coded by the *SLC24A5* gene. This gene is expressed in melanocytes and is found in the Golgi apparatus, an organelle that processes proteins and other molecules that are to be secreted. This protein appears to play a role in exchanging sodium and calcium ions across membranes.

In humans, nearly all individuals in African and East Asian populations have one allele at a site of *SLC24A5*, whereas nearly all individuals in European populations have a different allele. This nearly fixed difference is in contrast with nearly all other genes, where the differences in frequency across different populations are much more modest. Several lines of evidence point to the African/East Asian allele being ancestral and the European allele being derived. Moreover, the European allele shows signs of having been driven to high frequency by natural selection. In African Americans, which have a mixture of African and European ancestry, skin color variation is associated with the genotype at *SLC24A5*. However, this gene is not the only contributor to skin color. Moreover, the contribution of SLC24A5 to skin color is often dependent of alleles at other genes; gene interaction affects skin color.[61]

With time, more genes associated with skin color were found, such as another transport protein (*SLC45A*), a gene that affects the survival and proliferation of precursors to melanocytes (*KITLG*), and a gene that encodes an enzyme found mainly in melanocytes (*TYRP1*). Knowing the effects of variants at these and a few others is what enabled researchers to infer the skin color of Cheddar Man and other individuals for whom we have ancient DNA.

This knowledge also has allowed researchers to address when the genetic changes that led to lighter skin in Europeans started.[62] Molecular evolution studies of contemporary Europeans allow researchers to determine that the earliest change to lighter skin was at *KITLG* and occurred about 30,000 years ago. Next came changes in *SLC24A5* and *SLC45A2* and *TYRP1* around 11,000 to 19,000 years ago, well after the initial migrations of modern humans into Europe (60,000 years ago). These dates are before Cheddar Man existed; remember that there were other populations in Eurasia with lighter skin. Moreover, there likely was a long period where both dark and light alleles existed at intermediate frequencies in the same populations. The selection coefficient for the *SLC24A5* was quite large; individuals with some genotypes had a 10 percent or more fitness advantage over those with other genotypes. The selection coefficients for changes at *KITLG* were much smaller.

Until recently, most of the studies linking genes to skin color were mostly done either within Europe or comparing European and African populations. A major conclusion of these studies was that a few genes contribute a reasonably large proportion of the variation that we see in skin color. While skin color is not due to a single gene, it is not like height, where thousands of genes affect the trait, each with a miniscule effect.

In 2017, two studies looking within Africa changed how we view the genetic basis of skin color. Sarah Tishkoff's group took skin color measurements of more than 1,500 individuals in numerous populations across Africa.[63] They found considerable variation in skin color: many individuals, especially in southern and eastern Africa, had intermediate skin color—much lighter skin than in other parts of Africa.

After getting genomes from these individuals, Tishkoff's group found some genes previously implicated in skin color variation in Europe also contributed to variation in Africa. Notably, an allele of *SLC24A5* that leads to light skin color appears to have arisen (likely from migration from Europe) in East Asia, first appearing about 5,000 years ago. It has since increased to high frequency. They also found new genes that contribute to the variation in Africa. For instance, a gene called *MFSD12* is expressed more in individuals with light skin; the protein that comes from this gene is involved in the lysosome, the organelle that breaks down defective parts of cells. Another new link to skin color includes genes that function in repairing damage from ultraviolet radiation. Alleles that lead to lighter skin and those that lead to darker skin both existed in Africa before the origin of modern humans. Moreover, about half of the ancestral alleles are light alleles and half are dark.

Another group led by Alicia Martin and Carlos Bustamante at Stanford University took a closer look at variation in skin color in Nama and Khomani San, two groups with intermediate skin color who have lived in far southern area of Africa for several tens of thousands of years.[64] Notably, they found that heritability of skin color is extraordinarily high in these groups: an individual's skin color greatly resembles that of their parents, much more than in European populations. Another important finding is that numerous genes—many more than had been found before—contribute to skin color of these people.

As with the East Africans in the Tishkoff study, variants at *SLC24A5* that confer lighter skin are at high frequency in the Nama and Khomani San. There are several possibilities for their high frequency, including the intriguing prospect that the alleles first arose in southern Africa long before the migration of modern humans out of Africa.

Skin color evolution has been a dynamic process within Africa for an extraordinarily long time, and much of that variation still exists. Further study of skin color (and other traits) within Africa and their genetic basis will improve our understanding of the fullness of the genetic diversity existing within our species.

TROUBLESOME SCIENCE

Nicholas Wade, a veteran science journalist with a long career at the *New York Times*, published a controversial book called *Troublesome Inheritance* in 2014.[65] Reviewing the book for the *New York Times*, David Dobbs called it a "deeply flawed, deceptive, and dangerous book".[66] In his review, Dobbs noted that Wade declared that there are separate races of humans, comparable to subspecies of other animals, by misinterpreting the information coming from evolutionary geneticists. Dobbs also made a more damning criticism: Wade—without evidence—links genetic differences across different races to societal behavior. For instance, Wade links (again without evidence) reduced societal trust and increased violence to genetic variants in African Americans.

Evolutionary geneticists echoed the criticisms Dobbs laid out. A group of biologists led by Graham Coop responded to the review Dobbs wrote in a letter to the editor in the *New York Times*. Explicitly disavowing Wade, they concluded: "We reject Wade's implication that our findings substantiate his guesswork. They do not. We are in full agreement that there is no support for Wade's conjectures."[67] More than 140 evolutionary geneticists—including myself—signed on to that letter.[68] Many biologists and anthropologists have written responses, including a full-length book,[69] mostly decrying Wade's book.

Wade begins by stating "New analyses of the human genome establish that human evolution has been recent, copious and regional".[70] He presents this sentence as though he were bringing up a controversial point that few scientists were willing to accept. But on the whole, practically no evolutionary geneticist today would deny that human evolution is recent and copious. We have discussed several examples of recent human evolution in this book, including adaptations to high elevations and the rising frequency of lactase permanence (allowing more of us to partake in delicious ice cream and other dairy products). While we can debate how much of this evolution is regional, clearly some of it is.

Wade goes further than declaring recent, copious, and regional human evolution. He asserts that "analysis of genomes from around the world establishes that there is indeed a biological reality to race, despite the official statements to the contrary of leading social science organizations."[71] But as we have seen, it's not just the social scientists who object to the reality of human biological races; it's also the evolution-ary geneticists. The differences that we observe among continental groups are far smaller than the variation seen within populations.

Wade brings up the Lewontin study and the subsequent findings that partitioned genetic variation among and within populations. Here, Wade makes several errors. He notes that there is a "15% genetic difference between races"[72] and then says that this magnitude of difference is equivalent to that of subspecies with nonhuman ani-mals. Wade is referring to the proportion of the genetic variation that is due to differ-ences among groups. First, 15 percent is less than the typically accepted magnitude for subspecies; it's less than dog breeds. Second, 15 percent is not the proportion of the genetic variation among the groups typically considered major races. It's the part of the genetic variation among all populations. The proportion that is among the groups typically considered major races was about 6 percent in Lewontin's original finding; recent analysis of genomic data place this figure somewhat lower—about 4 percent, far below the cutoff for subspecies.

Wade also points to a 2005 study by Noah Rosenberg and his colleagues that found human populations could be clustered into five clusters whose boundaries roughly correspond to continents. Wade takes this study as support for biological races. But Rosenberg, who was one of the leaders of the letter to the *New York Times* criticizing Wade, and his colleagues are point blank about these clusters not corresponding to races: "Our evidence for clustering should not be taken as evidence of any particular concept of 'biological race.'"[73] Moreover, Rosenberg and colleagues also noted that they found considerable evidence for clines and that the clusters were due to "small discontinuous jumps in genetic distance for most population pairs on opposite sides of geographic barriers".[74] In other words, the differences across clusters are small.

The approach Rosenberg and his colleagues used to cluster human populations is one based on a statistical program developed by Jonathan Pritchard called STRUCTURE.[75] This program, which is commonly used in evolutionary genetics to look for population structure in organisms (not just humans), assigns individuals to groups based on their genotype. The number of such groups (K) can be set by the researchers or can be estab-lished by examining which value of K has the highest statistical support. The authors of STRUCTURE are clear in urging to be careful in interpreting K, on both statistical and biological grounds. K is sensitive to the genetic loci and populations examined.

After discussing the Rosenberg study, Wade then moves to subsequent work that used a similar approach. That investigation found seven clusters, two more than in the Rosenberg study, with the two additional clusters corresponding to the people from the Indian subcontinent and those living in the Middle East. One of the differences between this study and the earlier one by Rosenberg was that it was based on point changes in the DNA sequence (SNP—single nucleotide polymorphisms) instead of repeats. This difference is not as important as the second one: the latter study used more markers. Wade notes that "the more DNA markers that are used, whether tandem repeats or SNPs, the more subdivisions can be established in the human population".[76]

The last statement by Wade is correct. More markers allow for greater resolution, allowing more clusters to be found. But Wade glibly dismisses the additional clusters:

> It might be reasonable to elevate the Indian and Middle Eastern groups to the level of major races, making seven in all. But then many more subpopulations could be

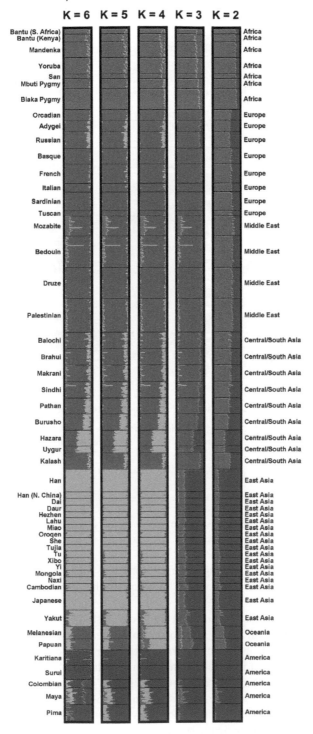

The program STRUCTURE assigns individuals to different groups. The figure shows partitioning with different number (K) of groups.

Source: From Figure 1 of Rosenberg, N. A., J. K. Pritchard, J. L. Weber, H. M. Cann, K. K. Kidd, L. A. Zhivotovsky, and M. W. Feldman. 2002. Genetic structure of human populations. *Science* 298: 2381–2385.

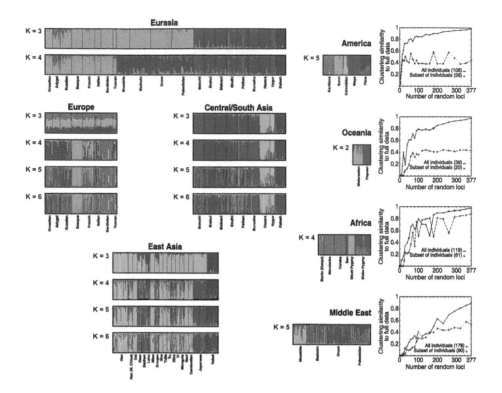

The program STRUCTURE can also find subdivisions within continental groups.

Source: From Figure 2 of Rosenberg, N. A., J. K. Pritchard, J. L. Weber, H. M. Cann, K. K. Kidd, L. A. Zhivotovsky, and M. W. Feldman. 2002. Genetic structure of human populations. *Science* 298: 2381–2385.

declared races, so to keep things simple, the five-race, continent-based scheme seems the most practical for most purposes.[77]

That's the problem—with enough markers one can find many small clusters. But numerous clusters apparently would undermine the narrative Wade wishes to establish.

As Dobbs noted, Wade does not just try to establish the reality of biological races in humans. He also with flimsy (or worse, no) evidence tries to link differences in societies to genetic differences among populations. For instance, after noting the adaptations that have evolved in Tibetans and other populations living at high elevations, Wade notes: "The adaptation of Jews to capitalism is another such evolutionary process, though harder to recognize because the niche to which Jews are adapted is one that has required a behavioral change, not a physical one."[78] The problem is that Wade has no support for his statements, statements that only serve to perpetuate harmful stereotypes.

CODA

The data show that humanity cannot be divided into separate biological races. There are no races, only clines. Yet the absence of races does not mean racism does not exist. The sad reality is that people are judged by the color of their skin. As Joe Graves at North Carolina A & T University bluntly and elegantly puts it: "Trayvon Martin was not killed due to the presence of any particular genetic polymorphism within his genome; rather it was due to long-standing socially defined racial stereotypes deeply embedded in the cultural psyche of this nation."[79]

The racial stereotypes and other insults have biological effects.[80] They alter stress responses and cause other physical harm in the individuals who have endured them and likely contribute to the very real and health disparities in groups traditionally viewed as races. Evolutionary geneticists have an obligation not to contribute to the perpetuation of these harmful stereotypes and to show why the science does not support them.

NOTES

1. Tishkoff is quoted in Saini (2019, p. 143).
2. Livingstone (1962, p. 280).
3. Reich (2018, p. 268).
4. Yeginsu and Zimmer (2018).
5. Saini (2019).
6. Brace et al. (2019).
7. Olalde et al. (2014).
8. In addition, this individual also had the ancestral allele for lactase persistence, suggesting he had lost the ability to digest large quantities of lactose during early childhood. He also had only five copies of the amylase genes that break down starches, suggesting that his ability to digest starch was poor. This is not surprising for a hunter–gatherer, as the alleles for eating a starch-rich diet and being able to tolerate large quantities of lactose-rich dairy products usually were not favored until agriculture arrived.
9. Reich is quoted in Saini (2019, p. 150).
10. Eligon et al. (2021).
11. I will refer to Americans who have relatively recent ancestors from sub-Saharan Africa as Black or African American interchangeably.
12. Comas (2020).
13. The quote is from Pitts (2017, p. 49). Pitts continues,

    > But I can't see where not facing it has helped us surmount it. To the contrary, it is lodged like a bone in the throat, sits astride virtually every aspect of our American lives, ever present even if unspoken. Ignoring it has not made it go away.

14. Saini (2019, chapter 2).
15. Okrent (2019, p. 57).
16. The influx of people from China to work in the gold mines and railroads during the middle of the 19th century sparked resentment amongst other workers. This race-driven resentment led to Congress passing the Chinese exclusion act of 1882, which barred most immigration from China and many other countries in Asia. Around the same time, the US Supreme Court ruled that immigrants from China and other countries in Asia could not become citizens.

17. The quote is from Okrent (2019, p. 186). Galton was Francis Galton, Darwin's cousin, who coined the term eugenics and was one of its leading proponents.
18. Okrent (2019, p. 212).
19. See Zimmer (2018) for details.
20. Kevles (2016).
21. Livingstone (1962).
22. Smith (2018).
23. Lewontin (1974).
24. Lewontin (1972).
25. Beyond the ABO and the Rh factor, there are numerous other proteins and glycoproteins that act as antigens.
26. Lewontin's seven groups were sub-Saharan Africans, Caucasians (which included people from Egypt, the Arabian peninsula, and the Near East, going as far east as Pakistan), East Asians, Oceanians (including those indigenous to Papua New Guinea and the Solomon Islands), people indigenous to South Asia (including the Tamils and the Todas), people indigenous to the Americas, and people indigenous to Australia.
27. Lewontin (1972, p. 397).
28. Edwards (2003).
29. Hochman (2021).
30. Rosenberg et al. (2002).
31. Norton et al. (2019).
32. See Templeton (2013) and Graves (2015).
33. Handley et al. (2007).
34. Pickrell and Reich (2014).
35. Reich (2018).
36. Pickrell and Reich (2014).
37. Background information on Neanderthals and Neanderthal DNA can be found in chapter 2 of Reich (2018).
38. A key advance was getting better-quality DNA from the toe of a Neanderthal that existed 50,000 years ago. This Neanderthal shared DNA at some sites with non-African populations but not with people currently living in sub-Saharan Africa. This and other findings are strong evidence that Neanderthals were exchanging a trickle of genes with Eurasian populations of modern humans.
39. Sankararaman et al. (2014).
40. Some genes that have come from Neanderthals appear to be adaptive. These include some that play roles in the immune system. Recall from Chapter 2 that an allele that increases the risk of severe Covid and an allele that protects against it both likely came from Neanderthals.
41. Wu et al. (1996).
42. Johnson and Lachance (2012).
43. Reich (2018).
44. Bhatia et al. (2014).
45. The amount of Neanderthal DNA found in the African populations (which were mostly from northern Africa) was on average about 17 million base pairs or about 0.5 to 0.6 percent of the genome. Europeans and East Asian populations had about 51 and 55 million base pairs (1.6 and 1.7 percent), respectively. See Chen et al. (2020) for details.
46. Jones (2019).
47. General information on Denisovans can be found in chapter 3 of Reich (2018).
48. The study examined three populations (the Baka, Bakola, and Bedzan) from present-day Cameroon, the Hadza from Tanzania, and the Sandawe from Tanzania. Details in Lachance et al. (2012).

49. Reich (2018).
50. They, too, regularly have been exchanging genes with other populations.
51. Skoglund et al. (2017).
52. Reich (2018).
53. See Jablonski and Chaplin (2010) for basic information on the selective pressures that have shaped skin color variation.
54. Curiously, as Jablonski and Chaplin (2010) have noted, Darwin dismissed the role of adaptation to physical features in the environment to explain the evolution of skin color and skin color variation in humans. This was one of the major errors he made with respect to evolution. Instead, Darwin emphasized the role of sexual selection to explain variation in skin color—as well as hair color and texture, among other features of humans.
55. Specifically, the UV-B type of ultraviolet radiation.
56. See Jablonski (2004) for basic information on the production of skin color.
57. Quillen et al. (2019).
58. Hoekstra et al. (2006).
59. Rogers et al. (2004).
60. Lamason et al. (2005).
61. Quillen et al. (2019).
62. Beleza et al. (2012).
63. Crawford et al. (2017b).
64. Martin et al. (2017).
65. Wade (2014).
66. Dobbs (2014).
67. Coop et al. (2014).
68. The letter and full list of co-signers can be found at https://cehg.stanford.edu/letter-from-population-geneticists. Accessed June 5, 2021.
69. DeSalle and Tattersall (2018).
70. Wade (2014, p. 2).
71. *Ibid.*, p. 4.
72. *Ibid.*, p. 120.
73. Rosenberg et al. (2005). The quote is on p. 9.
74. *Ibid.*, p. 1.
75. Pritchard et al. (2000). See Desalle and Tattersall (2018) for more discussion on STRUCTURE.
76. Wade (2014, p. 99).
77. *Ibid.*, p. 100.
78. *Ibid.*, p. 214.
79. Graves (2015, p. 1475).
80. Chae et al. (2014).

FURTHER READING

Reich, D. 2018. *Who We Are and How We Got Here: Ancient DNA and the New Science of the Human Past*. Pantheon Books.
Saini, A. 2019. *Superior: The Return of Race Science*. Beacon Press.

Epilogue
Lessons Learned and Lamentable Lacunae

This book has been considerably longer than I initially planned. In researching topics, I kept finding more stories to discuss. Moreover, the COVID-19 pandemic that broke out while I was writing necessitated a longer section on viruses than I had planned. The book has also, consequently, taken longer to write than I had envisioned. Despite the book's length, there are still many evolutionary applications that I have either neglected or given insufficient attention to.

The food section especially could be much longer. My original plans had included a chapter on how evolutionary principles have been applied to fermented food and drink. Fortunately, Rob Dunn and Monica Sanchez have written a very recent excellent book,[1] *Delicious*, in which they discuss how microbes have enriched our lives with fermentation. Some of my original chapter on fermentation survives in the "Blessed Are the Cheesemakers" subsection of Chapter 4, in which I discuss how microbes have been used in the processing of milk into cheese. Through the microbial processing, some of the lactic acid present in milk has been removed by the separation of the curds and whey and by the microbial fermentation. This lower lactic acid concentration may have allowed cheese to become a gateway to other dairy products. Cheesemakers continue to use the principles of evolution to choose the right microbes and the appropriate environments in making the desired cheese.[2] In addition, cheesemaking had been accidentally instrumental in the history of gene editing: several key findings about CRISPR gene editing were discovered by Rodolphe Barrangou and other researchers working at Danisco, a company that produces starter cultures of bacteria used in cheese production.[3] To protect Danisco's starter cultures from phages, Barrangou had been studying the evolutionary genetics of both the phages and the defense mechanisms the bacteria used to combat the phage, one of which was CRISPR.

Breeding of livestock, as in the plant breeding discussed in Chapter 8, has been an important evolutionary application. Sewall Wright, one of the founders of population genetics, spent the first decade of his research career at the US Department of Agriculture. While at the USDA, Wright examined the pedigrees of Shorthorn cattle and designed experiments to improve the breed.[4] Wright influenced and inspired Jay Lush at Iowa State, one of the early leaders in livestock breeding. Among other things, Lush formulated the breeder's equation, which states that the response to selection of a trait is equal to the selection differential (the difference between the mean of selected individuals and the mean of the general population) times the heritability (the proportion of the variation of a trait due to additive genetic factors). This breeder's equation, which remains a mainstay of both breeding programs and evolutionary biology, has been extended to account for multiple traits, some of which may be genetically correlated with each other. These correlations between traits—for instance, weight gain speed and fat accumulation—can constrain responses to artificial selection. Here, it might be difficult to obtain cattle that gain weight quickly but remain lean. Similarly, these correlated genetic responses can constrain evolution in natural settings.

One environmental concern about livestock production is that it contributes to global climate change. Methane, a potent greenhouse gas, is a by-product of the fermentation of bacteria in the rumen, the first stomach of cows, wherein bacteria breakdown the cellulose of grasses into compounds that cows can metabolize. Lately, cattle breeders have used various methods of selective breeding combined with data from GWAS mapping (see Chapter 5) to attempt to reduce the methane output of cows.[5] One intriguing finding is that cattle that have been selected to be more efficient feeders tend to emit less methane.

Chinook salmon caught after 2010 are about 8 percent smaller in body length as compared to those caught before 1990. This decline has caused individual salmon to decline in value by about 20 percent. Such size declines have occurred in many fishes and have been a concern of fisheries.[6] The extent to which these size declines are due to genetic changes arising from the selection imposed by fishing and how much fisheries-induced evolution affects the ability of fisheries to recover after overexploitation ends are important but still largely unaddressed empirical questions.[7] The research needed to address these questions is fundamentally evolutionary genetics.

More and more people are reducing their consumption of meat and fish—or eschewing it altogether—for economic, health, environmental, and moral ethical/animal welfare reasons. With this trend against meat has come the rise of meatless meat; that is, substances that look and taste and smell like meat but are not meat.[8] Most meat substitutes today are based on plant-derived food that mimics animal meat. However, cell-based meat, meaning meat grown from animal cells cultured in the lab, is likely to gain favor and should be more appealing than plant-based "meat" to those who currently eat meat. Indeed, the consulting firm Kearney recently estimated that a third of all meat consumed across the world will come from cultured cells by the year 2040.[9] While much of the genetic manipulation to improve this cultured meat will come from CRISPR and other high-tech, artificial selection is likely to be a useful tool.

Outside of the food section, I could have devoted more attention to some of the misapplications of evolution. One such misapplication is making facile inferences about human behavior—especially when trying to make normative arguments—from looking at other animals. The Canadian psychologist and author Jordan Peterson argued that like lobsters, humans are governed by dominance hierarchies and drives for dominance. Dominance, according to Peterson, is determined at least in part by the neurotransmitter serotonin in both; add a bit of serotonin to the defeated lobster and it will bounce back, he claims. Peterson exhorts his readers: "Look for your inspiration to the victorious lobster, with its 350 million years of practical wisdom. Stand up straight, with your shoulders back."[10]

The biology behind this glib advice is flawed not just because there are differences between the dominance hierarchies between humans and lobsters but also because choosing lobsters as the example of nature to emulate is cherry picking. We are, as marine biologist Bailey Steinworth pointed out responding to Peterson, evolutionarily quite distinct from lobsters. The common ancestor of lobsters and humans—likely a marine worm the size of a rice grain—does not have dominance hierarchies. Nor do many organisms that are more closely related to humans than they are to lobsters. The large differences of behavior seen across these organisms challenges the usefulness of singling out any one to make inferences about human behavior. Peterson is hardly alone in making such inferences; for instance, the aggression seen in chimpanzees' behavior has often been used to leap to conclusions about human behavior. The problem is that humans are equally related to bonobos, which have vastly different and less aggressive behavior than chimpanzees.[11]

What general lessons have been learned from examining the various types of applications of evolutionary biology? An important consideration underlying many of the applications is that environments matter. Moreover, the definition of environment has been extended to microenvironments. We saw that predisposition to inflammatory bowel disease is affected by the collections of microbes living in our digestive systems as well as by genetic variants (and their interaction). Similarly, we saw that the microenvironment substantially influences whether cancer will develop. Changes in the cellular microenvironment can alter the selective milieu, making cancer-predisposing genetic variants go from having a selective disadvantage to having a selective advantage. With this microenvironment change, cancer becomes much more likely. These topics are fundamentally in the province of ecology as well as evolution, albeit at the cellular level.

Another consideration is that in some applications, we want evolutionary change to occur, while in other cases, we want to hinder evolutionary change. Evolutionary change is desired in the case of marine organisms responding to changing environments due to climate change. In contrast, we want to hinder evolutionary change in the case of weeds evolving resistance or in coronavirus evolving new variants. Whether evolution is a feature or a bug, the same general principles come into play. For instance, large population sizes are conducive for evolution. Thus, large marine populations are more likely to adapt to climate change. Likewise, limiting the number of coronavirus infections is likely to hinder the evolution of new troublesome variants.

Costs of adaptations and tradeoffs are commonplace, though not universal. These will complicate applications where we desire evolution, such as sea urchins adapting to more acidic oceans. In contrast, tradeoffs and costs to adaptation can be put to use in cases where we want to prevent or even reverse evolution. For instance, we saw in Chapter 1 in Chan and Turner's phage therapy that a tradeoff led to a reversal of evolution in bacteria: as the bacteria evolved defenses against the phage, they lost their resistance to antibiotics. Similarly, we saw in Chapter 9 that many plants have evolved resistance to glyphosate and maintain that resistance in part because the costs are minimal. Whether evolution is a friend or a foe for a particular application, understanding the nature of these tradeoffs and costs is useful (if not essential). This often means examining the biochemical and/or mechanistic details of the tradeoff or cost. Partnerships between evolutionary biologists and those researchers who are more mechanistically inclined will often be useful here.

Breaking down walls between researchers in general can be beneficial. We saw in Chapter 4 how the lack of communication between people studying allergies and those studying type 1 diabetes and other autoimmune diseases hindered progress toward development of what is now the "old friends" hypothesis. Many more such connections can be found, some of which have been discussed here. For example, we find connections between cancer and pregnancy involving the invasiveness of the placenta. We also see connections between the spread of cancer and the spread of cane toads and other invasive species. Moreover, adaptive therapy used in managing cancer (Chapter 6) was inspired by the successes of integrative pest management used to control agricultural pests (Chapter 9).

Establishing and nurturing connections between researchers across different, apparently disconnected fields is not trivial. There are substantial inherent barriers, including insufficient rewards for those who engage in such research. Nonetheless, the potential rewards are substantial for such interdisciplinary research in evolutionary applications. Academia, governments, industries, and nonprofits all can play a role in fostering such connections.

NOTES

1. Dunn and Sanchez (2021).
2. See, for instance, Gallone et al. (2020).
3. Isaacson (2021).
4. Hill (2014). The exchange between breeding and evolutionary genetics was not a one-way street. Wright's work on cattle helped shape his views on evolution.
5. See, for instance, Hegarty et al. (2007) and Kenny et al. (2018).
6. Oke et al. (2020).
7. Hutchings and Kuparinen (2019, 2021).
8. Rubio et al. (2020).
9. Dolgin (2020).
10. Steinworth (2018).
11. I wrote more about the differences between chimpanzees and bonobos and the challenges involved in making inferences about human biology by looking at chimpanzees in my previous book. See Johnson (2007, chapter 8).

References

Abbasi, J. 2018. Anthony Fauci, MD: Working to end HIV/AIDS. *JAMA* 320: 327–329.

Abdelrahman, Z., M. Li, and X. Wang. 2020. Comparative review of SARS-CoV-2, SARS-CoV, MERS-CoV, and influenza a respiratory viruses. *Frontiers in Immunology* 11: 552909.

Abed, R., M. Brune, and D. R. Wilson. 2019. The role of the evolutionary approach in psychiatry. *World Psychology* 18: 370–371.

Abegglen, L. M., A. F. Caulin, A. Chan, K. Lee, R. Robinson, M. S. Campbell et al. 2015. Potential mechanisms for cancer resistance in elephants and comparative cellular response to DNA damage in humans. *JAMA* 314: 1850–1860.

Abu-Raddad, L. J., P. Patnaik, and J. B. Kublin. 2006. Dual infection with HIV and malaria fuels the spread of both diseases in Sub-Sahara Africa. *Science* 314: 1603–1606.

Adamowicz, S. J., P. M. Hollingsworth, S. Ratnasingham, and M. van der Bank. 2017. International barcode of life: Focus on big biodiversity in South Africa. *Genome* 60: 875–879.

Adler, L. S., R. E. Irwin, S. H. McArt, and R. L. Vanette. 2020. Floral traits affecting the transmission of beneficial and pathogenic pollinator-associated microbes. *Current Opinion in Insect Science* 44: 1–7.

Afify, A., J. F. Betz, O. Riabinina, C. Lahondere, and C. J. Potter. 2019. Commonly used insect repellents hide human odors from *Anopheles* mosquitoes. *Current Biology* 29: 3669–3680.

Aime, M. C., C. D. Bell, and A. W. Wilson. 2018. Deconstructing the evolutionary complexity between rust fungi (Puccinales) and their plant hosts. *Studies in Mycology* 89: 143–152.

Aitken, S. N. and M. C. Whitlock. 2013. Assisted gene flow to facilitate local adaptation to climate change. *Annual Review of Ecology, Evolution, and Systematics* 44: 367–388.

Aktipis, C. A. 2020. *The Cheating Cell: How Evolution Helps Us Understand and Treat Cancer*. Princeton University Press.

Aktipis, C. A., A. M. Boddy, G. Jansen, U. Hibner, M. E. Hochberg, C. C. Maley, and G. S. Wilkinson. 2015. Cancer across the tree of life: Cooperation and cheating in multicellularity. *Philosophical Transactions of the Royal Society B* 370: 20140219.

Albertson, R. C., W. Cresko, H. W. Detrich III, and J. H. Postlethwait. 2009. Evolutionary mutant models for human disease. *Trends in Genetics* 25: 74–81.

Aldenderfer, M. S. 2003. Moving up in the world: Archaeologists seek to understand how and when people came to occupy the Andean and Tibetan plateaus. *American Scientist* 91: 542–549.

Alford, R. A., G. P. Brown, L. Schwarzkopf, B. L. Phillips, and R. Shine. 2009. Comparisons through time and space suggest rapid evolution of dispersal behaviour in invasive species. *Wildlife Research* 36: 23–38.

Allen, K., J. Hanna, and C. Mossburg. 2018. Police used free genealogy database to track golden state killer suspect, investigator says. April 27, 2018. www.cnn.com/2018/04/26/us/golden-state-killer-dna-report. Accessed April 10, 2020.

Allen, R. C., R. Popat, S. P. Diggle, and S. P. Brown. 2014. Targeting virulence: Can we make evolution-proof drugs? *Nature Reviews Microbiology* 12: 300–308.

Alonge, M., X. Wang, M. Benoit, S. Soyk, L. Pereira, L. Zhang, H. Suresh, S. Ramakrishnan, F. Manumus, D. Ciren, Y. Levy et al. 2020. Major impacts of widespread structural variation on gene expression and crop improvement in tomato. *Cell* 182: 145–161.

Alonso-Curbelo, D., Y.-J. Ho, C. Burdziak, J. L. Maag, J. P. Morris IV., R. Chandwani, H.-A. Chen, K. M. Tsanov, F. M. Barriga, W. Luan et al. 2021. A gene-environment-induced epigenetic program initiates tumorigenesis. *Nature* 590: 642–648.

Amato, K. R., C. J. Yeoman, A. Kent, N. Righini, F. Carbonero, A. Estrada, H. R. Gaskins, R. M. Stumpf, S. Yildirin, M. Torralba et al. 2013. Habitat degradation impacts black howler monkeys (*Alouata pigra*) gastrointestinal tract. *The ISME Journal* 2013: 1–10.

Anderson, T. J. C., S. Nair, M. McDrew-White, I. H. Cheeseman, S. Nikhoma, F. Bilgic, R. McGreaady, E. Ashley, A. Pyae Phyo, N. J. White, and F. Nosten. 2016. Population parameters underlying an ongoing soft sweep in southeast Asia malaria parasites. *Molecular Biology and Evolution* 34: 131–144.

Angel, A. and J. Cooper. 2006. *A Review of the Impacts of Introduced Rodents on the Islands of Tristan da Cunha and Gough.* RSPB Research Report No. 17. RSPB.

Angier, N. 1988. *Natural Obsessions: The Search for the Oncogene.* Houghton Mifflin.

Angrist, M. 2010. *Here Is a Human Being: At the Dawn of Personal Genomics.* Harper Collins.

Anguita-Ruiz, A., C. M. Aguilera, and A. Gil. 2020. Genetics of lactose intolerance: Updated and online interactive world maps of phenotype and genotype frequencies. *Nutrients* 12: 2689.

Ariey, F., B. Witkowski, C. Amaratunga, J. Beghain, A.-C. Langlois, N. Khim, S. Kim, V. Dura, C. Bouchier, L. Ma et al. 2014. A molecular marker of artemisinin-resistant *Plasmodium falciparum* malaria. *Nature* 505: 50–56.

Armstrong, S. 2014. *P53: The Gene that Cracked the Cancer Code.* Bloomberg Signal.

Arney, K. 2020. *Rebel Cell: Cancer, Evolution, and the New Science of Life's Oldest Betrayal.* BenBella Books, Inc.

Arnold, M. L. 1997. *Natural Hybridization and Evolution.* Oxford University Press.

Arthur, A., A. Terman, A. Kurzontkowski, K. Keene, T. Babatunde, and K. Summers. 2018. Molecular evolution of genes associated with preeclampsia: Genetic conflict, antagonistic coevolution and signals of selection. *Journal of Evolutionary Medicine* 6: msq034.

Atkins, D. P., I. C. Trueman, C. B. Clarke, and A. D. Bradshaw. 1982. The evolution of lead tolerance by Festuca rubra on a motorway verge. *Environmental Pollution A* 27: 233–241.

Atukorala, A. D. S., V. Bhatia, and R. Ratnayake. 2019. Craniofacial skeleton of Mexican tetra (*Astyanax mexicanus*) as a bone disease model. *Developmental Dynamics* 248: 153–161.

Audent, J.-N., S. Ducatez, and L. Lefebvre. 2015. The town bird and the country bird: Problem-solving and immunocompetence vary with urbanization. *Behavioral Ecology* 27: 637–644.

Axelrod, R. 1984. *The Evolution of Cooperation.* Basic Books.

Axelrod, R. and W. D. Hamilton. 1981. The evolution of cooperation. *Science* 211: 1390–1396.

Azvolinsky, A. 2018. Genome collector: A profile of Charles Rotimi. *The Scientist.* October 1, 2018.

Bachem, C. W. B., H. J. van Eck, and M. E. de Vries. 2019. Understanding genetic load in potato for hybrid diploid breeding. *Molecular Plant* 12: 896–898.

Baik, L. S. and J. R. Carlson. 2021. The mosquito taste system and disease control. *Proceedings of the National Academy of Science.* doi: 10.1073/pnas2013076117.

Baker, H. G. and G. L. Stebbins (eds.). 1965. *The Genetics of Colonizing Species.* Academic Press.

Balakrishnan, V. S. 2018. Celebrated Brazilian bee scientist Warwick Kerr dies. *The Scientist.* October 2, 2018.

Barelli, C., D. Albanese, C. Donati, M. Pindo, C. Dallago, F. Rovero, D. Cavalieri, K. M. Tuohy, H. C. Hauffe, and C. De Filippo. 2015. Habitat fragmentation is associated to gut microbiota diversity of an endangered primate: Implications for conservation. *Scientific Reports* 5: 14862.

Barr, I. G., R. O. Donis, J. M. Katz, J. W. McCauley, T. Odagiri, H. Trusheim, T. F. Tsai, and D. E. Wentworth. 2018. Cell culture-derived influenza vaccines in the severe 2017–2018 epidemic season: A step towards improved influenza vaccine effectiveness. *NPJ Vaccines* 3: 44.

Barrett, L. G., M. Legros, N. Kumaran, D. Glassop, S. Raghu, and D. M. Gardiner. 2019a. Gene drives in plants: Opportunities and challenges for weed control and engineered resilience. *Proceedings of the Royal Society of London B* 286: 201915.

Barrett, R. D. H., S. Laurent, R. Mallarino, S. P. Pfeifer, C. Xu, M. Foll, K. Wakamatsu, J. S. Duke-Cohen, J. D. Jensen, and H. E. Hoekstra. 2019b. Linking a mutation to survival in wild mice. *Science* 363: 499–504.

Barrett, S. C. H. 2015. Foundations of invasion genetics: The baker and Stebbins legacy. *Molecular Ecology* 24: 1927–1941.

Basavaraju, S. V., M. E. Patton, K. Grimm, M. A. U. Rasheed, S. Lester, L. Mills, M. Stumpf, B. Freeman, A. Tamin, J. Harcourt et al. 2020. Serological testing of U. S. Blood donations to identify SARS-CoV-2-reactive antibodies December 2019–January 2020. *Clinical Infectious Disease* ciaa1785.

Batai, K., S. Hooker, and R. A. Kittles. 2021. Leveraging genetic ancestry to study health disparities. *American Journal of Physical Anthropology* 175: 363–375.

Bauch, H. 2001. Terror risk real. Thousands in U. S. will die, ex-presidential hopeful says. *Montreal Gazette*. September 5, 2001.

Baucom, R. G. 2019. Evolutionary and ecological insights from herbicide-resistant weeds: What have we learned about plant adaptation, and what is left to uncover? *The New Phytologist* 223: 68–82.

Baucom, R. G. and R. Busi. 2019. Commentary on "Evolutionary epidemiology predicts the emergence of glyphosate resistance in a major agricultural weed". *The New Phytologist* 223: 1056–1058.

Baucom, R. S., S.-M. Chang, J. M. Kniskern, M. D. Rausher, and J. R. Stinchcombe. 2011. Morning glory as a powerful model in ecological genomics: Tracing adaptation through both natural and artificial selection. *Heredity* 107: 377–385.

Beall, C. M., G. L. Cavalleri, L. Deng, R. C. Elston, Y. Gao, J. Knight, C. Li, J. C. Li, M. McCormack, H. E. Montgomery et al. 2010. Natural selection on EPAS1 (HIF2α) associated with low hemoglobin concentration. *Proceedings of the National Academy of Sciences* 107: 11459–11464.

Beaury, E. M., J. T. Finn, J. D. Corbin, V. Barr, and B. A. Bradley. 2020. Biotic resistance is ubiquitous across ecosystems of the United States. *Ecology Letters* 23: 476–482.

Beckett, K. M. 1997. The cystic fibrosis heterozygote advantage: A synthesis of ideas. *Anthropologica* 39: 147–158.

Bedford, T., A. L. Greinger, P. Roychoudhury, L. M. Starita, M. Famulare, M.-L. Huang, A. Nalla, G. Pepper, A. Reinhardt, H. Xie et al. 2020. Cryptic transmission of SARS-CoV-2 in Washington state. *Science* 370: 571–575.

Beleza, S., A. M. Santos, B. McEvoy, I. Alves, C. Martinho, E. Cameron, M. D. Shriver, E. J. Parra, and J. Rocha. 2012. The timing of pigmentation lightening in Europeans. *Molecular Biology* 30: 24–35.

Benedict, M. Q. 2021. Sterile insect technique: Lessons from the past. *Journal of Medical Entomology* 58: 1974–1979.

Benton, M. L., A. Abraham, A. L. LaBella, P. Abbot, A. Rokas, and J. A. Capra. 2021. The influence of evolutionary history on human health and disease. *Nature Reviews Genetics* 22: 269–283.

Berlocher, S. H. and J. L. Feder. 2002. Sympatric speciation in phytophagous insects: Moving beyond controversy? *Annual Review of Entomology* 47: 773–815.

Bernstein, L. 2020. 'Nobody has very clear answers for them': Doctors search for treatments for covid-19 long-haulers. *The Washington Post*. October 16, 2020.

Bestion, E., S. Jacob, L. Zinger, L. Di Gesu, M. Richard, J. White, and J. Cote. 2017. Climate warming reduces gut microbiota diversity in a vertebrate ectotherm. *Nature Ecology & Evolution* 1: 0161.

Bethke, P. C., D. A. Halterman, and S. Jansky. 2017. Are we getting better at using wild potato species in light of new tools? *Crop Science* 57: 1241–1258.

Betti, L. and A. Manica. 2018. Human variation in the shape of the birth canal is significant and geographically structured. *Proceedings of the Royal Society of London B* 285: 20181807.

Beyer, G., A. Habtezion, J. Werner, M. M. Lerch, and J. Mayerle. 2020. Chronic pancreatitis. *The Lancet* 396: 499–512.

Beyer, J., H. C. Trannum, T. Bakke, P. V. Hodson, and T. K. Collier. 2016. Environmental effects of the Deepwater Horizon oil spill: A review. *Marine Pollution Bulletin* 110: 28–51.

Bhatia, G., A. Tandon, N. Patterson, M. C. Aldrich, C. B. Ambrosone, C. Amos, E. V. Bandera, S. I. Berndt, L. Bernstein, W. J. Blot et al. 2014. Genome-wide scan of 29,141 African Americans finds no evidence of directional selection since admixture. *The American Journal of Human Genetics* 95: 437–444.

Bhuiyan, B. A. 2016. An overview of game theory and some applications. *Philosophy and Progress* LIX–LX: 112–128.

Biggerstaff, M., S. Cauchemez, C. Reed, M. Ganbhir, and L. Finelli. 2014. Estimates of the reproduction number for seasonal, pandemic, and zoonotic influenza: A systematic review of the literature. *BMC Biology* 14: 480.

BirdLife International. 2018. Diomedea dabbenena. *The IUCN Red List of Threatened Species* 2018: e.T22728364A132657527.

Birnbaum, J., S. Scharf, S. Schmidt, E. Jonscher, W. A. M. Hoeikmakers, S. Flemming, C. G. Toenhake, M. Schmitt, R. Sabitzki, B. Bergmann et al. 2020. A Kelch13-defned endocytosis pathway mediates artemisinin resistance in malarial parasites. *Nature* 367: 51–59.

Blackburn, T. M., C. Bellard, and A. Ricciardi. 2019. Alien versus native species as driver of recent extinction. *Frontiers in Ecology and the Environment* 17: 203–207.

Blanco-Melo, D., B. E. Nilsson-Payant, W.- C. Liu, S. Uhl, D. Hoagland, R. Moller, T. X. Jordan, K. Oishi, M. Panis, D. Sachs et al. 2020. Imbalanced host response to SARS-CoV-2 drives development of COVID-19. *Cell* 181: 1036–1045.

Blasser, M. J. 2014. *Missing Microbes: How the Overuse of Antibiotics is Fueling Our Modern Plagues*. Harper Collins.

Bleasdale, M., K. K. Richter, A. Janzen, S. Brown, A. Scott, J. Zech, S. Wilkin, K. Wang, S. Schiffels, J. Desideri et al. 2021. Ancient proteins provide evidence of dairy consumption in eastern Africa. *Nature Communications* 12: 632.

Bloomfield, S. F., G. A. W. Rook, E. A. Scott, F. Shanahan, R. Stanwell-Smith, and P. Turner. 2016. Time to abandon the hygiene hypothesis: New perspectives on allergic disease, the human microbiome, infectious disease prevention and the role of targeted hygiene. *Royal Society for Public Health* 136: 213–224.

Blumstein, D. T. 2008. Fourteen security lessons from antipredator behavior. Pp. 147–158 In *Natural Security: A Darwinian Approach to a Dangerous World*. R. Sagarin and T. Taylor (eds.). University of California Press.

Blumstein, D. T., J. Brucker, S. Shah, S. Patel, M. E. Alfaro, and B. Natterson-Horowitz. 2015. The evolution of capture myopathy in hooved animals: A model for human stress cardiomyopathy. *Evolution, Medicine, and Public Health*: 195–203.

Bock, D. G., C. Caseys, R. D. Cousens, M. A. Hahn, S. M. Heredia, S. Hubner, K. D. Whitney, and L. H. Rieseberg. 2015. What we still don't know about invasion genetics. *Molecular Ecology* 24: 2277–2297.

Boddy, A. M., L. M. Abegglen, A. P. Pessier, A. Aktipis, J. D. Schiffman, C. C. Maley, and C. Witte. 2020a. Lifetime cancer prevalence and life history traits in mammals. *Evolution, Medicine, and Public Health* 2020: 187–195.

Boddy, A. M., T. M. Harrison, and L. M. Abegglen. 2020b. Comparative oncology: New insights into an ancient disease. *iScience* 23: 10373.

Boddy, A. M., H. Kokko, F. Breden, G. S. Wilkinson, and C. A. Aktipis. 2015. Cancer susceptibility and reproductive trade-offs: A model of the evolution of cancer defenses. *Philosophical Transactions of the Royal Society* 370: 20140220.

Bosch, B., D. Bilton, P. Sosnay, K. S. Raraigh, D. Y. F. Mak, H. Ishiguro, V. Gulmans, M. Thomas, H. Cuppens, M. Amaral, and K. De Boeck. 2017. Ethnicity impacts the cystic fibrosis diagnosis: A note of caution. *Journal of Cystic Fibrosis* 16: 488–491.

Bosch, L., B. Bosch, K. De Boeck, T. Nawrot, I. Meyts, D. Vanneste, C. A. Le Bourlegat, J. Croda, and L. V. Riberio. 2017. Cystic fibrosis carriership and tuberculosis: Hints toward an evolutionary selective advantage based on data from the Brazilian territory. *BMC Infectious Diseases* 17: 340.

Bove, C. B., J. B. Ries, S. W. Davies, I. T. Westfield, J. Umbanhowar, and K. D. Castillo. 2019. Common Caribbean corals exhibit highly variable responses to future acidification and warming. *Proceedings of the Royal Society of London B* 286: 20182840.

Bower, D. S., C. K. Jennings, R. J. Webb, Y. Amepou, L. Schwarzkopf, L. Berger, R. A. Alford, A. Georges, D. T. McKnight, L. Carr et al. 2020. Disease surveillance of the amphibian chytrid fungus *Batrachochytrium dendrobatidis* in Papua New Guinea. *Conservation Science and Practice* 2: e256.

Boyle, E. A., Y. I. Li, and J. K. Pritchard. 2017. An expanded view of complex traits: From polygenic to omnigenic. *Cell* 169: 1177–1186.

Boylston, A. W. 2012. *Defying Providence: Smallpox and the Forgotten 18th Century Medical Revolution*. CreateSpace Independent Publishing.

Brace, S., Y. Diekmann, T. J. Booth, L. van Dorp, Z. Faltyskova, N. Rohland, S. Mallick, I. Olalde, M. Ferry, M. Michel et al. 2019. Ancient genomes indicate population replacement in early Neolithic Britain. *Nature Ecology & Evolution* 3: 765–771.

Bradshaw, L. D., S. P. Padgette, S. L. Kimball, and B. H. Wells. 1997. Perspectives on glyphosate resistance. *Weed Technology* 11: 189–198.

Brady, S. P., D. I. Bolnick, A. L. Angert, A. Gonzalez, R. D. H. Barrett, E. Crisko, A. M. Derry, C. G. Eckert, D. J. Fraser, G. F. Fussmann et al. 2019. Causes of maladaptation. *Evolutionary Applications* 12: 1229–1242.

Brady, S. P. and J. L. Richardson. 2017. Road ecology: Shifting gears toward evolutionary perspectives. *Frontiers in Ecology and Environment* 15: 91–98.

Brans, K. I. and L. De Meester. 2018. City life on fast lanes: Urbanization induces an evolutionary shift in the water flea *Daphnia. Functional Ecology* 32: 2225–2240.

Brans, K. I., M. Jansen, J. Vanoverberke, N. Tuzun, R. Stokes, and L. De Meester. 2017. The heat is on: Genetic adaptation to urbanization mediated by thermal tolerance and body size. *Global Change Biology* 23: 5218–5227.

Brans, K. I., R. Stokes, and L. De Meester. 2018. Urbanization drives genetic differentiation in physiology and structures the evolution of pace-of-life syndromes in the water flea *Daphnia magna. Proceedings of the Royal Society of London B* 285: 20180169.

Branswell, H. and A. Joseph. 2020. WHO declares the coronavirus outbreak a pandemic. *STAT News*. March 11, 2020.

Brauner, J. M., S. Mindermann, M. Sharma, D. Johnston, J. Salvatier, T. Gavenciak, A. B. Stephenson, G. Leech, G. Altman et al. 2021. Inferring the effectiveness of government interventions against COVID-19. *Science* 371: eabd9338.

Breed, M. D., E. Guzma'n-Novoa, and G. J. Hunt. 2004. Defense behavior of honey bees: Organization, genetics, and comparisons with other bees. *Annual Review of Entomology* 49: 271–298.

Breidenstein, E. B. M., B. K. Khaira, I. Wiegard, J. Overhage, and R. E. W. Hancock. 2008. Complex ciprofloxacin resistance revealed by screening a *Pseudomonas aeruginosa* library for altered susceptibility. *Antimicrobial Agents and Chemotherapy* 52: 4486–4491.

Brennan, R. S., A. D. Garrett, K. E. Huber, H. Hargarten, and M. H. Pespeni. 2019. Rare genetic variation and balanced polymorphisms are important for survival in global change conditions. *Proceedings of the Royal Society B* 286: 2019043.

Brenner, S. L., J. P. Jones, R. H. Rutanen-Whaley, W. Parker, M. V. Flinn, and M. P. Muehlenbein. 2015. Evolutionary mismatch and chronic psychological stress. *Journal of Evolutionary Medicine* 30: 32–44.

Brinkman-Van der Linden, E. C. M., N. Hurtado-Ziola, T. Hayakawa, L. Wiggleton, K. Benirschke, A. Varki, and N. Varki. 2007. Human-specific expression of Siglec-6 in the placenta. *Glycobiology* 17: 922–931.

Brook, C. E. and A. P. Dobson. 2014. Bats as 'special' reservoirs for emerging zoonotic pathogens. *Trends in Microbiology* 23: 172–180.

Brown, C. R. and M. Bomberger-Brown. 2013. Where has all the road kill gone? *Current Biology* 23: R233–R234.

Brown, G. P., C. Shilton, B. L. Phillips, and R. Shine. 2007. Invasion, stress, and spinal arthritis in cane toads. *Proceedings of the National Academy of Sciences* 104: 17698–17700.

Brown, H. C. 2017. Oysters on acid: How the ocean's declining PH will change the way we eat. *The New Food Economy*. November 28, 2017.

Browne, J. 1995. *Charles Darwin Vol. 1: Voyaging*. Jonathan Cape.

Bryson, B. 2019. *The Body: A Guide for Occupants*. Penguin Random House.

Buffenstein, R. 2005. The naked mole-rat: A new long-living model for human aging research. *Journal of Gerontology* 60A: 1369–1377.

Buisson, R., A. Langenbucher, D. Bowen, E. E. Kwan, C. H. Benes, L. Zou, and M. S. Lawrence. 2019. Passenger hotspot mutations in cancer driven by APOBEC3A and mesoscale genomic features. *Science* 364: eaaw2872.

Buiting, K. 2010. Prader–Willi syndrome and Angelman syndrome. *American Journal of Medical Genetics Part C (Seminars in Medical Genetics)* 154C: 365–376.

Burgess, M. R. and C. L. Sawyers. 2006. Treating imatinib-resistant leukemia: The next generation targeted therapies. *TheScientificJournal* 6: 918–930.

Burt, A. 2003. Site-specific selfish genes as tools for the control and genetic engineering of natural populations. *Proceedings of the Royal Society of London B* 270: 921–928.

Burt, A. and A. Crisanti. 2018. Gene drive: Evolved and synthetic. *ACS Chemical Biology* 13: 343–346.

Bush, G. L. 1969. Sympatric host race formation and speciation in frugivorous flies of the genus *Rhaglotis* (Diptera: Tephritidae). *Evolution* 23: 237–251.

Byars, S. G., D. Ewbank, D. R. Govindaraju, and S. C. Stearns. 2010. Natural selection in a contemporary human population. *Proceedings of the National Academy of Science* 107 (Suppl.): 1787–1792.

Byars, S. G., Q. Q. Huang, L.-A. Gray, A. Bakshi, S. Ripatti, G. Abraham, S. C. Stearns, and M. Inouye. 2017. Genetic loci associated with coronary artery disease harbor evidence of selection and antagonistic pleiotropy. *PLoS Genetics* 13: e1006328.

Byers, K. A., M. J. Lee, D. M. Patrick, and C. G. Himsworth. 2019. Rats about town: A systematic review of rat movements in urban ecosystems. *Frontiers in Ecology and Evolution* 7: 13.

Byrne, K. and R. A. Nichols. 1999. *Culex pipiens* in London Underground tunnels: Differentiation between surface and subterranean populations. *Heredity* 82: 7–15.

Byrnes, W. M. and S. A. Newman. 2014. Ernst Everett just: Egg and embryo as excitable systems. *Journal of Experimental Zoology (Molecular and Developmental Evolution)* 322B: 191–201.

Caballero, A., I. Bravo, and J. Wang. 2017. Inbreeding load and purging: Implications for the short-term survival and the conservation management of small populations. *Heredity* 118: 177–185.

Callaway, E. 2020. Making sense of coronavirus mutations. *Nature* 585: 174–178.

Campbell-Staton, S. C., K. W. Winchell, N. C. Rochette, J. Fredette, I. Maayan, R. M. Schweizer, and J. Catchen. 2020. Parallel selection on thermal physiology facilitates repeated adaptation of city lizards to urban heat islands. *Nature Ecology and Evolution* 4: 652–658.

Cannataro, V. L., S. G. Gaffney, and J. P. Townsend. 2018. *Journal of the National Cancer Institute* 110: 1171–1177.

Cannataro, V. L., J. D. Mandell, and J. P. Townsend. Preprint. Attribution of cancer origins to endogenous, exogenous, and actionable processes. Doi.org/10.1101/2020.10.24.352989. Published October 25, 2020.

Cannataro, V. L. and J. P. Townsend. 2018. Neutral theory and the somatic evolution of cancer. *Molecular Biology and Evolution* 35: 1308–1315.

Cannataro, V. L. and J. P. Townsend. 2019. Wagging the long tail of drivers of prostate cancer. *PLoS Genetics* 15: e1007820.

Caraco, Y., S. Blotnick, and M. Muskat. 2008. CYP2C9 genotype-guided warfarin prescribing enhances the efficacy and safety of anticoagulation: A prospective randomized controlled study. *Clinical Pharmacology and Therapeutics* 83: 460–470.

Caravaggi, A., R. J. Cuthbert, P. G. Ryan, J. Cooper, and A. L. Bond. 2019. The impacts of introduced house mice on the breeding success of nesting seabirds on Gough Island. *IBIS* 161: 648–661.

Carbone, M., S. T. Arron, B. Beutler, A. Bononi, W. Cavenee, J. E. Cleaver, C. M. Croce, A. D'Andrea, W. D. Foulkes, G. Gaudino et al. 2020. Tumour predisposition and cancer syndromes as models to study gene-environment interactions. *Nature Reviews Cancer* 20: 533–549.

Carbone, M., S. Emri, A. U. Dogan, I. Steele, M. Tuncer, H. I. Pass, and Y. I. Baris. 2007. A mesothelioma epidemic in Cappadocia: Scientific developments and unexpected social outcomes. *Nature Reviews Cancer* 7: 147–154.

Carey, A. F., G. Wang, C.-Y. Su, L. J. Zwiebel, and J. R. Carlson. 2010. Odorant reception in the malaria mosquito *Anopheles gambiae*. *Nature* 464: 66–71.

Cargill, M., S. J. Schrodi, M. Chang, V. E. Garcia, R. Brandon, K. P. Callis, N. Matsuami, K. G. Ardlie, D. Civello, J. J. Catanese et al. 2007. A large scale genetic association study confirms *IL12B* and leads to the identification of *IS23R* as psoriasis-risk genes. *The American Journal of Human Genetics* 80: 273–290.

Caroll, S. P., J. E. Loye, H. Dingle, M. Mathieson, T. R. Famula, and M. P. Zaklucki. 2005. And the beak shall inherit—evolution in response to invasion. *Ecology Letters* 8: 944–951.

Carrière, Y., J. A. Fabrick, and B. E. Tabashnik. 2016. Can pyramids and seed mixtures delay resistance to Bt crops? *Trends in Biotechnology* 34: 291–303.

Carson, R. 1955. *The Edge of the Sea.* Houghton Mifflin.

Carson, R. 1962. *Silent Spring.* Houghton Mifflin.

Carter, H. 2019. Mutation hotspots may not be drug targets. *Science* 364: 1228–1229.

Caruso, R., T. Mathes, E. C. Martens, N. Kamada, A. Nusrat, N. Inohara, G. Núñes. 2019. A specific gene-microbe interaction drives the development of Crohn's disease-like colitis in mice. *Science Immunology* 4: eaaw4341.

Cassini, M. H. 2020. A review of the critics of invasion biology. *Biological Reviews* 95: 1467–1478.

Caulin, A. F. and C. C. Maley. 2011. Peto's paradox: Evolution's prescription for cancer prevention. *Trends in Ecology and Evolution* 26: 175–182.

Ceredeira, A. S. and S. A. Karumanchi. 2012. Angiogenic factors in preeclampsia and related disorders. *Cold Spring Harbor Perspectives in Medicine* 2: a006585.

Chae, D. H., A. M. Nuru-Jeter, N. E. Alder, G. H. Brody, J. Lin, E. H. Blackburn, and E. S. Epel. 2014. Discrimination, racial bias, and telomerase length in African-American men. *American Journal of Preventative Medicine* 46: 103–111.

Chakraborty, R. and K. K. Kidd. 1991. The utility of DNA typing in forensic work. *Science* 254: 1735–1739.

Champer, J., E. Yang, E. Lee, J. Liu, A. G. Clark, and P. W. Messer. 2020. A CRISPR homing gene drive targeting a haplolethal gene removes resistance alleles and successfully spreads through a cage population. *Proceedings of the National Academy of the Sciences* 117" 24377–24383.

Chan, B. K., M. Sistrom, J. E. Wertx, K. E. Kortright, D. Narayan, and P. E. Turner. 2016. Phage selection restores antibiotic sensitivity in MDR *Pseudomonas aeruginosa*. *Scientific Reports* 6: 26717.

Chan, B. K., P. E. Turner, S. Kim, H. R. Mojiban, J. A. Elefteriades, and D. Narayan. 2018. Phage treatment of an aortic graft infected with *Pseudomonas aeruginosa*. *Evolution, Medicine, and Public Health* 2018: 60–66.

Chanishvili, N. 2012. Phage therapy—history from Twort and d'Herrelle through Soviet experiences to current approaches. *Advances in Virus Research* 83: 3–40.

Chapman, S. N., J. E. Pettay, V. Lummaa, and M. Lahdenpera. 2019. Limits to fitness benefits of prolonged reproductive lifespan in women. *Current Biology* 29: 645–650.

Chappell, B. 2020. Coronavirus: New York creates "containment area' around cluster in new Rochelle. *NPR*. March 10, 2020. https:www.npr.org/sections/health-shots/2020/03/10/814099444/new-york-creates-containment-area-around-cluster-in-new-rochelle. Accessed May 10, 2021.

Charr, J.-C., A. Garavito, C. Guyeux, D. Crouzillat, P. Desombes, C. Fournier, S. N. Ivy, E. N. Raharimalala, J.-J. Rakotomalala, P. Stoffelen et al. 2020. Complex evolutionary history of coffees revealed by full plastid genomes and 28,800 nuclear SNP analyses, with particular emphasis on *Coffea canephora* (Robusta coffee). *Molecular Phylogenetics and Evolution* 151 (online) doi: 10.1016.j.ympev.2020.106906.

Chatterjee, N., J. Shi, and M. Garcia-Closas. 2016. Developing and evaluating polygenic risk prediction models for stratified disease prevention. *Nature Reviews Genetics* 17: 392–406.

Chen, L., A. B. Wolf, W. Fu, L. Li, and J. M. Akey. 2020. Identifying and interpreting apparent Neanderthal ancestry in African individuals. *Cell* 180: 677–687.

Chen, Y. D. and L. L. Stelinski. 2017. Resistance management for Asian citrus psyllid, *Diaphorina citri* Kuwayama, in Florida. *Insects* 8: 103. doi: 10.3390/insects8030103.

Cheptou, P.-O., O. Carruem, S. Rouifed, and A. Cantarel. 2008. Rapid evolution of seed dispersal in an urban environment in the weed *Crepis sancta*. *Proceedings of the National Academy of Science* 105: 3796–3799.

Cheptou, P. O. and S. C. Lambrecht. 2020. Sidewalk plants as a model for studying adaptation to urban environments. Pp. 130–141 In *Urban Evolutionary Biology*. M. Sazulkin, J. Munshi South, and A. Charmantier (eds.). Oxford University Press.

Choudhary, P., B. N. Singh, H. Chakdar, and A. K. Saxena. 2021. DNA barcoding of phyto-pathogens for disease diagnostics and bio-surveillance. *World Journal of Microbiology and Biotechnology* 37: 54.

Chroback, U. 2019. Invasive earthworms are burrowing into boreal forests worldwide. *Popular Science*. December 12, 2019.

Clark, B. W., C. W. Matson, D. Jung, and R. T. Di Giulio. 2010. AHR2 mediates cardiac tetraogenesis of polycyclic aromatic hydrocarbons and PCB-126 in Atlantic killifish (*Fundulus heteroclitus*). *Aquatic Toxicology* 99: 232–240.

Claussnitzer, M. J. H. Cho, R. Collins, N. J. Cox, E. T. Dermitakis, M. E. Hurles, S. Kathiresan, E. E. Kenny, C. M. Lindgren, D. G. MacArthur et al. 2020. A brief history of human disease genetics. *Nature* 577: 179–189.

Cobey, S., D. B. Larremore, Y. H. Grad, and M. Lipstich. 2021. Concerns about SARS-CoV-2 evolution should not hold back efforts to expand vaccination. *Nature Review Immunology* 21: 330–335.

Cohen, J. M., D. J. Civitello, M. D. Venesky, T. A. McMahon, and J. R. Rohr. 2019. An interaction between climate change and infectious disease drove widespread amphibian declines. *Global Change Biology* 25: 927–937.

Cohen, J. M., M. D. Venesky, E. L. Sauer, D. J. Civitello, T. A. McMahon, E. A. Roznik, and J. R. Rohr. 2017. The thermal mismatch hypothesis explains outbreaks of an emerging infectious disease. *Ecology Letters* 20: 184–193.

Coissac, E., P. M. Hollingsworth, S. Lavergne, and P. Taberlet. 2016. From barcodes to genomes: Extending the concept of DNA barcoding. *Molecular Ecology* 25: 1423–1428.

Collins, A., D. Milbourne, L. Ramsay, R. Meyer, C. Chatot-Balandras, P. Oberhagemann, W. De Jong, C. Gebhardt, E. Bonnel, and R. Waugh. 1999. QTL for field resistance to late blight in potato are strongly correlated with maturity and vigour. *Molecular Breeding* 5: 387–398.

Collins, F. C. 1995. Positional cloning moves from perditional to traditional. *Nature Genetics* 9: 347–350.

Collins, R. A. and R. H. Cruickshank. 2013. The seven deadly sins of DNA barcoding. *Molecular Ecology Resources* 13: 969–975.

Comas, M. E. 2020. Trayvon Martin's Sanford killing launched black lives matter. City's racial reckoning isn't over. *Orlando Sentinel*. June 12, 2020.

Combs, M., K. Byers, C. Himsworth, and J. Munshi-South. 2019. Harnessing population genetics for pest management: Theory and application for urban rats. *Human-Wildlife Interactions* 13: 250–263.

Combs, M., E. E. Puckett, J. Richardson, D. Mims, and J. Munshi-South. 2018. Spatial population genomics of the brown rat (*Rattus norvegicus*) in New York City. *Molecular Ecology* 27: 83–98.

Comont, D., H. Hicks, L. Crook, R. Hull, E. Cocciantelli, J. Hadfield, D. Childs, R. Freckleton, and P. Neve. 2019. Evolutionary epidemiology predicts the emergence of glyphosate resistance in a major agricultural weed. *The New Phytologist* 223: 158401594.

Cook, L. M. 2003. The rise and fall of the carbonaria form of the peppered moth. *Quarterly Review of Biology* 78: 399–417.

Cook, L. M., B. S. Grant, L. J. Sacheri, and J. Mallet. 2012. Selective bird predation on the peppered moth: The last experiment of Michael Majerus. *Biological Letters* 8: 609–612.

Coop, G., M. B. Eisen, M. Przeworski, and N. Rosenberg. 2014. Letters: "A troublesome inheritance". *The New York Times*. August 8, 2014.

Cooper, D. S., A. J. Schultz, and D. T. Blumstein. 2020. Temporally separated data sets reveal similar traits of birds persisting in a United States megacity. *Frontiers in Ecology and Evolution* 8: 251.

Coronaviridae Study Group of the International Committee on Taxonomy of Viruses. 2020. The species Severe acute respiratory syndrome-related coronavirus: Classifying 2019-nCoV and naming it SARS-CoV-2. *Nature Microbiology* 5: 536–544.

Coss, S. 2016. *The Fever of 1721: The Epidemic that Revolutionized Medicine and American Politics.* Simon & Schuster.

Costa, R. A., I. R. Ferreira, H. A. Cintra, L. H. F. Gomes, and L. D. C. Guida. 2019. Genotype-phenotype relationships and endocrine findings in Prader–Willi syndrome. *Frontiers in Endocrinology* 10: 864.

Coyne, J. A. and H. A. Orr. 2004. *Speciation.* Sinauer Associates.

Crall, J. D., B. L. deBivort, B. Dey, and A. N. Ford Versypt. 2019. Social buffering of pesticides in bumblebees: Agent-based modeling of the effects of colony size and neonicotinoid exposure on behavior within nests. *Frontiers in Ecology and Evolution* 7: 51.

Crall, J. D., C. M. Switzer, R. L. Oppenheimer, A. N. Ford Versypt, B. Dey, A. Brown, M. Eyster, C. Guerin, N. E. Pierce, S. A. Combes, and B. L. de Bivort. 2018. Neonicotinoid exposure disrupts bumblebee nest behavior, social networks, and thermoregulation. *Science* 362: 683–686.

Crawford, J. E., J. M. Alves, W. Palmer, J. P. Day, M. Sylla, R. Ramasamy, S. N. Surendran, W. C. Black IV., A. Pain, and F. M. Jiggins. 2017a. Population genomics reveals that an anthropophilic population of *Aedes aegypti* mosquitoes in West Africa recently gave rise to American and Asian populations of this major disease vector. *BMC Biology* 15: 16.

Crawford, N. G., D. E. Kelly, M. E. B. Hansen, M. H. Beltrame, S. Fan, S. L. Bowman, E. Jewett, A. Ranciaro, S. Thompson, Y. Lo et al. 2017b. Loci associated with skin pigmentation identified African populations. *Science* 358: eaan8433.

Crespi, B. J. 2020. Why and how imprinted genes drive fetal programming. *Frontiers in Endocrinology* 10: 940.

Crespi, B. J. and M. C. Go. 2015. Diametrical diseases reflect evolutionary-genetic tradeoffs. *Evolution, Medicine, and Public Health* 2015: 216–253.

Croft, D. P., R. A. Johnstone, S. Ellis, S. Nattrass, D. W. Franks, L. J. N. Brent, S. Mazzi, K. C. Balcomb, J. K. B. Ford, and M. C. Cant. 2017. Reproductive conflict and the evolution of menopause in killer whales. *Current Biology* 27: 298–304.

Cromie, W. J. 2003. Genes in conflict: David Haig looks at internal genetic warfare. *The Harvard Gazette.* March 20, 2003.

Crow, J. F. 2000. The origins, patterns and implications of human spontaneous mutation. *Nature Reviews Genetics* 1: 40–47.

Crow, J. F. 2008. Just and unjust: E. E. Just (1883–1941). *Genetics* 179: 1735–1740.

Crow, J. F. 2010. Wright and fisher on inbreeding and genetic drift. *Genetics* 184: 609–611.

Cui, J., F. Li, and Z.-L. Shi. 2019. Origin and evolution of pathogenic coronaviruses. *Nature Reviews Microbiology* 17: 181–192.

Cuthbert, R. J., R. M. Wainless, A. Angel, M.-H. Burle, G. M. Hilton, H. Louk, P. Visser, J. W. Wilson, and P. G. Ryan. 2016. Drivers of predatory behavior of extreme size in house mice *Mus musculus* in Gough Island. *Journal of Mammalogy* 97: 533–544.

Danforth, B. N., S. Cardinal, C. Praz, E. A. B. Almeida, and D. Michez. 2013. The impact of molecular data on our understanding phylogeny and evolution. *Annual Review of Entomology* 58: 57–78.

Danforth, B. N., R. L. Minckley, and J. L. Neff. 2019. *The Solitary Bees: Biology, Evolution, Conservation.* Princeton University Press.

Davidson, E. H. 2001. *Genome Regulatory Systems: Development and Evolution.* Academic Press.

Davies, N. G., S. Abbott, R. C. Barnard, C. J. Jarvis, A. J. Kucharski, J. D. Munday, C. A. B. Pearson, T. W. Russell, D. C. Tully, A. D. Washburne et al. 2021. Estimated transmissibility and impact of SARS-CoV-2 lineage B.1.1.7 in England. *Science* 372: 147–158.

Davis, A. P., H. Chadburn, J. Moat, R. O'Sullivan, S. Hargreaves, and E. N. Lughadha. 2019. High extinction risk for wild coffee species and implications for coffee sector sustainability. *Science Advances* 5: eaav3473.

Davis, A. P., R. Gargiulo, M. F. Fay, D. Sarmu, and J. Haggar. 2020. Lost and found: *Coffea stenophylla* and *C. affinis*, the forgotten coffee crop species of West Africa. *Frontiers in Plant Science* 11: 618.

Dearlove, B., E. Lewitus, H. Bai, Y. Li, D. B. Reeves, M. G. Joyce, P. T. Scott, M. F. Amare, S. Vasan, N. L. Mitchell et al. 2020. A SARS-CoV-2 vaccine candidate would likely match all currently circulating variants. *Proceedings of the National Academy of Science* 117: 23652–23662.

DeBiasse, M. B. and M. W. Kelly. 2015. Plastic and evolved responses to global change: What can we learn from comparative transcriptomics? *Journal of Heredity* 107: 71–81.

De Boeck, K. 2020. Cystic fibrosis in the year 2020: A disease with a new face. *Acta Paediatrica* 109: 893–899.

DeCarlo, T. M., H. B. Harrison, L. Gajdzik, D. Alaguarda, R. Rodolfo-Metalpa, J. D'Olivio, D. Patalwala, and M. T. McCulloch. 2019. Acclimatization of massive reef-building corals to consecutive heat waves. *Proceedings of the Royal Society of London B* 286: 20190235.

DeGregori, J. 2018. *Adaptive Oncogenesis: A New Understanding of How Cancer Evolves Inside Us.* Harvard University Press.

De Jong, H. 2016. Impact of the potato on society. *American Journal of Potato Research* 93: 415–429.

DeLeo, D. M., S. Herrera, S. D. Lengyel, A. M. Quattrini, R. J. Kulathinal, and E. E. Cordes. 2018. Gene expression profiling reveals deep-sea coral response to the Deepwater Horizon oil spill. *Molecular Ecology* 27: 4066–4077.

DeLeo, D. M., D. V. Ruiz-Ramos, H. B. Baums, and E. E. Cordes. 2016. Response of deepwater corals to oil and chemical dispersant exposure. *Deep Sea Research* 129: 137–147.

Delph, L. F. and J. P. Demuth. 2016. Haldane's rule: Genetic bases and their empirical support. *Journal of Heredity* 107: 383–391.

DeSalle, R. and I. Tattersall. 2018. *Troublesome Science: The Misuse of Genetics and Genomics in Understanding Race.* Columbia University Press.

Desceux, N., A. Decourtye, and J.-M. Delpuesch. 2007. The sublethal effects of pesticides on beneficial arthropods. *Annual Review of Entomology* 52: 81–106.

De Vries, S., J. K. von Dahlen, C. Uhlmann, A. Schnake, T. Kloesges, and L. E. Rose. 2017. Signatures of selection and host-adapted gene expression of the *Phytophthora infestans* RNA silencing suppressor PSR2. *Molecular Plant Pathology* 18: 110–124.

De Weerdt, S. 2020. Fix what's broken: Drugs that target specific mutations in the protein at the root of cystic fibrosis have supercharged treatments for the disease and spurred the search for further therapies. *Nature* 583: S2–S4.

Diamond, S. A., L. D. Chick, A. Perez, S. A. Strickler, and R. A. Martin. 2018. Evolution of thermal tolerance and its fitness consequences: Parallel and non-parallel responses to urban heat islands across three cities. *Proceedings of the Royal Society of London B* 285: 20180036.

Diamond, S. A. and R. A. Martin. 2020. Evolutionary consequences of the urban heat island. Pp. 91–110 In *Urban Evolutionary Biology.* M. Szulkin, J. Munshi-South, and A. Charmantier (eds.). Oxford University Press.

Diamond, S. E. 2017. Evolutionary potential of upper thermal tolerance: Biogeographic patterns and expectations under climate change. The year in evolutionary biology. *Annals of the New York Academy of Sciences* 1389: 5–19.

Diamond, S. E., D. M. Sorger, J. Hulcr, S. L. Pelini, I. Del Toro, C. Hirsch, E. Oberg, and R. R. Dunn. 2011. Who likes it hot? A global analysis of the climatic, ecological, and evolutionary determinants of warming in ants. *Global Change Biology* 18: 448–456.

Diaz, S., J. Seetele, E. Brondizio, N. T. Ngo, M. Gue'ze, J. Agard, A. Arneth, P. Balvanera, K. Braumann et al. 2019. *Global Assessment Report on Biodiversity and Ecosystem Services*. IPBES.

Dickinson, E. 2012. *The Poems of Emily Dickinson*. Start Publishing.

Dilley, B. J., S. Schoombie, J. Schoombie, and P. G. Ryan. 2016. "Scalping" of albatross fledglings by introduced mice spreads rapidly at Marion Island. *Antarctic Science* 28: 73–80.

Dobbs, D. 2014. The fault in our DNA. *The New York Times*. July 10, 2014.

Dobson, A. P., S. L. Pimm, L. Hannah, L. Kaufman, J. A. Ahumada, A. W. Ando, A. Bernstein, J. Busch, P. Daszak, J. Engelmann et al. 2020. Ecology and economics for pandemic prevention. *Science* 369: 379–381.

Dobzhansky, T. 1973. Nothing in biology makes sense except in the light of evolution. *American Biology Teacher* 35: 125–129.

Dolezal, A. G., J. Carrillo-Tripp, T. M. Judd, W. A. Miller, B. C. Bonning, and A. L. Toth. 2019. Interacting stressors matter: Diet and virus infection in honeybee health. *Royal Society Open Science* 6: 181803.

Dolgin, E. 2020. Cell-based meat with a side of science. *Nature* 588: S64–S67.

Doney, S. C., M. Ruckelshaus, J. E. Duffy, J. P. Berry, F. Chan, C. A. English, H. M. Galindo, J. M. Grebmeier et al. 2012. Climate change impacts on marine ecosystems. *Annual Review of Marine Sciences* 4: 11–37.

Donohoe, D. R., N. Garge, X. Zhang, W. Sun, T. M. O'Connell, M. K. Bunger, and S. J. Buttman. 2011. The microbiome and butyrate regulate energy metabolism and autophagy in the mammalian colon. *Cell Metabolism* 13: 517–526.

Doudna, J. A. and S. H. Sternberg. 2017. *A Crack in Creation: Gene Editing and the Unthinkable Power to Control Evolution*. Mariner Books.

Dowd, K. 2020. After decades in the shadows, Joseph DeAngelo confesses he is the golden state killer. *SFGate*. June 29, 2020. www.sfgate.com/crime/article/Josepth-DeAngelo-earon-guilty-plea-hearing-15372192.php. Accessed April 14, 2021.

Driver, J. A. 2014. Inverse association between cancer and neurodegenerative diseases: Review of the epidemiological and biological evidence. *Biogerontology* 15: 547–557.

Dubinsky, B., A. Whitehead, J. T. Miller, C. D. Rice, and F. Galvez. 2013. Multitissue molecular, genomic, and developmental effects of the Deepwater Horizon oil spill on resident gulf killifish (*Fundulus grandis*). *Environmental Science & Technology* 47: 5074–5082.

Dubois, J. and P. Cheptou. 2016. Effects of fragmentation on plant adaptation to urban environments. *Philosophical Transactions of the Royal Society of London B* 372: 20160038.

Ducatez, S., J.-N. Audet, J. R. Rodriguez, L. Kayello, and L. Lefebvre. 2017. Innovativeness and the effects on urbanization on risk-taking behaviors in wild Barbados birds. *Animal Cognition* 20: 33–42.

Dugatkin, L. A. 2018. The silver fox domestication experiment. *Evolution: Education and Outreach* 11: 16.

Dugatkin, L. A. and L. W. Trut. 2017. *How to Tame a Fox (and Build a Dog)*. University of Chicago Press.

Duke, S. O. 2017. The history and current status of glyphosate. *Pest Management Science* 74: 1027–1034.

Dunbar, R. I. M. 1992. Neocortex size as a constraint on group size in primates. *Journal of Human Evolution* 22: 469–473.

Dunlap, D. W. 1995. Frank Lily, a geneticist, 65; Member of national AIDS panel. *The New York Times*. October 16, 1995.

Dunn, R. 2017. *Never Out of Season: How Having the Food We Want When We Want It Threatens Our Food Supply and Our Future*. Little, Brown and Company.

Dunn, R. and M. Sanchez. 2021. *Delicious: The Evolution of Flavor and How It Made Us Human*. Princeton University Press.

Dunsworth, H. M. 2018. There is no "obstetrical dilemma": Toward a braver medicine with fewer childbirth interventions. *Perspectives in Biology and Medicine* 61: 249–263.

Dunsworth, H. M., A. G. Warrener, T. Deacon, P. T. Eillison, and H. Pontzner. 2012. Metabolic hypothesis for human altriciality. *Proceedings of the National Academy of Sciences* 109: 15212–15216.

Edge, M. D. and G. Coop. 2019. How lucky was the genetic investigation in the Golden State Killer case? bioRxiv preprint. This version posted January 29, 2019.

Edge, M. D. and G. Coop. 2020. Attacks on genetic privacy via uploads to genealogical databases. *eLife* 9: e51810.

Edmands, S. 2007. Between a rock and a hard place: Evaluating the relative risks of inbreeding and outbreeding for conservation and management. *Molecular Ecology* 16: 463–475.

Edwards, A. W. F. 2003. Human genetic diversity: Lewontin's fallacy. *Bioessays* 25: 798–801.

Eilers, E. J., C. Kremen, S. Smith Greenleaf, A. K. Garber, and A.-M. Klein. 2011. Contribution of pollinator-mediated crops to nutrients in the human food supply. *PLoS One* 6: e21363.

Eilperin, J. 2015. Rafe Sagarin, who merged ecological thinking with national security dies. *The Washington Post.* June 4, 2015.

Eisenberg, D. T. A., C. W. Kuzawa, and M. G. Hayes. 2010. Worldwide allele frequencies of the human Apolipoprotein E gene: Climate, local adaptation, and evolutionary history. *American Journal of Physical Anthropology* 143: 100–111.

Eligon, J., T. Arango, S. Dewan, and N. Bogel-Burroughs. 2021. Derek Chauvin verdict brings a rare rebuke of police misconduct. *The New York Times.* April 20, 2021. www.nytimes.com/2021/04/20/us/george-floyd-chauvin-verdict.html. Accessed April 28, 2021.

Ellinghaus, D., L. Jostins, S. L. Spain, A. Coates, J. Bethune, B. Han, Y. R. Park, S. Raychaudhuri, J. G. Pouget, M. Hubenthal et al. 2016. Analysis of five chronic inflammatory diseases identifies 27 new associations and highlights disease-specific at shared loci. *Nature Genetics* 48: 510–521.

Ellis, J., P. Dodds, and T. Pryor. 2000. Structure, function and evolution of plant disease resistance genes. *Current Opinion in Plant Biology* 3: 278–284.

Ellis, S., D. W. Franks, S. Nattrass, T. E. Currie, M. A. Cant, D. Giles, K. C. Balcomb, and D. P. Croft. 2018. Analysis of ovarian activity reveal repeated evolution of post-reproductive lifespans in toothed whales. *Scientific Reports* 8: 12833.

Ellis, W. W. 2004. Terrorism in the United States: Revisiting the hart-Rudman commission. *Mediterranean Quarterly* 15: 23–37.

Embry, P. 2018. *Our Native Bees: North America's Endangered Pollinators and the Fight to Save Them.* Timber Press.

Emlen, D. J. 2008. The evolution of animal weapons. *Annual Review of Ecology, Evolution, and Systematics* 39: 387–411.

Emlen, D. J. 2014. *Animal Weapons: The Evolution of Battle.* Henry Holt and Company.

Emlen, D. J., J. Marangello, B. Ball, and C. W. Cunningham. 2005. Diversity in the weapons of sexual selection: Horn evolution in the beetle genus *Onthophagus* (Coleoptera: Scarabaeidae). *Evolution* 59: 1060–1084.

Emlen, D. J. and T. K. Phillips. 2006. Phylogenetic evidence for an association between tunneling behavior and the evolution of horns in dung beetles (Coleoptera: Scarbaeidae: Scarabaeinae). Coleopterists society monographs. Scarabaeoidea in the 21st century: A Festschrift Honoring Henry F. *Howden* 5: 47–56

Enard, D. and. D. A. Petrov. 2018. Evidence that RNA viruses drove the adaptive introgression between Neanderthals and modern humans. *Cell* 175: 360–375.

Engelhardt, S. C., P. Bergeson, A. Gagnon, L. Dillion, and F. Pelletier. 2019. Using geographical distance as a potential proxy for help in the assessment of the grandmother hypothesis. *Current Biology* 29: 651–656.

Erlich, Y., T. Shor, I. Pe'er, and S. Carmi. 2018. Identity inference of genomic data using long-range familial searches. *Science* 362: 690–694.

Ernst, S. G. 1997. A century of sea urchin development. *American Zoologist* 37: 250–259.

Estroup, A. and T. Guillemaud. 2010. Reconstructing routes of invasion using genetic data: Why, how and so what? *Molecular Ecology* 19: 4113–4130.

Evans, T. G., J. L. Padilla-Gamino, M. W. Kelly, M. H. Pespeni, F. Chan, B. A. Menge, B. Gaylord, T. M. Hill, A. D. Russell, S. R. Palumbi, E. Sanford, and G. E. Hoffman. 2015. Ocean acidification research in the 'post-genomic' era: Roadmaps from the purple sea urchin *Strongylocentrotus purpuratus*. *Comparative Biochemistry and Physiology Part A* 185: 33–42.

Evans, T. G., M. H. Pespeni, G. E. Hoffman, S. R. Palumbi, and E. Sanford. 2017. Transcriptomic responses to seawater acidification among sea urchin populations inhabiting a natural pH mosaic. *Molecular Ecology* 26: 2257–2275.

Fabry, V. 2008. Marine calcifiers in a high-CO2 ocean. *Science* 320: 1020–1022.

Fan, D., W. Fang, D-J. Li, J. Zhang, and F. Zhao. 2020. Association between ABO blood group system and COVID-19 susceptibility in Wuhan. *Frontiers in Cellular and Infection Microbiology* 10: 404.

Fang, Z., A. M. Gonzales, M. L. Durbin, K. K. T. Meyer, B. H. Miller, K. M. Volz, M. T. Clegg, and P. L. Morrell. 2013. Tracing the geographic origins of weedy *Ipomoea purpurea* in the southern United States. *Journal of Heredity* 104: 666–677.

Fea, M. P., R. P. Boisseau, D. J. Emlen, and G. I. Holwell. 2020. Cybernetic combatants support the importance of duels in the evolution of extreme weapons. *Proceedings of the Royal Society of London B* 287: 20200254.

Feder, A. F., S.-Y. Rhee, S. P. Holmes, R. W. Shafer, D. A. Petrov, and P. S. Pennings. 2016. More effective drugs lead to harder selective sweeps in the evolution of drug resistance. *eLife* 5: e10670.

Feder, J. L., S. B. Opp, B. Wlazlo, K. Reynolds, W. Go, and S. Spisak. 1994. Host fidelity is an effective premating barrier between sympatric races of apple maggot fly. *Proceedings of the National Academy of Science* 91: 7990–7994.

Ferguson-Smith, A. C. 2011. Genomic imprinting: The emergence of an epigenetic paradigm. *Nature Reviews Genetics* 12: 565–575.

Ferrarezi, R. S., C. I. Vincent, A. Urbaneja, and M. A. Machado. 2020. Editorial: Unravelling citrus huanglongbing disease. *Frontiers in Plant Science* 11: 609655.

Ferrer-Admetlla, A., M. Sikora, H. Laayouni, A. Esteve, F. Rouhinet, A. Blancher, F. Calafell, J. Bertranpetit, and F. Casals. 2009. A natural history of *FUT2* polymorphism in humans. *Molecular Biology and Evolution* 26: 1993–2003.

Finlay, B. B., K. R. Amato, M. Azad, M. J. Blaser, T. C. G. Bosch, H. Chu, M. G. Dominguez-Bello, S. D. Ehrlich, N. Elinav, N. Geva-Zatorsky et al. 2021. The hygiene hypothesis, the COVID pandemic, and the consequences for the human microbiome. *Proceedings of the National Academy of Sciences* 118: e2010217118.

Fischer, L. 2020. The real danger posed by coronavirus-infected mink. *Scientific American*. November 11, 2020.

Fortunato, A., A. Boddy, D. Mallo, A. Aktipis, C. C. Maley, and J. W. Pepper. 2017. Natural selection in cancer biology: From molecular snowflakes to trait hallmarks. *Cold Spring Harbor Perspectives in Medicine* 7: a029652.

Frank, S. A. 1995. George Price's contribution to evolution. *Journal of Theoretical Biology* 175: 373–388.

Frankham, R. 2015. Genetic rescue of small inbred populations: Meta-analysis reveals large and consistent benefits of gene flow. *Molecular Ecology* 24: 2610–2618.

Frankham, R., C. J. A. Bradshaw, and B. W. Brook. 2014. Genetics in conservation management: Revised recommendations for the 50/500 rules, red list criteria and population viability analysis. *Biological Conservation* 170: 56–63.

Franks, S. J., N. C. Kane, N. B. O'Hara, S. Tittes, and J. S. Rest. 2016. Rapid genome-wide evolution in *Brassica Rapa* populations following drought revealed by sequencing of ancestral and descendant gene pools. *Molecular Ecology* 25: 3622–3631.

Franks, S. J. and J. Munshi-South. 2014. Go forth, evolve and prosper: The genetic basis of adaptive evolution in an invasive species. *Molecular Ecology* 23: 2137–2140.

Franks, S. J. S. Sim, and A. E. Weis. 2007. Rapid evolution of flowering time by an annual plant in response to a climate fluctuation. *Proceedings of the National Academy of Sciences* 104: 1278–1282.

Frazier, I. 2017. Frogpocalpse now. *Outside Magazine*. March 23, 2017.

Freda, P. J., Z. M. Ali, N. Heter, G. J. Ragland, and T. J. Morgan. 2019. Stage-specific genotype-by-environment interactions for cold and heat hardiness in *Drosophila melanogaster*. *Heredity* 123: 479–491.

Freler, K., M. Meinshausen, A. Goly, M. Mengel, K. Lebek, S. D. Donner, and O. Hoegh-Guldberg. 2013. Limiting global warming to 2 degrees C is unlikely to save most coral reefs. *Nature Climate Change* 3: 165–170.

Fridley, J. D. and D. F. Sax. 2014. The imbalance of nature: Revisiting a Darwinian framework for invasion biology. *Global Ecology and Biogeography* 23: 1157–1166.

Fry, A. E., M. J. Griffiths, S. Auburn, M. Diakite, J. T. Forton, A. Green, A. Richardson, J. Wilson, M. Jallow, F. Sisay-Joof et al. 2008. Common variation in the ABO glycosyltransferase is associated with susceptibility to severe *Plasmodium falciparum* malaria. *Human Molecular Gene* 17: 567–578.

Fry, W. 2008. *Phytophthora infestans*: The plant (and R gene) destroyer. *Molecular Plant Pathology* 9: 385–402.

Fu, F., S. D. Kocher, and M. A. Nowak. 2015. The risk-return trade-off between solitary and eusocial reproduction. *Ecology Letters* 18: 74–84.

Fukuda, Y., R. Tingley, B. Grase, G. Webb, and K. Saalfeld. 2015. Long-term monitoring reveals declines in an endemic predator following invasion by an exotic prey. *Animal Conservation* 19: 75–87.

Fung, B. and C. Dewey. 2018. Justice department approves Bayer-Monsanto merger in landmark deal. *Washington Post*. May 29, 2018.

Gale, J. 2021. Bats in Laos caves harbor closest relatives to Covid-19 virus. Bloomberg. http://www.bloomberg.com/news/articles/2021-09-18/bats-in-laos-caves-harbor-closest-relatives-to-covid-19-virus. Accessed September 25, 2021.

Gallone, B., J. Steensels, and K. J. Verstrepen. 2020. Moulded by humans: The domestication of blue-veined cheese fungi. *Molecular Ecology* 29: 2517–2520.

Gampa, A., P. A. Engen, R. Shobar, and E. A. Mutlu. 2017. Relationships between gastrointestinal microbiota and blood group antigens. *Physiological Genetics* 49: 473–483.

Gantz, V. M. and E. Bier. 2014. The mutagenic chain reaction: A method for converting heterozygous to homozygous mutations. *Science* 348: 442–444.

Garcia, M. D., A. Nouwens, T. G. Lonhienne, and L. W. Guddat. 2017. Comprehensive understanding of acetohydroxyacid synthase inhibition by different herbicide families. *Proceedings of the National Academy of Science* 114: E1091–E1100.

Gardner, A., J. Alpedrinha, and S. A. West. 2012. Haplodiploidy and the evolution of eusociality: Split sex ratios. *The American Naturalist* 179: 240–256.

Gardner, H. L., J. M. Fenger, and C. A. London. 2016. Dogs as a model for cancer. *Annual Review of Animal Bioscience* 4: 199–222.

Gatenby, R. A. and R. J. Gillies. 2004. Why do cancers have high aerobic glycolysis? *Nature Reviews Cancer* 4: 891–899.

Gatenby, R. A. and R. J. Gillies. 2011. Of cancer and cave fish. *Nature Reviews Cancer* 11: 237–238.

Gatenby, R. A., A. S. Silva, R. J. Gillies, and B. R. Frieden. 2009. Adaptive therapy. *Cancer Research* 69: 4894–4904.

Geerts, A. N., J. Vanoverbeke, W. VanDoorslaer, H. Feuchtmayr, D. Atkinson, B. Moss, T. A. Davidson, C. D. Sayer, and L. De Meester. 2015. Rapid evolution of thermal tolerance in the water flea *Daphnia*. *Nature Climate Change* 5: 665–670.

George, A. L. 2013. The paleo diet is a paleo fantasy. *Slate*. April 7, 2013.

George, A. L. and R. L. Mayden. 2005. Species concepts and the endangered species act: How a valid biological definition of species enhances the legal protection of biodiversity. *Natural Resources Journal* 45: 369–407.

Gessen, M. 2008. *Blood Matters: From Inherited Illness to Designer Babies, How the World and I Found Ourselves in the Future of the Gene.* Harcourt.

Gevers, D., S. Kugathasan, L. A. Denson, Y. Va'zquez-Baeza, W. van Treuren, B. Ren, E. Schwager, D. Knights, S. J. Song, M. Yassour et al. 2014. The treatment-naïve microbiome in new-onset Crohn's disease. *Cell Host and Microbe* 15: 382–392.

Gibney, E., M. S. van Noorden, H. Ledford, D. Castlelvecchi, and M. Warren. 2018. 'Test-tube' evolution wins chemistry Nobel prize. *Nature* 562: 176.

Gibson, M. J. S. and L. C. Moyle. 2020. Regional differences in the abiotic environment contribute to genomic divergence within a wild tomato. *Molecular Ecology* 29: 2204–2217

Gilbert, M. T. P., A. Rambaut, G. Wlasiuk, T. J. Spira, A. E. Pitchenik, and M. Worobey. 2007. The emergence of HIV/AIDS in the Americas and Beyond. *Proceedings of the National Academy of Science* 104: 18566–18570.

Gluckman, P., A. Beedle, T. Buklijas, F. Low, and M. Hanson. 2016. *Principles of Evolutionary Medicine* (2nd ed.). Oxford University Press.

Gluckman, P. and M. Hanson. 2006. *Mismatch: Why Our World No Longer Fits Our Bodies.* Oxford University Press.

Gomulkiewicz, R., M. L. Thies, and J. J. Bull. 2021. Evading resistance to gene drives. *Genetics* 217: iyaa040.

Gould, S. J. 1989. *Wonderful Life: The Burgess Shale and the Nature of History.* W. W. Norton & Company.

Goulson, D. 2003. Effects of introduced bees on native ecosystems. *Annual Review of Ecology, Evolution, and Systematics* 34: 1–26.

Goulson, D., E. Nicholls, C. Botias, and E. L. Rotheray. 2015. Bee declines driven by combined stress from parasites, pesticides, and lack of flowers. *Science* 347: 1435–1444.

Graham, D. B. and B. J. Xavier. 2020. Pathway paradigms revealed from the genetics of inflammatory bowel disease. *Nature* 578: 527–539.

Graham, R. L. and R. S. Barle. 2010. Recombination, reservoirs, and modular spike: Mechanisms of coronavirus cross-species transmission. *Journal of Virology* 84: 3134–3146.

Grant, B. S. 1999. Fine tuning the peppered moth paradigm. *Evolution* 53: 980–984.

Grant, B. S. 2004. Allelic melanism in American and British peppered moths. *Journal of Heredity* 95: 97–102.

Grant, B. S. 2020. Vignette 10.3 industrial melanism: Genetic adaptation to pollution. In *Fundamentals of Ecotoxicology: The Science of Pollution.* M. C. Neman (eds.). CRC Press.

Grant, B. S. and R. J. Howlett. 1988. Background selection by the peppered moth (*Biston betularia* Linn.): Individual differences. *Biological Journal of the Linnean Society* 33: 217–232.

Grant, B. S. and L. L. Wiseman. 2002. Recent history of melanism in American peppered moths. *Journal of Heredity* 93: 86–90.

Graves, J. L. Jr. 2015. Why the nonexistence of biological races does not mean the nonexistence of racism. *American Behavioral Science* 59: 1474–1495.

Gray, K. J., R. Saxena, and S. A. Karuamanchi. 2018. Genetic predisposition to preeclampsia is conferred by fetal DNA variants near *FLT1*, a gene involved in the regulation of angiogenesis. *American Journal of Obstetrics and Gynecology* 218: 211–218.

Gray, M. M., M. D. Paramenter, C. A. Hogan, I. Ford, R. J. Cuthbert, P. G. Ryan, K. W. Bowman, and B. A. Payseur. 2015. Genetics of rapid and extreme size evolution in island mice. *Genetics* 201: 213–228.

Gray, M. M., D. Wegmann, R. J. Haasl, M. A. White, S. I. Gabriel, J. B. Searle, R. J. Cuthbert, P. G. Ryan, and B. A. Payseur. 2014. Demographic history of a recent invasion of house mice on the isolated island of Gough. *Molecular Ecology* 23: 1923–1939.

Greenfield, K. T. 2009. *China Syndrome: The First Story of the 21st Century's First Great Epidemic.* Harper Collins.

Gregoric, A. and I. Planinc. 2005. The control of *Varroa destructor* in honey bees using the thymol-based acaricide—Apiguard. *American Bee Journal* 145: 672–675.

Gregory, T. R. 2008. Evolution as fact, theory, and path. *Evolution: Education and Outreach* 1: 46–52.

Griffiths, J. S., T.-C. Francis Pan, and M. W. Kelly. 2019. Differential responses to ocean acidification between populations of *Balanophyllia elegans* corals from high and low upwelling environments. *Molecular Ecology* 28: 2715–2730.

Grubaugh, N. D., W. P. Hanage, and A. L. Rasmussen. 2020. Making sense of mutation: What D614G means for the COVID-19 pandemic remains unclear. *Cell* 182: 794–795.

Grubaugh, N. D., J. T. Ladner, P. Lemey, O. G. Pybus, A. Rambaut, E. C. Holmes, and K. G. Andersen. 2019. Tracing virus outbreaks in the twenty-first century. *Nature Microbiology* 4: 10–19.

Gurdasani, D., I. Barroso, E. Zeggini, and M. S. Sandhu. 2019. Genomics of disease risk in globally diverse populations. *Nature Reviews Genetics* 20: 520–535.

Gurven, M. D. ands D. E. Lieberman. 2020. WEIRD bodies: Mismatch, medicine, and missing diversity. *Evolution and Human Behavior* 41: 330–340.

Gussow, A. B., N. Auslander, G. Faure, Y. I. Wolf, F. Zeng, and E. V. Koonin. 2020. Genomic determinants of pathogenicity in SARS-CoV-2 and other human coronaviruses. *Proceedings of the National Academy of Sciences* 117: 15193–15199.

Gutaker, R. M., C. L. Weiss, D. Ellis, N. L. Anglin, S. Knapp, J. L. Fernández-Alonso, S. Prat, and H. Burbano. 2019. The origins and adaptation of European potatoes reconstructed from historical genomes. *Nature Ecology & Evolution* 3: 1093–1101.

Guthrie, J. R. 2007. Darwinian Dickinson: The scandalous rise and noble fall of the common clover. *The Emily Dickinson Journal* 16: 73–91.

Haig, D. 1993. Genetic conflicts in human pregnancy. *Quarterly Review of Biology* 68: 495–532.

Haig, D. 2000. The kinship theory of genomic imprinting. *Annual Review of Ecology and Systematics* 31: 9–32.

Haig, D. 2014. Maternal-fetal conflict, genomic imprinting and mammalian vulnerabilities to cancer. *Philosophical Transactions of the Royal Society of London B* 370: 20140178.

Haig, D. 2019. Cooperation and conflict in human pregnancy. *Current Biology* 29: R455–R458.

Haig, D. 2020. Commentary on the exchange between Boddy et al. and Wagner et al.: Malignancy, placentation and litter size. *Evolution, Medicine, and Public Health* 2020: 217–218.

Haig, D. and M. Westoby. 1989. Parent-specific gene expression and the triploid endosperm. *The American Naturalist* 134: 147–155.

Hamilton, W. D. 1964. The genetical evolution of social behavior I & II. *Journal of Theoretical Biology* 7: 1–52.

Hamilton, W. D. 1972. Altruism and related phenomena, mainly in social insects. *Annual Review of Ecology and Systematics* 3: 193–232.

Hammer, S. C., A. M. Knight, and F. H. Arnold. 2017. Design and evolution of enzymes for non-natural chemistry. *Current Opinion in Green and Sustainable Chemistry* 7: 23–30.

Hamon, P., C. E. Grover, A. P. Davis, J.-J. Rakatomala, N. E. Raharimalala, V. A. Albert, H. L. Sreeenath, P. Stoffelen, S. E. Mitchell, E. Couturon et al. 2017. Genotyping-by-sequencing provides the first well-resolved phylogeny for coffee (*Coffea*) and insights into the evolution of caffeine content in its species. *Molecular Phylogenetics and Evolution* 109: 351–361.

Hanahan, D. A. and R. A. Weinberg. 2000. The hallmarks of cancer. *Cell* 100: 57–70.

Hanahan, D. A. and R. A. Weinberg. 2011. Hallmarks of cancer: The next generation. *Cell* 144: 646–673.

Handley, L. J. L., A. Manica, J. Goudet, and F. Balloux. 2007. Going the distance: Human population genetics in a clinal world. *Trends in Genetics* 23: 432–439.

Hangartner, S. and A. A. Hoffmann. 2016. Evolutionary potential of multiple measures of upper thermal tolerance in *Drosophila melanogaster*. *Functional Ecology* 30: 442–452.

Hardy, K., J. Brand-Miller, K. D. Brown, M. G. Thomas, and L. Copeland. 2015. The importance of dietary carbohydrate in human evolution. *The Quarterly Review of Biology* 90: 251–268.

Hardy, N. B., C. Kaczvinsky, G. Bird, and B. B. Normark. 2020. What we don't know about diet-breadth evolution in herbivorous insects. *Annual Review of Ecology, Evolution, and Systematics* 51: 103–122.

Hardy, N. B., D. A. Peterson, L. Ross, and J. A. Rosenheim. 2018. Does a plant-eating insect's diet govern the evolution of insecticide resistance? Comparative tests of the pre-adaptation hypothesis. *Evolutionary Applications* 11: 739–747.

Harman, O. 2010. *The Price of Altruism: George Price and the Search for the Origins of Kindness*. W. W. Norton & Company.

Harris, S. and J. M. V. Rayner. 1986. A discriminant analysis of the current distribution of urban foxes (*Vulpes vulpes*) in Britain. *Journal of Animal Ecology* 55: 605–611.

Hartl, D. L. and A. G. Clark. 2007. *Principles of Population Genetics* (4th ed.). Sinauer Associates, Inc.

Hassan, J. 2020. Man named William Shakespeare, one of the first to get Pfizer vaccine, sets off pun cascade. *Washington Post*. December 8, 2020.

Hawkes, K., J. F. O'Connell, N. G. Blurton Jones, H. Alvarez, and E. L. Charnov. 1998. Grandmothering, menopause, and the evolution of human life histories. *Proceedings of the National Academy of Sciences* 95: 1336–1339.

Hawkes, K. and K. R. Smith. 2010. Do women stop early? Similarities in fertility declines in humans and chimpanzees. *Annals of the New York Academy of Sciences* 1204: 43–53.

He, W., S. Neil, H. Kulkami, E. Wright, B. K. Agan, V. C. Marconi, M. J. Dolan, R. A. Weiss, and S. K. Ahuja. 2008. Duffy antigen receptor for chemokines mediates trans-infection of HIV-1 from red blood cells to target cells and affects HIV-AIDS. *Cell Host & Microbe* 4: 52–62.

Head, I. M., D. M. Jones, and W. F. M. Roling. 2006. Marine microorganisms make a meal of oil. *Nature Reviews Microbiology* 4: 173–182.

Hedrick, P. W. 2013. Adaptive introgression in animals: Examples and comparison to new mutation and standing variation as sources of adaptive variation. *Molecular Ecology* 22: 4606–4618.

Hegarty, R. S., J. P. Goopy, R. M. Herd, and B. McCorkell. 2007. Cattle selected for lower residual feed intake have reduced daily methane production. *Journal of Animal Science* 85: 1487–1486.

Heinen-Kay, J. L. and M. Zuk. 2019. When does sexual signal exploitation lead to signal loss? *Frontiers in Ecology and Evolution* 7: 255.

Henderson, D. A. 2009. *Smallpox: The Death of a Disease: The Inside Story of Eradicating a Worldwide Killer*. Prometheus.

Hendry, A. P. 2017. *Eco-evolutionary Dynamics*. Princeton University Press.

Hensely, S. 2020. FDA analysis of Moderna COVID-19 vaccine finds it effective and safe. *NPR*. December 15, 2020.

Herbert, P. D. N., A. Cywinska, S. L. Ball, and J. R. deWaard. 2003. Biological identifications through DNA barcodes. *Proceedings of the Royal Society of London B* 270: 313–321.

Herbert, P. D. N. and T. R. Gregory. 2005. The promise of DNA barcoding for taxonomy. *Systematic Biology* 54: 852–859.

Herbert, P. D. N., M. Y. Stoecke, T. S. Zeimak, C. M. Francis. 2004. Identification of birds through DNA barcodes. *PLOS Biology* 2: e312.

Hermisson, J. and P. S. Pennings. 2005. Soft sweeps: Molecular population genetics of adaptation from standing genetic variation. *Genetics* 169: 2335–2352.

Hill, M. O., D. B. Roy, and K. Thompson. 2002. Hemeroby, urbanity and ruderality: Bioindicators of disturbance and human impact. *Journal of Applied Ecology* 39: 708–720.

Hill, W. G. 2014. Applications of population genetics to the animal breeding, from wright, fisher, and lush to genomic prediction. *Genetics* 196: 1–16.

Hillis, D. M., J. J. Bull, M. E. White, M. R. Badgett, and I. J. Molineux. 1992. Experimental phylogenetics: Generation of a known phylogeny. *Science* 255: 589–595.

Hillis, D. M., C. H. Heller, S. D. Hacker, D. W. Hall, M. J. Laskowski, and D. E. Savada. 2020. *Life: The Science of Biology* (12th ed.). Macmillian.

Hochman, A. 2021. Janus-faced race: Is race biological, social, or mythical? *American Journal of Physical Anthropology* 175: 453–464.

Hoekstra, H., R. J. Hirschmann, R. A. Bundey, P. A. Insel, and J. P. Crossland. 2006. A single amino acid mutation contributes to adaptive beach mouse color pattern. *Science* 313: 101–104.

Holmes, E. C., S. A. Goldstein, A. L. Rasmussen, D. L. Robertson, A. Crits-Christoph, J. O. Wertheim, S. J. Anthony, W. S. Barclay, M. F. Boni, P. C. Doherty, et al. 2021. The origins of SARS-CoV-2: An critical review. *Cell* 184: 4848–4856.

Horton, R. 2020. *The COVID-19 Catastrophe: What's Gone Wrong and How to Stop it Happening Again*. Polity Press.

Hottes, A. K., P. L. Freddolino, A. Khare, Z. N. Donnell, J. C. Liu, and S. Tavazoie. 2013. Bacterial adaptation through loss of function. *PLoS Genetics* 9: e1003617.

Hrdy, S. B. 2009. *Mothers and Others: The Evolutionary of Mutual Understanding*. Harvard University Press.

Hubbell, S. 1988. *A Book on Bees: And How to Keep Them*. Houghton Mifflin.

Hubbell, S. 2001. *Shrinking the Cat: Genetic Engineering Before We Knew about Genes*. Houghton Mifflin Harcourt.

Hueber, W., D. D. Patel, T. Dryja, A. M. Wright, I. Koroleva, G. Bruin, C. Antoni, Z. Draelos, M. H. Gold, the Psoriasis Study Group et al. 2010. Effects of AIN457, a fully human antibody to interleukin-17A, on psoriasis, rheumatoid arthritis, and uveitis. *Science Translational Medicine* 2: 52ra72.

Huetra-Sánchez, X. Jin, Asan, Z. Bianba, B. M. Peter, N. Vinckerbosch, Y. Liang, X. Yi, M. He, M. Somel et al. 2010. Altitude adaptation in Tibetans caused by introgression of Denisovan-like DNA. *Nature* 512: 194–197.

Hufbauer, R. A. and G. R. Roderick. 2005. Microevolution in biological control: Mechanisms, patterns, and processes. *Biological Control* 35: 227–239.

Humphreys, A. M., R. Govaerts, S. Z. Ficinski, E. Nic Lughadha, and M. S. Vorontsova. 2019. Global dataset shows geography and life form predict modern plant extinction and rediscovery. *Nature Ecology & Evolution* 3: 1043–1047.

Hutchings, J. A. and A. Kuparinen. 2019. Implications of fisheries-induced evolution for population recovery: Refocusing the science and refining its communication. *Fish and Fisheries* 21: 453–464.

Hutchings, J. A. and A. Kuparinen. 2021. Throwing down a genomic gauntlet on fisheries-induced evolution. *Proceedings of the National Academy of Sciences* 118: e2105319118.

Hylander, B. L. and E. A. Repasky. 2019. Temperature as a modulator of the gut microbiome: What are the implications and opportunities for thermal medicine? *International Journal of Hyperthermia* 36 (Suppl.): 83–89.

Immink, R. G. H., D. Pose, S. Ferrario, F. Ott, K. Kaufmann, F. L. Valentim, S. de Folter, F. van der Wal, A. D. J. van Dijk, M. Schmid, and G. C. Angenent. 2012. Characterization of SOC1's central role in flowering by the identification of its upstream and downstream regulators. *Plant Physiology* 160: 433–449.

Isaacson, W. 2021. *The Code Breaker: Jennifer Doudna, Gene Editing, and the Future of the Human Race.* Simon & Schuster.

Isberg, S., S. A. Balaguera-Reina, and J. P. Ross. 2017. Crocodylus johnstoni. *The IUCN Red List of Threatened Species* 2017: e.T46589A3010118.

Jablonski, N. G. 2004. The evolution of human skin and skin color. *Annual Review of Anthropology* 33: 585–623.

Jablonski, N. and G. Chaplin. 2010. Human skin pigmentation as an adaptation to UV radiation. *Proceedings of the National Academy of Sciences. Proceedings of the National Academy of Sciences* 107: 8962–8968.

Jacobs, W. B., D. R. Kaplan, and F. D. Miller. 2006. The p53 family in nervous system development and disease. *Journal of Neurochemistry* 97: 1571–1584.

Jamniczky, H. J. C., C. Rolian, P. N. Gonzalez, C. C. Powell, T. E. Parsons, F. L. Bookstein, and B. Hallgrimsson. 2010. Rediscovering Waddington in the post-genomic age. *Bioessays* 32: 553–558.

Jansky, S. H., A. O. Charkowski, D. S. Douches, G. Gusmini, C. Richael, P. C. Berke, D. M. Spooner, R. G. Novy, H. De Jong, W. S. De Jong et al. 2016. Reinventing potato as a diploid inbred line-based crop. *Crop Science* 56: 1412–1422.

Jensen, J. D. and M. Lynch. 2020. Considering mutational meltdown as a potential SARS-CoV-2 treatment strategy. *Heredity* 124: 619–620.

Jensen, M. P., C. A. Allen, T. Eguchi, I. P. Bell, E. L. LaCaella, W. A. Hilton, C. A. M. Hof, and P. H. Dutton. 2018. Environmental warming and feminization of one of the largest sea turtle populations in the world. *Current Biology* 28: 154–159.

Jeschke, P., R. Nauen, M. Schindler, and A. Elbert. 2010. Overview of the status and global strategy for neonicotinoids. *Journal of Agricultural and Food Chemistry* 59: 2897–2908.

Jing, W., S. Zhao, J. Liu, and M. Liu. 2020. ABO blood groups and hepatitis B virus infection: A systematic review and meta-analysis. *BMJ Open* 10: e034114.

Joh, E. E. 2011. DNA theft: Recognizing the crime of nonconsensual genetic collection and testing. *Boston University Law Review* 91: 665–700.

Johnson, D. 2009. Darwinian selection in asymmetric warfare: The natural advantages of insurgents and terrorists. *Washington Academy of Sciences* 93: 89–112.

Johnson, M. T. J., C. M. Prashad, M. Lavoignat, and H. S. Saini. 2018. Contrasting the effects of natural selection, genetic drift and gene flow on urban evolution in white clover (*Trifolium repens*). *Proceedings of the Royal Society of London B* 285: 20181019.

Johnson, N. A. 1987. The effect of relatedness among foundresses on the sex ratios of their broods in the parasitoid wasp, *Nasonia vitripennis*. Undergraduate honors thesis. College of William and Mary.

Johnson, N. A. 2007. *Darwinian Detectives: Revealing the Natural History of Genes and Genomes*. Oxford University Press.

Johnson, N. A. 2008. Hybrid incompatibility and speciation. *Nature Education* 1 (1): 20.

Johnson, N. A. 2010. Hybrid incompatibility genes: Remnants of genomic battlefield. *Trends in Genetics* 26: 317–325.

Johnson, N. A. 2016. Evolution and agriculture II. Evolutionary applications to breeding. Pp. 25–31 In *Encyclopedia of Evolutionary Biology*. R. M. Kliman (ed.), Vol. 2. Academic Press.

Johnson, N. A. 2018. "Heredity's history and hope" Review of C. Zimmer. 2018. She has her mother's laugh: The powers, perversions, and potential of heredity. *Evolution* 72: 1733–1735.

Johnson, N. A. 2020. Let bees be bees. Review of Seeley, T. D. 2019. The lives of bees: The untold story of honey bees in the wild. *Evolution: Education and Outreach* 13: 17.

Johnson, N. A. and J. Lachance. 2012. The genetics of sex chromosomes: Evolution and implications for hybrid incompatibility. This year in evolutionary biology. *Annals of the New York Academy of Sciences* 1256: e1–e22.

Johnson, N. A., D. C. Lahti, and D. T. Blumstein. 2012. Combating the assumption of evolutionary progress: Lessons from the decay and loss of traits. *Evolution: Education and Outreach* 5: 128–138.

Johnson, N. A. and A. H. Porter. 2000. Rapid speciation via parallel, directional selection on regulatory genetic pathways. *Journal of Theoretical Biology* 205: 527–542.

Johnson, W. E., D. P. Onorato, M. E. Roelke, E. D. Lloyd, M. Cunningham, R. C. Belden, R. McBride, D. Jansen, M. Lotz et al. 2010. Genetic restoration of the Florida panther. *Science* 329: 1641–1645.

Johnstone, R. A. and M. A. Cant. 2019. Evolution of menopause. *Current Biology* 29: R112–R115.

Jolie, A. 2013. My medical choice. *The New York Times*. May 14, 2013.

Jones, A. 2021. Record drop in US cancer death rate. *The Scientist*. January 13, 2021. www.the-scientist.com/news/news-opinion/record-drop-in-us-cancer-death-rate-68551. Accessed March 7, 2021.

Jones, N. 2018. Redrawing the map: How the world's climate zones are shifting. *Yale Environment* 360. October 22, 2018.

Jones, N. 2019. New studies reveal the history of Denisova cave. *The Leakey Foundation*. February 1, 2019. www.leakerfoundation.org/new-studies-denosva-cave/

Jordan, D. and D. Mills. 2021. Past, present, and future of DNA typing for analyzing human and non-human forensic samples. *Frontiers in Ecology and Evolution* 9: 646130.

Jorgensen, A. L., C. Prince, G. Fitzgerald, A. Hanson, J. Downing, J. Reynolds, J. E. Zhang, A. Alfirevic, and M. Pirmohamed. 2019. Implementation of genotype-guided dosing of warfarin with point-of-care genetic testing in three UK clinics: A matched cohort study. *BMC Medicine* 17: 76.

Jorgensen, K. M., T. Wassermann, P. O. Jensen, W. Hengzuang, S. Molin, N. Hoiby, and O. Ciofu. 2013. Sublethal ciprofloxacin treatment leads to rapid development of high-level ciprofloxacin resistance during long-term experimental evolution of Pseudomonas aeruginosa. *Antimicrobial Agents and Chemotherapy* 57: 4215–4221.

Jove', V., Z. Gong, F. J. H. Hol, Z. Zhao, T. R. Sorrells, M. Prakash, C. S. McBride, and L. B. Vosshall. 2020. Sensory discrimination of blood and floral nectar by *Aedes aegypti* mosquitoes. *Neuron* 108: 1163–1180.

Joye, S. M., A. Bracco, T. M. Ozgokmen, J. P. Chanton, M. Crosell, I. R. MacDonald, E. E. Cordes, J. P. Montoya, and U. Passow. 2016. The Gulf of Mexico ecosystem, six years after the Macondo oil well blowout. *Deep-Sea Research* 129: 4–19.

Juhasz, A. 2011. *Black Tide: The Devastating Impact of the Gulf Oil Spill*. Wiley.

Just, E. E. 1933. Cortical cytoplasm and evolution. *The American Naturalist* 67: 20–29.

Kahn, J. 2020. The gene drive dilemma: We can alter entire species, but should we? *The New York Times*. January 8, 2020. www.nytimes.com/2020/01/08/magazine/gene-drive-mosquitoes.html.

Kang, H.-J. and Z. Rosenwaks. 2018. p53 and reproduction. *Fertility and Sterility* 109: 39–43.

Kansas State University. 2016. Left uncontrolled, weeds would cost billions in economic losses every year. *ScienceDaily*. May 16, 2016.

Katella, K. 2021. 5 things to know about the Delta variant. Yale Medicine. September 24, 2021. https://www.yalemedicine.org/5-things-to-know-delta-variant-covid. Accessed September 25, 2021.

Katju, V., L. B. Packard, L. Bu, P. D. Kneightly, and U. Bergthorsson. 2015. Fitness decline in spontaneous mutation accumulation lines of *Caenorhabditis elegans* populations with varying effective population sizes. *Evolution* 69: 104–116.

Katju, V., L. B. Packard, and P. D. Kneightly. 2018. Fitness decline under osmotic stress in *Caenorhabditis elegans* populations subjected to spontaneous mutation accumulation at varying population sizes. *Evolution* 72: 1000–1008.

Katz, B. 2019. Voracious purple sea urchins are ravaging kelp forests on the west coast. *Smithsonian*. October 25. 2019.

Kelly, D. A. 2016. Evolutionary medicine II. Use of the comparative method and the animal model. Pp. 65–68 In *Encyclopedia of Evolutionary Biology*. R. M. Kliman (ed.), Vol. 2. Academic Press.

Kelly, M. W., M. B. DeBiasse, V. A. Villela, H. L. Roberts, and C. F. Cecola. 2016. Adaptation to climate change: Trade-offs among responses to multiple stressors in an intertidal crustacean. *Evolutionary Applications* 9: 1147–1155.

Kelly, M. W. and G. E. Hoffman. 2013. Adaptation and the physiology of ocean acidification. *Functional Ecology* 27: 980–990.

Kelly, M. W., J. L. Padilla-Gamino, and G. E. Hoffman. 2013. Natural variation and the capacity to adapt to ocean acidification in the keystone sea urchin *Strongylocentrotus purpuratus*. *Global Change Biology* 19: 2536–2546.

Kelly, M. W. and B. Thapa. 2016. Evolution and responses to climate change. In *Encyclopedia of Evolutionary Biology*. R. M. Kliman (ed.). PAGE. Academic Press.

Kemp. S. A., D. A. Collier, R. P. Datir, I. A. T. M. Ferreira, S. Gayed, A. Jahun, M. Hosmillo, C. Rees-Spear, P. Micochova, I. U. Lumb et al. 2021. SARS-CoV-2 evolution during treatment of chronic infection. *Nature* 592: 277–282.

Kenny, D. A., C. Fitzsimons, S. M. Waters, and M. McGee. 2018. Invited review: Improving the feed efficiency of beef cattle—the current state of the art and future challenges. *Animal* 12: 1815–1826.

Kevles, D. J. 2016. The history of eugenics. *Issues in Science and Technology* 32: 45–49.

Kim, M. S., K. P. Patel, A. K. Teng, A. J. Berens, and J. Lachance. 2018. Genetic disease risks can be misestimated across global populations. *BMC Genome Biology* 19: 179.

Kimmig, S., E. J. Beninde, M. Brandt, A. Schleimer, S. Kramer-Schadt, H. Hiofer et al. 2020. Beyond the landscape: Resistance modelling infers physical and behavioral gene flow barriers to a mobile carnivore across a metropolitan area. *Molecular Ecology* 29: 466–484.

Kinch, M. 2018. *Between Hope and Fear: A History of Vaccines and Human Immunity*. Pegasus Books.

Kindstedt, P. S. 2012. *Cheese and Culture: A History of Cheese and Its Place in Western Civilization*. Chelsea Green Publishing.

King, E. A., J. W. Davis, and J. F. Degner. 2019. Are drug targets with genetic support twice as likely to be approved? Revised estimates of the impact of genetic support for drug mechanisms on the probability of drug approval. *PLoS Genetics* 15: e1008489.

King, G. M., J. E. Kosta, T. C. Hazen, and P. A. Sobecky. 2015. Microbial responses to the Deepwater Horizon oil spill: From coastal wetlands to the deep sea. *Annual Review of Marine Science* 7: 377–401.

Kirkby, J. 2010. "[W]e thought Darwin had thrown 'the Redeemer' away": Darwinizing with Emily Dickinson. *The Emily Dickinson Journal* XIX: 1–29.

Klatt, Björn K., Andrea Holzschuh, Catrin Westphal, Yann Clough, Inga Smit, Elke Pawelzik, and Teja Tscharntke. 2014. Bee pollination improves crop quality, shelf life and commercial value. *Proceedings of the Royal Society of London B: Biological Sciences* 281 (1775): 20132440.

Klee, H. J. and D. M. Tieman. 2018. The genetics of fruit flavour preferences. *Nature Review Genetics* 19: 347–356.

Knight, C. A., N. A. Molinari, and D. A. Petrov. 2005. The large genome constraint hypothesis: Evolution, ecology, and phenotype. *Annals of Botany* 95: 177–190.

Knipling, E. F. 1955. Possibilities of insect control or eradication through the use of sexually sterile males. *Journal of Economic Entomology* 48: 459–462.

Knipling, E. F. 1959. Sterile-male method of population control. *Science* 130: 902–904.

Kohlbert, E. 2014. *The Sixth Extinction: An Unnatural History*. Henry Holt & Co.

Kolata, G. 1991. Critic of 'genetic fingerprint' tests tells of pressure to withdraw paper. *New York Times*. December 20, 1991.

Kolata, G. 1992. Theory tying AIDS to polio vaccine is discounted. *The New York Times*. October 23, 1992.

Konin, W. E. 2019. Robust evidence of insect declines. *Nature* 574: 641–642.

Korber, B., W. M. Fischer, S. Gnanakaran, H. Yoon, J. Theiler, W. Abfalterer, N. Hengarner, E. E. Giorgi, T. Bhattacharya, B. Foley et al. 2020. Tracking changes in SARS-CoV-2 spike: Evidence that D614G increases infectivity of the COVID-19 virus. *Cell* 182: 812–827.

Kozlovsky, D. Y., E. A. Weissgerber, and V. V. Pravsudov. 2017. What makes specialized food-caching mountain chickadees successful city slickers? *Proceedings of the Royal Society of London B* 284: 20162613.

Kreiner, J. M., J. R. Stinchcombe, and S. I. Wright. 2018. Population genomics of herbicide resistance: Adaptation via evolutionary rescue. *Annual Review of Plant Biology* 69: 611–635.

Kress, W. J., C. Garcia-Robiedo, M. Uriate, and D. L. Erickson. 2015. DNA barcodes for ecology, evolution, and conservation biology. *Trends in Ecology and Evolution* 30: 25–35.

Kroeker, K. J., R. L. Kordas, R. N. Crim, and G. G. Singh. 2010. Meta-analysis reveals negative yet variable effects of ocean acidification on marine organisms. *Ecology Letters* 13: 1419–1434.

Kruglyak, L. 1999. Prospects for whole-genome linkage disequilibrium mapping of common disease genes. *Nature Genetics* 22: 139–144.

Kuester, A., S.-M. Chang, and R. S. Baucom. 2015. The geographic mosaic of herbicide resistance in the common morning glory, *Ipomoea purpurea*: Evidence for resistance hotspots and low genetic differentiation across the landscape. *Evolutionary Applications* 8: 821–833.

Kuester, A., E. Fall, S.-M. Chang, and R. S. Baucom. 2017. Shifts in outcrossing rates and changes to floral traits associated with the evolution of herbicide resistance in the common morning glory. *Ecology Letters* 20: 41–49.

Kulaberoglu, Y., B. Bhushan, F. Hadi, S. Chakrabarti, W. T. Khaled, K. S. Rankin, E. St. John Smith, and D. Frankel. 2019. The material properties of naked mole-rat hyaluronan. *Scientific Reports* 9: 6632.

Kurzrock, R. and F. J. Giles. 2015. Precision oncology for patients with advanced cancer: The challenges of malignant snowflakes. *Cell Cycles* 14: 2219–2221.

Kushimoto, K. S., S. Gando, D. Saitoh, T. Mayumi, H. Ogura, S. Fujishima, T. Araki, H. Ikeda, J. Kotani, Y. Miki et al. 2013. The impact of temperature abnormalities on the disease susceptibility and outcome in patients with severe sepsis: An analysis from a multicenter, prospective survey of severe sepsis. *Critical Care* 17: R271.

Lachance, J. and S. A. Tishkoff. 2013. Population genomics of human adaptation. *Annual Review of Ecology, Evolution, and Systematics* 44: 123–144.

Lachance, J., B. Vermont, C. C. Elbers, B. Ferwerda, A. Froment, J.-M. Bodo, G. Lema, W. Fu., T. B. Nyambo, T. R. Rebbeck et al. 2012. Evolutionary history and adaptation from high-coverage whole-genome sequences of diverse African hunter-gatherers. *Cell* 150: 457–469.

Lamason, R. L., M.-A. P. K. Mohideen, J. R. Mest, M. J. Jurynec, X. Mao, V. R. Humphreville, J. E. Humbert, S. Sinha, J. L. Moore, P. Jagadeeswaran, W. Zhao et al. 2005. SLC24A5, a putative cation exchanger, affects pigmentation in zebrafish and humans. *Science* 310: 1782–1786.

Lambert, J. 2019. Living near your grandmother has evolutionary benefits. *NPR*. February 7, 2019.

Lambert, M. P. and C. M. Donihue. 2020. Urban biodiversity management using evolutionary tools. *Nature Ecology and Evolution* 4: 903–910.

Landau, E. 2021. How much did grandmothers influence human evolution? Scientists debate the evolutionary benefits of menopause. *Smithsonian Magazine*. January 4, 2021.

Lander, E. S. and B. Budowie. 1994. DNA fingerprinting dispute laid to rest. *Nature* 371: 735–738.

Lander, E. S. and N. J. Schork. 1994. Genetic dissection of complex traits. *Science* 265: 2037–2049.

Larson, B. M. H. 2005. The war of the roses: Demilitarizing invasion biology. *Frontiers in Ecology and Environment* 3: 495–500.

Latinne, A., B. Hu, K. J. Olival, G. Zhu, L. Zhang, H. Li, A. A. Chmura, H. E. Field, C. Zambrana-Torrelio, J. H. Epstein et al. 2020. Origin and cross-species transmission of bat coronaviruses in China. *Nature Communications* 11: 4235.

Laundre', J. W., L. Hernandez, and W. J. Ripple. 2010. The landscape of fear: Ecological implications of being afraid. *The Open Ecology Journal* 3: 1–7.

Leddy, T. 2007. Review of R. Quacchia. Julia Morgan, architect and the creation of the Asilomar conference. *The Journal of Aesthetics and Art Criticism* 65: 432–434.

Ledford, H. 2019. World's largest plant survey reveals alarming extinction rate. *Nature* 570: 148–149.

Lee, C. E. and G. W. Gelembiuk. 2008. Evolutionary origins of invasive populations. *Evolutionary Applications* 1: 427–448.

Lee, C. E., J. L. Remfert, T. Opgenorth, K. M. Lee, E. Stanford, J. W. Connolly, J. Kim, and S. Tomke. 2017. Evolutionary responses to crude oil from the Deepwater Horizon oil spill by the copepod *Eurytemora affinis*. *Evolutionary Applications* 10: 813–828.

Lees, R. S., J. R. L. Gilies, J. Hendrichs, M. J. B. Vreysen, and K. Bourtzis. 2015. Back to the future: The sterile insect technique against mosquito disease vectors. *Current Opinion in Insect Science* 10: 156–162.

Lenski, R. E. and G. J. Velicer. 2000. Games microbes play. *Selection* 1: 89–93.

Lents, N. H. 2018. *Human Errors: A Panorama of Our Glitches, From Pointless Bones to Broken Genes*. Mariner Books.

Letko, M., S. N. Seifert, K. J. Olival, R. K. Plowright, and V. J. Munster. 2020. Bat-borne virus diversity, spillover and emergence. *Nature Reviews Microbiology* 18: 461–471.

Lewontin, R. C. 1972. The apportionment of human diversity. Pp. 381–398 In *Evolutionary Biology*. T. Dobzhansky, M. K. Heckt, and W. C. Steere (eds.), Vol. 6. Appleton-Century-Crofts.

Lewontin, R. C. 1974. *The Genetic Basis of Evolutionary Change*. Columbia University Press.

Lewontin, R. C. and D. L. Hartl. 1991. Population genetics in forensic DNA testing. *Science* 254: 1745–1750.

Lewontin, R. C. and J. L. Hubby. 1966. A molecular approach to the study of genic heterozygosity in natural populations. II. Amount of variation in natural populations of *Drosophila pseudoobscura*. *Genetics* 54: 595–609.

Lewontin, R. C. and K. Kojima. 1960. The evolutionary dynamics of complex polymorphisms. *Evolution* 14: 458–472.

Li, G., M. Yuan, H. Li, C. Deng, Q. Wang, Y. Tang, H. Zhang, W. Yu, Q. Xu, Y. Zou et al. 2021. Safety and efficacy of artemisinin-piperaquine for treatment of COVID-19: An open-label, non-randomised and controlled trial. *International Journal of Antimicrobial Agents* 57: 106216.

Li, J., X. Hong, S. Mesiano, L. J. Muglia, X. Wang, M. Synder, D. K. Stevenson, and G. M. Shaw. 2018. Natural selection has differentiated the progesterone receptor among human populations. *The American Journal of Human Genetics* 103: 45–57.

Li, P., J.-Z. Jiang, X.-F. Wan, Y. Hua, L. Li, J. Zhou, X. Wang, F. Hou, J. Chen, J. Zhou, and J. Chen. 2020a. Are pangolins the intermediate host of the 2019 novel coronavirus (SARS-CoV-2)? *PLoS Pathogens* 16: e1008421.

Li, Q., X. Guan, P. Wu, X. Wang, L. Zhou, Y. Tong, R. Ren, K. S. M. Leung, E. H. Y. Lau, J. Y. Wong, X. Xing et al. 2020b. Early transmission dynamics in Wuhan, China, of novel coronavirus-infected pneumonia. *New England Journal of Medicine* 382: 1199–1207.

Li, W., Z. Shi, M. Yu, W. Ren, C. Smith, J. H. Epstein, H. Wang, G. Crameri, Z. Hu, H. Zhang et al. 2005. Bats are natural reservoir of SARS-like coronaviruses. *Science* 310: 676–679.

Lieberman, D. E. 2013. *The Story of the Human Body: Evolution, Health, and Disease*. Random House.

Lindo, J., R. Haas, C. Hofman, M. Apata, M. Moraga, R. A. Verdugo, J. T. Watson, C. V. Llave, D. Witonsky, C. Beall et al. 2018. The genetic prehistory of the Andean highlands 7000 BP through European contact. *Science Advances* 2018: eaau4921.

Linksvayer, T. A. and M. J. Wade. 2005. The evolutionary origin and elaboration of sociality in the aculeate Hymenoptera: Maternal effects, sib-social effects, and heterochrony. *The Quarterly Review of Biology* 80: 317–336.

Lipton, E., D. E. Sanger, M. Haberman, M. D. Shear, M. Mazzetti, and J. E. Barnes. 2020. He could have seen what was coming: Behind Trump's failure on the virus. *The New York Times*. April 11, 2020 (Updated October 2, 2020).

Littleford-Colquhoun, B. L., L. Weyrich, N. Kent, and C. H. Frere. 2019. City life alters the gut microbiome and stable isotope profiling of the eastern water dragon (*Intellagama lesueurii*). *Molecular Ecology* 28: 4592–4607.

Liu, T., L. Chen, X. Sun, Y. Wang, S. Li, X. Yin, X. Wang, C. Ding, H. Li, and W. Di. 2014. Progesterone receptor PROGINS and +331G/A polymorphisms confer susceptibility to ovarian cancer: A meta-analysis based on 17 studies. *Tumor Biology* 35: 2427–2436.

Livingstone, F. B. 1962. On the non-existence of human races. *Current Anthropology* 3: 279–281.

Lloyd, M. M., A. D. Makukhov, and M. H. Pespeni. 2016. Loss of genetic diversity as a consequence of selection in response to high pCO2. *Evolutionary Applications* 9: 1124–1132.

Lobo, I. 2008. Pleiotropy: One gene can affect multiple traits. *Nature Education* 1: 10.

Locke, B. 2016. Natural *Varroa* mite-surviving *Apis mellifera* honeybee populations. *Apidologie* 47: 467–482.

Longo, A. V., R. C. Fleischer, and K. R. Lips. 2019. Double trouble: Co-infections of chytrid fungi will severely impact widely distributed newts. *Biological Invasions* 21: 2233–2245.

Lopez-Vaamonde, C., L. Sire, B. Rasmussen, R. Rougerie, C. Wieser, A. A. Allaoui, J. Minet, J. R. deWaard, T. Decaens, and D. C. Lee. 2019. DNA barcodes reveal deeply neglected diversity and numerous invasions of micromoths in Madagascar. *Genome* 62: 108–121.

Losos, J. B. 2017. *Improbable Destinies: Fate, Chance and the Future of Evolution*. Riverhead Press.

Losos, J. B. and C. J. Schneider. 2009. Anolis lizards. *Current Biology* 19: 316–8.

Lownes, A. E. 1947. Charles Darwin to Charles Harrison Blackley (An Early Chapter on Pollen Allergy). *Isis* 37: 21–24.

Lu, J., T. Tang, H. Tang, J. Huang, S. Shi, and C.-I. Wu. 2006. The accumulation of deleterious mutations in rice genomes: A hypothesis on the cost of domestication. *Trends in Genetics* 22: 126–131.

Luan, Z., Y. Zhang, T. Lu, Y. Ruan, H. Zhang, J. Han, L. Li, W. Sun, L. Wang, W. Yue, and D. Zhang. 2011. Positive association of the human STON2 gene with schizophrenia. *NeuroReport* 22: 288–293.

Luo, Y. 2020. Neanderthal DNA raises risk of severe COVID. *Nature* 587: 552–553.

Lynas, M. 2018. *Seeds of Science: Why We Got It So Wrong on GMOs*. Bloomsbury Sigma.

Lynch, M., J. Conery, and R. Burger. 1995. Mutational meltdowns in sexual populations. *Evolution* 49: 1067–1080.

Mahley, R. W. 2016. Apolipoprotein E: From cardiovascular disease to neurological disorders. *Journal of Molecular Medicine* 94: 739–746.

Mahmood, S. S., D. Levy, R. S. Vasan, and T. Wang. 2014. The Framingham Heart Study and the epidemiology of cardiovascular disease: A historical approach. *The Lancet* 383: 999–1008.

Majerus, M. 2009. Industrial melanism in the peppered moth, *Biston betularia*: An excellent teaching example of Darwinian evolution in action. *Evolution Education and Outreach* 2: 63–74.

Malaria Genomic Epidemiology Network. 2014. Reappraisal of known malaria resistance loci in a large multicenter study. *Nature Genetics* 46: 1197–1204.

Manari, A. K., B. H. Funke, H. L. Rehm, M. S. Olessen, B. A. Maron, P. Szolovits, D. M. Margulies, J. Loscalzo, and I. S. Kohane. 2016. Genetic misdiagnoses and the potential for health disparities. *The New England Journal of Medicine* 375: 655–665.

Manley, R., B. Templeton, T. Doyle, D. Gates, S. Hedges, M. Boots, and L. Wilfert. 2019. Knock-on community impacts of a novel vector: Spillover of emerging DWV-B from Varroa-infested honeybees to wild bumblebees. *Ecology Letters* 22: 1306–1315.

Mann, C. C. 2005. *1491: New Revelations of the Americas Before Columbus*. Alfred A. Knopf.

Mann, C. C. 2018. *The Wizard and the Prophet: Two Remarkable Scientists and Their Dueling Visions to Shape Tomorrow's World*. Alfred A. Knopf.

Mansbach, R. A., S. Chakraborty, K. Nguyen, D. C. Montefiori, B. Korber, and S. Gnanakaran. 2021. The SARS-CoV-2 spike variant D614G favors an open conformational state. *Science Advances* 7: eabf3671.

Marapana, D. and A. F. Cowman. 2020. Uncovering the ART of antimalarial resistance. *Nature* 367: 22–23.

Marinić, M. and V. J. Lynch. 2020. Relaxed constraint and functional divergence of the progesterone receptor (PGR) in the human-stem lineage. *PLoS Genetics* 16: e1008666.

Marshe, V. S., I. Gorbovskaya, S. Kanji, M. Kish, and D. J. Muller. 2019. Implications of APOE genotyping for late-onset Alzheimer's disease (LOAD) risk estimation: A review of the literature. *Journal of Neural Transmission* 126: 65–85.

Martin, R. A., L. D. Chick, A. R. Yilmaz, and S. E. Diamond. 2019. Evolution, not transgenerational plasticity, explains the adaptive divergence of acorn ant thermal tolerance across an urban-rural temperature cline. *Evolutionary Applications* 12: 1678–1687.

Martin, R. A., M. Lin, J. M. Granka, J. W. Myrick, X. Liu, A. Sockrell, E. G. Atkinson, C. J. Werely, M. Moller, M. S. Sandhu et al. 2017. An unexpectedly complex architecture for skin pigmentation in Africans. *Cell* 171: 1340–1353.

Martin, S. J. and L. E. Bretell. 2019. Deformed wing virus in honeybees and other insects. *Annual Review of Virology* 6: 12.1–12.21.

Masel, J. and D. E. L. Promislow. 2016. Answering evolutionary questions: A guide for mechanistic biologists. *Bioessays* 38: 704–711.

Matthews, W. A., D. A. Sumner, and T. Hanon. 2019. Contributions of the U. S. honey industry to the U.S. economy. https://aic.ucdavis.edu/wp-content/uploads/2019/02/HONEY-COMPLETE_DRAFT_FEBRUARY-11-2009.pdf. Accessed May 18, 2021.

Maurano, M. T., S. Ramaswami, P. Zappile, D. Dimartino, L. Boytard, A. M. Ribeiro-dos-Santos, N. A. Vulpescu, G. Westby, G. Shen, X. Feng et al. 2020. Sequencing identifies multiple early introductions of SARS-CoV-2 to the New York City region. *Genome Research*. doi: 10.1101/gr.266676.120.

May, R. M. 2011. Why worry about how many species and their loss? *PLoS Biology* 9: e1001130.

Maynard Smith, J. 1976. Evolution and the theory of games: In situations characterized by conflict of interest, the best strategy to adopt depends on what others are doing. *American Scientist* 64: 41–45.

Maynard Smith, J. and J. Haigh 1974. The hitch-hiking effect of a favorable gene. *Genetical Research Cambridge* 23: 23–35.

Maynard Smith, J. and G. R. Price. 1973. The logic of animal conflict. *Nature* 246: 15–18.

Mayr, E. 1963. *Animal Species and Evolution*. Belknap Press.

McArt, S. H., H. Koch, R. E. Irwin, and L. S. Adler. 2014. Arranging the bouquet of disease floral traits and the transmission of plant and animal pathogens. *Ecology* 17: 624–636.

McBride, C. S. 2016. Genes and odors underlying the recent history of mosquito preference for humans. *Current Biology* 26: R41–R46.

McBride, C. S., F. Baier, A. B. Omondi, S. A. Spitzer, J. Lutomiah, R. Sang, R. Ignell, and L. B. Vosshall. 2014. Evolution of mosquito preference for humans linked to an odorant receptor. *Nature* 515: 222–227.

McCann, S., M. J. Greenlees, D. Newell, and R. Shine. 2014. Rapid acclimation to cold allows the cane to invade montane areas within its Australian range. *Functional Ecology* 28: 1166–1174.

McCook, S. and J. Vandermeer. 2015. The big rust and the red queen: Long-term perspectives on coffee rust research. *Phytopathology* 105: 1164–1173.

McCullough, D. 1977. *The Path Between the Seas: The Creation of the Panama Canal*. Simon & Schuster.

McDonald, C. A., A. V. Longo, K. R. Lips, and K. R. Zamudio. 2020. Incapacitating effects of fungal coinfection in a novel pathogen system. *Molecular Ecology* 29: 3173–3186.

McDonald, J. H. and M. Kreitman. 1991. Adaptive protein evolution at the *Adh* locus in *Drosophila*. *Nature* 351: 652–654.

McFarland, C. D., K. S. Korolev, G. V. Kryukov, S. R. Sunyaev, and L. A. Mirny. 2013. Impact of deleterious passenger mutations on cancer progression. *Proceedings of the National Academy of Sciences* 110: 2910–2915.

McGinnis, R., V. Steinthorsdoittir, N. O. Williams, G. Thorleifsson, S. Shooter, S. Hjartardottir, S. Bumpstead, L. Stefansdottir, L. Hildyard, J. K. Sigurdsson et al. 2017. Variants at the fetal genome near *FLT1* are associated with risk of preeclampsia. *Nature Genetics* 49: 1255–1263.

McKenna, M. 2020. Coffee rust is going to ruin your morning. *The Atlantic*. September 16, 2020.

McKinney, M. L. 2002. Urbanization, biodiversity, and conservation. *BioScience* 52: 883–890.

McNamara, M. 2018. *I'll Be Gone in the Dark: One Woman's Obsessive Search for the Golden State Killer*. HarperCollins.

Melnyk, A., A. Wong, and R. Kassen. 2014. The fitness costs of antibiotic resistance. *Evolutionary Applications* 8: 273–283.

Meredith, S. 2018. Two thirds of global population will live in cities by 2050, UN says. *CNBC*. May 17, 2018. www.cnbc.co/2018.05.17/two-thirds-of-global-population-will-live-in-cities-by-2050-un-says-html. Accessed August 4, 2020.

Metcalf, J. L. 2019. Estimating the postmortem interval using microbes: Knowledge gaps and a path to technology adoption. *Forensic Science International: Genetics* 38: 211–218.

Metcalf, J. L., D. O. Carter, and R. Knight. 2016. Microbiology of death. *Current Biology* 26: R561–R563.

Metcalf, J. L., L. Wegner Parfrey, A. Gonzalez, C. L. Lauber, D. Knights, G. Ackermann, G. C. Humphrey, M. J. Gebert, W. Van Truren, D. Berg-Lyons et al. 2013. A microbial clock provides an accurate estimate of the postmortem interval in a mouse model system. *eLife* 2: e101044.

Metzker, M. L., D. P. Mindell, X.-M. Liu, R. G. Ptak, R. A. Gibbs, and D. M. Hillis. 2002. *Proceedings of the National Academy of Science* 99: 14292–14297.

Mikheyev, A. S., M. M. Y. Tin, J. Arora, and T. D. Seeley. 2015. Museum samples reveal rapid evolution by wild honey bees exposed to a novel parasite. *Nature Communications* 6: 7991.

Mills, M. 2017. Rapid evolution of resistance to parasitism in biological control. *Proceedings of National Academy of Science* 114: 3792–3794.

Mittan, C. S. and K. R. Zamudio. 2019. Rapid adaptation to cold tolerance in the invasive cane toad *Rhinella marina*. *Conservation Physiology* 7: FILL.

Mohr, S. E. 2018. *First in Fly: Drosophila Research and Biological Discovery*. Harvard University Press.

Monosson, E. 2018. *Viral Rescue. When Antibiotics Fail, Could Phage Therapy Succeed? The Germ's Eye View of Infection Might Open Up Revolutionary Treatments*. Aeon.

Montalalvo, D. 2015. Insects feast on plants, endangering crops and costing billions. *CNBC*. May 9, 2015. www.cnbc.com/2015/insects-feast-on-plants-endangering-crops-and-costing-billions.html.

Moore, T. and D. Haig. 1991. Genomic imprinting in mammalian development: A parental tug-of-war. *Trends in Genetics* 7: 15–19.

Mora, C., D. P. Tittensor, S. Adi, A. G. B. Simpson, and B. Worm. 2011. How many species are there on Earth and in the ocean? *PLoS Biology* 9: e1001127.

Moradall, M. F., S. Ghodes, and B. H. A. Rehm. 2017. *Pseudomonas aeruginosa* lifestyle: A paradigm for adaptation, survival, and persistence. *Frontiers in Cellular and Infectious Microbiology* 7: 39.

Morens, D. M. and A. S. Fauci. 2007. The 1918 influenza pandemic: Insights for the 21st century. *The Journal of Infectious Diseases* 195: 1018–1028.

Morens, D. M. and A. S. Fauci. 2020. Emerging pandemic diseases: How we got to COVID-19. *Cell* 182: 1077–1092.

Moritz, C. and C. Cicero. 2004. DNA barcoding: Promise and pitfalls. *PLoS Biology* 2: e354.

Morley, J. W., R. L. Selden, R. J. Latour, T. L. Frolicher, R. J. Seagraves, and M. L. Pinsky. 2018. Projecting shifts in thermal habitat for 686 species on the North American continental shelf. *PLoS One* 13: e0196127.

Morse, G. E. and B. B. Normark. 2006. A molecular phylogenetic study of armoured scale insects (Hemiptera: Diaspididae). *Systematic Entomology* 31: 338–349.

Motsinger-Reil, A. A., E. Jorgenson, M. V. Relling, D. L. Kroetz, R. Weinshilboum, N. J. Cox, and D. M. Roden. 2013. Genome-wide association studies in pharmacogenomics: Successes and lessons. *Pharmacogenetics and Genomics* 23: 383–394.

Motuhi, S-E., M. Mehiri, C. E. Payri, S. La Barre, and S. Bach. 2016. Marine natural products from New Caledonia—a review. *Marine Drugs* 14: 58–117.

Moyers, B. T., P. L. Morrell, and J. K. McKay. 2018. Genetic costs of domestication and crop improvement. *Journal of Heredity* 109: 103–116.

Mueller, J. C., J. Partecke, B. J. Hatchwell, K. J. Gaston, and K. L. Evans. 2013. Candidate gene polymorphisms for behavioral adaptations during urbanization in blackbirds. *Molecular Ecology* 22: 3629–3637.

Mukherjee, S. 2010. *The Emperor of All Maladies: A Biography of Cancer*. Scribner.

Murthy, A., Y. Li, M. Reichelt, A. K. Katakam, R. Noubade, M. Rose-Girma, J. DeVoss, L. Diehl, R. R. Graham, and M. van Lookeren Campagne. 2014. A Crohn's disease variant in *Atg16l1* enhances its degradation by caspase 3. *Nature* 506: 456–462.

Nacci, D. E., D. Champlin, and S. Jayaraman. 2010. Adaptation of the estuarine fish *Fundulus heteroclitus* (Atlantic killifish) to polychlorinated biphenyls (PCBs). *Estuaries and Coasts* 33: 853–864.

Nacci, D. E., D. Proestou, D. Champlin, J. Martinson, and E. R. Watts. 2016. Genetic basis for rapidly evolved tolerance in the wild: Adaptation to toxic pollutants by an estuarine fish species. *Molecular Ecology* 25: 5467–5482.

Nair, S., X. Li, G. A. Arya, M. McDew-White, M. Ferrari, F. Nosten, and T. J. C. Anderson. 2018. Fitness costs and the rapid spread of kelch130 C580Y substitutions conferring artemisinin resistance. *Antimicrobial Agents and Chemotherapy* 62: e00605–18.

Natterson-Horowitz, B. and K. Bowers. 2012. *Zoobiquity: What Animals Can Teach Us About Health and the Science of Healing*. Knopf.

Nazzi, F., S. P. Brown, D. Annoscia, F. Del Piccolo, G. Di Prisco, P. Varricchio, G. Della Vedova, F. Cattonaro, E. Caprio, and F. Pennacchio. 2012. Synergetic parasite-pathogen interactions mediated by host immunity can drive collapse of honeybee colonies. *PLoS Pathogens* 8: e1002735.

Nelson, M. R., H. Tipney, J. L. Painter, J. Shen, P. Nicoletti, Y. Shen, A. Floratos, P. C. Sham, M. J. Li, J. Wang et al. 2015. The support of human genetic evidence for approved drug indications. *Nature Genetics* 47: 856–860.

Nesse, R. M. 2008. Evolution: Medicine's most basic science. *The Lancet* 372: S21–S27.

Nesse, R. M. 2011. Ten questions for evolutionary studies of disease vulnerability. *Evolutionary Applications* 4: 264–271.

Nesse, R. M. 2019. *Good Reasons for Bad Feelings: Insights from the Frontier of Evolutionary Psychiatry*. Dutton.

Nesse, R. M. and G. C. Williams. 1994. *Why We Get Sick: The New Science of Darwinian Medicine*. Vintage Books.

Ngor, L., E. C. Palmer-Young, R. B. Nevarez, K. A. Russell, L. Leger, J. Giacomini, M. S. Pinilla-Gallego, R. E. Irwin, and Q. S. McFrederick. 2020. Cross-infectivity of honey and bumble-associated parasites across three bee families. *Parasitology* 147: 1290–1304.

Nishkimi, M. and K. Yagi. 1991. Molecular basis for the deficiency in humans of gulonolactone oxidase, a key enzyme for ascorbic acid, a key enzyme for ascorbic acid. *American Journal of Clinical Nutrition* 54: 1203S–1208S.

Noble, C., J. Min, J. Olejarz, J. Buchthal, A. Chavez, A. L. Smidler, E. A. DeBenedictis, G. M. Church, M. A. Nowak, and K. M. Esvelt. 2019. Daisy-chain gene drive for the alteration of local populations. *Proceedings of the National Academy of Sciences* 116: 8275–8282.

Nolte, M. J., P. Jing, C. N. Dewey, and B. A. Payseur. 2020. Giant island mice exhibit widespread gene expression changes in key metabolic organs. *Genome Biology and Evolution* 12: 1277–1301.

Noor, M. A. F. 2018. *Live Long and Evolve: What Star Trek Can Teach Us about Evolution, Genetics, and Life on Other Worlds*. Princeton University Press.

Normark, B. B. and N. A. Johnson. 2011. Niche explosion. *Genetica* 139: 551–564.

Norton, H. L., E. E Quillen, A. W. Bingham, L. N. Pearson, and H. Dunsworth. 2019. Human races are not like dog breeds: Refuting a racist analogy. *Evolution: Education and Outreach* 12: 17.

NPR. 2019. Privacy and DNA tests. NPR. November 8, 2019. www.npr.org/2019/11/09/777888000/privacy-and-dna-tests. Accessed April 11, 2021.

Nyholt, D. R., C.-E. Yu, and P. M. Visscher. 2010. On Jim Watson's *ApoE* status: Genetic information is hard to hide. *European Journal of Human Genetics* 17: 147–149.

O'Brien, S. J., D. E. Wildt, and M. Bush. 1986. The cheetah in genetic peril. *Scientific American* 254: 84–95.

O'Hanlon, S. J., A. Rieux, R. A. Farrer, G. M. Rosa, B. Waldman, A. Bataille, T. A. Kosch, K. A. Murray, B. Brankovics, M. Fumagalli et al. 2018. Recent Asian origin of chytrid fungi causing global amphibian declines. *Science* 360: 621–627.

O'Hara, N. B., J. S. Rest, and S. J. Franks. 2016. Increased susceptibility to fungal disease accompanies adaptation to drought in *Brassica Rapa*. *Evolution* 70: 241–248.

Oke, K. B., C. J. Cunningham, P. A. H. Westley, M. L. Baskett, S. M. Carlson, J. Clark, A. P. Hendry, V. A. Karatayev, N. W. Kendall, J. Kieble et al. 2020. Recent declines in salmon body size impact ecosystems and fisheries. *Nature Communications* 11: 4155.

Okrent, D. 2019. *The Guarded Gate: Bigotry, Eugenics and the Law That Kept Two Generations of Jews, Italians, and Other European Immigrants Out of America*. Scribner.

Olalde, O., M. Allentoft, F. Sánchez-Quinto, G. Santpere, C. W. K. Chiang, M. DeGiorgio, J. Prado-Martinez, J. A. Rodrigeuz, S. Ramussen, J. Quilez et al. 2014. Derived immune and ancestral pigmentation alleles in a 7,000-year-old Mesolithic European. *Nature* 507: 225–228.

Olson, M. V. 1999. When less is more: Gene loss as an engine of evolution. *American Journal of Human Genetics* 64: 18–23.

Onion, R. 2016. "Inexplicable, terrible, and capricious": A history of scurvy, the mysterious disease that haunted the exploration. *Slate*. December 8, 2016.

Orth, M., P. Metzger, S. Gerum, J. Mayerle, G. Schneider, C. Belka, M. Schnurr, and K. Lauber. 2019. Pancreatic ductal adenocarcinoma: Biological hallmarks, current status, and future perspectives of combined modality treatment approaches. *Radiation Oncology* 14: 141.

Otto, S. P., T. Day, J. Arino, C. Colijn, J. Dushoff, M. Li, S. Mechai, G. Van Domsealaar, J. Wu, D. J. D. Earn, and N. H. Ogden. 2021. The origins and potential future of SARS-CoV-2 variants of concern in the evolving COVID-19 pandemic. *Current Biology* 31: R918–R929.

Oziolor, E. M., B. Dubansky, W. W. Burggren, and C. W. Matson. 2016. Cross-resistance in gulf killifish (*Fundulus grandis*) populations resistant to dioxin-like compounds. *Aquatic Toxicology* 175: 222–231.

Oziolor, E. M., N. M. Reid, S. Yair, K. M. Lee, S. Guberman VerPloeg, P. C. Bruns, J. R. Shaw, A. Whitehead, and C. W. Matson. 2019. Adaptive introgression enables evolutionary rescue from extreme environmental pollution. *Science* 364: 455–457.

Paabo, S. 2014. *Neanderthal Man: In Search of Lost Genomes*. Basic Books.

Paigen, K. and P. M. Petkov. 2018. PRDM9 and its role in genetic recombination. *Trends in Genetics* 34: 291–300.

Palmer-Young, E. C., I. W. Farrell, L. S. Adler, N. J. Milano, P. A. Egan, R. J. Junker, R. E. Irwin, and P. C. Stevenson. 2019. Chemistry of floral rewards: Intra- and inter-specific variability of nectar and pollen secondary metabolites across taxa. *Ecological Monographs* 89: e0135.

Palmer-Young, E. C., B. M. Sadd, and L. S. Adler. 2016. Evolution of resistance to single and combined floral phytochemicals by a bumble bee parasite. *Journal of Evolutionary Biology* 30: 300–312.

Palmer-Young, E. C., B. M. Sadd, P. C. Stevenson, R. E. Irwin, and L. S. Adler. 2016. Bumble bee parasite strains vary in resistance to phytochemicals. *Scientific Reports* 6: 37087.

Palumbi, S. R., T. G. Evans, M. H. Pespeni, and G. N. Somero. 2019. Present and future adaptation of marine species assemblages: DNA-based insights into climate change from studies of physiology, genomics, and evolution. *Oceanography* 32: 70–81.

Parker-Pope, T. 2021. Can the Covid vaccine protect me against virus variants? *The New York Times*. April 15, 2021. www.nytimes.com/2021/04/15/well/live/covid-variants-vaccine.html. Accessed May 12, 2021.

Parmesan, C. and G. Yohe. 2003. A globally coherent fingerprint of climate change impacts across natural systems. *Nature* 421: 37–42.

Parsons, K. J., A. Rigg, A. J. Conith, A. C. Kitchener, S. Harris, and H. Zhu. 2020. Skull morphology diverges between urban and rural populations of red foxes mirroring patterns of domestication and macroevolution. *Proceedings of the Royal Society of London B* 287: 20200763.

Payseur, B. A. and P. Jing. 2021. Genomic targets of positive selection in giant mice from Gough Island. *Molecular Biology and Evolution* 38: 911–926.

Penn, E. A. 2000. Biological warfare in eighteenth-century North America: Beyond Jeffrey Amherst. *The Journal of American History* 86: 1552–1580.

Pennings, P. S., S. Kryazhimskiy, and J. Wakeley. 2014. Loss and recovery of genetic diversity in adapting populations of HIV. *PLoS Genetics* 10: e1004000.

Pennisi, E. 2019. Billions of North American birds have vanished. *Science* 365: 1228–1229.

Pepin, J. 2011. *The Origin of AIDS*. Cambridge University Press.

Pespeni, M. H., B. T. Barney, and S. R. Palumbi. 2013. Differences in growth in the regulation of growth and biomineralization genes revealed through long-term common-garden acclimation and experimental genomics in the purple sea urchin. *Evolution* 67: 1901–1914.

Pespeni, M. H., F. Chan, B. A. Menge, and S. R. Palumbi. 2013. Signs of adaptation to local pH conditions across an environmental mosaic in the California current ecosystem. *Integrative and Comparative Biology* 33: 857–870.

Peterson, D., N. B. Hardy, G. E. Morse, I. C. Stocks, A. Okusu, and B. B. Normark. 2015. Phylogenetic analysis reveals positive correlations between adaptations to diverse hosts in a group of pathogen-like herbivores. *Evolution* 69: 2785–2792.

Pfennig, K. S. 2019. How to survive in a human-dominated world. *Science* 364: 433–434.

Phillips, B. L., G. P. Brown, M. Greenlees, J. K. Webb, and R. Shine. 2007. Rapid expansion of the cane toad (*Bufo marinus*) invasion front in tropical Australia. *Austral Ecology* 32: 169–176.

Phillips, B. L. and R. Shine. 2004. Adapting to an invasive species: Toxic cane toads induce morphological change in Australian snakes. *Proceedings of the National Academy of Science* 101: 17150–17155.

Phillips, B. L. and R. Shine. 2006. An invasive species induces rapid adaptive change in a native predator: Cane toads and black snakes in Australia. *Proceedings of the Royal Society of London B* 273: 1545–1550.

Phillips, B. L., R. Shine, and R. Tingley. 2016. The genetic backburn: Using rapid evolution to halt invasions. *Proceedings of the Royal Society of London B* 283: 20153037.

Phillips, C. 2018. The golden state killer investigation and the nascent field of forensic genealogy. *Forensic Science International: Genetics* 36: 186–188.

Phipps, E. A., R. Thadhani, T. Benzing, and S. A. Karumanchi. 2019. Pre-eclampsia: Pathogenesis, novel diagnostics, and therapies. *Nature Reviews Nephrology* 15: 275–289.

Pickrell, J. K. and D. Reich. 2014. Towards a new history and geography of human genes informed by ancient DNA. *Trends in Genetics* 30: 377–389.

Pierce, G. B. and W. C. Speers. 1988. Tumors as caricatures of the process of tissue renewal: Prospects for therapy by directing differentiation. *Cancer Research* 48: 1996–2004.

Pinker, S. 2018. *Enlightenment Now. The Case for Reason, Science, Humanism, and Progress.* Penguin Books.

Pitts, L. 2017. *Racism in America: Cultural Codes and Color Lines in the 21st Century.* Herald Books.

Plumer, B. 2019a. Humans are speeding extinction and altering the natural world at an 'unprecedented' rate. *New York Times.* May 6, 2019.

Plumer, B. 2019b. The world's oceans are in danger, a major climate change report warns. *The New York Times.* September 25, 2019.

Pollan, M. 2004. *The Omnivore's Dilemma: A Natural History of Four Meals.* Penguin Books.

Poolman, E. M. and A. P. Galvani. 2007. Evaluating candidate agents of selective pressure for cystic fibrosis. *Journal of the Royal Society Interface* 4: 91–98.

Posvai, M., D. Gulisija, J. B. Munro, J. C. Silva, and C. E. Lee. 2020. Rapid evolution of genome-wide gene expression and plasticity during saline to freshwater invasions by the copepod *Eurytemora affinis* species complex. *Molecular Ecology* 29: 4835–4856.

Powell, A. L. T., C. V. Nguyen, T. Hill, K. Lam Cheng, R. Figueroa-Balderas, H. Ashrati, C. Pons, R. Fernández-Munoz, A. Vicente, and J. Lopez-Baltazar. 2012. *Uniform ripening* encodes a *Golden 2-like* transcription factor regulating tomato fruit chloroplast development. *Science* 336: 1711–1714.

Powles, S. B. 2008. Evolved glyphosate-resistant weeds around the world: Lessons to be learned. *Pest Management Science* 64: 360–365.

Powles, S. B., D. F. Lorraine-Colwill, J. J. Bellow, and C. Preston. 1998. Evolved resistance to glyphosate in rigid ryegrass (*Lolium rigidum*) in Australia. *Weed Science* 46: 604–607.

Pringle, P. 2008. *The Murder of Nikolai Vavilov: The Story of Stalin's Persecution of One of the Great Scientists of the Twentieth Century.* Simon & Schuster.

Pritchard, J. K., M. Stephens, and P. Donnelly. 2000. Inference of population structure using multilocus genotype data. *Genetics* 155: 945–959.

Provine, W. B. 1986. *Sewall Wright and Evolutionary Biology.* University of Chicago Press.

Puckett, E. E., J. Park, M. Combs, M. J. Blum, J. E. Bryant, A. Caccone, F. Costa, E. E. Deinum, A. Esther, C. G. Himsworth et al. 2016. Global population structure of the brown rat (*Rattus norvegicus*). *Proceedings of the Royal Society of London B* 283: 20161762.

Puckett, E. E., E. Sherrat, M. Combs, E. J. Carlen, W. Harcourt-Smith, and J. Munshi-South. 2020. Variation in brown rat cranial shape shows directional selection over 120 years in *New York City, Ecology and Evolution* 10: 4739–4748.

Pyšek, P., J. Cuda, P. Smilauer, H. Skalova, Z. Chumova, C. Lambertini, M. Lucaniva, H. Rysava, P. Travnicek, K. Semberova, and L. A. Meyerson. 2019. Competition among native and invasive *Phragmites australis* populations: An experimental test of the effects of the invasion status, genome size, and ploidy level. *Ecology and Evolution* 10: 1106–1118.

Pyšek, P., P. E. Hulme, D. Simberloff, S. Bacher, T. M. Blackburn, J. T. Carlton, W. Dawson, F. Essl, L. C. Foxcroft, P. Genovesi et al. 2020. Scientists' warning on invasive species. *Biological Reviews* 95: 1511–1534.

Pyšek, P., H. Skalove, J. Cuda, W.-Y, Guo, J. Suda, J. Dolezal, O. Kauzal, C. Lambertini, M. Lucanova, T. Mandakova et al. 2018. Small genome separates native and invasive populations in an ecologically important cosmopolitan grass. *Ecology* 99: 79–90.

Quammen, D. 1996. The *Song of the Dodo: Island Biogeography in an Age of Extinction*. Scribner.

Quammen, D. 2012. *Spillover: Animal Infections and the Next Human Pandemic*. W. W. Norton & Company.

Quammen, D. 2018. *The Tangled Tree: A Radical New History of Life*. Simon & Schuster.

Quammen, D. 2020. The sobbing pangolin. *The New Yorker*. August 24, 2020. Pp. 26–31.

Quillen, E. E., N. L. Norton, E. J. Parra, F. Lona-Durazo, K. C. Ang, F. M. Illiescu, L. N. Pearson, M. D. Shriver, T. Lasisi, O. Gokcumen et al. 2019. Shades of complexity: New perspectives on the evolution and genetic architecture of human skin. *American Journal of Physical Anthropology* 168: 4–26.

Rapoport, A., D. A. Seale, and A. M. Colman. 2015. Is tit-for-tat the answer? On the conclusions drawn from Axelrod's tournaments. *PLoS One* 10: e0134128.

Rasko, D. A. and V. Sperandio. 2010. Anti-virulence strategies to combat bacteria-mediated disease. *Nature Reviews Drug Discovery* 9: 117–129.

Rautiala, P., H. Helantera, and M. Puurtinen. 2019. Extended haplodiploidy hypothesis. *Evolution Letters* 3: 263–270.

Razifard, H., A. Ramos, A. L. Della Valle, C. Bodary, E. Goetz, E. J. Manser, X. Li, L. Zhang, S. Visa, D. Tieman, E. can der Knaap, and A. L. Caiedo. 2020. Genomic evidence for complex domestication history of the cultivated tomato in Latin America. *Molecular Biology* 37: 1118–1132.

Reboliar, E. A., E. Martinez-Ugalde, and A. H. Orta. 2020. The amphibian skin microbiome and its protective role against chytridiomycosis. *Herpetologica* 76: 167–177.

Rehm, H. L. 2017. Evolving health care through personalized genomics. *Nature Reviews Genetics* 18: 259–267.

Reich, D. 2018. *Who We Are and How We Got Here: Ancient DNA and the New Science of the Human Past*. Pantheon Books.

Reid, N. M., D. A. Preston, B. W. Clark, W. C. Warren, J. K. Colbourne, J. K. Shaw, S. I. Karchner, M. E. Hahn, D. Nacci, M. F. Oleksiak, D. L. Crawford, and A. Whitehead. 2016. The genomic landscape of rapid repeated evolutionary adaptation to toxic pollution in wild fish. *Science* 354: 1305–1308.

Reidenbach, K. R., S. Cook, M. Bertone, R. E. Harbach, B. M. Wiegmann, and N. J. Besansky. 2009. Phylogenetic analysis and temporal diversification of mosquitoes (Diptera: Culicidae) based on nuclear genes and morphology. *BMC Evolutionary Biology* 9: 298.

Reilly, J. R., D. R. Artz, D. Biddinger, K. Bobiwashi, N. K. Boyle, C. Brittain, J. Brokaw, J. W. Campbell, J. Daniels, E. Elle et al. 2020. Crop production in the USA is frequently limited by a lack of pollinators. *Proceedings of the Royal Society of London B* 287: 20200922.

Resnick, D. N., J. Losos, and J. Travis. 2019. From low to high gear: There has been a paradigm shift in our understanding of evolution. *Ecology Letters* 22: 233–244.

Rice, W. R. 2013. Nothing in genetics makes sense except in the light of genomic conflict. *Annual Review of Ecology, Evolution, and Systematics* 44: 217–237.

Richardson, J. L., M. K. Burak, C. Hernandez, J. M. Shirvell, C. Mariani, T. S. A. Carvalho-Pereira, A. C. Pertile, J. A. Panti-May, G. G. Pedra, S. Serrano et al. 2017. Using fine-scale spatial genetics of Norway rats to improve control efforts and reduce leptospirosis in urban slum environments. *Evolutionary Applications* 10: 323–337.

Rinker, D. C., N. K. Specian, S. Zhao, and J. G. Gibbons. 2019. Polar bear evolution is marked by rapid changes in gene copy number in response to dietary shift. *PNAS* 116: 13446–13451.

Risch, N. and K. Merikangas. 1996. The future of genetic studies of complex human diseases. *Science* 273: 1516–1517.

Riskin, D. 2014. *Mother Nature is Trying to Kill You: A Lively Tour Through the Dark Side of the Natural World.* Simon & Schuster.

Ristaino, J. B. and D. H. Pfister. 2016. "What a painfully interesting subject": Charles Darwin's studies of potato late blight. *BioScience* 66: 1035–1045.

Rivera-Marchand, B., D. Oskay, and T. Giray. 2012. Gentle Africanized bees on an oceanic island. *Evolutionary Applications* 5: 746–756.

Robertson, M., A. Schrey, A. Shayter, C. J. Moss, and C. Richards. 2017. Genetic and epigenetic variation in *Spartina alterniflora* following the Deepwater Horizon oil spill. *Evolutionary Applications* 10: 792–801.

Roelke, M. E., J. S. Martenson, and S. J. O'Brien. 1993. The consequences of demographic reduction and genetic depletion in the endangered Florida panther. *Current Biology* 3: 340–350.

Rogers, A. R., D. Iltis, and S. Wooding. 2004. Genetic variation at the MC1R locus and the time since loss of human body hair. *Molecular Biology and Evolution* 45: 105–108.

Rogers, R. L. and M. Slatkin. 2017. Excess of genomic defects in a wooly mammoth on Wrangel Island. *PLoS Genetics* 13: e1006601.

Rohner, N. 2018. Cavefish as an evolutionary mutant model system for human disease. *Developmental Biology* 441: 355–357.

Rohr, J. R., A. P. Dobson, P. T. J. Johnson, A. M. Kilpatrick, S. H. Paul, T. R. Raffel, D. Ruiz-Moreno, and M. B. Thomas. 2011. Frontiers in climate change-disease research. *Trends in Ecology and Evolution* 26: 270–277.

Rolfes, M. A., B. Flannery, J. R. Chung, A. O'Halloran, S. Gang, E. A. Belongia et al. 2019. Effects of the Influenza vaccination in the United States during the 2017–8 influenza season. *Clinical Infectious Disease.* doi: 10:1093/cid/ciz075.

Rollins, L. A., M. F. Richardson, and R. Shine. 2015. A genetic perspective on rapid evolution in cane toads (*Rhinella marina*). *Molecular Ecology* 24: 2264–2276.

Romero, P. A. and F. H. Arnold. 2009. Exploring protein fitness landscapes by directed evolution. *Nature Reviews Molecular and Cell Biology* 10: 868–876.

Romiguier, J., S. A. Cameron, S. H. Woodward, B. J. Fischman, L. Keller, and C. J. Praz. 2016. Phylogenomics controlling for base composition bias reveals a single origin of eusociality in corbiculate bees. *Molecular Biology and Evolution* 33: 670–678.

Romiguier, J., J. Lourenco, P. Gayral, N. Faivre, L. A. Weinert, S. Ravel, M. Ballenghen, V. Cahais, A. Bernard, E. Loire, L. Keller, and N. G. Galtier. 2014. Population genomics of eusocial insects: The costs of a vertebrate-like effective population size. *Journal of Evolution Biology* 27: 593–603.

Rommens, J. M., M. C. Iannuzzi, B.-S. Kerem, M. L. Drumm, G. Melmer, M. Dean, R. Rozmahel, J. L. Cole, D. Kennedy, N. Hidaka et al. 1989. Identification of the cystic fibrosis gene: Chromosome walking and jumping. *Science* 245: 1059–1066.

Rook, G. A. W. 2007. The hygiene hypothesis and the increasing prevalence of chronic inflammatory disorders. *Royal Society of Tropical Medicine and Hygiene* 101: 1072–1074.

Rook, G. A. W., V. Adams, J. Hunt, R. Palmer, R. Martinelli, and L. R. Brunet. 2004. Mycobacteria and other environmental organisms as immunomodulators for immunoregulatory disorders. *Springer Seminars in Immunopathology* 25: 237–255.

Rook, G. A. W. and L. R. Brunet. 2005. Microbes, immunoregulation, and the gut. *Gut* 54: 317–320.

Rose, N. H., M. Sylla, A. Badolo, J. Lutomiah, D. Ayala, O. B. Aribodor, N. Ibe, J. Akorli, S. Otoo, J.-P. Mutebi et al. 2020. Climate and urbanization drive mosquito preference for humans. *Current Biology* 30: 3570–3579.

Rosenberg, K. V., A. M. Dokter, P. J. Blancher, J. R. Sauer, A. C. Smith, P. A. Smith, J. C. Stanton, A. Panjabi, L. Helft, M. Parr, and P. P. Marra. 2019. Decline of the North American avifauna. *Science* 366: 120–124.

Rosenberg, N. A., S. Mahajan, S. Ramachandran, C. Zhao, J. K. Pritchard, and M. W. Feldman. 2005. Clines, clusters, and the effect of study design on the inference of human population. *PLoS Genetics* 1: e70.

Rosenberg, N. A., J. K. Pritchard, J. L. Weber, H. M. Cann, K. K. Kidd, L. A. Zhivotovsky, and M. W. Feldman. 2002. Genetic structure of human populations. *Science* 298: 2381–2385.

Ross-Ibarra, J. 2004. The evolution of recombination under domestication: A test of two hypotheses. *The American Naturalist* 163: 105–112.

Rowe-Rowe, D. T. and J. E. Crafford. 1992. Density, body size, and reproduction of the feral house mice on Gough Island. *South African Journal of Zoology* 27: 1–5.

Roy, D., S. Tomo, P. Purohit, and P. Setia. 2021. Microbiome in death and beyond: Current vista and future trends. *Frontiers in Ecology and Evolution* 9: 630397.

Roy, H. and E. Wajnberg. 2008. From biological control to invasion: The ladybird *Harmonia axyridis* as a model species. *BioControl* 53: 1–4.

Rubio, N. R., N. Xiang, and D. L. Kaplan. 2020. Plant-based and cell-based approaches to meat production. *Nature Communications* 11: 6276.

Rumer, K. M., J. Uyensishi, M. C. Hoffman, B. M. Fisher, and V. D. Winn. 2012. Siglec-6 expression is increased in placentas from pregnancies complicated by preterm preeclampsia. *Reproductive Sciences* 20: 646–653.

Rundlof, M., G. S. K. Anderson, R. Bommarco, I. Fries, V. Hederstrom, L. Herbertsson, O. Jonasson, B. K. Klatt, T. R. Pedersen, J. Yourstone, and H. G. Smith. 2015. Seed coating with a neonicotinoid insecticide negatively affects wild bees. *Nature* 521: 77–80.

Sabatine, M. S. 2019. PCSK9 inhibitors: Clinical evidence and implementation. *Nature Reviews Cardiology* 16: 155–165.

Sabeti, P. 2008. Natural selection: Uncovering mechanisms of evolutionary adaptation to infectious disease. *Nature Education* 1: 13.

Sagarin, R. 1994. Students care far too much about grades. *New York Times*. Letter to the editor. June 12, 1994.

Sagarin, R. 2012. *Learning from the Octopus: How Secrets from Nature Can Help Us Fight Terrorist Attacks, Natural Disasters, and Diseases*. Basic Books.

Sagarin, R., D. T. Blumstein, and G. P. Dieti. 2016. Security, evolution and. Pp. 10–15 In *Encyclopedia of Evolution*. R. M. Kliman (ed.), Vol. 4. Academic Press.

Saini, A. 2019. *Superior: The Return of Race Science*. Beacon Press.

Salewski, V. and C. Watt. 2016. Bergmann's rule: A biophysiological rule examined in birds. *Oikos* 126: 161–172.

Samavati, L. and B. D. Uhal. 2020. ACE2, much more than just a receptor for SARS-COV-2. *Frontiers in Cellular and Infection Microbiology* 10: 317.

Sanford, M. and K. McKeage. 2015. Secukinumab: First global approval. *Drugs* 75: 329–338.

Santini, L., M. González-Suárez, D. Russo, A. Gonzalez-Voyer, A. von Hardenberg, and L. Ancilloto. 2019. One strategy does not fit all: Determinants of urban adaptation in mammals. *Ecology Letters* 22: 365–376.

Sankararaman, S., S. Mallick, M. Dannemann, K. Prufer, J. Kelso, S. Paabo, N. Patterson, and D. Reich. 2014. The genomic landscape of Neanderthal ancestry in present-day humans. *Nature* 507: 354–357.

Sann, M., O. Niehuis, R. S. Peters, C. Mayer, A. Kozlov, L. Podsiadloski, S. Bank, K. Meusemann, B. Misof, C. Bleidorn, and M. Ohl. 2018. Phylogenomic analysis of Apoidea sheds new light on the sister group of bees. *BMC Evolutionary Biology* 18: 71.

Sano, T., S. Morita, R. Tominaga, M. Masuda, Y. Tomita, T. Yasutsune, and H. Yasui. 2002. Adaptive immunity is severely impaired by open-heart surgery. *Japanese Journal of Thoracic and Cardiovascular Surgery* 50: 201–205.

Savile, C. K., J. M. Janey, E. C. Mundorff, J. C Moore, S. Tam et al. 2010. Biocatalytic asymmetric synthesis of chiral amines from ketones applied to sitagliptin manufacture. *Science* 329: 305–310.

Savolainen, V., R. S. Cowan, A. P. Vogler, G. R. Roderick, and R. Lane. 2005. Toward writing the encyclopedia of life: An introduction to DNA barcoding. *Philosophical Transactions of the Royal Society of London B* 360: 1805–1811.

Sax, D. F. and J. H. Brown. 2000. The paradox of invasion. *Global Ecology and Biogeography* 9: 363–371.

Sayol, F., D. Soi, and A. L. Pigot. 2020. Brain size and life history interact to predict urban tolerance in birds. *Frontiers in Ecology and Evolution* 8: 58.

Scaduto, D. L., J. M. Brown, W. C. Haaland, D. J. Zwickl, D. M. Hillis, and M. L. Metzker. 2010. Source identification in two criminal cases using phylogenetic analysis of HIV-1 DNA sequences. *Proceedings of the National Academy of Science* 107: 21242–21247.

Schiffman, J. and L. Abegglen. 2017. What elephants teach us about cancer prevention. *The Conversation.* May 1, 2017. www.theconversation.com/what-elephants-teach-us-about-cancer-prevention-76237

Schilthuizen, M. 2018. *Darwin Comes to Town: How the Urban Jungle Drives Evolution.* Picador.

Schlee, B. C., F. Pasmans, L. F. Skerratt, L. Berger, A. Martel, W. Beukema, A. A. Acevado, P. A. Burrowes, T. Carvalho, A. Catenazzi et al. 2019. Amphibian fungal panzootic causes catastrophic and ongoing loss of biodiversity. *Science* 363: 1459–1463.

Schmidt, A., D. M. Morales-Prieto, J. Pastuscheck, K. Froelich, and U. R. Markert. 2015. Only humans have human placentas: Molecular differences between mice and humans. *Journal of Reproductive Immunology* 108: 65–71.

Schneider, S. S., G. DeGrandi-Hoffmann, and D. R. Smith. 2004. The African honey bee: Factors contributing to a successful biological invasion. *Annual Review of Entomology* 49: 351–376.

Schwartz, J. 2020. Robert May, an uncontainable "big picture' scientist, dies at 84. *New York Times.* May 11, 2020.

Schweinfurth, M. R. and M. Taborsky. 2020. Rats play tit-for-tat instead of integrating social experience over multiple interactions. *Proceedings of the Royal Society of London B* 287: 20192423.

Scudellari, M. 2017. Cleaning up the hygiene hypothesis. *Proceedings of the National Academy of Science* 114: 1433–1436.

Scudellari, M. 2019. Hijacking evolution: Gene-drive technology could alter the genome of a species. *Nature* 571: 160–162.

Seabrook, J., 2007. Sowing for apocalypse: The quest for a global seed bank. *The New Yorker.* August 27, 2007. Pp. 60–71.

Seeley, T. D. 2010. *Honeybee Democracy.* Princeton University Press.

Seeley, T. D. 2019. *The Lives of Bees: The Untold Story of the Honey Bee in the Wild.* Princeton University Press.

Seeley, T. D., J. W. Nowicke, M. Meselson, J. Guillemin, and P. Akratanakul. 1985. Yellow rain. *Scientific American* 253: 128–137.

Segerstrale, U. 2013. *Nature's Oracle: The Life and Work of W. D. Hamilton.* Oxford University Press.

Segurel, L. and C. Bon. 2017. On the evolution of lactase persistence in humans. *Annual Review of Genomics and Human Genetics* 31: 297–319.

Sengupta, S. 2017. How a seed bank, almost lost in Syria, could help feed a warming world. *The New York Times.* October 13, 2017.

Sethuraman, A., F. J. Janzen, D. W. Weistock, and J. J. Obrycki. 2020. Insight from population genomics to enhance and sustain biological control of insect pests. *Insects* 11: 462.

Shah, S. 2010. *The Fever: How Malaria Has Ruled Humankind for 500,000 Years.* Sarah Crichton Books.

Sharma, V., S. Verma, E. Seranova, S. Sarkar, and D. Kumar. 2018. Selective autophagy in infection and disease. *Frontiers in Cell and Developmental Biology* 6: 147.

Sharp, P. M. and B. H. Hahn. 2010. The evolution of HIV-1 and the origin of AIDS. *Proceedings of the Royal Society B* 365: 2487–2494.

Shendure, J., G. M. Findley, and M. W. Snyder. 2019. Genomic medicine—progress, pitfalls, and promise. *Cell* 177: 45–57.

Sheppard, C. A. 2004. Benjamin Dann Walsh: Pioneer entomologist and proponent of Darwinian theory. *Annual Review of Entomology* 49: 1–25.

Shi, J., Z. Wen, G. Zhong, H. Yang, C. Wang, B. Huang, R. Liu, X. He, L. Shauai, Z. Sun et al. 2020. Susceptibility of ferrets, cats, digs, and other domesticated animals to SARS-coronavirus 2. *Science* 368: 1016–1020.

Shiits, R. 1987. *And the Band Planned On: Politics, People, and the AIDS Epidemic.* St. Martin's Press.

Shine, R. 2010. The ecological impact of the invasive cane toad (*Bufo marinus*) in Australia. *The Quarterly Review of Biology* 85: 253–291.

Shriner, D. and C. N. Rotimi. 2018. Whole-genome-sequence-based haplotypes reveal single origin of the sickle allele during the Holocene Wet Phase. *The American Journal of Human Genetics* 102: 547–556.

Sievert, L. L. 2011. The evolution of post-reproductive life: Adaptationist scenarios. Pp. 149–170 In *Reproduction and Aging.* C. G. N. Mascie-Taylor and L. Rosetta (eds.). Cambridge University Press.

Sievert, L. L. 2016. Human Life Histories, Evolution and. Pp. 236–241 In *Encyclopedia of Evolutionary Biology.* R. M. Kliman (ed.), Vol. 2. Academic Press.

Silva, D. N., V. Va'rzea, O. Salgueiro Paulo, and D. Batisa. 2018. Population genomic footprints of host adaptation, introgression and recombination in coffee leaf rust. *Molecular Plant Pathology* 19: 1742–1753.

Silver, N. 2012. *The Signal and the Noise: Why Most Predictions Fail—But Some Don't.* Penguin.

Simberloff, D., L. Souza, M. A. Nunez, M. N. Barrios-Garcia, and W. Bunn. 2012. The natives are restless, but not often and mostly when disturbed. *Ecology* 93: 598–607.

Singh, R., S. Chandrashekharappa, S. R. Bodduluri, B. V. Baby, B. Hegde, N. G. Kotla, A. A. Hiwale, T. Saiyed, P. Patel, M. Vijay-Kumar et al. 2019. Enhancement of the gut barrier integrity by a microbial through the Nrf2 pathway. *Nature Communications* 10: 89.

Skoglund, P., J. C. Thompson, M. E. Prendergast, A. Mittnik, M. Hajdinjak, T. Salie, N. Rohland, S. Mallick, A. Peltzer, A. Heinze et al. 2017. Reconstructing prehistoric African population structure. *Cell* 171: 59–71.

Slatkin, M. 2008. Linkage disequilibrium—understanding the evolutionary past and mapping the medical future. *Nature Reviews Genetics* 9: 477–485.

Smallegange, R. C., N. O. Verhulst, and W. Takken. 2010. Sweaty skin: An invitation to bite? *Trends in Parasitology* 27: 143–148.

Smith, A. R., W. T. Wcislo, and S. O'Donnell. 2007. Survival and productivity benefits to social nesting in the sweat bee *Megalopta genalis. (Hymenoptera: Halcitidae) Behavioral Ecology and Sociobiology* 61: 1111–1120.

Smith, K. R., H. A. Hanson, G. P. Mineau, and S. S. Buys. 2012. Effects of BRCA1 and BRCA2 mutations on female fertility. *Proceedings of the Royal Society B* 279: 1389–1395.

Smith, S. L. 2018. Single (sub)species then and now: An examination of the nonracial perspective of C. Loring Brace. *American Journal of Physical Anthropology* 165: 104–125.

Snell-Rood, E. C. and N. Wick. 2013. Anthropogenic environments exert variable selection on cranial capacity in mammals. *Proceedings of the Royal Society of London B* 280: 20131384.

Snodgrass, R. C. 1956. *The Anatomy of the Honey Bee.* Cornell University Press.

Sol, D., O. Lapiedra, and S. Ducatez. 2020. Cognition and adaptation to urban environments. Pp. 253–267 In *Urban Evolutionary Biology.* M. Szulkin, J. Munshi-South, and A. Charmantier (eds.). Oxford University Press.

Somarelli, J. A. 2021. The hallmarks of cancer as ecologically-driven phenotypes. *Frontiers in Ecology and Evolution* 9: 226.

Somarelli, J. A., A. M. Boddy, H. L. Gardner, S. B. Dewitt, J. Tuohy, K. Megquier, M. U. Sheth, S. D. Hsu, J. L. Thorne, C. A. London, and M. C. Eward. 2020a. Improving cancer drug discovery by studying cancer across the tree of life. *Molecular Biology and Evolution* 37: 11–17.

Somarelli, J. A., H. Gardner, V. L. Cannataro, E. F. Gunady, A. M. Boddy, N. A. Johnson, J. N. Fisk, S. G. Gaffney et al. 2020b. Molecular biology and evolution of cancer: From discovery to action. *Molecular Biology and Evolution* 37: 320–326.

Somers, J. 2020. Immune disorder: The breakdown of cellular society in COVID-19. *The New Yorker.* November 9, 2020. Pp. 26–30.

Song, H.-D., C.-C. Tu, G.-W. Zhang, K. Zheng, L.-C. Lei, Q.-X. Chen, Y.-W. Gao, H.-Q. Zhou, H. Xiang, H.-J. Zheng et al. 2005. Cross-host evolution of severe acute respiratory syndrome coronavirus in palm civet and human. *Proceedings of the National Academy of Science* 102: 2430–2435.

Souza, R. O., M. A. Del Lama, M. Cervini, N. Mortari, T. Eltz, Y. Zimmermann, C. Bach, B. J. Brosi, S. Suni, J. J. G. Quezada-Euán, and R. J. Paxton. 2010. Conservation genetics of neotropical pollinators revisited: Microsatellite analysis suggests that diploid males are rare in orchid bees. *Evolution* 64: 3318–3326.

Span, P. 1994. The gene team. *The Washington Post.* December 14, 1994.

Spinney, L. 2017. *Pale Rider: The Spanish Flu of 1918 and How It Changed the World.* Public Affairs.

Spooner, D. M., M. McLean, G. Ramsay, R. Waugh, and G. J. Bryan. 2005. A single domestication for potato based on multilocus amplified fragment length polymorphism genotyping. *Proceedings of the National Academy of Sciences* 102: 14694–14699.

Srinivas, S. 2015. Oysters, clams and scallops face high risk from ocean acidification: New study finds. *The Guardian.* February 23, 2015.

St. John, P. 2020. The untold story of how the golden state killer was found: A covert operation and private DNA. *Los Angeles Times.* December 8, 2020. Accessed April 10, 2021.

St-Pierre, J., M.-F. Hivert, P. Perron, P. Poirer, S.-P. Guay, D. Brisson, and L. Bouchard. 2012. IGF2 methylation is a modulator of newborn's fetal growth and development. *Epigenetics* 7: 1125–1132.

Stammnitz, M. R., T. H. H. Coorens, K. C. Gori, D. Hayes, B. Fu, J. Wang, D. E. Martin-Herranz, L. B. Alexandrov, A. Baez-Ortega, S. Barthorpe et al. 2018. The origins and vulnerabilities of two transmissible cancers in Tasmanian devils. *Cancer Cell* 33: 607–619.

Steinworth, B. 2018. Jordan Peterson needs to reconsider the lobster. *The Washington Post.* June 4, 2018.

Stensmyr, M. S. 2020. Mosquito Biology: How a quest for water spawned a thirst for blood. *Current Biology* 30: R1046–R1048.

Stern, D. B. and C. E. Lee. 2020. Evolutionary origins of genomic adaptation in an invasive copepod. *Nature Ecology and Evolution* 4: 1084–1094.

Stiemsma, L. T., L. A. Reynolds, S. E. Turvey, and B. B. Finlay. 2015. The hygiene hypothesis: Current perspectives and future therapies. *ImmunoTargets and Therapy* 4: 143–157.

Stoeckle, M. Y. and D. S. Thaler. 2014. Barcoding works in practice but not in (neutral) theory. *PLoS One* 9: e100755.

Stott, R. 2013. *Darwin's Ghosts: The Secret History of Evolution.* Spiegel & Grau.

Strachan, D. P. 1989. Hay fever, hygiene, and household size. *British Medical Journal* 299: 1259–1260.

Strachan, D. P. 2000. Family size, infection and atopy: The first decade of the "hygiene hypothesis". *Thorax* 55 (Suppl 1): S2–S10.

Stratton, J. A., M. J. Nolte, and B. A. Payseur. 2021. Evolution of boldness and exploratory behavior in giant mice from Gough Island. Unpublished ms.

Sturcke, J. 2009. Amphibious warfare: Australians offered prizes in cane toad cull. *The Guardian.* March 26, 2009. www.theguardian.com/world/2009.australia-cane-toad-cull. Accessed October 11, 2020.

Sullivan, R. 2008. *Rats: Observations on the History & Habitat of the City's Most Unwanted Inhabitants.* Bloomsbury.

Sultan, S. E., T. Horagn-Kobelski, L. M. Nichols, C. E. Riggs, and R. K. Waples. 2013. A resurrection study reveals rapid adaptive evolution within populations of an invasive plant. *Evolutionary Applications* 6: 266–278.

Sunday, J. M., P. Calosi, S. Dupont, P. L. Munday, J. H. Stillman, and T. B. H. Reusch. 2014. Evolution in an acidifying ocean. *Trends in Ecology and Evolution* 29: 117–125.

Szukin, M., J. Munshi-South, and A. Charmantier. 2020. Introduction. Pp. 1–12 In *Urban Evolutionary Biology.* M. Szukin, J. Munshi-South, and A. Charmantier (eds.). Oxford University Press.

Tabashnik, B. E., T. Brévault, and Y. Carrière. 2013. Insect resistance to Bt crops: Lessons from the first billion acres. *Nature Biotechnology* 31: 510–521.

Tabashnik, B. E. and Y. Carrière. 2017. Surge in insect resistance to transgenic crops and prospects for sustainability. *Nature Biotechnology* 35: 926–935

Tabashnik, B. E., T. Dennehy, and Y. Carrière. 2005. Delayed resistance to transgenic cotton in pink bollworm. *Proceedings of the National Academy of Science* 102: 15389–15393.

Tabashnik, B. E., M. S. Sisterson, P. C. Ellswood, T. J. Dennehy, L. Antilla, L. Liesner, M. Whitlow, R. T. Staten, J. A. Fabrick, G. C. Unnithan et al. 2010. Suppressing resistance to *Bt* cotton with sterile insect releases. *Nature Biotechnology* 28: 1304–1309.

Takahashi, J., T. Ayabe, M. Mitsuhata, I. Shimizu, and M. Ono. 2008. Diploid male production in a rare and locally distributed bumble, *Bombus florilegus* (Hymenoptera,
Apidae). *Insect Society* 55: 43–50.

Takeuchi, F., R. McGinnis, S. Bourgeois, C. Barnes, N. Eriksson, N. Soranzo, P. Whittaker,
V. Ranganath, V. Kumanduri, W. McLaren et al. 2009. A genome-wide association
study confirms *VKORC1, CYP2C9*, and *CYP4F2* as principal genetic determinants of
warfarin dose. *PLoS Genetics* 5: e1000433.

Tam, V., N. Patel, M. Turcotte, Y. Boss, G. Pare, and D. Meyre. 2019. Benefits and limitations
of genome-wide association studies. *Nature Reviews Genetics* 20: 467–484.

Tang, M. L. K., A.-L. Ponsonby, F. Orsini, D. Tey, M. Robinson, E. L. Su, P. Licciardi, W.
Burks, and S. Donath. 2014. Administration of a probiotic with peanut oral immunotherapy: A randomized trial. *Journal of Allergy and Clinical Immunology* 135: 737–744.

Tanner, J. C., E. Swanger, and M. Zuk. 2019. Sexual signal loss in field crickets maintained
despite strong sexual selection favoring singing males. *Evolution* 73: 1482–1489.

Temmam, S., K. Vongphayloth, E. B. Salazar, S. Munier, M. Bonomi, B. Regnault, B.
Douangboubpha, Y. Karami, D, Chretien, D. Sanamxay et al. 2021. Coronaviruses
with a SARS-CoV-2-like receptor-binding domain allowing ACE2-mediated entry into
human cells isolated from bats of Indochinese peninsula. Preprint. Research Square.
Posted September 17, 2021. Accessed September 25, 2021.

Templeton, A. R. 2013. Biological races in humans. *Studies in History and Philosophy of
Biological and Biomedical Sciences* 44: 262–271.

Teyssier, A., E. Mattysen, N. Salleh Hudin, L. de Neve, J. White, and L. Lens. 2020. Diet contributes to urban-induced alterations in gut microbiota: Experimental evidence from a
wild passerine. *Proceedings of the Royal Society of London* 297: 20192182.

Teyssier, A., L. O. Rouffaer, N. S. Hudin, D. Strubbe, E. Matthysen, L. Lens, and J. White.
2018. Inside the guts of the city: Urban-induced alterations of the gut microbiota.
Science of the Total Environment 612: 1276–1286.

Thakkar, N., R. Burstein, H. Hu, P. Selvaraj, and D. Klein. 2020. Social distancing and mobility reductions have reduced COVID-19 transmission in King County, WA. IDM.

The Severe Covid-19 GWAS Group. 2020. Genome-wide association study of severe Covid-19
with respiratory failure. *The New England Journal of Medicine* 383: 1522–1534.

Thebault, R. 2019. The largest bird in North America was nearly wiped out. Here's how it
fought its way back. *Washington Post.* July 22, 2019.

Thomas, F., M. Giraudeau, F. Renaud, B. Ujvari, B. Roche, P. Pujori, M. Raymond, J.-F.
Lemaitre, and A. Alvergne. 2019. Can postfertile life stages evolve as an anti-cancer
mechanism? *PLoS Biology* 17: e3000565.

Thompson, K. A., M. Renaudin, and M. T. J. Johnson. 2016. Urbanization drives the evolution
of parallel clines in plant populations. *Proceedings of the Royal Society of London B*
283: 20162180.

Thompson, K. A., L. H. Rieseberg, and D. Schulter. 2018. Speciation and the city. *Trends in
Ecology and Evolution* 33: 815–825.

Tian, J., C. Wang, J. Xia, L. Wu, G. Xu, W. Wu, D. Li, W. Qin, Q. Chen et al. 2019. Teosinte
ligule allele narrows plant architecture and enhances high-density maize yields. *Science*
365: 658–664.

Tieman, D., P. Bliss, L. M. McIntyre, A. Blandon-Ubeda, D. Bles, A. Z. Odabasi, G. R.
Rodriguez, E. van der Knapp, M. G. Taylor, C. Goulet et al. 2012. The chemical interactions underlying tomato flavor preference. *Current Biology* 22: 1035–1039.

Tieman, D., G. Zhu, M. F. R. Resende Jr., T. Lin, C. Nguyen, D. Bies, J. L. Rambla, K. S. O.
Beltran, M. Taylor et al. 2017. A chemical genetic roadmap to improved tomato flavor.
Science 355: 391–394.

Tingley, R., G. Ward-Fear, L. Schwarzkopf, M. J. Greenlees, B. L. Phillips, G. Brown, S. Clulow, J. Webb, R. Capon, A. Sheppard et al. 2017. New weapons in the toad toolkit: A review of methods to control and mitigate the biodiversity impacts of invasive cane toads (*Rhinella marina*). *The Quarterly Review of Biology* 92: 123–149.

Tishkoff, S. A., F. A. Reed, A. Ranciaro, B. F. Voight, C. C. Babbitt, J. S. Silverman, K. Powell et al. 2007. Convergent adaptation of human lactase persistence in Africa and Europe. *Nature Genetics* 39: 31–40.

Tiwari, S., R. S. Mann, M. E. Rogers, and L. L. Stelinski. 2011. Insecticide resistance in field populations of Asian citrus psyllid in Florida. *Pest Management Science* 67: 1258–1268.

Tizard, J., S. Patel, J. Waugh, E. Tavares, T. Bergmann, B. Gill, J. Norman, L. Christidis, P. Scofield, O. Haddrath, A. Baker, D. Lambert, and C. Millar. 2019. DNA barcoding a unique avifauna: An important tool for evolution, systematics and evolution. *BMC Evolutionary Biology* 19: 52.

Tollis, M., E. Ferris, M. S. Campbell, V. K. Harris, S. M. Rupp, T. M. Harrison, W. K. Kiso, D. L. Schmitt, M. M. Garner, C. A. Aktipis et al. Preprint. Elephant genomes elucidate disease defenses and other traits. BioRxiv. http://www.biorxiv.org/content/1 0.1101/2020.05.29.124396v2. Version posted October 26, 2020. Accessed September 25, 2021.

Tollis, M., J. Robbins, A. E. Webb, L. F. K. Kuderna, A. F. Caulin, J. D. Garcia, M. Be'rube', N. Pourmand, T. Marques-Bonet, M. J. O'Connell, P. J. Palsboll, and C. C. Maley. 2019. Return to the sea, get huge, beat cancer: An analysis of cetacean genomes including an assembly for the humpback whale (*Megaptera novaeangliae*). *Molecular Biology and Evolution* 36: 1746–1763.

Tomasetto, F., J. M. Tylianakis, M. Reale, S. Wratten, and S. L. Goldson. 2017. Intensified agriculture favors evolved resistance to biological control. *Proceedings of the National Academy of Science* 114: 3885–3890.

Trevelline, B. K., S. S. Fontaine, B. K. Hartup, and K. D. Wolf. 2019. Conservation biology needs a microbial renaissance: A call for the consideration of host-associated microbiota in wildlife management practices. *Proceedings of the Royal Society of London B* 286: 20182448.

Trivedi, B. P. 2020. *Breath from Salt: A Deadly Genetic Disease, a New Era in Science, and the Patients and Families Who Changed Medicine Forever.* BenBella Books, Inc.

Trumble, B. C., J. Stieglitz, A. D. Blackwell, H. Allayee, B. Beheim, C. E. Finch, M. Gurven, and H. Kaplan. 2017. Apolipoprotein E4 is associated with improved cognitive function in Amazonian forager-horticulturalists with a high parasite burden. *The FASEB Journal* 31: 1508–15115.

Tseng, A., J. Seet, and E. J. Phillips. 2014. The evolution of three decades of antiretroviral therapy: Challenges, triumphs and the promise of the future. *British Journal of Clinical Pharmacology* 79: 182–192.

Tu, Y. 2016. Artemisinin—a gift from traditional Chinese medicine to the world (Nobel lecture). *Angewandte Chemie International Edition* 55: 10210–10226.

Tunnicliffe, V., K. T. Davies, D. A. Buterfield, R. W. Embley, J. M. Rose, and W. W. Chadwich Jr. 2009. Survival of mussels in extremely acidic waters of a submarine volcano. *Nature Geoscience* 2: 344–348.

Valenti, L., S. Villa, G. Baselli, R. Temporiti, A. Bandera, L. Scudeller, and D. Prati. 2020. Association of ABO blood group and secretor phenotype with severe COVID-19. *Transfusion* 2020: 1–3.

van Alphen, J. J. M. and B. J. Fernhout. 2020. Natural selection, selective breeding, and the evolution of resistance of honeybees (*Apis mellifera*) against *Varroa*. *Zoological Letters* 6: 6.

Van Dorn, L., D. Richard, C. C. S. Tan, L. P. Shaw, M. Acman, and F. Balloux. 2020. No evidence for increased transmission from recurrent mutations in SARS-CoV-2. *Nature Communications* 11: 5986.

Van Etten, M. L., A. Kuester, S.-M. Chang, and R. S. Baucom. 2016. Fitness costs of herbicide resistance across natural populations of the common morning glory, *Ipomea purpurea*. *Evolution* 70: 2199–2210.

Van Etten, M. L., K. M. Lee, S.-M. Chang, and R. S. Baucom. 2020. Parallel and non-parallel genomic responses contribute to herbicide resistance in *Ipomoea purpurea*, a common agricultural weed. *PLoS Genetics* 16: e1008593.

Vanbergen, A. J. and the Insect Pollinators Initiative. 2013. Threats to an ecosystem service: Pressure on pollinators. *Frontiers in Ecology and Environment* 11: 251–259.

Vandepitte, K., T. De Meyer, K. Helsen, K. Van Acker, I. Roldán-Ruiz, J. Mergeay, and O. Honnay. 2014. Rapid genetic adaptation precedes the spread of an exotic plant species. *Molecular Ecology* 23: 2157–2164.

van't Hof, A. E., N. Edmonds, M. Dalikova, F. Marec, and I. J. Saccheri. 2011. Industrial melanism has in British peppered moths has a singular and recent mutational history. *Science* 332: 958–960.

Vassiliadis, D. and M. A. Dawson. 2021. Mutation alters injury response to drive cancer. *Nature* 590: 557–558.

Vazquez, J. M. and V. J. Lynch. 2021. Pervasive duplication of tumor suppressors in Afrotherians during the evolution of large bodies and reduced cancer risk. *eLife* 10: e65041.

Vazquez, J. M., M. Sulak, S. Chigurupati, and V. J. Lynch. 2018. A zombie *LIF* gene in elephants is upregulated by TP53 to induce apoptosis in response to DNA damage. *Cell Reports* 24: 1765–1776.

Venkatesan, S., C. Swanton, B. S. Taylor, and J. F. Costello. 2017. Treatment-induced mutagenesis and selective pressures sculpt cancer evolution. *Cold Spring Harbor Perspectives* 7: a026617.

Vergara, C., Y. J. Tsai, A. V. Grant, N. Rafaels, T. Hand, M. Stockton, M. Campbell, D. Mercado, M. Faruque, G. Dunston et al. 2008. Gene encoding Duffy antigen/receptor for chemokines is associated with asthma and IgE in three populations. *American Journal of Respiratory and Critical Care Medicine* 178: 1017–1022.

Verhulst, N. O., A. Umanets, B. T. Weldegergis, J. P. A. Maas, T. M. Visser, M. Dicke, H. Smidt, and W. Takken. 2018. Do apes smell like humans? The role of skin bacteria and volatiles of primates in mosquitoes. *The Journal of Experimental Biology* 221: 1–11.

Vikram, V., B. P. Malikarjuna Swamy, S. Dixit, R. Singh, B. P. Singh, B. Miro, A. Henry, N. K. Singh, and A. Kumar. 2015. Drought susceptibility of modern rice varieties: An effect of linkage of drought tolerance with undesirable traits. *Scientific Reports* 5: 14799.

Visscher, P. M., S. Macgregor, B. Benyamin, G. Zhu, S. Gordon, S. Medland, W. G. Hill, J. J. Hottenga et al. 2007. Genome partitioning of genetic variation for height from 11, 214 sibling pairs. *American Journal of Human Genetics* 81: 1104–1112.

Visscher, P. M., N. R. Wray, Q. Zhang, P. Sklar, M. I. McCarthy, M. A. Brown, and J. Yang. 2017. 10 years of GWAS discovery: Biology, function, and translation. *The American Journal of Human Genetics* 101: 5–22.

Wachter, K. W. 1997. Between Zeus and the Salmon: Introduction. Pp. 1–16 In *Between Zeus and the Salmon: The Biodemography of Longevity*. K. W. Wachter and C. E. Finch (eds.). National Academy Press.

Wade, N. 2014. *A Troublesome Inheritance: Genes, Race, and Human History*. The Penguin Press.

Wake, D. B. and V. T. Vredenburg. 2008. Are we in the midst of the sixth mass extinction? A view from the world of amphibians. *Proceedings of the National Academy of Science* 105: 11466–11473.

Wall, J. D. and J. K. Pritchard. 2003. Haplotype blocks and linkage disequilibrium in the human genome. *Nature Review Genetics* 4: 587–597.

Wallace-Wells, D. 2020. We had the vaccine the whole time. *New York Magazine.* December 7, 2020.

Walsh, B. D. 1864. On phytophagic varieties and phytophagic species. *Proceedings of the Entomological Society of Philadelphia* 3: 403–430.

Walsh, B. D. 1866. Imported Insects; the gooseberry saw-fly. *Practical Entomology* 1: 117–125.

Walsh, B. D. 1867. The apple-worm and the apple-maggot. *American Journal of Horticulture* 2: 338–343.

Wamsley, L. 2020. As U. S. reaches 250,000 deaths from COVID-19, a long winter is coming. *NPR.* November 18, 2020. https:www,npr.org/coronavirus-live-updates/2020/11/18/935930352/as-u-s-reaches-250–000-deaths-from-covid-19-a-long-winter-is-coming.

Wang, J., C. Xu, Y. K. Wong, Y. Li, F. Liao, T. Jiang, and Y. Tu. 2019. Artemisinin, the magic drug discovered from traditional Chinese medicine. *Engineering* 5: 32–39.

Wang, L. F. and B. T. Eaton. 2007. Bats, civets and the emergence of SARS. *CTMI* 315: 325–344.

Wanless, R. M., P. G. Ryan, R. Altwegg, A. Angel, J. Cooper, R. Cuthbert, and G. M. Hilton. 2016. From both sides: Dire demographic consequences of carnivorous mice and longlining for the Critically Endangered Tristan albatrosses on Gough Island. *Biological Conservation* 142: 1710–1718.

Wapner, J. 2019. What happens when a jury grapples with perplexing science. *Slate.* March 25, 2019.

Warrener, A. W., K. L. Lewton, H. Pontzer, and D. E. Lieberman. 2015. A wider pelvis does not increase locomotor costs in humans, with implications for the evolution of childbirth. *PLoS One* 10: e0118903.

Watanabe, K. 2015. Potato genetics, genomics, and applications. *Breeding Science* 65: 53–68.

Watanabe, M. E. 2008. Colony collapse disorder: Many suspects, no smoking gun. *BioScience* 58: 384–388.

Wayne, M. L. and B. M. Bolker. 2015. *Infectious Disease: A Very Short Introduction.* Oxford University Press.

Webster, B. 1975. Ancient farmers knew pesticides. *The New York Times.* December 1, 1975.

Weir, A. 2014. *The Martian.* Broadway Books.

Welland, N. and C. Zimmer. 2020. Pfizer's vaccine offers strong protection after first dose. *New York Times.* December 8, 2020.

Werren, J. H. 1997. Biology of *Wolbachia. Annual Review of Entomology* 42: 587–609.

Werren, J. H., L. Baldo, and M. E. Clark. 2008. *Wolbachia*: Master manipulators of invertebrate biology. *Nature Reviews Microbiology* 6: 741–775.

Whitehead, A., B. W. Clark, N. M. Reid, M. E. Hahn, and D. Nacci. 2017. When evolution is the solution to pollution: Key principles, and lessons from rapid repeated adaptation of killifish (*Fundlus heteroclitus*) populations. *Evolutionary Applications* 10: 762–783.

Whitehead, A., B. Dubansky, C. Bodinier, T. L. Garcia, S. Miles, C. Pilley, V. Raghunathan, J. L. Roach, N. Walker, R. B. Walter, C. D. Rice, and F. Galvez. 2012. Genomic and physiological footprint of the Deepwater Horizon oil spill on resident marsh fishes. *Proceedings of the National Academy of Sciences* 109: 20298–20302.

Whittaker, K. and V. Vredenburg. 2011. An overview of chytridiomycosis. www.amphibianweb.org/chytrid/chytridiomycosis.html. Accessed October 30, 2020.

Wickens, K., J. Crane, N. Pearce, and R. Beasley. 1999. The magnitude of the effect of smaller family sizes on the increase in the prevalence of asthma and hay fever in the United Kingdom and New Zealand. *Journal of Allergy and Clinical Immunology* 104: 554–558.

Wilczek, A. M., M. D. Cooper, T. M. Korves, and J. Schmitt. 2014. Lagging adaptation to warming climate in *Arabidopsis thaliana*. *Proceedings of the National Academy of Sciences* 111: 7906–7913.

Wilfert, L., J. Gadau, and P. Schmid-Hempel. 2007. Variation in genomic recombination rates among animal taxa and the case of social insects. *Heredity* 98: 189–197.

Wilfert, L., G. Long, H. C. Legget, P. Schmid-Hempel, R. Butlin, S. J. M. Martin, and M. Boots. 2016. Deformed wing virus is a recent global epidemic in honeybees driven by *Varroa* mites. *Science* 351: 594–597.

Williams, G. C. 1957. Pleiotropy, natural selection, and the evolution of senescence. *Evolution* 11: 398–411.

Williams, G. C. 1966. *Adaptation and Natural Selection: A Critique of Some Current Evolutionary Thought*. Princeton University Press.

Williams, G. C. and R. M. Nesse. 1991. The dawn of Darwinian medicine. *The Quarterly Review of Biology* 66: 1–22.

Williams, T. N. and S. L. Thein. 2018. Sickle cell anemia and its phenotypes. *Annual Review of Genomics and Human Genetics* 19: 113–147.

Wilson, E. O. 2005. Kin selection as the key to altruism: Its rise and fall. *Social Research* 72: 159–166.

Wilson, E. O. 2016. *Half Earth: Our Planet's Fight for Life*. Liveright.

Wilson Sayres, M. A. and K. D. Makova. 2011. Genome analyses substantiate mutation bias in many species. *Bioessays* 93: 938–945.

Wilson Sayres, M. A., C. Venditti, M. Pagel, and K. D. Makova. 2011. Do variations in substitution rates and male mutation bias correlate with life-history traits? A study of 32 mammalian genomes. *Evolution* 65: 2800–2815.

Winchell, K. M., A. C. Battles, and T. Y. Moore. 2020. Terrestrial locomotor evolution in urban environments. Pp. 197–216 In *Urban Evolutionary Biology*. M. Szulkin, J. Munshi-South, and A. Charmantier (eds.). Oxford University Press

Winchell, K. M., K. P. Schliep, D. L. Mahler, and L. J. Revell. 2020. Phylogenetic signal and evolutionary correlates of urban tolerance in a widespread neotropical clade. *Evolution* 74: 1274–1288.

Winegard, T. 2019. *The Mosquito: A Human History of Our Deadliest Predator*. Penguin Random House.

Witt, J. D. S., D. L. Threloff, and P. D. N. Herbert. 2006. DNA barcoding reveals extraordinary cryptic diversity in an amphipod genus: Implications for desert spring conservation. *Molecular Ecology* 15: 3073–3082.

Witt, K. E. and E. Huerta-Sánchez. 2019. Convergent evolution in human and domesticate adaptation to high-altitude environments. *Philosophical Transactions of the Royal Society of London B* 374: 20180235.

Wolfe, L. M. 2002. Why alien invaders succeed? Support for the escape-from-enemy hypothesis. *The American Naturalist* 160: 705–711.

Wolfe, N. 2011. *The Viral Storm: The Dawn of a New Pandemic Age*. Time Books.

Wong, A., N. Rodrigue, and R. Kassen. 2012. Genomics of adaptation during experimental evolution of the opportunistic pathogen *Pseudomonas aeruginosa*. *PLoS Genetics* 8: e1002928.

Woodhead, A. J., C. C. Hicks, A. V. Norstrom, G. J. Williams, N. A. J. Graham. 2019. Coral reef ecosystem services in the Anthropocene. *Functional Ecology* 33: 1023–1034.

Worobey, M., M. Gemmel, D. E. Teuwen, T. Haselkorn, K. Kunstman, M. Bunce, J.-J. Muyembe, J.-J. M. Kabongo, R. M. Kalengayi, E. Van Marck et al. 2008. Direct evidence of extensive diversity of HIV-1 in Kinshasa by 1960. *Nature* 455: 661–665.

Worobey, M., J. Pekar, B. B. Larsen, M. I. Nelson, V. Hill, J. B. Joy, A. Rambaut, M. A. Suchard, J. O. Wethelem, and P. Lemey. 2020. The emergence of SARS-CoV-2 in Europe and North America. *Science* 370: 564–570.

Wright, L. B., M. J. Schoemaker, M. E. Jones, A. Ashworth, and A. J. Swerdlow. 2018. Breast cancer risk in relation to history of preeclampsia and hyperemesis gravidarum: Prospective analysis in the generations study. *International Journal of Cancer* 143: 782–792.

Wright, R. O. 2000. *Nonzero: The Logic of Human Destiny*. Pantheon.

Wright, R. O., H. Hu, E. K. Silverman, S. W. Tsaih, J. Schwartz, D. Bellinger, E. Palazuelos, S. T. Weiss, and M. Hernandez-Avila. 2003. Apolipoprotein E genotype predicts 24-month Bayley scales infant development score. *Pediatric Research* 54: 819–825.

Wright, S. 1969. *Evolution and the Genetics of Populations. II. The Theory of Gene Frequencies*. University of Chicago Press.

Wrotek, S., E. K. LeGrand, A. Dzialuk, and J. Alcock. 2021. Let fever do its job: The meaning of fever in the pandemic era. *Evolution, Medicine, & Public Health* 2021: 26–35. doi:10.193/emph/eoaa044.

Wu, C.-I., N. A. Johnson, and M. F. Palopoli. 1996. Haldane's rule and its legacy: Why are there so many sterile males? *Trends in Ecology and Evolution* 11: 281–284.

Wu, S., S. Powers, W. Zhu, and Y. A. Hannun. 2016. Substantial contribution of extrinsic factors to cancer development. *Nature* 529: 47.

Xiricostas, Z. A., S. E. Everingham, and A. T. Moles. 2020. The sex with the reduced sex chromosome: A comparison across the tree of life. *Biological Letters* 16: 20190867.

Xu, S. and X. Cao. 2010. Interleukin-17 and its expanding biological functions. *Cellular and Molecular Immunology* 7: 164–174.

Yang, W. S. Kandula, M. Huyng, S. K. Greene, G. VanWye, H. T. Chan, E. McGibbon, A. Yeung, D. Olson, A. Fine, and J. Shaman. 2021. Estimating the infection: Fatality risk of SARS-CoV-2 in New York City during the spring 2020 pandemic wave: A model-based analysis. *The Lancet* 21: 203–212.

Yeginsu, C. and C. Zimmer. 2018. 'Cheddar Man', Britain's oldest skeleton, had dark skin, DNA shows. *The New York Times*. February 7, 2018. www.nytimes.com/2018/02/07/world/europe/uk-cheddar-man-skeleton-skin.html

Yong, E. 2016. *I Contain Multitudes: The Microbes Within Us and a Grander View of Life*. Ecco.

Yoshizawa, M., A. Settle, M. C. Hermosura, L. J. Tuttle, N. Cetaro, C. N. Passow, and S. E. McGaugh. 2018. The evolution of a series of behavioral traits is associated with autism-risk genes in cavefish. *BMC Evolutionary Biology* 18: 89.

Young, N. D. 1996. QTL mapping and quantitative disease resistance in plants. *Annual Review of Phytopathology* 34: 479–501.

Yu, Y., Z. Zhao, X. Zheng, J. Zhou, W. Kong, P. Wang, W. Bai, H. Zheng, H. Zhang, J. Li et al. 2018. A selfish genetic element confers non-Mendelian inheritance in rice. *Science* 360: 1130–1132.

Yuan, J. S., P. J. Tranel, and C. N. Stewart Jr. 2007. Non-target-site herbicide resistance as a family business. *Trends in Plant Science* 12: 6–13.

Zagorski, N. 2006. Profile of Alec J. Jeffreys. *Proceedings of the National Academy of Sciences* 103: 8918–8920.

Zayed, A. and L. Packer. 2005. Complementary sex determination substantially increases extinction proneness of haplodiploid population. *Proceedings of the National Academy of Sciences* 102: 10742–10746.

Zeberg, H., J. Kelso, and S. Paabo. 2020. The Neandertal progesterone receptor. *Molecular Biology and Evolution* 37: 2655–2660.

Zeberg, H. and S. Paabo. 2020. The major genetic risk factor for severe COVID-19 is inherited from Neanderthals. *Nature* 587: 610–612.

Zentner, G. E. and M. J. Wade. 2017. The promise and peril of CRISPR gene drives. *Bioessays* 39: 1799109.

Zhang, C., P. Wang, D. Tang, Z. Yang, F. Lu, J. Qi, N. R. Tawari, Y. Shang, C. Li, and S. Huang. 2019. The genetic basis of inbreeding depression in potato. *Nature Genetics* 51: 374–378.

Zhang, Y.-Z. and E. C. Holmes. 2020. A genomic perspective on the origin and emergence of SARS-CoV-2. *Cell* 181:

Zheng, X., D. Zhang, Y. Li, C. Yang, Y. Win, X. Liang, Y. Liang, X. Pan, L. Hu, Q. Sun et al. 2019. Incompatible and sterile insect techniques combined eliminate mosquitoes. *Nature* 572: 56–61.

Zimmer, C. 2016. A virus, fished out of a lake, may have saved a man's life—and advanced science. *STAT.* December 7, 2016. www.statnews.com/2016/12/07/virus-bacteria-phage-therapy/. Accessed May 10, 2021.

Zimmer, C. 2018. *She Has Her Mother's Laugh: The Powers, Perversions, and Potential of Heredity.* Dutton Press.

Zipkin, E. F., G. V. DiRenzo, J. M. Ray, S. Rossman, and K. R. Lips. 2020. Tropical snake diversity collapses after widespread amphibian loss. *Science* 367: 814–816.

Zug, G. R. and P. B. Zug. 1979. The marine toad, *Bufo marinus*: A natural history resume' of the native populations. *Smithsonian Contributions to Zoology* 284: 1–58.

Zuk, M. 2013. *Paleofantasy: What Evolution Really Tells Us about Sex, Diet, and How We Live.* W. W. Norton & Company.

Zuk, M., L. W. Simmons, and J. R. Rotenberry. 2006. Silent night: Adaptive disappearance of a sexual signal in a parasitized population of field crickets. *Biology Letters* 2: 521–524.

Index

Note: Chemicals best known by their abbreviations are listed under that abbreviation. Genes are usually listed under the gene abbreviation in italics and the protein product spelled out in parenthesis. For people whose primary link to this work is one of their books, that book is listed in parenthesis.